FORMULA WEIGHTS

$AgBr$	187.78	$K_2Cr_2O_7$	294.19
$AgCl$	143.32	$K_3Fe(CN)_6$	329.26
Ag_2CrO_4	331.73	$K_4Fe(CN)_6$	368.36
AgI	234.77	$KHC_8H_4O_4$ (phthalate)	204.23
$AgNO_3$	169.87	$KH(IO_3)_2$	389.92
$AgSCN$	165.95	K_2HPO_4	174.18
Al_2O_3	101.96	KH_2PO_4	136.09
$Al_2(SO_4)_3$	342.15	$KHSO_4$	136.17
As_2O_3	197.84	KI	166.01
B_2O_3	69.62	KIO_3	214.00
$BaCO_3$	197.35	KIO_4	230.00
$BaCl_2 \cdot 2H_2O$	244.28	$KMnO_4$	158.04
$BaCrO_4$	253.33	KNO_3	101.11
$Ba(IO_3)_2$	487.14	KOH	56.11
$Ba(OH)_2$	171.35	$KSCN$	97.18
$BaSO_4$	233.40	K_2SO_4	174.27
Bi_2O_3	465.96	$La(IO_3)_3$	663.62
CO_2	44.01	$Mg(C_9H_6ON)_2$	312.62
$CaCO_3$	100.09	$MgCO_3$	84.32
CaC_2O_4	128.10	$MgNH_4PO_4$	137.32
CaF_2	78.08	MgO	40.31
CaO	56.08	$Mg_2P_2O_7$	222.57
$CaSO_4$	136.14	$MgSO_4$	120.37
$Ce(HSO_4)_4$	528.40	MnO_2	86.94
CeO_2	172.12	Mn_2O_3	157.87
$Ce(SO_4)_2$	332.25	Mn_3O_4	228.81
$(NH_4)_2Ce(NO_3)_6$	548.23	$Na_2B_4O_7 \cdot 10H_2O$	381.37
$(NH_4)_4Ce(SO_4)_4 \cdot 2H_2O$	632.55	$NaBr$	102.90
Cr_2O_3	151.99	$NaC_2H_3O_2$	82.04
CuO	79.54	$Na_2C_2O_4$	134.00
Cu_2O	143.08	$NaCl$	58.44
$CuSO_4$	159.60	$NaCN$	49.01
$Fe(NH_4)_2(SO_4)_2 \cdot 6H_2O$	392.14	Na_2CO_3	105.99
FeO	71.85	$NaHCO_3$	84.01
Fe_2O_3	159.69	$Na_2H_2EDTA \cdot 2H_2O$	372.24
Fe_3O_4	231.54	Na_2O_2	77.98
HBr	80.92	$NaOH$	40.00
$HC_2H_3O_2$ (acetic acid)	60.05	$NaSCN$	81.07
$HC_7H_5O_2$ (benzoic acid)	122.12	Na_2SO_4	142.04
HCl	36.46	$Na_2S_2O_3 \cdot 5H_2O$	248.18
$HClO_4$	100.46	NH_4Cl	53.49
$H_2C_2O_4 \cdot 2H_2O$	126.07	$(NH_4)_2C_2O_4 \cdot H_2O$	142.11
H_5IO_6	227.94	NH_4NO_3	80.04
HNO_3	63.01	$(NH_4)_2SO_4$	132.14
H_2O	18.015	$(NH_4)_2S_2O_8$	228.20
H_2O_2	34.01	NH_4VO_3	116.98
H_3PO_4	98.00	$Ni(C_4H_7O_2N_2)_2$	288.92
H_2S	34.08	$PbCrO_4$	323.18
H_2SO_3	82.08	PbO	223.19
H_2SO_4	98.08	PbO_2	239.19
HgO	216.59	$PbSO_4$	303.25
Hg_2Cl_2	472.09	P_2O_5	141.94
$HgCl_2$	271.50	Sb_2S_3	339.69
KBr	119.01	SiO_2	60.08
$KBrO_3$	167.01	$SnCl_2$	189.60
KCl	74.56	SnO_2	150.69
$KClO_3$	122.55	SO_2	64.06
KCN	65.12	SO_3	80.06
K_2CrO_4	194.20	$Zn_2P_2O_7$	304.68

ANALYTICAL CHEMISTRY

SAVITA PATEL

ANALYTICAL CHEMISTRY

Third Edition

Douglas A. Skoog
Professor Emeritus, Stanford University

Donald M. West
San Jose State University

SAUNDERS COLLEGE PUBLISHING
Philadelphia

SAUNDERS COLLEGE PUBLISHING
West Washington Square
Philadelphia, PA 19105

Acquiring Editor: THOM GORMAN
Managing Editor: JEAN SHINDLER
Production Manager: LULA SCHWARTZ
Design Supervisor: RENEE DAVIS
Text Design: RUTH RILEY
Cover Designer: FRED CHARLES

Library of Congress Cataloging in Publication Data

Skoog, Douglas Arvid, 1918-
  Analytical chemistry.

  Includes bibliographical references and index.
  1. Chemistry, Analytic—Quantitative. I. West,
Donald M., joint author. II. Title.
QD101.2.S55   1979   545   78-14543
ISBN 0-03-044416-0

345   074   10987

PREFACE

The third edition of *Analytical Chemistry* is intended—as were its predecessors—to present a manageable body of information on the subject for an audience with educational goals in the natural sciences, health services, engineering, and other fields that make use of quantitative information. As before, the emphasis is upon principles. It remains our view that it is by means of principles that an understanding of the topic will develop. Likewise, the pervading role of equilibrium is stressed in that it provides a guide to such an understanding. The introduction of new material without removal or condensation of an equivalent amount of material deemed less important is a temptation against which we have strived.

Changes in format have been made with the dual aims of making the information more readily accessible and minimizing overlap in the presentation. It has been our experience that an appreciable number of students benefit from a review of concepts that carry over from the welter of topics presented in the freshman course. Accordingly, for those who need it, Chapter 2 is devoted to methods of expressing mass and concentration in chemical terms, as well as to the use of these terms to solve simple problems in stoichiometry. Chapter 3 provides a similar review of equilibrium concepts.

The order in which the subject matter is presented can be altered to suit the preferences of the instructor. Volumetric analysis (Chapters 7–13) can be considered ahead of gravimetric methods (Chapters 5, 6). Consideration of the problems associated with the evaluation of analytical data (Chapter 4) can be timed to coincide with submission of the first laboratory results. Potentiometric methods are discussed in a separate chapter (14); at the discretion of the instructor, other electroanalytical methods (Chapter 15) can be introduced or omitted entirely. Similarly, optical methods (Chapters 16, 17) can be treated in whatever detail seems useful. Analytical separations are considered briefly in Chapter 18.

Laboratory instructions have been combined in the form of two separate chapters (19, 20). Ample references to supporting material in earlier chapters accompany these directions.

A new set of problems has been prepared; answers have been provided for approximately half of them. A solutions manual is also available.

We are again indebted to Professor Alfred Armstrong of the College of William and Mary for his painstaking review of the manuscript and his many helpful criticisms.

Stanford, California
San Jose, California
September, 1978

Douglas A. Skoog
Donald M. West

CONTENTS

Preface   v

1 Introduction   1

Choice of Methods for an Analysis   2

2 A Review of Some Elementary Concepts   4

The Chemical Composition of Solutions   4
Units of Weight and Concentration   8
Stoichiometric Relationships   15
Mathematical Operations Associated with Equilibrium Calculations   18

3 A Review of Simple Equilibrium Constant Calculations   24

The Equilibrium State   24
Equilibrium Constant Expressions   26
Common Types of Equilibrium Constant Expressions   27

4 The Evaluation of Analytical Data   46

Definition of Terms   47
Detection and Correction of Determinate Errors   52
Indeterminate Error   56
Propagation of Errors in Computation   74
Significant Figure Convention   77

5 The Solubility of Precipitates — 87

Effect of Competing Equilibria on the Solubility of Precipitates — 88

Separation of Ions by Control of the Concentration of the Precipitating Reagent — 99

Effect of Electrolyte Concentration on Solubility — 103

Additional Variables That Affect the Solubility of Precipitates — 112

Rate of Precipitate Formation — 112

6 Gravimetric Analysis — 117

Calculation of Results from Gravimetric Data — 117

Properties of Precipitates and Precipitating Reagents — 122

Applications of the Gravimetric Method — 134

7 An Introduction to Volumetric Methods of Analysis — 143

Definition of Some Terms — 143

Reactions and Reagents Used in Volumetric Analysis — 144

End Points in Volumetric Methods — 146

Calculations Associated with Titrimetric Methods — 148

8 Precipitation Titrations — 167

Titration Curves for Precipitation Reactions — 167

Applications of Precipitation Titrations — 180

9 Theory of Neutralization Titrations — 186

Indicators for Acid-Base Titrations — 186

Titration Curves for Strong Acids or Strong Bases — 190

Titration Curves for Weak Acids or Weak Bases — 193

Titration Curves for Weak Bases — 206

Titration Curves for a Mixture of a Weak and Strong Acid (or a Weak and Strong Base) — 206

Titration Curves for Polyfunctional Acid-Base Systems — 208

Titration Curves for Polyfunctional Bases 217

Composition of Solutions of a Polybasic Acid as a
Function of pH 218

10 Applications of Neutralization Titrations 227

Reagents for Neutralization Reactions 227

Typical Applications of Neutralization Titrations 233

11 Complex Formation Titrations 247

Titrations with Inorganic Complexing Reagents 249

Titrations with Aminopolycarboxylic Acids 250

12 Theory of Oxidation-Reduction Titrations 268

Fundamentals of Electrochemistry 270

Oxidation-Reduction Titrations 295

Oxidation-Reduction Indicators 306

13 Application of Oxidation-Reduction Titrations 313

Auxiliary Reagents 313

Applications of Standard Oxidants 317

Application of Reductants 335

14 Potentiometric Methods 350

Potential Measurement 350

Reference Electrodes 354

Metal Indicator Electrodes 357

Membrane Indicator Electrodes 359

Direct Potentiometric Measurements 371

Potentiometric Titrations 376

15 Additional Electroanalytical Methods 389

Behavior of Cells during Current Passage 389

Electrodeposition 394

Coulometry 397

Polarography 408

Amperometric Titrations with One Microelectrode 421

Amperometric Titrations with Two Polarized
Microelectrodes 425

16 Absorptiometric Methods of Analysis 433

Properties of Electromagnetic Radiation 433

Generation of Electromagnetic Radiation 437

Quantitative Aspects of Absorption Measurements 438

Components of Instruments for Absorption
Measurements 443

The Absorption Process 460

Applications of Absorption Spectroscopy to
Qualitative Analysis 466

Quantitative Analysis by Absorption Measurements 467

Analytical Errors in Absorption Measurements 471

17 Atomic Spectroscopy 482

Flame Spectroscopy 483

Atomic Absorption Spectroscopy 486

Flame Emission Spectroscopy 494

18 Analytical Separations 497

Separation by Precipitation 497

Extraction Methods 498

Chromatographic Separations 502

**19 Chemicals, Apparatus, and Unit Operations for
Analytical Chemistry** 526

Choosing and Handling Chemicals and Reagents 526

Cleaning and Marking Laboratory Ware 528

Evaporation of Liquids 529

The Measurement of Mass 530

Equipment and Manipulations Associated with
Weighing 543

Equipment and Manipulations for Filtration and
Ignition   **547**
The Measurement of Volume   **555**
The Laboratory Notebook   **568**

20 Selected Methods of Analysis   571

1	Gravimetric Methods	**571**
2	Precipitation Titrations with Silver Ion	**578**
3	Neutralization Titrations	**583**
4	Complex Formation Titrations	**590**
5	Titrations with Potassium Permanganate	**593**
6	Titrations with Cerium (IV)	**600**
7	Titration with Potassium Dichromate	**602**
8	Iodimetric Titrations	**603**
9	Iodometric Methods of Analysis	**605**
10	Titrations with Potassium Bromate	**608**
11	Titrations with Potassium Iodate	**609**
12	Potentiometric Titrations	**611**
13	Electrogravimetric Methods	**614**
14	Coulometric Titrations	**615**
15	Voltammetry	**616**
16	Methods Based upon Absorption of Radiation	**618**

Answers to Problems   622

Appendixes   641

1 Use of Exponential Numbers   643

2 Logarithms   646

3 The Quadratic Equation   651

4 Solution of Higher Order Equations   652

5 Simplification of Equations by Neglect of Terms   654

6 Some Standard and Formal Electrode Potentials   656

7 Solubility Product Constants   660

8 Dissociation Constants for Acids   663

9 Dissociation Constants for Bases   665

Index   667

ANALYTICAL CHEMISTRY

1 introduction

A quantitative analysis provides numerical information concerning the quantity of some species (the *analyte*) in a measured amount of matter (the *sample*). The results of an analysis are expressed in such relative terms as parts of analyte per hundred (the percent), per thousand, per million, or perhaps per billion of the sample. Other terms include the weight (or volume) of analyte per unit volume of sample and the mole fraction.

Applications of chemical analyses are to be found everywhere in an industrialized society. For example, measurement of the parts per million of hydrocarbons, nitrogen oxides, and carbon monoxide in exhaust gases defines the effectiveness of automotive smog control devices. Determination of the concentration of ionized calcium in blood serum is important in the diagnosis of hyperparathyroidism in human patients. The nitrogen content of breakfast cereals and other foods can be directly related to their protein content and thus their nutritional qualities. Periodic quantitative analyses during the production of steel permit the manufacture of a product having a desired strength, hardness, ductility, or corrosion resistance. The continuous analysis for mercaptans in the household gas supply assures the presence of an odorant to warn of dangerous leaks in the gas distribution system. The analysis of soils for nitrogen, phosphorus, potassium, and moisture throughout the growing season enables the farmer to tailor fertilization and irrigation schedules to meet plant needs efficiently and economically.

In addition to practical applications of the types just cited, quantitative analytical data are at the heart of research activity in chemistry, biochemistry, biology, geology, and the other sciences. Thus, for example, much of what is known of the mechanisms by which chemical reactions occur has been learned through kinetic studies employing quantitative measurements of the rates at which reactants are used up or products appear. Recognition that the conduction of nerve signals in animals and the contraction or relaxation of muscles involve the transport of sodium and potassium ions across membranes was the result of quantitative measurements for these ions on each side of such membranes. Studies concerned with the mechanisms by which gases are transported in blood have required methods for continuously monitoring the concentration of oxygen, carbon dioxide, and other species within a living organism. An understanding of the behavior of semiconductor devices has required the development of methods for the quantitative determination of impurities in pure silicon and germanium in the range of 1×10^{-6} to $1 \times 10^{-10}\%$. Recognition that the amounts of various minor elemental constituents in obsidian samples permit identification and location of their sources has enabled archeologists to trace prehistoric trade routes for tools and weapons fashioned from these materials.

For many investigators in chemistry and biochemistry, as well as some of the biological sciences, the acquisition of quantitative information represents a significant fraction of their experimental efforts. Analytical procedures, then, are among the important tools employed by such scientists in pursuit of their research goals. Both an understanding of the basis of the quantitative analytical process and the competence and confidence to perform analyses are therefore prerequisites for research in these fields. The role of analytical chemistry in the education of chemists and biochemists is analogous to the role of calculus and matrix algebra for aspiring theoretical physicists or the role of ancient languages in the education of scholars of classics.

Choice of Methods for an Analysis

The chemist or scientist who needs analytical data is frequently confronted with an array of methods which could be used to provide the desired information. Such considerations as speed, convenience, accuracy, availability of equipment, number of analyses, amount of sample that can be sacrificed, and concentration range of the analyte must all be considered; the success or failure of an analysis is often critically dependent upon the proper selection of method. Because no generally applicable rules exist, the choice of method is a matter of judgment. Such decisions are difficult; the ability to make them comes only with experience.

This text presents many of the common unit operations associated with chemical analyses and includes a variety of methods for the final measurement of analytes. Both theory and practical detail are treated. Mastery of this material will permit students to perform many useful analyses and will provide them with the background from which they can develop the judgment necessary for the prudent choice of analytical methods.

2

a review of some elementary concepts

Most quantitative analytical measurements are performed on solutions of the sample. The study of analytical chemistry therefore makes use of solution concepts with which students should already be familiar. The purpose of this chapter and Chapter 3 is to review the most important of these concepts.

The Chemical Composition of Solutions

Although both aqueous and organic solvents are widely used in chemical analysis, the former is more commonly encountered. Our discussion will therefore focus on the behavior of solutes in water; reactions in nonaqueous polar media will be considered only briefly.

ELECTROLYTES

Electrolytes are solutes that ionize in a solvent to produce an electrically conducting medium. *Strong electrolytes* are ionized completely—or nearly

TABLE 2-1
Classification of Electrolytes

STRONG ELECTROLYTES	WEAK ELECTROLYTES
1. The inorganic acids HNO_3, $HClO_4$, H_2SO_4,[a] HCl, HI, HBr, $HClO_3$, $HBrO_3$	1. Many inorganic acids such as H_2CO_3, H_3BO_3, H_3PO_4, H_2S, H_2SO_3
2. Alkali and alkaline-earth hydroxides	2. Most organic acids
3. Most salts	3. Ammonia and most organic bases
	4. Halides, cyanides, and thiocyanates of Zn, Cd, and Hg

[a] H_2SO_4 is completely dissociated into HSO_4^- and H_3O^+ ions and for this reason is classified as a strong electrolyte. However, it should be noted that the HSO_4^- ion is a weak electrolyte because it is only partially dissociated.

so—whereas *weak electrolytes* undergo partial ionization only. Table 2-1 summarizes the common strong and weak electrolytes in aqueous media.

ACIDS AND BASES

For analytical chemists the most useful acid-base concept is that proposed independently by Brønsted and Lowry in 1923. According to the Brønsted–Lowry view, *an acid is any substance that is capable of donating a proton*; *a base is any substance that can accept a proton.*

It is important to recognize that the acidic character of a substance manifests itself only in the presence of a proton acceptor or base; similarly, basic behavior requires the presence of an acid. Many solvents act as proton acceptors or donors and thus induce acidic or basic behavior of the solutes dissolved in them. For example, when nitrous acid is dissolved in water, the solvent acts as a proton acceptor and thus behaves as a base:

$$HNO_2 + H_2O \rightleftharpoons NO_2^- + H_3O^+ \tag{2-1}$$

$$\text{acid}_1 \quad \text{base}_2 \quad \text{base}_1 \quad \text{acid}_2$$

On the other hand, when ammonia is dissolved in water, the solvent provides a proton and is thus an acid:

$$NH_3 + H_2O \rightleftharpoons NH_4^+ + OH^- \tag{2-2}$$

$$\text{base}_1 \quad \text{acid}_2 \quad \text{acid}_1 \quad \text{base}_2$$

Water is the classic example of an *amphiprotic* solvent because it exhibits both acidic and basic properties, depending on the solute. Other useful amphiprotic solvents include methyl alcohol, ethyl alcohol, and anhydrous acetic acid. When nitrous acid or ammonia is dissolved in one of these, reactions similar to those shown by Equations 2-1 and 2-2 occur. With methanol, for example, we may write

$$HNO_2 + CH_3OH \rightleftharpoons NO_2^- + CH_3OH_2^+ \qquad \textbf{(2-3)}$$

and

$$NH_3 + CH_3OH \rightleftharpoons NH_4^+ + CH_3O^- \qquad \textbf{(2-4)}$$

Conjugate Acids and Bases.   After an acid has donated a proton, the species that remains is capable of accepting a proton to re-form the original acid. In Equation 2-1, for example, it is seen that nitrite ion is the product of the acidic action of nitrous acid; nitrite ion, however, can behave as a base and accept a proton from a suitable donor. This reaction occurs to a small extent when sodium nitrite is dissolved in water.

$$NO_2^- + H_2O \rightleftharpoons HNO_2 + OH^-$$

$$\text{base}_1 \qquad \text{acid}_2 \qquad \text{acid}_1 \qquad \text{base}_2$$

Thus every Brønsted–Lowry acid is paired with a corresponding base called its *conjugate base*, and every Brønsted–Lowry base is paired with a *conjugate acid*. In Equation 2-1 nitrite ion is seen to be the conjugate base of nitrous acid; the hydronium ion, H_3O^+, is the conjugate acid of the base water. Note also (Equation 2-2) that reaction between the base ammonia and the acid water results in the formation of the conjugate acid ammonium ion and the conjugate base hydroxide ion, respectively.

Autoprotolysis.   Amphiprotic solvents undergo self-ionization or *autoprotolysis* to form a pair of ionic species. Autoprotolysis is an acid-base reaction, as illustrated by the following equations:

$$\text{acid}_1 + \text{base}_2 \quad \rightleftharpoons \quad \text{base}_1 + \text{acid}_2$$

$$H_2O + H_2O \quad \rightleftharpoons \quad OH^- + H_3O^+$$

$$CH_3OH + CH_3OH \quad \rightleftharpoons \quad CH_3O^- + CH_3OH_2^+$$

$$HCOOH + HCOOH \rightleftharpoons HCOO^- + HCOOH_2^+$$

$$NH_3 + NH_3 \quad \rightleftharpoons \quad NH_2^- + NH_4^+$$

The positive ion formed by the autoprotolysis of water is called the *hydronium* ion, the proton being covalently bonded to the parent molecule by one of the unshared electron pairs of the oxygen. Higher hydrates such as $H_5O_2^+$ and $H_7O_3^+$ also exist, but they are significantly less stable than H_3O^+. No unhydrated hydrogen ions appear to exist in aqueous solutions.

To emphasize the high stability of the singly hydrated proton many chemists use the notation H_3O^+ when writing equations for reactions in which the proton is a participant. As a matter of convenience others use H^+

to symbolize the proton, whatever its actual degree of hydration may be. This notation has the advantage of simplifying the writing of equations that require the proton for balance.

STRENGTHS OF ACIDS OR BASES

Figure 2-1 shows the reactions of a few common acids with water. The first entry, hydrochloric acid, is termed a *strong acid* because its reaction is sufficiently complete to leave essentially no HCl molecules, as such, in the solvent. The remaining acids are *weak acids*, which react incompletely to give solutions that contain significant quantities of both the parent acid and the conjugate base. Note that acids may be cationic, anionic, or electrically neutral.

Strongest acid $\qquad$ $HCl + H_2O \rightleftharpoons H_3O^+ + Cl^-$ $\qquad$ Weakest base

$$Al(H_2O)_6^{3+} + H_2O \rightleftharpoons H_3O^+ + AlOH(H_2O)_5^{2+}$$

$$HC_2H_3O_2 + H_2O \rightleftharpoons H_3O^+ + C_2H_3O_2^-$$

$$H_2PO_4^- + H_2O \rightleftharpoons H_3O^+ + HPO_4^{2-}$$

Weakest acid $\qquad$ $NH_4^+ + H_2O \rightleftharpoons H_3O^+ + NH_3$ $\qquad$ Strongest base

FIGURE 2-1   Relative strengths of some common weak acids and their conjugate bases.

The acids shown in Figure 2-1 become progressively weaker from the top to the bottom of the list. Thus, for the purposes of classification, hydrochloric acid is completely dissociated; in contrast, only a few thousandths of a percent of the ammonium ions in an ammonium chloride solution are converted to ammonia molecules. It is also important to note that ammonium ion, the weakest acid, forms the strongest conjugate base of the group. That is, NH_3 has a much stronger affinity for protons than any base above it in Figure 2-1.

The extent of reaction between a solute acid (or base) and a solvent is critically dependent upon the tendency of the latter to donate or accept protons. Thus, for example, perchloric, hydrochloric, and hydrobromic acids are all classified as strong acids in water. If anhydrous acetic acid, a poorer proton acceptor than water, is used *as the solvent*, only perchloric acid undergoes complete dissociation and remains a strong acid; the process can be expressed by the equation

$$\underset{\text{acid}_1}{HClO_4} + \underset{\text{base}_2}{HC_2H_3O_2} \rightleftharpoons \underset{\text{base}_1}{ClO_4^-} + \underset{\text{acid}_2}{H_2C_2H_3O_2^+}$$

Because they undergo only partial dissociation, hydrochloric and hydrobromic acids are weak acids in glacial acetic acid.

Units of Weight and Concentration

In the laboratory the mass of a substance is ordinarily determined in such metric units as the kilogram (kg), the gram (g), the milligram (mg), the microgram (μg), the nanogram (ng), or the picogram (pg).[1] For chemical calculations, however, it is more convenient to employ units that express the weight relationship or *stoichiometry* among reacting species in terms of small whole numbers. The gram formula weight, the gram molecular weight, and the gram equivalent weight are employed in analytical work for this purpose. These terms are often shortened to the formula weight, the molecular weight, and the equivalent weight.

CHEMICAL FORMULAS, FORMULA WEIGHTS, AND MOLECULAR WEIGHTS

An *empirical formula* expresses the simplest combining ratio of atoms in a substance; the *chemical formula*, on the other hand, specifies the number of atoms in a molecule. The two are identical unless experimental evidence indicates that the fundamental aggregate is actually some multiple of the empirical formula. For example, the chemical formula for hydrogen is H_2 because the gas exists as diatomic molecules under ordinary conditions. In contrast, Ne serves adequately to describe the composition of neon, which is observed to be monatomic.

The entity expressed by the chemical formula may or may not actually exist. For example, no evidence has been found for the existence of sodium chloride molecules, as such, in the solid state or in an aqueous solution. Rather, this substance consists of sodium ions and chloride ions, no one of which can be shown to be in simple combination with any other single ion. Nevertheless, the formula NaCl is convenient for stoichiometric accounting and is therefore used. It is also necessary to note that the chemical formula is frequently that of the principal species only. Thus, for example, water in the liquid state contains small amounts of such entities as H_3O^+, OH^-, H_4O_2 and undoubtedly others, in addition to H_2O. Here the chemical formula of H_2O is that for the predominant species and is perfectly satisfactory for chemical accounting; it is, however, only an approximation of the actual composition of the real substance.

The *gram formula weight* (gfw) is the summation of the atomic weights in grams for all the atoms in the chemical formula of a substance. Thus the gram formula weight for H_2 is 2.016 (2×1.008) g; for NaCl it is 58.44 ($35.45 + 22.99$) g. The definition for the gram formula weight carries with it no inference concerning the existence or nonexistence of the substance for which it has been calculated.

---

[1] The relationship among these units is 10^{-3} kg = 1 g = 10^3 mg = 10^6 μg = 10^9 ng = 10^{12} pg.

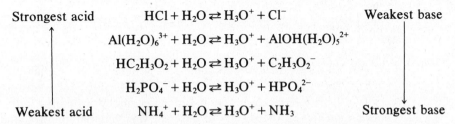

We shall employ the term *gram molecular weight* (gmw) rather than gram formula weight to indicate that we are concerned with a real chemical species. Thus, the gram molecular weight of H_2 is also its gram formula weight, 2.016 g. It would be incorrect to assign a gram molecular weight to sodium chloride in the solid state (or in aqueous solution) because the *species* NaCl does not exist in these states. It is perfectly proper to assign gram molecular weights to Na^+ (22.99 g) and Cl^- (35.45 g) because these are real chemical entities (in the strict sense these should be called gram ionic weights rather than gram molecular weights, although this terminology is seldom encountered).

One molecular weight of a species contains 6.02×10^{23} particles of that species; this quantity is frequently referred to as the *mole*.[2] In a similar way the formula weight represents 6.02×10^{23} units of a substance, whether real or not, represented by the chemical formula.

EXAMPLE

A 25.0-g sample of H_2 contains

$$\frac{25.0 \text{ g}}{2.016 \text{ g/mole}} = 12.4 \text{ moles of } H_2$$

$$12.4 \text{ moles} \times \frac{6.02 \times 10^{23} \text{ molecules}}{\text{mole}} = 7.47 \times 10^{24} \text{ molecules } H_2$$

The same weight of NaCl contains

$$\frac{25.0 \text{ g}}{58.44 \text{ g/fw}} = 0.428 \text{ fw NaCl}$$

which corresponds to 0.428 mole Na^+ and 0.428 mole Cl^-.

Let us further distinguish between the formula weight and the molecular weight by considering exactly 1 fw of water which by definition weighs 18.015 g. Such a quantity contains slightly less than 1 mole of the species H_2O because of the existence of H_3O^+, OH^-, H_4O_2, and such other species as may be present.

Laboratory quantities are frequently more conveniently expressed in terms of *milliformula weights* (mfw) or *millimoles* (mmole); these

---

[2] In the International System (SI) of Units, proposed by the International Bureau of Weights and Measures, the only chemical unit for amount of substance is the *mole*. The mole is defined as the quantity of a material that contains as many elementary entities (these may be atoms, ions, electrons, ion-pairs, or molecules, and must be explicitly defined) as there are atoms of carbon in exactly 0.012 kg of carbon-12 (that is, Avogadro's number). It seems probable that a shift to SI units will ultimately occur. It is equally important to have an understanding of the units upon which the present chemical literature is based, even though these may eventually disappear.

represent, respectively, $\frac{1}{1000}$ of the gram formula weight and gram molecular weight.

CONCENTRATION OF SOLUTIONS

Several methods are employed to describe the concentration of solutions.

Formality or Formal Concentration. The *formality*, F, gives the number of formula weights of a solute contained in 1 liter of solution. The term also expresses the number of milliformula weights per milliliter of solution.

EXAMPLE

Exactly 4.57 g of $BaCl_2 \cdot 2H_2O$ (gfw = 244) are dissolved in sufficient water to give exactly 250 ml. What are the formal concentrations of $BaCl_2 \cdot 2H_2O$ and Cl^- in the resulting solution?

The number of milliformula weights of solute is

$$\text{no. mfw } BaCl_2 \cdot 2H_2O = \frac{4.57 \text{ g } BaCl_2 \cdot 2H_2O}{0.244 \text{ g } BaCl_2 \cdot 2H_2O/mfw} = 18.73$$

$$F = \frac{\text{no. mfw } BaCl_2 \cdot 2H_2O}{ml} = \frac{18.73 \text{ mfw}}{250 \text{ ml}} = 0.0749 \text{ mfw/ml}$$

$$\text{no. mfw } Cl^- = 2 \times \text{no. mfw } BaCl_2 \cdot 2H_2O$$

Therefore

$$F_{Cl^-} = 2 \times F_{BaCl_2 \cdot 2H_2O} = 2 \times 0.0749 = 0.1498 \text{ mfw/ml}$$

Molarity or Molar Concentration. The *molarity*, M, expresses the number of moles of a solute per liter of solution or the number of millimoles per milliliter. Often the molar concentration of a given species is symbolized by placing the chemical formula for the species in square brackets. For example, $[SO_4^{2-}]$ is the symbol for the molar concentration of sulfate ion in a solution.

As shown in the following examples, the formal and the molar concentrations of some solutions may be identical; in others they will be quite different.

EXAMPLE

Calculate the formal and the molar solute concentrations in (1) an aqueous solution that contains 2.30 g of ethanol, C_2H_5OH (gfw = 46.1), in 3.50 liters, and (2) an aqueous solution that contains 285 mg of trichloroacetic acid, Cl_3CCOOH (gfw = 163), in 10.0 ml of aqueous solution (the acid is 73% ionized in water).

1. $F = \dfrac{2.30 \text{ g C}_2\text{H}_5\text{OH}}{46.1 \text{ g C}_2\text{H}_5\text{OH/fw}} \times \dfrac{1}{3.50 \text{ liters}} = 0.0143 \text{ fw/liter}$

In an aqueous solution of ethanol the only solute species present in any significant quantity is C_2H_5OH. Therefore

$$M_{HA} = [HA] = \frac{(285 \times 0.27) \text{ mg HA}}{163 \text{ mg HA/mmole}} \times \frac{1}{10.0 \text{ ml}} = 0.047 \text{ mmole/ml}$$

or

$$[C_2H_5OH] = 0.0143$$

2. Employing HA as the symbol for Cl_3CCOOH, we write

$$F = \frac{285 \text{ mg HA}}{163 \text{ mg HA/mfw}} \times \frac{1}{10 \text{ ml}} = 0.175 \text{ mfw/ml}$$

Because all but 27% of the Cl_3CCOOH is dissociated as H_3O^+ and Cl_3CCOO^-, the molar concentration of Cl_3CCOOH is given by

$$M_{HA} = [HA] = \frac{(285 \times 0.27) \text{ mg HA}}{163 \text{ mg HA/mmole}} \times \frac{1}{10.0 \text{ ml}} = 0.047 \text{ mmole/ml}$$

The molarity of H_3O^+ as well as Cl_3CCOO^- will be equal to the formal concentration of the acid minus the concentration of undissociated acid; that is,

$$M_{H_3O^+} = [H_3O^+] = [A^-] = 0.175 - 0.047 = 0.128 \text{ mmole/ml}$$

It is important to appreciate that the practice of restricting the terms "mole" and "molarity" to a real species and its solution is not universally followed. Thus many chemists employ the terms "formal concentration" and "molar concentration" interchangeably; others do not use the former term at all. The solution just considered is then described as having an analytical or total acid concentration of $0.175\ M$. When specifying the concentration of the undissociated acid, statements such as "$0.047\ M$ in the undissociated acid" or "a species concentration of $0.047\text{-}M$ acid" are employed.

The foregoing example reveals that quantitative information concerning the fate of a solute is needed before the molar concentration of its solution can be specified. In contrast, the formal concentration can be established from the specifications for preparation of the solution and the formula weight of the solute.

EXAMPLE

Describe the preparation of 2.00 liters of $0.100\text{-}F$ Na_2CO_3 from the solid.

$$\text{no. fw Na}_2\text{CO}_3 \text{ needed} = 0.100 \frac{\text{fw Na}_2\text{CO}_3}{\text{liter}} \times 2.00 \text{ liters} = 0.200$$

$$\text{g Na}_2\text{CO}_3 = 0.200 \text{ fw} \times 106 \frac{\text{g Na}_2\text{CO}_3}{\text{fw}} = 21.2$$

Therefore dissolve 21.2 g of the Na_2CO_3 in water, and dilute to exactly 2.00 liters.

EXAMPLE

Describe the preparation of 1.50 liters of 0.100-M Na^+ from pure Na_2CO_3.

$$\text{no. fw } Na_2CO_3 \text{ needed} = 0.100 \frac{\text{mole } Na^+}{\text{liter}} \times 1.50 \text{ liters} \times \frac{1 \text{ fw } Na_2CO_3}{2 \text{ moles } Na^+}$$

$$= 0.0750$$

$$\text{g } Na_2CO_3 = 0.0750 \text{ fw } Na_2CO_3 \times 106 \frac{\text{g } Na_2CO_3}{\text{fw}} = 7.95$$

Dissolve 7.95 g Na_2CO_3 in water, and dilute to 1.50 liters.

Normality or Normal Concentration. This specialized method for expressing concentration is based upon the number of equivalents of solute that are contained in a liter of solution. Normality and equivalent weight are defined in Chapter 7.

Titer. Titer defines concentration in terms of the weight of some species with which a unit volume of the solution reacts. The applications of titer are considered in Chapter 7.

p-Values. Frequently scientists express the concentration of a species in a dilute solution in terms of its *p-value* or *p-function*, where *the p-value is defined as the negative logarithm (to the base 10) of the molar concentration of that species.* Thus for the species X

$$pX = -\log [X]$$

As shown by the following examples, p-values offer the advantage of providing concentration information in terms of small numbers that can be written in the integer form.

EXAMPLE

Calculate the p-values for each ion in a solution that is 2.00×10^{-3} F in NaCl and 5.40×10^{-4} F in HCl.

$$pH = -\log [H_3O^+] = -\log (5.40 \times 10^{-4})$$

To obtain the logarithm we follow the procedure shown in Appendix 2. Thus

$$pH = -\log 5.4 - \log 10^{-4}$$

$$= -0.73 - (-4) = 3.27$$

To obtain pNa we write

$$pNa = -\log(2.00 \times 10^{-3}) = -\log 2.0 - \log 10^{-3}$$

$$= -0.30 - (-3.00) = 2.70$$

The chloride ion concentration is given by the sum of the two concentrations; that is,

$$[Cl^-] = 2.00 \times 10^{-3} + 5.40 \times 10^{-4} = 2.00 \times 10^{-3} + 0.540 \times 10^{-3}$$

$$= 2.54 \times 10^{-3}$$

$$pCl = -\log 2.54 \times 10^{-3} = 2.60$$

EXAMPLE

Calculate the molar concentration of Ag^+ in a solution having a pAg of 6.372.

$$pAg = -\log[Ag^+] = 6.372$$

Here we must obtain the antilogarithm of 6.372. Employing the procedure shown in Appendix 2, we write

$$\log[Ag^+] = -6.372 = -7.000 + 0.628$$

$$[Ag^+] = \text{antilog}(-7.000) \times \text{antilog}(0.628)$$

$$= 10^{-7} \times 4.246 = 4.25 \times 10^{-7}$$

It is noteworthy that the p-value for a species becomes negative when its concentration is greater than unity. For example, in a 2.0-F solution of HCl

$$[H_3O^+] = 2.0$$

$$pH = -\log 2.0 = -0.30$$

Density and Specific Gravity. The *density* of a substance measures its mass per unit volume, whereas the *specific gravity* is the ratio of its mass to that of an equal volume of water at 4°C. In the metric system, density has units of kilograms per liter or grams per milliliter. Specific gravity, on the other hand, is dimensionless and thus not tied to any particular system of units; for this reason it is widely used in describing items of commerce. Because water at 4°C has a density of exactly 1.00 g/ml, and because we shall be employing the metric system throughout the text, density and specific gravity will be used interchangeably.

Parts per Million; Parts per Billion. For very dilute solutions it is convenient to express concentrations in terms of parts per million:

$$ppm = \frac{\text{weight of solute}}{\text{weight of solution}} \times 10^6$$

For even more dilute solutions 1×10^9 rather than 1×10^6 is employed in the foregoing equation; the results are then given as parts per billion (ppb). The term parts per thousand (ppt) is also used.

If the solvent is water and the quantity of solute is so small that the density of the solution is essentially 1.00 g/ml,

$$\text{ppm} = \frac{\text{mg solute}}{10^6 \text{ mg water}} = \frac{\text{mg solute}}{\text{liter solution}}$$

EXAMPLE

What is the molarity of K^+ in a solution that contains 63.3 ppm of $K_3Fe(CN)_6$? Because the solution is so dilute, it is safe to assume that the density of the solution is 1.00 g/ml. Therefore

$$63.3 \text{ ppm } K_3Fe(CN)_6 = 63.3 \text{ mg } K_3Fe(CN)_6/\text{liter}$$

$$M = \frac{63.3 \text{ mg } K_3Fe(CN)_6}{\text{liter}} \times 10^{-3} \frac{g}{mg} \times \frac{1}{329 \text{ g } K_3Fe(CN)_6/\text{fw}} \times \frac{3 \text{ fw } K^+}{1 \text{ fw } K_3Fe(CN)_6}$$

$$= 5.77 \times 10^{-4}$$

Percentage of Concentration. Chemists frequently express concentrations in terms of percentage (parts per hundred). Unfortunately this practice can be a source of ambiguity because of the many ways the percentage composition of a solution can be expressed. Common methods include

$$\text{weight percent (w/w)} = \frac{\text{wt of solute}}{\text{wt of soln}} \times 100$$

$$\text{volume percent (v/v)} = \frac{\text{volume of solute}}{\text{volume of soln}} \times 100$$

$$\text{weight-volume percent (w/v)} = \frac{\text{wt of solute, g}}{\text{volume of soln, ml}} \times 100$$

It should be noted that the denominator in each of these expressions refers to the *solution* rather than to the solvent. Moreover, the first two expressions do not depend on the units employed (provided, of course, that there is consistency between numerator and denominator), whereas units must be defined for the third. Of the three expressions only weight percentage has the virtue of being temperature-independent.

Weight percentage is frequently used to express the concentration of commercial aqueous reagents; thus nitric acid is sold as a 70% solution, which means that the reagent contains 70 g of HNO_3 per 100 g of solution.

Weight-volume percentage is often employed to indicate the composition of dilute aqueous solutions of solid reagents; thus 5% aqueous silver nitrate *usually* refers to a solution that is prepared by dissolving 5 g of silver nitrate in sufficient water to give 100 ml of solution.

Thus

$$\text{no. mfw AgNO}_3 = \frac{2 \text{ mfw AgNO}_3}{\text{mfw Na}_2\text{CO}_3} \times \text{no. mfw Na}_2\text{CO}_3$$

$$\frac{0.0669 \text{ mfw AgNO}_3}{\text{ml AgNO}_3} \times V = \frac{2 \text{ mfw AgNO}_3}{\text{mfw Na}_2\text{CO}_3} \times \frac{0.348 \text{ g Na}_2\text{CO}_3}{0.106 \text{ g Na}_2\text{CO}_3/\text{mfw Na}_2\text{CO}_3}$$

$$V = 2 \times \frac{0.348 \text{ g}}{0.106 \text{ g/mfw}} \times \frac{1}{0.0669 \text{ mfw/ml}}$$

$$= 98.1 \text{ ml AgNO}_3$$

EXAMPLE

What weight of Ag_2CO_3 is formed after mixing 25.0 ml of 0.200-F $AgNO_3$ with 50.0 ml of 0.0800-F Na_2CO_3?

We must first determine which of the reactants is present in the stoichiometrically lesser amount because this species will limit the quantity of silver carbonate that can be produced:

$$\text{no. mfw AgNO}_3 = 25.0 \text{ ml} \times 0.200 \text{ mfw/ml} = 5.00$$

$$\text{no. mfw Na}_2\text{CO}_3 = 50.0 \text{ ml} \times 0.0800 \text{ mfw/ml} = 4.00$$

$$\text{no. mfw AgNO}_3 \text{ required} = 2 \times \text{no. mfw Na}_2\text{CO}_3$$

Thus the reaction is limited by the number of milliformula weights of $AgNO_3$, and

$$\text{no. mfw Ag}_2\text{CO}_3 = \frac{5.00}{2} = 2.50 \text{ mfw}$$

$$w = 2.50 \text{ mfw} \times \frac{0.276 \text{ g Ag}_2\text{CO}_3}{\text{mfw}} = 0.690 \text{ g}$$

Mathematical Operations Associated with Equilibrium Calculations

Throughout the study of analytical chemistry no concept is more pervasive than that of chemical equilibrium. The notion that reactions are never entirely complete will recur time and again throughout this text, as will discussion of the variables that affect the condition of equilibrium in a chemical system. For the present it is sufficient to note that the equilibrium law and its consequences are of such importance that it will be convenient to introduce the topic several times in the contexts of solubility, neutralization processes, complex formation reactions, and oxidation-reduction reactions. In these treatments it is assumed that the reader is familiar with manipulating exponential numbers, using logarithms, and solving

To avoid uncertainty it is necessary to specify explicitly the type of percentage composition that has been used. If this information is lacking, the user is forced to decide intuitively which of the several types is involved.

EXAMPLE

Describe the preparation of 100 ml of 6.0-F HCl from the concentrated reagent. The label on the bottle states that the specific gravity of the reagent is 1.18 and that it is 37% HCl.

Generally the percentages employed in describing commercial reagents are weight-weight. Therefore

$$\frac{\text{g HCl}}{\text{ml concd soln}} = \frac{1.18 \text{ g soln}}{\text{ml soln}} \times \frac{37 \text{ g HCl}}{100 \text{ g soln}} = 0.437$$

$$\text{g HCl required} = 100 \text{ ml} \times \frac{6.00 \text{ mfw HCl}}{\text{ml}} \times \frac{0.0365 \text{ g}}{\text{mfw HCl}} = 21.9$$

$$\text{ml concd soln} = \frac{21.9 \text{ g HCl}}{0.437 \text{ g HCl/ml concd soln}} = 50.2$$

Dilute 50 ml of the concentrated reagent to a volume of about 100 ml.

Solution-Diluent Volume Ratios. The composition of a dilute solution is sometimes specified in terms of the volume of a more concentrated reagent and the volume of solvent to be used in diluting it. The volume of the former is separated from that of the latter by a colon. Thus a 1:4 HCl solution contains four volumes of water for each volume of concentrated hydrochloric acid taken. This method of notation is frequently ambiguous in that the concentration of the original reagent solution is not always obvious to the reader; the use of formal concentrations is preferable.

Stoichiometric Relationships

A balanced chemical equation is a statement of the combining ratios (in formula weights) that exist between reacting substances and their products. Thus the equation[3]

$$2\text{NaI(aq)} + \text{Pb(NO}_3)_2\text{(aq)} = \text{PbI}_2\text{(s)} + 2\text{NaNO}_3\text{(aq)}$$

[3] Here it is advantageous to depict the reaction in terms of chemical compounds. If we wish to focus on reaction species, the net ionic representation is preferable:

$$2\text{I}^-\text{(aq)} + \text{Pb}^{2+}\text{(aq)} = \text{PbI}_2\text{(s)}$$

indicates that 2 fw of sodium iodide combine in aqueous solution with 1 fw of lead nitrate to produce 1 fw of solid lead iodide and 2 fw of aqueous sodium nitrate.[4]

Experimental measurements are never obtained directly in terms of formula weight; instead they have units such as grams, milligrams, liters, or milliliters. To convert such measurements to the weight of some other substance, the chemist first transforms the raw data into units of formula weights, then takes account of the stoichiometry that is involved, and finally reconverts to ordinary metric units. These transformations, which are of fundamental importance to analytical chemistry, are summarized in Tables 2-2 and 2-3. These definitions—alone or in combination—will provide the solution to any stoichiometric problem; the ability to manipulate them must be cultivated. In setting up equations, it is helpful to supply units for all quantities that possess dimensions; the best proof that a correct relationship has been generated is agreement between the units that appear on the two sides of the equal sign.

TABLE 2-2
Expression of Weight in Chemical Units

CHEMICAL UNIT	WEIGHT OF UNIT IN GRAMS GIVEN BY	METHOD OF CONVERSION FROM METRIC UNITS TO CHEMICAL UNITS
Formula weight (fw)	gfw	$\text{no. fw} = \dfrac{\text{gram of substance}}{\text{gfw}}$
Milliformula weight (mfw)	$\dfrac{\text{gfw}}{1000}$	$\text{no. mfw} = \dfrac{\text{gram of substance}}{\text{gfw}/1000}$
Mole	gmw	$\text{no. mole} = \dfrac{\text{gram of species}}{\text{gmw}}$
Millimole	$\dfrac{\text{gmw}}{1000}$	$\text{no. mmole} = \dfrac{\text{gram of species}}{\text{gmw}/1000}$
Equivalent (eq)	eq wt	$\text{no. eq} = \dfrac{\text{gram of substance}}{\text{eq wt}}$
Milliequivalent (meq)	$\dfrac{\text{eq wt}}{1000}$	$\text{no. meq} = \dfrac{\text{gram of substance}}{\text{eq wt}/1000}$

[4] Chemists frequently include information about the physical state of substances in equations; thus (g), (l), (s), and (aq) refer to gaseous, liquid, solid, and aqueous solution states, respectively.

TABLE 2-3
Expression of Concentration in Chemical Units

CHEMICAL TERM FOR CONCENTRATION	METHOD OF CALCULATION FROM CHEMICAL UNITS OF WEIGHT	METHOD OF CALCULATION FROM M... UNITS OF W...
Formality, F	$F = \dfrac{\text{no. fw}}{\text{liter of soln}}$	$F = \dfrac{\text{gram solute}}{\text{liter of soln} \times \text{g...}}$
	$= \dfrac{\text{no. mfw}}{\text{ml of soln}}$	$= \dfrac{\text{gram solut...}}{\text{ml of soln} \times \text{gfw...}}$
Molarity, M	$M = \dfrac{\text{no. mole}}{\text{liter of soln}}$	$M = \dfrac{\text{gram solute}}{\text{liter of soln} \times \text{gm...}}$
	$= \dfrac{\text{no. mmole}}{\text{ml of soln}}$	$= \dfrac{\text{gram solute}}{\text{ml of soln} \times \text{gmw...}}$
Normality, N	$N = \dfrac{\text{no. eq}}{\text{liter of soln}}$	$N = \dfrac{\text{gram solute}}{\text{liter of soln} \times \text{eq }...}$
	$= \dfrac{\text{no. meq}}{\text{ml of soln}}$	$= \dfrac{\text{gram solute}}{\text{ml of soln} \times \text{eq w...}}$

EXAMPLE

Calculate the weight in grams w of $AgNO_3$ (gfw = 170) required to ... 2.33 g of Na_2CO_3 (gfw = 106) to Ag_2CO_3 (gfw = 276).

$$\text{no. fw } AgNO_3 \text{ required} = 2 \times \text{no. fw } Na_2CO_3$$

From Table 2-2 we substitute

$$\frac{w}{170 \text{ g } AgNO_3/\text{fw } AgNO_3} = \frac{2 \text{ fw } AgNO_3}{\text{fw } Na_2CO_3} \times \frac{2.33 \text{ g } Na_2CO_3}{106 \text{ g } Na_2CO_3/\text{fw}}$$

$$w = \frac{170 \times 2 \times 2.33}{106} = 7.47 \text{ g } AgNO_3$$

EXAMPLE

How many milliliters of 0.0669-F $AgNO_3$ will be needed to conve... of pure Na_2CO_3 to Ag_2CO_3?

Because the volume V in milliliters is desired, it will be more con... base our calculations on milliformula weights than on formula...

second-order and higher algebraic equations. A brief review of each of these techniques will be found in the following sections of the appendix:

Appendix 1: Use of Exponential Numbers
Appendix 2: Use of Logarithms
Appendix 3: Solution of Quadratic Equations
Appendix 4: Solution of Higher Order Equations by Systematic Approximation
Appendix 5: Simplification of Equations by Neglecting Certain Terms

PROBLEMS*

*1. How many milliformula weights are contained in
   (a) 14.2 g of PbO_2?
   (b) 126 μg of KIO_4?
   (c) 500 ml of 0.130-F Na_2SO_4?
   (d) 5.00 liters of 2.00×10^{-3} F $KMnO_4$?
   (e) 100 ml of an aqueous solution containing 2.64 ppm NH_3?

2. How many milliformula weights are contained in
   (a) 18.3 mg of SnO_2?
   (b) 10.0 g of dry ice (CO_2)?
   (c) 2.00 kg of $Na_2B_4O_7 \cdot 10 H_2O$?
   (d) 5.20 ml of 2.75-F acetic acid?
   (e) 20.0 liters of 0.0125-F $K_2Cr_2O_7$?

*3. How many milligrams are contained in
   (a) 5.0 moles of CCl_4?
   (b) 1.96 mfw of acetic acid (CH_3COOH)?
   (c) 20.0 fw of H_2SO_4?
   (d) 10.8 ml of 0.116-F sucrose (gfw = 342)?
   (e) 30.2 liters of 6.0-F H_3PO_4?

4. How many grams are contained in
   (a) 1.11 moles of K_2SO_4?
   (b) 67.3 mfw of I_2?
   (c) 225 ml of 0.100-F methanol (CH_3OH)?
   (d) 2.50 liters of 1.50-F KOH?
   (e) 2.00 ml of 1.00×10^{-5} F KCl?

*5. A solution was prepared by dissolving 273 mg of $Al_2(SO_4)_3$ in dilute HCl and diluting to exactly 500 ml. Calculate

* Answers to problems or parts of problems marked with an asterisk are to be found at the end of the book.

(a) the formal concentration of $Al_2(SO_4)_3$.
(b) the molar concentration of Al^{3+}.
(c) the formal concentration of SO_4^{2-}.
(d) the weight-volume percentage of $Al_2(SO_4)_3$.
(e) the number of millimoles of Al^{3+} in 25.0 ml of the solution.
(f) the parts per million of Al^{3+} in the solution.
(g) the pAl of the solution.
(h) the pSO$_4$ of the solution.

$ppm = \dfrac{mg}{l}$

6. A solution was prepared by dissolving 432 mg of $K_3Fe(CN)_6$ in water and diluting to 1500 ml. Calculate
   (a) the formal concentration of $K_3Fe(CN)_6$.
   (b) the molar concentration of K^+.
   (c) the weight-volume percentage of $K_3Fe(CN)_6$.
   (d) the molar concentration of $Fe(CN)_6^{3-}$.
   (e) the number of millimoles of K^+ in 100.0 ml of the solution.
   (f) the parts per million of K^+ in the solution.
   (g) the pK of the solution.
   (h) the pFe(CN)$_6$ of the solution.

$ppm = \dfrac{mg}{kg}$

$ppm = \dfrac{mg\ solute}{1000\ g\ solvent}$

*7. Average sea water contains 1.06×10^4 ppm of Na^+ and 380 ppm of K^+.
   (a) Calculate the molar concentration of each of these ions if the density of seawater is 1.02 g/ml.
   (b) Calculate the pK and pNa of the solution.

8. Average human blood serum contains 2.7 mg Mg^{2+} and 10 mg Ca^{2+} per 100 ml.
   (a) Calculate the molar concentration of each of these species.
   (b) Calculate the pMg and pCa of the solution.

9. Calculate the p-value of each of the indicated ions in the following solutions:
   *(a) Na^+, SO_4^{2-}, and OH^- in a solution that was 0.0200 F in Na_2SO_4 and 0.0100 F in NaOH.
   (b) Mg^{2+}, Ca^{2+}, and Cl^- in a solution that was 1.30×10^{-3} F in $MgCl_2$ and 2.20×10^{-4} F in $CaCl_2$.
   *(c) H^+, NO_3^-, and Zn^{2+} in a solution that was 1.50 F in HNO_3 and 0.200 F in $Zn(NO_3)_2$.
   (d) Pb^{2+}, Cd^{2+}, and NO_3^- in a solution that was 0.300 F in $Pb(NO_3)_2$ and 0.400 F in $Cd(NO_3)_2$.
   *(e) H^+, Ba^{2+}, and ClO_4^- in a solution that was 1.00×10^{-4} F in $Ba(ClO_4)_2$ and 7.32×10^{-5} F in $HClO_4$.
   (f) K^+, OH^-, and $Fe(CN)_6^{4-}$ in a solution that was 1.12×10^{-6} F in $K_4Fe(CN)_6$ and 3.12×10^{-5} F in KOH.

10. Calculate the molar hydrogen ion concentration of a solution having a pH of
    *(a) 12.12.   *(e) 3.75.
    (b) 1.97.   (f) 8.06.
    *(c) 6.43.   *(g) −2.17.
    (d) 9.86.   (h) −0.32.

second-order and higher algebraic equations. A brief review of each of these techniques will be found in the following sections of the appendix:

Appendix 1: Use of Exponential Numbers
Appendix 2: Use of Logarithms
Appendix 3: Solution of Quadratic Equations
Appendix 4: Solution of Higher Order Equations by Systematic Approximation
Appendix 5: Simplification of Equations by Neglecting Certain Terms

PROBLEMS*

*1. How many milliformula weights are contained in
   (a) 14.2 g of PbO_2?
   (b) 126 μg of KIO_4?
   (c) 500 ml of 0.130-F Na_2SO_4?
   (d) 5.00 liters of 2.00×10^{-3} F $KMnO_4$?
   (e) 100 ml of an aqueous solution containing 2.64 ppm NH_3?

2. How many milliformula weights are contained in
   (a) 18.3 mg of SnO_2?
   (b) 10.0 g of dry ice (CO_2)?
   (c) 2.00 kg of $Na_2B_4O_7 \cdot 10 H_2O$?
   (d) 5.20 ml of 2.75-F acetic acid?
   (e) 20.0 liters of 0.0125-F $K_2Cr_2O_7$?

*3. How many milligrams are contained in
   (a) 5.0 moles of CCl_4?
   (b) 1.96 mfw of acetic acid (CH_3COOH)?
   (c) 20.0 fw of H_2SO_4?
   (d) 10.8 ml of 0.116-F sucrose (gfw = 342)?
   (e) 30.2 liters of 6.0-F H_3PO_4?

4. How many grams are contained in
   (a) 1.11 moles of K_2SO_4?
   (b) 67.3 mfw of I_2?
   (c) 225 ml of 0.100-F methanol (CH_3OH)?
   (d) 2.50 liters of 1.50-F KOH?
   (e) 2.00 ml of 1.00×10^{-5} F KCl?

*5. A solution was prepared by dissolving 273 mg of $Al_2(SO_4)_3$ in dilute HCl and diluting to exactly 500 ml. Calculate

* Answers to problems or parts of problems marked with an asterisk are to be found at the end of the book.

(a) the formal concentration of $Al_2(SO_4)_3$.

(b) the molar concentration of Al^{3+}.

(c) the formal concentration of SO_4^{2-}.

(d) the weight-volume percentage of $Al_2(SO_4)_3$.

(e) the number of millimoles of Al^{3+} in 25.0 ml of the solution.

(f) the parts per million of Al^{3+} in the solution.

(g) the pAl of the solution.

(h) the pSO$_4$ of the solution.

$ppm = \dfrac{mg}{l}$

6. A solution was prepared by dissolving 432 mg of $K_3Fe(CN)_6$ in water and diluting to 1500 ml. Calculate

(a) the formal concentration of $K_3Fe(CN)_6$.

(b) the molar concentration of K^+.

(c) the weight-volume percentage of $K_3Fe(CN)_6$.

(d) the molar concentration of $Fe(CN)_6^{3-}$.

(e) the number of millimoles of K^+ in 100.0 ml of the solution.

(f) the parts per million of K^+ in the solution.

(g) the pK of the solution.

(h) the pFe(CN)$_6$ of the solution.

$ppm = \dfrac{mg}{kg}$

$ppm = \dfrac{mg\ solute}{1000\ g\ solvent}$

*7. Average sea water contains 1.06×10^4 ppm of Na^+ and 380 ppm of K^+.

(a) Calculate the molar concentration of each of these ions if the density of seawater is 1.02 g/ml.

(b) Calculate the pK and pNa of the solution.

8. Average human blood serum contains 2.7 mg Mg^{2+} and 10 mg Ca^{2+} per 100 ml.

(a) Calculate the molar concentration of each of these species.

(b) Calculate the pMg and pCa of the solution.

9. Calculate the p-value of each of the indicated ions in the following solutions:

*(a) Na^+, SO_4^{2-}, and OH^- in a solution that was 0.0200 F in Na_2SO_4 and 0.0100 F in NaOH.

(b) Mg^{2+}, Ca^{2+}, and Cl^- in a solution that was 1.30×10^{-3} F in $MgCl_2$ and 2.20×10^{-4} F in $CaCl_2$.

*(c) H^+, NO_3^-, and Zn^{2+} in a solution that was 1.50 F in HNO_3 and 0.200 F in $Zn(NO_3)_2$.

(d) Pb^{2+}, Cd^{2+}, and NO_3^- in a solution that was 0.300 F in $Pb(NO_3)_2$ and 0.400 F in $Cd(NO_3)_2$.

*(e) H^+, Ba^{2+}, and ClO_4^- in a solution that was 1.00×10^{-4} F in $Ba(ClO_4)_2$ and 7.32×10^{-5} F in $HClO_4$.

(f) K^+, OH^-, and $Fe(CN)_6^{4-}$ in a solution that was 1.12×10^{-6} F in $K_4Fe(CN)_6$ and 3.12×10^{-5} F in KOH.

10. Calculate the molar hydrogen ion concentration of a solution having a pH of

*(a) 12.12.   *(e) 3.75.

(b) 1.97.   (f) 8.06.

*(c) 6.43.   *(g) −2.17.

(d) 9.86.   (h) −0.32.

To avoid uncertainty it is necessary to specify explicitly the type of percentage composition that has been used. If this information is lacking, the user is forced to decide intuitively which of the several types is involved.

EXAMPLE

Describe the preparation of 100 ml of 6.0-F HCl from the concentrated reagent. The label on the bottle states that the specific gravity of the reagent is 1.18 and that it is 37% HCl.

Generally the percentages employed in describing commercial reagents are weight-weight. Therefore

$$\frac{\text{g HCl}}{\text{ml concd soln}} = \frac{1.18 \text{ g soln}}{\text{ml soln}} \times \frac{37 \text{ g HCl}}{100 \text{ g soln}} = 0.437$$

$$\text{g HCl required} = 100 \text{ ml} \times \frac{6.00 \text{ mfw HCl}}{\text{ml}} \times \frac{0.0365 \text{ g}}{\text{mfw HCl}} = 21.9$$

$$\text{ml concd soln} = \frac{21.9 \text{ g HCl}}{0.437 \text{ g HCl/ml concd soln}} = 50.2$$

Dilute 50 ml of the concentrated reagent to a volume of about 100 ml.

Solution-Diluent Volume Ratios. The composition of a dilute solution is sometimes specified in terms of the volume of a more concentrated reagent and the volume of solvent to be used in diluting it. The volume of the former is separated from that of the latter by a colon. Thus a 1:4 HCl solution contains four volumes of water for each volume of concentrated hydrochloric acid taken. This method of notation is frequently ambiguous in that the concentration of the original reagent solution is not always obvious to the reader; the use of formal concentrations is preferable.

Stoichiometric Relationships

A balanced chemical equation is a statement of the combining ratios (in formula weights) that exist between reacting substances and their products. Thus the equation[3]

$$2NaI(aq) + Pb(NO_3)_2(aq) = PbI_2(s) + 2NaNO_3(aq)$$

---

[3] Here it is advantageous to depict the reaction in terms of chemical compounds. If we wish to focus on reaction species, the net ionic representation is preferable:

$$2I^-(aq) + Pb^{2+}(aq) = PbI_2(s)$$

indicates that 2 fw of sodium iodide combine in aqueous solution with 1 fw of lead nitrate to produce 1 fw of solid lead iodide and 2 fw of aqueous sodium nitrate.[4]

Experimental measurements are never obtained directly in terms of formula weight; instead they have units such as grams, milligrams, liters, or milliliters. To convert such measurements to the weight of some other substance, the chemist first transforms the raw data into units of formula weights, then takes account of the stoichiometry that is involved, and finally reconverts to ordinary metric units. These transformations, which are of fundamental importance to analytical chemistry, are summarized in Tables 2-2 and 2-3. These definitions—alone or in combination—will provide the solution to any stoichiometric problem; the ability to manipulate them must be cultivated. In setting up equations, it is helpful to supply units for all quantities that possess dimensions; the best proof that a correct relationship has been generated is agreement between the units that appear on the two sides of the equal sign.

TABLE 2-2
Expression of Weight in Chemical Units

CHEMICAL UNIT	WEIGHT OF UNIT IN GRAMS GIVEN BY	METHOD OF CONVERSION FROM METRIC UNITS TO CHEMICAL UNITS
Formula weight (fw)	gfw	$\text{no. fw} = \dfrac{\text{gram of substance}}{\text{gfw}}$
Milliformula weight (mfw)	$\dfrac{\text{gfw}}{1000}$	$\text{no. mfw} = \dfrac{\text{gram of substance}}{\text{gfw}/1000}$
Mole	gmw	$\text{no. mole} = \dfrac{\text{gram of species}}{\text{gmw}}$
Millimole	$\dfrac{\text{gmw}}{1000}$	$\text{no. mmole} = \dfrac{\text{gram of species}}{\text{gmw}/1000}$
Equivalent (eq)	eq wt	$\text{no. eq} = \dfrac{\text{gram of substance}}{\text{eq wt}}$
Milliequivalent (meq)	$\dfrac{\text{eq wt}}{1000}$	$\text{no. meq} = \dfrac{\text{gram of substance}}{\text{eq wt}/1000}$

[4] Chemists frequently include information about the physical state of substances in equations; thus (g), (l), (s), and (aq) refer to gaseous, liquid, solid, and aqueous solution states, respectively.

TABLE 2-3

Expression of Concentration in Chemical Units

CHEMICAL TERM FOR CONCENTRATION	METHOD OF CALCULATION FROM CHEMICAL UNITS OF WEIGHT	METHOD OF CALCULATION FROM METRIC UNITS OF WEIGHT
Formality, F	$F = \dfrac{\text{no. fw}}{\text{liter of soln}}$	$F = \dfrac{\text{gram solute}}{\text{liter of soln} \times \text{gfw}}$
	$= \dfrac{\text{no. mfw}}{\text{ml of soln}}$	$= \dfrac{\text{gram solute}}{\text{ml of soln} \times \text{gfw}/1000}$
Molarity, M	$M = \dfrac{\text{no. mole}}{\text{liter of soln}}$	$M = \dfrac{\text{gram solute}}{\text{liter of soln} \times \text{gmw}}$
	$= \dfrac{\text{no. mmole}}{\text{ml of soln}}$	$= \dfrac{\text{gram solute}}{\text{ml of soln} \times \text{gmw}/1000}$
Normality, N	$N = \dfrac{\text{no. eq}}{\text{liter of soln}}$	$N = \dfrac{\text{gram solute}}{\text{liter of soln} \times \text{eq wt}}$
	$= \dfrac{\text{no. meq}}{\text{ml of soln}}$	$= \dfrac{\text{gram solute}}{\text{ml of soln} \times \text{eq wt}/1000}$

EXAMPLE

Calculate the weight in grams w of $AgNO_3$ (gfw = 170) required to convert 2.33 g of Na_2CO_3 (gfw = 106) to Ag_2CO_3 (gfw = 276).

$$\text{no. fw } AgNO_3 \text{ required} = 2 \times \text{no. fw } Na_2CO_3$$

From Table 2-2 we substitute

$$\frac{w}{170 \text{ g } AgNO_3/\text{fw } AgNO_3} = \frac{2 \text{ fw } AgNO_3}{\text{fw } Na_2CO_3} \times \frac{2.33 \text{ g } Na_2CO_3}{106 \text{ g } Na_2CO_3/\text{fw}}$$

$$w = \frac{170 \times 2 \times 2.33}{106} = 7.47 \text{ g } AgNO_3$$

EXAMPLE

How many milliliters of 0.0669-F $AgNO_3$ will be needed to convert 0.348 g of pure Na_2CO_3 to Ag_2CO_3?

Because the volume V in milliliters is desired, it will be more convenient to base our calculations on milliformula weights than on formula weights.

Thus

$$no. \ mfw \ AgNO_3 = \frac{2 \ mfw \ AgNO_3}{mfw \ Na_2CO_3} \times no. \ mfw \ Na_2CO_3$$

$$\frac{0.0669 \ mfw \ AgNO_3}{ml \ AgNO_3} \times V = \frac{2 \ mfw \ AgNO_3}{mfw \ Na_2CO_3} \times \frac{0.348 \ g \ Na_2CO_3}{0.106 \ g \ Na_2CO_3/mfw \ Na_2CO_3}$$

$$V = 2 \times \frac{0.348 \ g}{0.106 \ g/mfw} \times \frac{1}{0.0669 \ mfw/ml}$$

$$= 98.1 \ ml \ AgNO_3$$

EXAMPLE

What weight of Ag_2CO_3 is formed after mixing 25.0 ml of 0.200-F $AgNO_3$ with 50.0 ml of 0.0800-F Na_2CO_3?

We must first determine which of the reactants is present in the stoichiometrically lesser amount because this species will limit the quantity of silver carbonate that can be produced:

$$no. \ mfw \ AgNO_3 = 25.0 \ ml \times 0.200 \ mfw/ml = 5.00$$

$$no. \ mfw \ Na_2CO_3 = 50.0 \ ml \times 0.0800 \ mfw/ml = 4.00$$

$$no. \ mfw \ AgNO_3 \ required = 2 \times no. \ mfw \ Na_2CO_3$$

Thus the reaction is limited by the number of milliformula weights of $AgNO_3$, and

$$no. \ mfw \ Ag_2CO_3 = \frac{5.00}{2} = 2.50 \ mfw$$

$$w = 2.50 \ mfw \times \frac{0.276 \ g \ Ag_2CO_3}{mfw} = 0.690 \ g$$

Mathematical Operations Associated with Equilibrium Calculations

Throughout the study of analytical chemistry no concept is more pervasive than that of chemical equilibrium. The notion that reactions are never entirely complete will recur time and again throughout this text, as will discussion of the variables that affect the condition of equilibrium in a chemical system. For the present it is sufficient to note that the equilibrium law and its consequences are of such importance that it will be convenient to introduce the topic several times in the contexts of solubility, neutralization processes, complex formation reactions, and oxidation-reduction reactions. In these treatments it is assumed that the reader is familiar with manipulating exponential numbers, using logarithms, and solving

*11. Calculate the formal concentration of $(NH_4)_2SO_4$ and the molar concentration of NH_4^+ in a 20.0% (w/w) $(NH_4)_2SO_4$ solution that has a specific gravity of 1.12.

12. Calculate the formal concentration of K_2CrO_4 and the molar concentration of K^+ in a 38.0% (w/w) K_2CrO_4 solution that has a specific gravity of 1.37.

*13. Describe the preparation of
   (a) 250 ml of 9.0% (w/v) aqueous ethylene glycol.
   (b) 250 ml of 9.0% (w/w) aqueous ethylene glycol.
   (c) 250 ml of 9.0% (v/v) aqueous ethylene glycol.

14. Describe the preparation of
   (a) 1.00 liter of 20.0% (w/v) aqueous methanol.
   (b) 1.00 liter of 20.0% (w/w) aqueous methanol.
   (c) 1.00 liter of 20.0% (v/v) aqueous methanol.

*15. Describe the preparation of
   (a) 1.00 liter of a $1.00 \times 10^{-2}\ F$ solution of I_2 in CCl_4.
   (b) 250 ml of 0.125-M NH_4^+ from pure $(NH_4)_2SO_4$.
   (c) 100 ml of 0.100-F $K_2Cr_2O_7$ from a 1.09-F solution of the salt.
   (d) 250 ml of 1.00% (w/v) $AgNO_3$ from a 0.500-F solution of $AgNO_3$.
   (e) 3.00 liters of a solution containing 10.0 ppm of chloride from pure $BaCl_2 \cdot 2H_2O$.

16. Describe the preparation of
   (a) 600 ml of 0.300-F $BaCl_2$ from solid $BaCl_2 \cdot 2H_2O$.
   (b) 1.75 liters of 0.200-M K^+ from solid $K_4Fe(CN)_6$.
   (c) 100 ml of 0.100-F $AgNO_3$ from a solution that was 0.330 F in the salt.
   (d) 600 ml of a 0.200% solution (w/v) of $KMnO_4$ from a 0.137-F solution of $KMnO_4$.
   (e) 1.00 liter of a solution containing 2.5 ppm of Na^+ from a solution that was $2.00 \times 10^{-3}\ F$ in Na_2SO_4.

*17. Describe the preparation of
   (a) 750 ml of 0.225-F $K_2Cr_2O_7$ from the solid.
   (b) 50.0 liters of a solution that is 0.200 F in K_2SO_4 from the solid.
   (c) 250 ml of 1.00% (w/v) Na_2CO_3 from a 1.20-F solution of Na_2CO_3.
   (d) 20.0 liters of a solution that is 0.0202 M in Br^- from a 0.242-F solution of $AlBr_3$.

18. Describe the preparation of
   (a) 2.00 liters of a solution that is 0.0150 F in $K_3Fe(CN)_6$ from solid $K_3Fe(CN)_6$.
   (b) 400 ml of a solution that is 0.0150 M in K^+ from solid $K_3Fe(CN)_6$.
   (c) 500 ml of a solution that is 0.0300 M in K^+ from a 0.700-F solution of K_2HPO_4.
   (d) 500 ml of a solution that is 0.0200 F in $BaCl_2$ from a 1.25-F solution of $BaCl_2 \cdot 2H_2O$.

*19. Concentrated HNO_3 has a specific gravity of 1.42 and is 69% (w/w) HNO_3.
  (a) How many grams of HNO_3 are contained in 750 ml of this reagent?
  (b) Describe the preparation of 2.00 liters of a solution having a concentration of about 0.25 F in HNO_3 from the concentrated reagent.

20. Concentrated HCl has a specific gravity of 1.185 and is 36.5% (w/w) in HCl.
  (a) How many milliliters of HCl gas (measured at S.T.P.) are contained in 10.0 liters of the reagent?
  (b) Describe how 6.0 liters of approximately 0.15-F HCl should be prepared from the concentrated reagent.

*21. Describe the preparation of 1.00 liter of 6.0-F H_3PO_4 from the commercial reagent that is 85% (w/w) H_3PO_4 and has a specific gravity of 1.69.—

22. Describe the preparation of 200 ml of 6.0-F H_2SO_4 from the concentrated reagent that is 95% (w/w) H_2SO_4 and has a specific gravity of 1.84.

*23. Lanthanum ion reacts with iodate ion to form a slightly soluble precipitate having the formula $La(IO_3)_3$ and a gram formula weight of 664.
  (a) How many grams of KIO_3 are required to react completely with 3.65 g of $La(NO_3)_3$ (gfw = 325)?
  (b) What weight of KIO_3 will react with 3.00 fw of $La(NO_3)_3$?
  (c) How many grams of $La(NO_3)_3$ are required to react completely with 1.67 g of KIO_3?
  (d) How many grams of $La(IO_3)_3$ are formed when 1.00 g of $La(NO_3)_3$ is mixed with 2.00 g of KIO_3?
  (e) How many grams of $La(IO_3)_3$ are formed when 10.0 ml of 0.200-F $NaIO_3$ are mixed with 12.0 ml of 0.100-F $La(NO_3)_3$?

24. Silver ion reacts with arsenate ion to give the sparingly soluble Ag_3AsO_4 (gfw = 463).
  (a) What weight of $AgNO_3$ will react completely with 4.00 g of Na_3AsO_4 (gfw = 208)?
  (b) What weight of Ag_3AsO_4 can be formed from 1.00 fw of Na_3AsO_4?
  (c) What weight of Ag_3AsO_4 will form when 1.05 g of $AgNO_3$ are mixed with 3.33 g of Na_3AsO_4?
  (d) What weight of Ag_3AsO_4 will form when 2.00 liters of 0.163-F $AgNO_3$ are mixed with 0.500 liter of 0.101-F Na_3AsO_4?
  (e) What volume of 0.600-F $AgNO_3$ is needed to react completely with 7.25 g of Na_3AsO_4?

*25. How much of the substance in the second column is needed to react completely with the indicated amount of substance in the first column?
  (a) 12.0 mfw H_2SO_4 (a) g KOH
  (b) 3.50 mfw KOH   (b) ml of 0.075-F H_2SO_4
  (c) 0.700 g $Ba(OH)_2$ (c) ml of 0.100-F HCl
  (d) 18.6 ml of 0.303-F $BaCl_2$ (d) ml of 0.125-F $AgNO_3$
  (e) 1.5 ml HCl, sp gr = 1.12, (e) ml of 0.0200-F $Ba(OH)_2$
      % HCl (w/w) = 24.0

26. How much of the substance in the second column is needed to react completely with the indicated amount of substance in the first column?

   (a) 16.3 mfw HCl   (a) g $Ba(OH)_2$

   (b) 16.3 mfw $Ba(OH)_2$ (b) ml 0.100-F HCl

   (c) 1.62 g $Pb(NO_3)_2$ (c) g KCl

   (d) 3.25 ml $HClO_4$, sp gr = 1.60,   (d) g $Ca(OH)_2$
        % $HClO_4$ (w/w) = 70.0

   (e) 21.0 ml of 0.700-F H_2SO_4 (e) ml of 0.567-F KOH

3

a review of simple equilibrium constant calculations

Most reactions that are useful for chemical analysis proceed rapidly to a state of *chemical equilibrium* in which reactants and products exist in constant and predictable ratios. A knowledge of these ratios, under various experimental conditions, often permits the chemist to decide whether or not a reaction is suitable for analytical purposes and to choose conditions which will minimize the error associated with an analysis.

Equilibrium constant expressions are algebraic equations that relate the concentrations of reactants and products in a chemical reaction to one another by means of a numerical quantity called an *equilibrium constant*. A chemist must know how to derive useful information from equilibrium constants; it is the purpose of this chapter to review the calculations that are applicable to systems in which a single equilibrium predominates. Methods for treating complex systems involving several equilibria will be dealt with in later chapters.

The Equilibrium State

For purposes of discussion consider the equilibrium

$$2Fe^{3+} + 3I^- \rightleftharpoons 2Fe^{2+} + I_3^-$$ (3-1)

The rate of this reaction and the extent to which it proceeds to the right can be readily judged by observing the orange-red color imparted to the solution by the triiodide ion (at low concentrations the other three participants in the reaction are essentially colorless). If, for example, 2 mfw of iron(III) are added to a liter of solution containing 3 mfw of potassium iodide, color appears instantaneously; within a second or less the color intensity becomes constant with time, showing that the triiodide concentration has become invariant.

A solution of identical color intensity (and hence triiodide concentration) can be produced by adding 2 mfw of iron(II) to a liter of solution containing 1 mfw of triiodide ion. Here an immediate decrease in color is observed as a result of the reaction

$$2Fe^{2+} + I_3^- \rightleftharpoons 2Fe^{3+} + 3I^- \tag{3-2}$$

Many other combinations of the four reactants could be employed to yield solutions indistinguishable from the two just described.

The foregoing examples illustrate that the concentration relationships at chemical equilibrium (that is, the *position of equilibrium*) are independent of the route by which the equilibrium state is achieved. On the other hand, it is readily shown that these relationships are altered by the application of stress to the system—for example, by changes in temperature, in pressure (if one of the reactants or products is a gas), or in the total concentration of one of the reactants. These effects can be predicted qualitatively from the *principle of Le Châtelier*, which states that the position of chemical equilibrium will always shift in a direction that tends to relieve the effect of an applied stress. Thus an increase in temperature will alter the concentration relationships in the direction that tends to absorb heat; an increase in pressure favors those participants that occupy the smaller total volume. Of particular importance in an analysis is the effect of introducing an additional amount of a participating species to the reaction mixture; here the resulting stress is relieved by a shift in equilibrium in a direction that partially consumes the added substance. Thus for the equilibrium we have been considering, the addition of iron(III) would cause an increase in color as more triiodide ion and iron(II) are formed; the addition of iron(II) would have a reverse effect. An equilibrium shift brought about by changing the amount of one of the participants is called a *mass-action effect*.

If it were possible to examine a system at the molecular level, it would be found that interactions among the species present continue unabated even after equilibrium is achieved. The observed constant concentration relationship is thus the consequence of an equality in the rates of the forward and reverse reactions; that is, chemical equilibrium is a dynamic state.

Equilibrium Constant Expressions

The influence of concentration (or pressure, if the species are gases) on the position of a chemical equilibrium is conveniently described in quantitative terms by means of an equilibrium constant expression. Such expressions are readily derived from thermodynamic theory; they are of great practical importance because they permit the chemist to predict the direction and the completeness of a chemical reaction. It is important to note, however, that equilibrium constant expressions yield no information concerning reaction rates.

The equilibrium constant expression for a reaction can take one of two forms—an exact form (called the *thermodynamic expression*) or an approximate form that is applicable to a limited set of conditions only. Approximate equilibrium constant expressions are more frequently encountered because they are more convenient to use; to be sure, calculations based upon them will be subject to a degree of uncertainty.

In an approximate equilibrium constant expression it is assumed that concentration is the sole factor that governs the influence of any ion (or molecule) upon the condition of equilibrium, that is, that each species acts independently of its neighbors. For example, the influence of each iron(III) ion upon the equilibrium state in Equation 3-1 is assumed to be the same in very dilute solutions (where the ion is most likely surrounded by neutral water molecules) as it is in very concentrated solutions (where other charged species are in close proximity). A solution (or a gas) in which the ions or molecules act independently of one another is termed *ideal* or *perfect*. Truly ideal solutions and ideal behavior are seldom encountered in the laboratory.

Consider the generalized equation for a chemical equilibrium

$$m\text{M} + n\text{N} \rightleftharpoons p\text{P} + q\text{Q} \qquad (3\text{-}3)$$

where the capital letters represent the formulas of participating chemical species and the italic letters are the small integers required to balance the equation. Thus the equation states that m moles of M react with n moles of N to form p moles of P and q moles of Q. The approximate equilibrium constant expression for this reaction is

$$K = \frac{[\text{P}]^p[\text{Q}]^q}{[\text{M}]^m[\text{N}]^n} \qquad (3\text{-}4)$$

where the letters in brackets represent the molar concentrations of dissolved solutes or partial pressures (in atmospheres) if the reacting substances are gases.

The letter K in Equation 3-4 is a temperature-dependent, numerical constant called the *equilibrium constant*. By convention the concentrations of the products, as the equation is *written*, are always placed in the numerator and the concentrations of the reactants in the denominator.

Note also that each concentration is raised to a power that is identical to the integer that accompanies the formula of that species in the balanced equation describing the equilibrium.

Equation 3-4 can be readily derived from thermodynamic concepts by assuming that the participants in a reaction exhibit ideal behavior.[1]

Equation 3-4 is useful for determining the approximate equilibrium composition of a dilute solution. More exact results require the use of thermodynamic equilibrium constant expressions, which are considered in Chapter 5. The approximate form suffices to provide the chemist with adequate information for many purposes.

Common Types of Equilibrium Constant Expressions

In this section we shall consider the common types of equilibria encountered in analyses and describe some of the characteristics of the corresponding equilibrium constant expressions.

DISSOCIATION OF WATER

Aqueous solutions always contain small amounts of hydronium and hydroxide ions as a consequence of the dissociation reaction

$$2H_2O \rightleftharpoons H_3O^+ + OH^- \tag{3-5}$$

An equilibrium constant for this reaction can be formulated as shown in Equation 3-4, namely,

$$K = \frac{[H_3O^+][OH^-]}{[H_2O]^2}$$

In dilute aqueous solutions, however, the concentration of water is large compared with the concentration of solutes and can be considered to be invariant. That is, the quantity of water in each liter of a dilute aqueous solution (approximately 55.6 fw) is enormous compared with the amount formed or lost through any equilibrium shift. Therefore $[H_2O]$ in Equation 3-5 can be taken as constant, and we may write

$$K[H_2O]^2 = K_w = [H_3O^+][OH^-] \tag{3-6}$$

where the new K_w is given the special name the *ion product constant for water*.

At 25°C the ion product constant for water has a numerical value of 1.01×10^{-14} (for convenience we will normally use the approximation $K_w \cong 1.00 \times 10^{-14}$). Table 3-1 shows the dependence of this constant upon temperature.

---

[1] See L. K. Nash, *Elements of Classical and Statistical Thermodynamics*, Book 1. Reading, Mass.: Addison-Wesley, 1970, pp. 124–130.

TABLE 3-1
Variation of K_w with Temperature

TEMPERATURE, °C	K_w
0	0.114×10^{-14}
25	1.01×10^{-14}
50	5.47×10^{-14}
100	49×10^{-14}

The ion product constant for water permits the ready calculation of the hydronium and hydroxide ion concentrations of aqueous solutions.

EXAMPLE

Calculate the hydronium and hydroxide ion concentration of pure water at 25°C and at 100°C.

Because OH^- and H_3O^+ are formed from the dissociation of water only, their concentrations must be equal; that is,

$$[H_3O^+] = [OH^-]$$

Substitution into Equation 3-6 gives

$$[H_3O^+]^2 = [OH^-]^2 = K_w$$

$$[H_3O^+] = [OH^-] = \sqrt{K_w}$$

At 25°C

$$[H_3O^+] = [OH^-] = \sqrt{1.00 \times 10^{-14}} = 1.00 \times 10^{-7}$$

At 100°C

$$[H_3O^+] = [OH^-] = \sqrt{49 \times 10^{-14}} = 7.0 \times 10^{-7}$$

EXAMPLE

Calculate the hydronium and hydroxide ion concentrations in 0.200-F aqueous NaOH.

Sodium hydroxide is a strong electrolyte, and its contribution to the hydroxide ion concentration of this solution will be 0.200 mole/liter. As in the previous example, hydroxide ions and hydronium ions are also formed *in equal amounts* from the dissociation of water. Therefore we may write

$$[OH^-] = 0.200 + [H_3O^+]$$

where $[H_3O^+]$ accounts for the hydroxide ions contributed by the solvent. The concentration of OH^- from the water will be small when compared with 0.200; therefore

$$[OH^-] \cong 0.200$$

We can then employ Equation 3-6 to calculate the hydronium ion concentration

$$[H_3O^+] = \frac{1.00 \times 10^{-14}}{0.200} = 5.00 \times 10^{-14}$$

Note that the approximation

$$[OH^-] = 0.200 + 5.00 \times 10^{-14} = 0.200$$

will cause no significant error.

EQUILIBRIUM INVOLVING SLIGHTLY SOLUBLE IONIC SOLIDS

When an aqueous solution is saturated with a sparingly soluble salt, one or more equilibria will be established. To illustrate, the equilibrium that results when water is saturated with solid silver chloride can be described by the equation

$$AgCl(s) \rightleftharpoons Ag^+ + Cl^-$$

It is important to note that the existence of this equilibrium requires that there be an excess of *solid* silver chloride in contact with the saturated solution.

Application of Equation 3-4 to this equilibrium yields the expression

$$K = \frac{[Ag^+][Cl^-]}{[AgCl]_s} \qquad (3-7)$$

The term $[AgCl]_s$ is the concentration of AgCl *in the solid phase* and is a constant because the concentration of a compound in a pure solid is invariant. Thus Equation 3-7 can be rewritten in the form

$$K[AgCl]_s = K_{sp} = [Ag^+][Cl^-] \qquad (3-8)$$

where the new constant formed by combining the original equilibrium constant with the invariant AgCl concentration is called the *solubility product constant* and is given the symbol K_{sp}.

It is of utmost importance to keep in mind that a solubility product expression such as Equation 3-8 applies only to saturated solutions that are in contact with an excess of undissolved solid.

Appendix 7 contains solubility product constants for numerous inorganic salts. Typical uses of these constants are illustrated in the following examples; further applications are considered in Chapters 5 and 8.

EXAMPLE

How many grams of barium iodate can be dissolved in 500 ml of water at 25°C?

The solubility product constant for $Ba(IO_3)_2$ is 1.57×10^{-9} (Appendix 7). Thus

$$Ba(IO_3)_2(s) \rightleftharpoons Ba^{2+} + 2IO_3^-$$

and

$$[Ba^{2+}][IO_3^-]^2 = 1.57 \times 10^{-9} = K_{sp}$$

It is seen from the equation describing the equilibrium that 1 mole of Ba^{2+} is produced from each formula weight of $Ba(IO_3)_2$ that dissolves. Thus

$$\text{formal solubility of } Ba(IO_3)_2 = [Ba^{2+}]$$

The equation also indicates that the iodate concentration is twice that for barium ion, that is,

$$[IO_3^-] = 2[Ba^{2+}]$$

Substituting the latter into the equilibrium constant expression gives

$$[Ba^{2+}](2[Ba^{2+}])^2 = 1.57 \times 10^{-9}$$

or

$$[Ba^{2+}] = \left(\frac{1.57 \times 10^{-9}}{4}\right)^{1/3} = 7.3 \times 10^{-4}$$

and

$$\text{solubility} = 7.3 \times 10^{-4} \, F$$

To obtain the solubility of $Ba(IO_3)_2$ in grams per 500 ml, we write

$$\text{solubility} = 7.3 \times 10^{-4} \frac{\text{mfw } Ba(IO_3)_2}{\text{ml}} \times 500 \text{ ml} \times 0.487 \frac{\text{g}}{\text{mfw } Ba(IO_3)_2}$$

$$= 0.178 \text{ g}$$

where 0.487 is the milliformula weight of $Ba(IO_3)_2$.

The common ion effect predicted from the Le Châtelier principle is demonstrated by the following examples.

EXAMPLE

Calculate the formal solubility of $Ba(IO_3)_2$ in a solution that is 0.0200 F in $Ba(NO_3)_2$.

In this example the solubility is not directly related to $[Ba^{2+}]$ but can be described in terms of $[IO_3^-]$; that is,

$$\text{solubility of } Ba(IO_3)_2 = \tfrac{1}{2}[IO_3^-]$$

Here barium ions arise from two sources: $Ba(NO_3)_2$ and $Ba(IO_3)_2$. The contribution from the former is 0.0200 M while that from the latter is equal to the formal solubility or $\frac{1}{2}[IO_3^-]$. Thus

$$[Ba^{2+}] = 0.0200 + \tfrac{1}{2}[IO_3^-]$$

Substitution of these quantities into the solubility product expression yields

$$(0.0200 + \tfrac{1}{2}[IO_3^-])[IO_3^-]^2 = 1.57 \times 10^{-9}$$

Because the exact solution for $[IO_3^-]$ will involve a cubic equation, it is worthwhile to seek an approximation that will simplify the algebra (see Appendix 5). The small numerical value for K_{sp} suggests that the solubility of $Ba(IO_3)_2$ is not large; therefore it is reasonable to suppose that the barium ion concentration derived from the solubility of $Ba(IO_3)_2$ is small with respect to that from the $Ba(NO_3)_2$. That is,

$$0.0200 + \tfrac{1}{2}[IO_3^-] \cong 0.0200$$

The original equation then simplifies to

$$0.0200[IO_3^-]^2 = 1.57 \times 10^{-9}$$

$$[IO_3^-] = 2.80 \times 10^{-4}$$

The assumption that

$$(0.0200 + \tfrac{1}{2} \times 2.80 \times 10^{-4}) \cong 0.0200$$

does not appear to cause serious error because the second term is only about 0.7% of 0.0200. Ordinarily we shall consider an assumption of this type to be acceptable if the discrepancy is less than 5 to 10%. Therefore

$$\text{solubility of } Ba(IO_3)_2 = \tfrac{1}{2}[IO_3^-] = \tfrac{1}{2} \times 2.80 \times 10^{-4} = 1.40 \times 10^{-4} \ F$$

If we compare this result with the solubility of barium iodate in pure water, we see that the presence of a small concentration of the common ion has lowered the formal solubility of $Ba(IO_3)_2$ by a factor of about 5.

EXAMPLE

Calculate the solubility of $Ba(IO_3)_2$ in the solution that results when 200 ml of $0.0100\text{-}F$ $Ba(NO_3)_2$ are mixed with 100 ml of $0.100\text{-}F$ $NaIO_3$.

We must first establish whether either reactant will be present in excess. The amounts available are

$$\text{no. mfw } Ba^{2+} = 200 \text{ ml} \times 0.0100 \text{ mfw/ml} = 2.00$$

$$\text{no. mfw } IO_3^- = 100 \text{ ml} \times 0.100 \text{ mfw/ml} = 10.00$$

If formation of $Ba(IO_3)_2$ is complete,

$$\text{excess } IO_3^- = 10.00 - 2 \times 2.00 = 6.00 \text{ mfw}$$

Thus

$$F_{IO_3^-} = \frac{6.00 \text{ mfw}}{300 \text{ ml}} = 0.0200$$

As in the first example,

$$\text{formal solubility of } Ba(IO_3)_2 = [Ba^{2+}]$$

Here, however,

$$[IO_3^-] = 0.0200 + 2[Ba^{2+}]$$

where $2[Ba^{2+}]$ represents the contribution of iodate from the solubility of the precipitate. We can obtain a provisional answer by making the assumption that $2[Ba^{2+}]$ is small with respect to 0.0200. Then

$$[IO_3^-] \cong 0.0200$$

Thus

$$\text{solubility of } Ba(IO_3)_2 = [Ba^{2+}] = \frac{K_{sp}}{[IO_3^-]^2}$$

$$= \frac{1.57 \times 10^{-9}}{(0.0200)^2} = 3.9 \times 10^{-6}$$

The approximation used in this calculation is clearly reasonable.

Note that the results from the last two examples demonstrate that the presence of excess iodate is more effective in decreasing the solubility of $Ba(IO_3)_2$ than is an equal excess of barium ion.

DISSOCIATION OF WEAK ACIDS AND BASES

When a weak acid or base is dissolved in water, partial dissociation occurs. Thus for nitrous acid we may write

$$HNO_2 + H_2O \rightleftharpoons H_3O^+ + NO_2^- \qquad K_a = \frac{[H_3O^+][NO_2^-]}{[HNO_2]}$$

where K_a is the *acid dissociation constant* for nitrous acid. In an analogous way the *basic dissociation constant* for ammonia is given by

$$NH_3 + H_2O \rightleftharpoons NH_4^+ + OH^- \qquad K_b = \frac{[NH_4^+][OH^-]}{[NH_3]}$$

Note that a concentration term for water ($[H_2O]$) does not appear in either equation; as with the ion product constant for water, the solvent concentration is assumed to be large and constant and is thus incorporated in the equilibrium constants K_a and K_b. Dissociation constants for a number of common weak acids and bases are found in Appendixes 8 and 9.

Relationship between Dissociation Constants for Conjugate Acid-Base Pairs. Consider the dissociation constant expressions for ammonia and its conjugate acid, ammonium ion.

$$NH_3 + H_2O \rightleftharpoons NH_4^+ + OH^- \qquad K_b = \frac{[NH_4^+][OH^-]}{[NH_3]}$$

and

$$NH_4^+ + H_2O \rightleftharpoons NH_3 + H_3O^+ \qquad K_a = \frac{[NH_3][H_3O^+]}{[NH_4^+]}$$

Multiplication of the two equilibrium constant expressions together gives

$$K_a K_b = \frac{[\cancel{NH_3}][H_3O^+]}{[\cancel{NH_4^+}]} \times \frac{[\cancel{NH_4^+}][OH^-]}{[\cancel{NH_3}]} = [H_3O^+][OH^-]$$

But

$$[H_3O^+][OH^-] = K_w$$

Therefore

$$K_a K_b = K_w \qquad \qquad (3\text{-}9)$$

This relationship is general for all conjugate acid-base pairs. Most tables of dissociation constants do not list both the acid and base dissociation constants for such pairs because it is so easy to calculate one from the other by Equation 3-9.

EXAMPLE

What is K_b for the reaction

$$CN^- + H_2O \rightleftharpoons HCN + OH^-$$

Examination of Appendix 9 (dissociation constants for bases) reveals no entry for CN^-. In Appendix 8, however, K_a for HCN is found to have a value of 2.1×10^{-9}. Thus

$$K_b = \frac{[HCN][OH^-]}{[CN^-]} = \frac{K_w}{K_{HCN}}$$

$$K_b = \frac{1.00 \times 10^{-14}}{2.1 \times 10^{-9}} = 4.8 \times 10^{-6}$$

Applications of Acid Dissociation Constants. Acid dissociation constants permit the calculation of the hydronium or hydroxide ion concentrations of solutions of weak acids and their conjugate bases.

For example, the equilibrium established when the weak acid HA is dissolved in water may be written as

$$HA + H_2O \rightleftharpoons H_3O^+ + A^- \qquad K_a = \frac{[H_3O^+][A^-]}{[HA]}$$

In addition, hydronium ions result from the equilibrium

$$2H_2O \rightleftharpoons H_3O^+ + OH^-$$

Ordinarily the hydronium ions produced from the first reaction will suppress the dissociation of water to such an extent that the concentration of hydronium and hydroxide ions from this source can be considered to be negligible. Under these circumstances we see from the acid dissociation equilibrium that one A^- ion is produced with each H_3O^+ ion; that is,

$$[H_3O^+] \cong [A^-] \tag{3-10}$$

Furthermore, the sum of the molar concentrations of the weak acid and its conjugate base must equal the formal concentration of the acid because the solution contains no other species that contributes A^-. Thus

$$F_{HA} = [A^-] + [HA] \tag{3-11}$$

Substituting Equation 3-10 into 3-11 and rearranging give

$$[HA] = F_{HA} - [H_3O^+] \tag{3-12}$$

When $[A^-]$ and $[HA]$ are replaced by Equations 3-10 and 3-12, the equilibrium expression becomes

$$\frac{[H_3O^+]^2}{F_{HA} - [H_3O^+]} = K_a \tag{3-13}$$

Equation 3-13 can be rearranged to the form

$$[H_3O^+]^2 + K_a[H_3O^+] - K_a F_{HA} = 0$$

The solution to this quadratic equation is (see Appendix 3)

$$[H_3O^+] = \frac{-K_a + \sqrt{(K_a)^2 + 4K_a F_{HA}}}{2} \tag{3-14}$$

It is frequently possible to assume that $[H_3O^+]$ is much smaller than F_{HA}; that is, when $[H_3O^+] \ll F_{HA}$, Equation 13-12 becomes

$$[HA] \cong F_{HA}$$

Substituting this relationship and Equation 3-10 into the expression for K_a and rearranging yield

$$[H_3O^+] = \sqrt{K_a F_{HA}} \tag{3-15}$$

The magnitude of the error introduced by the assumption that $[H_3O^+] \ll F_{HA}$ will increase as the formal concentration of acid becomes smaller and the acid dissociation constant becomes larger. This statement is supported by the data in Table 3-2. Note that the error introduced by the assumption is about 0.5% when the ratio F_{HA}/K_a is 10^4. The error increases to about 1.6% when the ratio is 10^3, to about 5% when it is 10^2, and to about 17% when it is 10. Figure 3-1 illustrates the effect graphically. It is

TABLE 3-2
Errors Introduced by Assuming H_3O^+ Concentration Small Relative to F_{HA} in Equation 3-13

VALUE OF K_a	VALUE OF F_{HA}	VALUE FOR $[H_3O^+]$ USING ASSUMPTION	VALUE FOR $[H_3O^+]$ BY MORE EXACT EQUATION	PERCENT ERROR
1.00×10^{-2}	1.00×10^{-3}	3.16×10^{-3}	0.92×10^{-3}	244
	1.00×10^{-2}	1.00×10^{-2}	0.62×10^{-2}	61
	1.00×10^{-1}	3.16×10^{-2}	2.70×10^{-2}	17
1.00×10^{-4}	1.00×10^{-4}	1.00×10^{-4}	0.62×10^{-4}	61
	1.00×10^{-3}	3.16×10^{-4}	2.70×10^{-4}	17
	1.00×10^{-2}	1.00×10^{-3}	0.95×10^{-3}	5.3
	1.00×10^{-1}	3.16×10^{-3}	3.11×10^{-3}	1.6
1.00×10^{-6}	1.00×10^{-5}	3.16×10^{-6}	2.70×10^{-6}	17
	1.00×10^{-4}	1.00×10^{-5}	0.95×10^{-5}	5.3
	1.00×10^{-3}	3.16×10^{-5}	3.11×10^{-5}	1.6
	1.00×10^{-2}	1.00×10^{-4}	9.95×10^{-5}	0.5
	1.00×10^{-1}	3.16×10^{-4}	3.16×10^{-4}	0.0

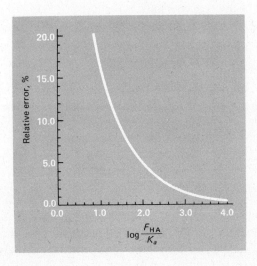

FIGURE 3-1 Relative error resulting from the assumption that $F_{HA} - [H_3O^+] \cong F_{HA}$ in Equation 3-13.

noteworthy that the hydronium ion concentration from the approximate solution becomes equal to or greater than the formality of the acid itself when the ratio is one or smaller. The approximation leads to meaningless results under these circumstances.

In general it is good practice to make the simplifying assumption and to obtain a trial value for $[H_3O^+]$ that may be compared with F_{HA} in Equation 3-13. If the trial value alters $[HA]$ by an amount smaller than the allowable error in the calculation, the solution may be considered satisfactory. Otherwise the quadratic equation must be solved to give a better value for $[H_3O^+]$.

nitrous acid HNO2

EXAMPLE

Calculate the hydronium ion concentration of an aqueous 0.120-F nitrous acid solution. The principal equilibrium in this solution is   *HNO2*

$$HNO_2 + H_2O \rightleftharpoons H_3O^+ + NO_2^-$$

In Appendix 8 we find

$$K_a = \frac{[H_3O^+][NO_2^-]}{[HNO_2]} = 5.1 \times 10^{-4}$$

Thus

$$[H_3O^+] = [NO_2^-]$$

$$[HNO_2] = 0.120 - [H_3O^+]$$

and

$$\frac{[H_3O^+]^2}{0.120 - [H_3O^+]} = 5.1 \times 10^{-4}$$

If we now assume $[H_3O^+] << 0.120$, we find

$$[H_3O^+] = \sqrt{0.120 \times 5.1 \times 10^{-4}} = 7.8 \times 10^{-3}$$

We must now examine the assumption that $(0.0120 - 0.0078) \cong 0.012$; a difference of about 7% is involved. The relative error in $[H_3O^+]$ will be smaller than this figure, however, as we can see by calculating $\log F_{HA}/K_a = 2.4$. Referring to Figure 3-1, we see that an error of about 3% results. If a more accurate figure were needed, solution of the quadratic equation would yield a value of 7.6×10^{-3} M.

EXAMPLE

Calculate the hydronium ion concentration of a solution which is 2.0×10^{-4} F in aniline hydrochloride $C_6H_5NH_3Cl$. In aqueous solution, dissociation to Cl^- and $C_6H_5NH_3^+$ is complete. The weak acid $C_6H_5NH_3^+$

dissociates as follows:

$$C_6H_5NH_3^+ + H_2O \rightleftharpoons C_6H_5NH_2 + H_3O^+ \qquad K_a = \frac{[H_3O^+][C_6H_5NH_2]}{[C_6H_5NH_3^+]}$$

Inspection of Appendix 8 reveals no entry for $C_6H_5NH_3^+$, but Appendix 9 gives a basic dissociation constant for aniline, $C_6H_5NH_2$; that is,

$$C_6H_5NH_2 + H_2O \rightleftharpoons C_6H_5NH_3^+ + OH^- \qquad K_b = 3.94 \times 10^{-10}$$

Thus Equation 3-9 is used to obtain a value for K_a:

$$K_a = \frac{1.00 \times 10^{-14}}{3.94 \times 10^{-10}} = 2.54 \times 10^{-5}$$

Proceeding as in the previous example,

$$[H_3O^+] = [C_6H_5NH_2]$$

$$[C_6H_5NH_3^+] = 2.0 \times 10^{-4} - [H_3O^+]$$

If we now assume that $[H_3O^+] \ll 2.0 \times 10^{-4}$ and substitute the simplified value for $[C_6H_5NH_2]$ into the dissociation constant expression,

$$\frac{[H_3O^+]^2}{2.0 \times 10^{-4}} = 2.54 \times 10^{-5}$$

$$[H_3O^+] = 7.1 \times 10^{-5}$$

Comparison of 7.1×10^{-5} with 2.0×10^{-4} suggests that a significant error exists in this value for $[H_3O^+]$ (using Figure 3-1, we find that this error is greater than 20%). Thus unless only an approximation is needed, it is necessary to use the more nearly exact expression

$$\frac{[H_3O^+]^2}{2.0 \times 10^{-4} - [H_3O^+]} = 2.54 \times 10^{-5}$$

which rearranges to

$$[H_3O^+]^2 + 2.54 \times 10^{-5}[H_3O^+] - 5.08 \times 10^{-9} = 0$$

and

$$[H_3O^+] = \frac{-2.54 \times 10^{-5} + \sqrt{(2.54 \times 10^{-5})^2 + 4 \times 5.08 \times 10^{-9}}}{2}$$

$$[H_3O^+] = 6.0 \times 10^{-5}$$

Application of Dissociation Constants for Weak Bases. The techniques discussed in previous sections are readily adapted to the calculation of the hydroxide ion concentration in solutions of weak bases.

Aqueous ammonia is basic by virtue of the reaction

$$NH_3 + H_2O \rightleftharpoons NH_4^+ + OH^-$$

Here the predominant species has been clearly demonstrated to be NH_3.

Nevertheless, such solutions are sometimes called ammonium hydroxide, the terminology being vestigial from the time when the substance NH_4OH rather than NH_3 was believed to be the undissociated form of the base. Application of the mass law to this equilibrium yields the expression

$$K_b = \frac{[NH_4^+][OH^-]}{[NH_3]}$$

The use of NH_3 in the denominator is preferable to the historical NH_4OH; in any event the magnitude of K_b remains unchanged.

EXAMPLE

Calculate the hydronium ion concentration of a 0.075-F NH_3 solution. The predominant equilibrium in this solution is

$$NH_3 + H_2O \rightleftharpoons NH_4^+ + OH^-$$

From the table of basic dissociation constants (Appendix 9),

$$\frac{[NH_4^+][OH^-]}{[NH_3]} = 1.76 \times 10^{-5} = K_b$$

The equation for the equilibrium indicates that

$$[NH_4^+] = [OH^-]$$

and

$$[NH_4^+] + [NH_3] = F_{NH_3} = 0.075$$

If we substitute $[OH^-]$ for $[NH_4^+]$ in the second of these equations and rearrange, we find that

$$[NH_3] = 0.075 - [OH^-]$$

Substituting these quantities into the dissociation constant expression yields

$$\frac{[OH^-]^2}{7.5 \times 10^{-2} - [OH^-]} = 1.76 \times 10^{-5}$$

If we assume that $[OH^-] << 7.5 \times 10^{-2}$, the equation then simplifies to

$$[OH^-]^2 \cong 7.5 \times 10^{-2} \times 1.76 \times 10^{-5}$$

and

$$[OH^-] = 1.15 \times 10^{-3}$$

Upon comparing the calculated value for $[OH^-]$ with 7.5×10^{-2}, we see that the error in $[OH^-]$ will be less than 2%. If needed, a better value for $[OH^-]$ could be obtained by solving the quadratic equation.

Finally, then,

$$[H_3O^+] = \frac{K_w}{[OH^-]} = \frac{1.00 \times 10^{-14}}{1.15 \times 10^{-3}} = 8.7 \times 10^{-12}$$

sodium hypochlorite NaOCl

EXAMPLE

Calculate the hydroxide ion concentration of a 0.010-F sodium hypo- $\boxed{NaOCl}$ chlorite solution.

The equilibrium between OCl^- and water is

$$OCl^- + H_2O \rightleftharpoons HOCl + OH^-$$

for which

$$K_b = \frac{[HOCl][OH^-]}{[OCl^-]}$$

Appendix 9 does not contain a value for K_b; an examination of Appendix 8, however, shows that the acid dissociation constant of HOCl is 3.0×10^{-8}. Therefore, employing Equation 3-9, we write

$$K_b = \frac{K_w}{K_a} = \frac{1.00 \times 10^{-14}}{3.0 \times 10^{-8}} = 3.3 \times 10^{-7}$$

Proceeding as in the previous example,

$$[OH^-] = [HOCl]$$

$$[OCl^-] + [HOCl] = 0.010$$

or

$$[OCl^-] = 0.010 - [OH^-] \cong 0.010$$

That is, we assume $[OH^-] \ll 0.010$. Substitution into the equilibrium constant expression gives

$$\frac{[OH^-]^2}{0.010} = 3.3 \times 10^{-7}$$

$$[OH^-] = 5.7 \times 10^{-5}$$

The error resulting from the approximation is clearly small.

COMPLEX FORMATION

An analytically important class of reactions involves the formation of soluble complex ions. Two examples are

$$Fe^{3+} + SCN^- \rightleftharpoons Fe(SCN)^{2+} \qquad K_f = \frac{[Fe(SCN)^{2+}]}{[Fe^{3+}][SCN^-]}$$

$$Zn(OH)_2(s) + 2OH^- \rightleftharpoons Zn(OH)_4^{2-} \qquad K_f = \frac{[Zn(OH)_4^{2-}]}{[OH^-]^2}$$

where K_f is called the *formation constant* for the complex.[2] Note that the second constant applies only to a saturated solution that is in contact with the sparingly soluble zinc hydroxide. Note also that no concentration term for the zinc hydroxide appears in the formation constant expression because its concentration in the solid is invariant and is included in the constant K_f. In this regard the treatment is analogous to that described for solubility product expressions.

An application involving the use of formation constants is given in the following example.

EXAMPLE

The average person can see the red color imparted by $FeSCN^{2+}$ to an aqueous solution if the concentration of the complex is $6 \times 10^{-6} M$ or greater. What minimum concentration of KSCN would be required to make it possible to detect 1 ppm of iron(III) in a natural water sample if the formation constant for the complex is 1.4×10^2?

$$Fe^{3+} + SCN^- \rightleftharpoons FeSCN^{2+} \qquad K_f = 1.4 \times 10^2$$

or

$$\frac{[FeSCN^{2+}]}{[Fe^{3+}][SCN^-]} = 1.4 \times 10^2$$

To convert the minimum detectable concentration of iron(III) to a formal concentration we recall (p. 14) that 1 ppm corresponds to 1 mg/liter. Thus

$$F_{Fe^{3+}} = \frac{1\ mg}{liter} \times \frac{1}{1000\ mg/g} \times \frac{1}{55.8\ g/fw}$$

$$= 1.8 \times 10^{-5}\ fw/liter$$

The detection limit for the complex is stated to be

$$[FeSCN^{2+}] = 6 \times 10^{-6}$$

At equilibrium the iron(III) will exist in two forms, Fe^{3+} and $FeSCN^{2+}$.

$$F_{Fe^{3+}} = 1.8 \times 10^{-5} = [Fe^{3+}] + [FeSCN^{2+}]$$

$$[Fe^{3+}] = 1.8 \times 10^{-5} - [FeSCN^{2+}] = 1.8 \times 10^{-5} - 6 \times 10^{-6}$$

$$= 1.2 \times 10^{-5}$$

We now substitute these concentrations into the formation constant expression in order to determine the SCN^- concentration that will be

---

[2] Equilibria involving complex ions are sometimes described in terms of dissociation reactions and *instability constants*; the latter are the reciprocals of formation constants. For example,

$$Fe(SCN)^{2+} \rightleftharpoons Fe^{3+} + SCN^- \qquad K_{inst} = \frac{1}{K_f} = \frac{[Fe^{3+}][SCN^-]}{[FeSCN^{2+}]}$$

required; that is,

$$\frac{6 \times 10^{-6}}{1.2 \times 10^{-5}[SCN^-]} = 1.4 \times 10^2$$

$$[SCN^-] = 0.0036$$

Thus if the test solution is made $0.0036\,F$ (or greater) in KSCN, a detectable amount of $FeSCN^{2+}$ will form, provided total iron(III) concentration is 1 ppm or greater.

OXIDATION-REDUCTION

Equilibrium constants for oxidation-reduction reactions can be formulated in the usual way. For example,

$$6Fe^{2+} + Cr_2O_7^{2-} + 14H_3O^+ \rightleftharpoons 6Fe^{3+} + 2Cr^{3+} + 21H_2O$$

$$K = \frac{[Fe^{3+}]^6[Cr^{3+}]^2}{[Fe^{2+}]^6[Cr_2O_7^{2-}][H_3O^+]^{14}}$$

As in earlier examples, no term for the concentration of water is needed.

Tables of equilibrium constants for oxidation-reduction reactions are not generally available because these constants are readily derived from more fundamental constants called *standard electrode potentials*. Derivations of this kind are treated in detail in Chapter 12.

DISTRIBUTION OF A SOLUTE BETWEEN TWO IMMISCIBLE LIQUIDS

Another important type of equilibrium involves the distribution of a solute between two liquid phases. For example, if an aqueous solution of iodine is shaken with an immiscible organic solvent such as hexane or chloroform, a portion of the solute is extracted into the organic layer. Ultimately an equilibrium is established between the two phases which can be described by

$$I_2(aq) \rightleftharpoons I_2(org) \qquad K = \frac{[I_2]_{org}}{[I_2]_{aq}}$$

The equilibrium constant for this reaction, K, is often called a *distribution* or *partition coefficient*.

As we shall see in Chapter 18, distribution equilibria are of vital importance in understanding many separation processes.

STEPWISE EQUILIBRIA

Many weak electrolytes associate or dissociate in a stepwise manner, and equilibrium constants can be written for each step. For example, when ammonia is added to a solution containing silver ions, at least two equilibria involving the two species are established.

$$Ag^+ + NH_3 \rightleftharpoons AgNH_3^+ \qquad K_1 = \frac{[AgNH_3^+]}{[Ag^+][NH_3]}$$

$$AgNH_3^+ + NH_3 \rightleftharpoons Ag(NH_3)_2^+ \qquad K_2 = \frac{[Ag(NH_3)_2^+]}{[AgNH_3^+][NH_3]}$$

Here K_1 and K_2 are stepwise formation constants for the two complexes. The two constants can be multiplied together to give an overall formation constant for the reaction

$$Ag^+ + 2NH_3 \rightleftharpoons Ag(NH_3)_2^+ \qquad \beta_2 = K_1K_2 = \frac{[Ag(NH_3)_2^+]}{[Ag^+][NH_3]^2}$$

Overall formation constants of this type are often symbolized as β_n, where n corresponds to the number of moles of complexing species that combine with 1 mole of cation. Thus, for example,

$$Cd^{2+} + 3CN^- \rightleftharpoons Cd(CN)_3^- \qquad \beta_3 = K_1K_2K_3 = \frac{[Cd(CN)_3^-]}{[Cd^{2+}][CN^-]^3}$$

$$Cd^{2+} + 4CN^- \rightleftharpoons Cd(CN)_4^{2-} \qquad \beta_4 = K_1K_2K_3K_4 = \frac{[Cd(CN)_4^{2-}]}{[Cd^{2+}][CN^-]^4}$$

where K_1, K_2, K_3, and K_4 are the corresponding stepwise constants.

Many common acids and bases undergo stepwise dissociation. For example, the following equilibria exist in an aqueous solution of phosphoric acid:

$$H_3PO_4 + H_2O \rightleftharpoons H_2PO_4^- + H_3O^+ \qquad K_1 = \frac{[H_3O^+][H_2PO_4^-]}{[H_3PO_4]}$$

$$= 7.11 \times 10^{-3}$$

$$H_2PO_4^- + H_2O \rightleftharpoons HPO_4^{2-} + H_3O^+ \qquad K_2 = \frac{[H_3O^+][HPO_4^{2-}]}{[H_2PO_4^-]}$$

$$= 6.34 \times 10^{-8}$$

$$HPO_4^{2-} + H_2O \rightleftharpoons PO_4^{3-} + H_3O^+ \qquad K_3 = \frac{[H_3O^+][PO_4^{3-}]}{[HPO_4^{2-}]}$$

$$= 4.2 \times 10^{-13}$$

As illustrated by H_3PO_4, numerical values of K_n for an acid or a base become smaller with each successive dissociation step. Equilibria of this type are considered in Chapter 9.

PROBLEMS

1. Generate solubility product expressions for
   *(a) $CuCN$.   (d) $Zn(OH)_2$.
   (b) $PbSO_4$.   *(e) $Al(OH)_3$.
   *(c) Ag_2CO_3.   (f) $Th(OH)_4$.

2. For each of the compounds in Problem 1, write an equation relating its formal solubility s to its solubility product constant K_{sp}.

3. Write the equilibrium constant expressions, and give the numerical values for each for the following reactions:
   *(a) the basic dissociation of ethylamine, $C_2H_5NH_2$.
   (b) the acidic dissociation of hydrogen cyanide, HCN.
   *(c) the acidic dissociation of pyridine hydrochloride, C_5H_5NHCl.

4. Write equilibrium constant expressions for each of the following reactions:
   *(a) the formation of $AgCl_2^-$ from $AgCl$.
   (b) the formation of CuI_2^- from CuI.
   *(c) the formation of $Cd(NH_3)_4^{2+}$ from NH_3 and Cd^{2+}.
   (d) the formation of AlF_6^{3-} from Al^{3+} and F^-.
   *(e) the dissociation of H_3AsO_4 to H_3O^+ and AsO_4^{3-}.
   (f) the reaction of CO_3^{2-} with H_2O to give H_2CO_3 and OH^-.

5. Calculate the solubility product constants for each of the following substances. Solubility data for each are given in parentheses.
   *(a) TlI (2.5×10^{-4} fw/liter) $TlI(s) \rightleftharpoons Tl^+ + I^-$
   (b) $SrSO_4$ (0.0104 g/100 ml) $SrSO_4(s) \rightleftharpoons Sr^{2+} + SO_4^{2-}$
   *(c) $Cu(IO_3)_2$ (1.09 mg/ml) $Cu(IO_3)_2(s) \rightleftharpoons Cu^{2+} + 2IO_3^-$
   (d) Ag_2SeO_3 (6.2×10^{-6} mfw/ml) $Ag_2SeO_3(s) \rightleftharpoons 2Ag^+ + SeO_3^{2-}$
   *(e) $Pb(IO_3)_2$ (0.024 mg/ml) $Pb(IO_3)_2(s) \rightleftharpoons Pb^{2+} + 2IO_3^-$
   (f) $Cu(OH)_2$ (1.11×10^{-2} fw/liter) $Cu(OH)_2(s) \rightleftharpoons Cu^{2+} + 2OH^-$
   *(g) $Ce(OH)_3$ (5.2×10^{-6} mfw/ml) $Ce(OH)_3(s) \rightleftharpoons Ce^{3+} + 3OH^-$
   (h) $La(IO_3)_3$ (460 mg/liter) $La(IO_3)_3 \rightleftharpoons La^{3+} + 3IO_3^-$

*6. Calculate the weight in grams of $PbCl_2$ that will dissolve in 100 ml of
   (a) H_2O.
   (b) 0.50-F KCl.
   (c) 0.50-F $Pb(NO_3)_2$.

7. The solubility product constant for $Ce(IO_3)_3$ is 3.2×10^{-10}. Calculate the weight of $Ce(IO_3)_2$ that will dissolve in 100 ml of
   (a) H_2O.
   (b) 0.50-F $NaIO_3$.
   (c) 0.50-F $Ce(NO_3)_3$.

*8. The solubility product for Tl_2CrO_4 is 9.8×10^{-13}. What CrO_4^{2-} concentration is required to lower the concentration of Tl^+ in a solution to 1.00×10^{-6} M?

9. What hydroxide concentration is required to lower the Fe^{3+} concentration of an $Fe_2(SO_4)_3$ solution to 1.00×10^{-9} F?

10. What is the hydroxide ion concentration of a saturated solution of
    *(a) $Pb(OH)_2$? 4×10^{-6} —→ Ans 7.9×10^{-6}
    (b) $Mg(OH)_2$?

*11. What will be the La^{3+} concentration in a solution prepared by mixing 50.0 ml of 0.0500-F La^{3+} with
    (a) 50.0 ml of water?
    (b) 50.0 ml of 0.150-F IO_3^-?
    (c) 50.0 ml of 0.300-F IO_3^-?
    (d) 50.0 ml of 0.050-F IO_3^-?

12. The solubility product constant for K_2PtCl_6 is 1.1×10^{-5}.

$$K_2PtCl_6 \rightleftharpoons 2K^+ + PtCl_6^{2-}$$

What is the K^+ concentration of a solution prepared by mixing 50.0 ml of 0.400-F KCl with
    (a) 50.0 ml of 0.100-M $PtCl_6^{2-}$?
    (b) 50.0 ml of 0.200-M $PtCl_6^{2-}$?
    (c) 50.0 ml of 0.400-M $PtCl_6^{2-}$?

13. At 25°C what are the molar H_3O^+ and OH^- concentrations of
    *(a) 0.0400-F HOCl?
    (b) 0.200-F acetic acid?
    *(c) 0.0500-F ethylamine?
    (d) 0.100-F hydroxylamine?
    *(e) 0.120-F sodium formate?
    (f) 0.0860-F sodium benzoate?
    *(g) 0.200-F hydroxylamine hydrochloride?
    (h) 0.0500-F methylamine hydrochloride?

14. What is the hydronium ion concentration of water at 0°C?

15. At 25°C what is the hydronium ion concentration of
    *(a) 0.100-F trichloroacetic acid? $Cl_3CCOOH + H_2O \rightleftharpoons Cl_3CCOO^- + H^+$
    *(b) 0.300-F sodium chloroacetate? $Na/ClCH_2COO + H_2O \rightleftharpoons ClCH_2COOH + OH^-$
    (c) 0.0100-F ethylamine?
    (d) 0.0100-F ethylamine hydrochloride?
    *(e) 1.00×10^{-4} F pyridine hydrochloride?
    (f) 0.200-F sulfamic acid?

16. Calculate the hydronium ion concentration of
    *(a) a 0.0200-F solution of NH_4NO_3.
    (b) a 0.0200-F solution of $NaNO_2$.

*17. Mercury(II) forms a soluble, neutral complex with Cl^- having the formula $HgCl_2$. The formation constant of the complex has a value of 1.7×10^{13}. Calculate the concentrations of the three species in a solution prepared by mixing 10 ml of 0.0200-F Hg^{2+} with

(a) 10 ml of 0.016-F Cl^-.

(b) 10 ml of 0.040-F Cl^-.

(c) 10 ml of 0.066-F Cl^-.

*18. Calculate the molar concentration of $Cu(SCN)_2^-$ in a solution that is saturated with CuSCN and has the following formal concentration of KSCN, given that the formation constant for the reaction $CuSCN + SCN^- \rightleftharpoons Cu(SCN)_2^-$ is 4.8×10^{-4}:

(a) 0.00100.

(b) 0.0100.

(c) 0.100.

19. The formation constant for $Pb(OH)_3^-$ is

$$Pb(OH)_2(s) + OH^- \rightleftharpoons Pb(OH)_3^- \qquad K_f = 5 \times 10^{-2}$$

Calculate the concentration of $Pb(OH)_3^-$ in a saturated $Pb(OH)_2$ solution that is

(a) 0.00300-F in NaOH.

(b) 0.300-F in NaOH.

4
the evaluation of
analytical data

Every physical measurement is subject to a degree of uncertainty that, at best, can only be decreased to an acceptable level. The magnitude of this uncertainty is often difficult to determine and requires additional effort, ingenuity, and good judgment on the part of the scientist. Nevertheless, it is a task that cannot be neglected because an analysis of totally unknown reliability is worthless. On the other hand, a result that is not highly accurate may be of great use provided that the limits of probable error affecting it can be set with a high degree of certainty. Unfortunately there exists no simple, generally applicable means by which the quality of an experimental result can be assessed with absolute certainty; indeed, the work expended in evaluating the reliability of data is frequently comparable to the effort that went into obtaining them. This effort may involve a study of the literature to profit from the experience of others, the calibration of equipment, additional experiments specifically designed to provide clues to possible errors, and statistical analysis of the data. It should be recognized, however, that none of these measures is infallible. Ultimately the scientist can only make a *judgment* as to the probable accuracy of a measurement; with experience, judgments of this kind tend to become harsher and less optimistic.

A direct relationship exists between the accuracy of an analytical measurement and the time and effort expended in its acquisition. A tenfold increase in reliability may require hours, days, or perhaps weeks of added labor. As a first step, then, the experienced scientist establishes how reliable the results of an analysis must be; this consideration will determine the amount of time and effort that must be invested in the analysis. Careful thought at the outset of an investigation often provides major savings in time and effort. *It cannot be too strongly emphasized that a scientist cannot afford to waste time in the indiscriminate pursuit of the ultimate in accuracy when such is not needed.*

In this chapter we consider the types of errors encountered in analyses, methods for their recognition, and techniques for estimating and reporting their magnitude.

Definition of Terms

The chemist generally repeats the analysis of a given sample two to five times. The individual results for such a set of replicate measurements will seldom be exactly the same; it thus becomes necessary to select a central "best" value for the set. Intuitively, the added effort of replication can be justified in two ways. First, the central value of the set ought to be more reliable than any of the individual results; second, the variations among the results ought to provide some measure of reliability in the "best" value that is chosen.

Either of two quantities, the *mean* or the *median*, may serve as the central value for a set of measurements.

THE MEAN AND MEDIAN

The *mean*, *arithmetic mean*, and *average* ($\bar{x}$) are synonymous terms for the numerical value obtained by dividing the sum of a set of replicate measurements by the number of individual results in the set.

The *median* of a set is that result about which all others are equally distributed, half being numerically greater and half being numerically smaller. If the set consists of an odd number of measurements, selection of the median may be made directly; for a set containing an even number of measurements, the average of the central pair is taken.

EXAMPLE

Calculate the mean and median for 10.06, 10.20, 10.08, 10.10.

$$\text{mean} = \bar{x} = \frac{10.06 + 10.20 + 10.08 + 10.10}{4} = 10.11$$

Because the set contains an even number of measurements, the median is the average of the middle pair:

$$\text{median} = \frac{10.08 + 10.10}{2} = 10.09$$

Ideally the mean and median should be numerically identical; more often than not, however, this condition is not realized, particularly when the number of measurements in the set is small.

PRECISION

The term *precision* is used to describe the reproducibility of results. It can be defined as the agreement between the numerical values of two or more measurements that have been made *in an identical fashion*. Several methods exist for expressing the precision of data.

Absolute Methods for Expressing Precision. The *deviation from the mean* $(x_i - \bar{x})$ is a common method for describing precision and is simply the numerical difference, *without regard to sign*, between an experimental value and the mean for the set of data that includes the value. To illustrate, suppose that a chloride analysis has yielded the following results:

SAMPLE	PERCENT CHLORIDE	DEVIATION FROM MEAN $\lvert x_i - \bar{x} \rvert$	DEVIATION FROM MEDIAN
x_1	24.39	0.077	0.03
x_2	24.19	0.123	0.17
x_3	24.36	0.047	0.00
	3|72.94	3|0.247	3|0.20
	$\bar{x} = 24.313 \cong 24.31$	avg $= 0.082 \cong 0.08$	avg $= 0.067 \cong 0.07$
	$w = x_{max} - x_{min} = 24.39 - 24.19 = 0.20$		

The mean value for the data is 24.31%; the deviation of the second result from the mean is 0.12%. The average deviation from the mean is 0.08%. Note that we have carried three figures to the right of the decimal point in the calculation for the average deviation even though the individual data are given to only two. The mean and the average deviation from the mean are then rounded to the proper number of figures *after the computation is complete*. This practice is generally worthwhile because it tends to minimize rounding errors.

Precision can also be reported in terms of *deviation from the median*. In the preceding example, deviations from 24.36 would be recorded, as shown in the last column of the table.

The *spread* or *range* (*w*) in a set of data is the numerical difference between the highest and lowest result; it is also a measure of precision. In the previous example the spread is 0.20% chloride.

Two other measures of precision are the *standard deviation* and the *variance*. These terms will be defined in a later section of this chapter.

Relative Methods for Expressing Precision. We have thus far calculated precision in absolute terms. It is often more informative, however, to indicate the precision relative to the mean (or the median) in terms of percentage or as parts per thousand. For example, for sample x_1 in the earlier example,

$$\text{relative deviation from mean} = \frac{0.077 \times 100}{24.31} = 0.32 = 0.3\%$$

Similarly, the average deviation of the set from the median can be expressed as

$$\text{relative average deviation from median} = \frac{0.067 \times 1000}{24.36} = 2.8 = 3 \text{ ppt}$$

ACCURACY

The term *accuracy* denotes the nearness of a measurement to its accepted value and is expressed in terms of *error*. Note the fundamental difference between this term and precision. Accuracy involves a comparison with respect to a true or accepted value; precision compares a result with other measurements made in the same way.

The accuracy of a measurement is often described in terms of *absolute error*, E, which can be defined as the difference between the observed value x_i and the accepted value x_t.

$$E = x_i - x_t \tag{4-1}$$

The accepted value may itself be subject to considerable uncertainty; as a consequence it is frequently difficult to arrive at a realistic estimate for the error of a measurement.

Returning to the previous example, suppose that the accepted value for the percentage of chloride in the sample is 24.36%. The absolute error of the mean is thus $24.31 - 24.36 = -0.05\%$ chloride; here we ordinarily retain the sign of the error to indicate whether the result is high or low.

Often a more useful quantity than the absolute error is the relative error, which is expressed as a percentage or in parts per thousand of the accepted value. Thus the chloride analysis we have been considering,

$$\text{relative error} = -\frac{0.05 \times 100}{24.36} = -0.21 = -0.2\%$$

$$\text{relative error} = -\frac{0.05 \times 1000}{24.36} = -2.1 = -2 \text{ ppt}$$

PRECISION AND ACCURACY OF EXPERIMENTAL DATA

The precision of a measurement is readily determined by performing replicate experiments under identical conditions. Unfortunately an estimate of the accuracy is not equally available because this quantity requires sure knowledge of the very information that is being sought, namely, the true value. It is tempting to ascribe a direct relationship between precision and accuracy. The danger of this approach is illustrated in Figure 4-1, which summarizes the results of an analysis for nitrogen in two pure compounds by four analysts. The dots give the absolute errors of replicate measurements for each sample and each analyst. Note that analyst 1 obtained relatively high precision and high accuracy. Analyst 2, on the other hand, had poor precision but good accuracy. The results from analyst 3 are of a kind that is by no means uncommon; the precision is excellent, but a significant error exists in the numerical average for the data. The scientist also encounters a situation similar to that recorded by analyst 4 whereby both precision and accuracy are poor.

The behavior illustrated by Figure 4-1 can be rationalized by assuming that experimental measurements are afflicted by two general

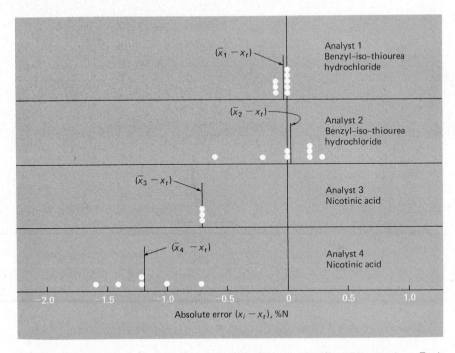

FIGURE 4-1   Absolute errors in nitrogen analyses by a micro Kjeldahl procedure. Each vertical line labeled $(\bar{x}_i - x_t)$ is the absolute deviation of the mean of the set from the true value. [Data taken from C. O. Willits and C. L. Ogg, *J. Ass. Offic. Anal. Chem.*, **32**, 561 (1949).]

types of uncertainty and that the effect of one of these types is not revealed by the precision of the measurements.

CLASSES OF ERRORS

The uncertainties that arise in a chemical analysis and are responsible for the behavior illustrated in Figure 4-1 may be classified into two broad categories, *determinate errors* and *indeterminate* or *random errors*. At the outset it should be pointed out that it is frequently difficult or impossible to be certain of the category in which a given error belongs; indeed, assignment to one or the other may be largely a subjective judgment. Nevertheless, the classification is useful for discussing analytical errors.

Determinate Errors. Determinate errors are those that have a definite value and an assignable cause; in principle (but not always in practice) these errors can be measured and accounted for. A determinate error is ordinarily unidirectional; that is, it will cause all of a series of replicate analyses to be either high or low. Thus the last two sets of data for nicotinic acid shown in Figure 4-1 appear to be affected by a negative determinate error, and the probable source of this error can be identified. The first step in this analysis requires the oxidation of the sample with concentrated sulfuric acid, a process that ordinarily converts the nitrogen present to ammonium sulfate. It has been found, however, that compounds containing a pyridine ring (such as nicotinic acid) are incompletely oxidized unless special precautions are taken; low results are the consequence. It is highly likely that the negative errors [$(\bar{x}_3 - x_t)$ and $(\bar{x}_4 - x_t)$] shown in Figure 4-1 are determinate and attributable to this incomplete oxidation.

Indeterminate Errors. Indeterminate errors result from extending a system of measurements to its maximum. The sources of these errors cannot be positively identified, and the magnitudes of individual indeterminate errors are not measurable. The most important consequence of indeterminate errors is that they cause the data from replicate measurements to fluctuate in a *random* manner, in some instances causing high results and in others low.

The scatter of the individual errors about mean values ($x_n - \bar{x}$) in Figure 4-1 is a direct indication of indeterminate-type uncertainties. Larger indeterminate errors appear to be associated with the work of analysts 2 and 4 than with that of analysts 1 and 3.

SOURCES OF DETERMINATE AND INDETERMINATE ERRORS

It is not possible to list all conceivable causes of determinate and indeterminate errors. We can, however, recognize that limitations to both precision and accuracy can be traced to three general sources: (1) *instrument uncertainties* that are attributable to imperfections in measuring devices, (2) *method uncertainties* that arise from nonideal chemical or physical behavior of analytical systems, and (3) *personal uncertainties* that are caused by physical or psychological limitations of experimenters. Examples of these three sources appear in the sections that follow.

Detection and Correction of Determinate Errors

SOURCES OF DETERMINATE ERRORS

Instrument Errors. All measuring devices are potential sources of determinate errors. For example, pipets, burets, and volumetric flasks frequently deliver or contain volumes slightly different from those indicated by their graduations. These differences may arise from such sources as their use at temperatures that differ significantly from the calibration temperature, distortions in the container walls due to heating while drying, errors in the original calibration, or contaminants on the inner surfaces of the containers. Many determinate errors of this type are readily eliminated by calibration.

Measuring devices powered by electricity are commonly subject to determinate errors. Examples include decreases in the voltage of battery-operated power supplies with use, increased resistance in circuits because of dirty electrical contacts, vibration, and currents induced from 110-V power lines. Again, these errors are detectable and correctable; most are unidirectional.

Method Errors. Determinate errors are often introduced from nonideal chemical or physical behavior of the reagents and reactions upon which an analysis is based. Such sources of nonideality include the slowness of some reactions, the incompleteness of others, the instability of some species, the nonspecificity of most reagents, and the possible occurrence of side reactions which interfere with the measurement process. For example, in a gravimetric analysis the chemist is confronted with the problem of isolating the element to be determined as a solid of the greatest possible purity. If this solid is not washed sufficiently, the precipitate will be contaminated with foreign substances and have a spuriously high weight. On the other hand, the washing needed to remove these contaminants may cause weighable quantities to be lost owing to the solubility of the precipitate; here a negative determinate error will result. In either event the accuracy of the procedure is limited by a method error associated with the analysis.

A method error frequently encountered in volumetric analysis is due to the volume of reagent, in excess of theoretical, which is required to cause an indicator to undergo the color change that signals completion of the reaction. In this case the ultimate accuracy of the analysis is limited by the very phenomenon that makes the determination possible.

Errors inherent in a method are probably the most serious of determinate errors because they are the most likely to remain undetected.

Personal Errors. Many measurements require personal judgments. Examples include the position of a pointer between two scale divisions, the color of a solution at the end point in a volumetric analysis, the level of a liquid with respect to a graduation in a pipet, or the relative intensity of two light beams. Judgments of this type are often subject to systematic, unidirectional uncertainties. For example, one person may read a pointer consistently high, another may be slightly slow in activating a timer, and a third may be less sensitive to color changes and thus tend to employ excess reagent in a volumetric analysis. Color blindness or other physical handicaps often increase the probability of determinate personal errors.

A near-universal source of personal error is prejudice or bias. Most of us, no matter how honest, have a natural tendency to estimate scale readings in a direction that improves the precision in a set of results or causes the results to fall closer to a preconceived notion of the true value for the measurement. Scientists must actively fight this tendency; it does not suffice to assume that bias occurs only in others.

Gross Mistakes. Errors of this type arise from arithmetic mistakes, transposition of numbers in recording data, reading a scale backward, reversing a sign, or using a wrong scale. In some instances this type of error will affect only a single result; in others, such as using the wrong scale of an instrument, an entire set of replicate measurements may be affected. Errors of this type are ordinarily the consequence of carelessness and can be eliminated by self-discipline.

EFFECT OF DETERMINATE ERROR UPON THE RESULTS OF AN ANALYSIS

Determinate errors may be classified as being either *constant* or *proportional*. The magnitude of a constant error is independent of the size of the quantity measured. On the other hand, proportional errors increase or decrease in proportion to the size of the sample taken for analysis.

Constant Errors. For any given analysis a constant error will become more serious as the size of the quantity measured decreases. This problem is illustrated by the solubility losses that attend the washing of a precipitate.

EXAMPLE

Suppose that 0.50 mg of precipitate is lost as a result of being washed with 200 ml of wash liquid. If 500 mg of precipitate are involved, the relative error due to solubility loss will be $-(0.50 \times 100/500) = -0.1\%$. Loss of the same quantity from 50 mg of precipitate will result in a relative error of -1.0%.

The amount of reagent required to bring about the color change in a volumetric analysis is another example of constant error. This volume, usually small, remains the same regardless of the total volume of reagent required for the titration. Again, the relative error will be more serious as the total volume decreases. Clearly one way of minimizing the effect of constant error is to use as large a sample as is consistent with the method at hand.

Proportional Errors. Interfering contaminants in the sample, if not eliminated somehow, will cause an error of the proportional variety. For example, a method widely employed for the analysis of copper involves reaction of copper(II) ion with potassium iodide; the quantity of iodine produced in the reaction is then measured. Iron(III), if present, will also liberate iodine from potassium iodide. Unless steps are taken to prevent this interference, the analysis will yield erroneously high results for the percentage of copper because the iodine produced will be a measure of the sum of the copper and iron in the sample. The magnitude of this error is fixed by the *fraction* of iron contamination and will produce the same relative effect regardless of the size of sample taken for analysis. If the sample size is doubled, for example, the amount of iodine liberated by both the copper and the iron contaminant will also be doubled. Thus the reported percentage of copper will be independent of sample size.

DETECTION OF INSTRUMENTAL AND PERSONAL ERRORS

Instrumental errors are usually found and corrected by calibration procedures. Indeed, periodic recalibration of equipment is always desirable because the response of most instruments changes with time owing to wear, corrosion, or mistreatment.

Most personal errors can be minimized by care and self-discipline. Thus most scientists develop the habit of always rechecking instrument readings, notebook entries, and calculations. Errors that result from a physical handicap can usually be avoided by a judicious choice of method, provided, of course, that the handicap is recognized.

DETECTION OF METHOD ERRORS

Method errors are particularly difficult to detect. Identification and compensation for systematic errors of this type may take one or more of the courses described in the following paragraphs.

Analysis of Standard Samples. A method may be tested for determinate error by analysis of a synthetic sample whose overall composition is known and closely approximates that of the material for which the analysis is intended. Great care must go into the preparation of standard samples to ensure that the concentration of the constituent to be determined is known with a high degree of certainty. Unfortunately the preparation of a sample whose composition truly resembles that of a complex natural substance is often difficult, if not impossible. Moreover, the problem is compounded by the requirement that the exact concentration of a particular constituent be known as a result of the method of preparation. These problems are frequently so imposing as to prevent the use of this approach.

Several hundred common substances that have been carefully analyzed for one or more constituents are available from the National Bureau of Standards. These standard materials are valuable for the testing of analytical procedures for accuracy.[1]

Independent Analysis. Parallel analysis of a sample by a method of established reliability that differs from the one under investigation is of particular value where samples of known purity are not available. In general the independent method should not resemble the one under study to minimize the possibility that some common factor in the sample will have an equal effect on both methods.

Blank Determinations. Constant errors affecting physical measurements can be frequently evaluated with a blank determination, in which all steps of the analysis are performed in the absence of a sample. The result is then applied as a correction to the actual measurement. Blank determinations are of particular value in exposing errors that are due to the introduction of interfering contaminants from reagents and vessels employed in an analysis. They also enable the analyst to correct titration data for the volume of reagent needed to cause an indicator to change color at the end point of a volumetric analysis.

[1] See the current edition of NBS Circular 552 for a description of available samples and their prices. For a description of the Reference Material Program of the NBS see J. P. Cali, *Anal. Chem.*, **48**, 802A (1976).

Variation in Sample Size. As was demonstrated in the example on page 54, a constant error has a decreasing effect on a result as the size of the measurement increases. This fact can be helpful in detecting constant errors in a method. Here the sample size is varied as widely as possible. In the presence of a constant error, the results will be found to increase or decrease systematically with sample size.

Indeterminate Error

As suggested by its name, indeterminate error arises from uncertainties in a measurement that are unknown and not controlled by the scientist. The effect of such uncertainties is to produce a *random* scatter of results for replicate measurements such as those for the four sets of data shown in Figure 4-1.

Table 4-1 illustrates the effect of indeterminate error upon the relatively simple process of calibrating a pipet. The procedure involves determining the weight of water (to the nearest milligram) delivered by the pipet. The temperature of the water must be measured to establish its

TABLE 4-1
Replicate Measurements from the Calibration of a 10-ml Pipet

TRIAL	VOLUME OF WATER DELIVERED, ml	TRIAL	VOLUME OF WATER DELIVERED, ml	TRIAL	VOLUME OF WATER DELIVERED, ml
1	9.990	9	9.988	17	9.977
2	9.985	10	9.976	18	9.982
3	9.973	11	9.981	19	9.974
4	9.980	12	9.974	20	9.985
5	9.982	13	9.970[b]	21	9.987
6	9.988	14	9.989	22	9.982
7	9.993[a]	15	9.981	23	9.979
8	9.970[b]	16	9.985	24	9.988

Mean volume = 9.9816 = 9.982 ml
Average deviation from mean = 0.0052 ml = 0.005 ml
Spread = 9.993 − 9.970 = 0.023 ml
Standard deviation = 0.0065 ml = 0.006 ml

[a] Maximum value
[b] Minimum value

density. The experimental weight can then be converted to the volume delivered by the pipet.

The data in Table 4-1 are typical of those that might be obtained by an experienced and competent worker who performs the weighings to the nearest milligram (which corresponds to 0.001 ml), with every effort being made to recognize and eliminate determinate errors. Even so, the average deviation from the mean of the 24 measurements is ±0.005 ml, and the spread is 0.023 ml. This dispersion among the data is the direct consequence of indeterminate error.

Variations among replicate results such as those in Table 4-1 can be rationalized by assuming that any measurement is affected by numerous small and individually undetectable instrument, method, and personal uncertainties caused by uncontrolled variables in the experiment. The cumulative effect of such uncertainties will be likewise variable. Ordinarily they tend to cancel one another and thus exert a minimal effect. Occasionally, however, they can act in concert to produce a relatively large positive or negative error. Sources for uncertainties in this calibration process might include such visual judgments as the liquid level with respect to the etch mark on the pipet, the mercury level in the thermometer, and the position of an indicator with respect to a scale in the balance (all personal uncertainties). Other sources include variation in the drainage time, the angle of the pipet as it drains (both method uncertainties), and temperature, and thus volume change, resulting from the way the pipet is handled (an instrument uncertainty). Undoubtedly numerous other uncertainties exist in addition to the ones cited. It is clear that many small and uncontrolled variables accompany even as simple an experiment as a pipet calibration. Although we are unable to detect the influence of any one of these uncertainties, their cumulative effect is an indeterminate error that accounts for the scatter of data about the mean.

THE DISTRIBUTION OF DATA FROM REPLICATE MEASUREMENTS

In contrast to determinate errors, indeterminate errors cannot be eliminated from measurements. Furthermore, the scientist cannot ignore their existence on the basis of their small size. For example, it would probably be safe to assume that the average value of the 24 measurements in Table 4-1 is closer to the true volume delivered by the pipet than any of the individual data. Suppose, however, that only a duplicate calibration had been performed and that by chance these measurements had corresponded to trials 1 and 7; the average of these two values, 9.992, differs by 0.010 ml from the mean of the 24 measurements. Note also that the average deviation of these two measurements *from their own mean* is only ±0.0015 ml. On the basis of this small average deviation from the mean, an overly optimistic idea of the magnitude of the indeterminate error might arise. Serious consequences would result if the user of the pipet needed to deliver volumes known, let us say, to the nearest ± 0.002 ml. Here failure to recognize the true magnitude of the indeterminate error would create a totally false sense of security with respect to the performance of the pipet. If this pipet were employed for 1000 measurements it can be shown that two to three of the transfers would, with high probability, differ from the mean of 9.982 ml by as much as 0.02 ml; more than 100 would differ by 0.01 ml or greater, despite every precaution on the part of the user.

To develop a qualitative grasp of the way small uncertainties affect the outcome of replicate measurements, let us first consider an imaginary situation in which just four uncertainties are the cause of indeterminate error. We shall specify that each of these uncertainties has an equal probability of occurring and can affect the final result in only one of two ways, namely, to cause it to be in error by plus or minus a fixed amount,

TABLE 4-2

Possible Ways Four Equal-sized Uncertainties U_1, U_2, U_3, and U_4 Can Combine

COMBINATIONS OF UNCERTAINTIES	MAGNITUDE OF INDETERMINATE ERROR	RELATIVE FREQUENCY OF ERROR
$+U_1 + U_2 + U_3 + U_4$	$+4U$	1
$-U_1 + U_2 + U_3 + U_4$		
$+U_1 - U_2 + U_3 + U_4$	$+2U$	4
$+U_1 + U_2 - U_3 + U_4$		
$+U_1 + U_2 + U_3 - U_4$		
$-U_1 - U_2 + U_3 + U_4$		
$+U_1 + U_2 - U_3 - U_4$		
$+U_1 - U_2 + U_3 - U_4$		
$-U_1 + U_2 - U_3 + U_4$	0	6
$-U_1 + U_2 + U_3 - U_4$		
$+U_1 - U_2 - U_3 + U_4$		
$+U_1 - U_2 - U_3 - U_4$		
$-U_1 + U_2 - U_3 - U_4$	$-2U$	4
$-U_1 - U_2 + U_3 - U_4$		
$-U_1 - U_2 - U_3 + U_4$		
$-U_1 - U_2 - U_3 - U_4$	$-4U$	1

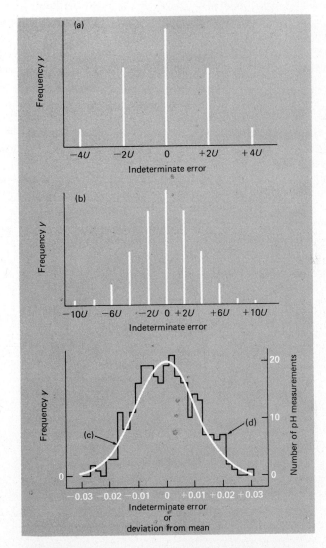

FIGURE 4-2 Theoretical distribution of indeterminate error arising from (a) 4 uncertainties, (b) 10 uncertainties, (c) a very large number of uncertainties. Curve (c) shows the normal error or Gaussian distribution. Curve (d) is an experimental distribution curve that might be obtained by plotting the deviations from the mean for about 250 replicate pH measurements against the number of times each deviation was observed.

U. Further, we shall stipulate that the magnitude of *U* is the same for each of the four uncertainties.

Table 4-2 shows all of the possible ways the four uncertainties can combine to give the indicated indeterminate errors. We note that there is only one way in which the maximum positive error of $4U$ can arise, compared with four combinations that lead to a positive error of $2U$, and six combinations that result in zero error. A similar relationship exists for negative indeterminate errors. This ratio of $6:4:1$ is a measure of the probability for an error of each size; if we made sufficient measurements, a frequency distribution of errors such as that shown in Figure 4-2a would be expected. Figure 4-2b shows the distribution for ten equal-sized uncertainties. Again, we see that the most frequent occurrence is zero error, while the maximum error of $10U$ would occur only occasionally (about once in 500 measurements).

If the foregoing arguments are extended to a very large number of uncertainties of smaller and smaller size, it can be demonstrated that the continuous distribution curve shown in Figure 4-2c will result. This bell-shaped curve is called a *Gaussian* or *normal error curve*.[2] Its properties include (1) a maximum frequency in occurrence of zero indeterminate error, (2) a symmetry about this maximum indicating that negative and positive errors occur with equal frequency, and (3) an exponential decrease in frequency as the magnitude of the error increases. Thus a small indeterminate error will occur much more often than a very large one.

Numerous *empirical* observations have shown that indeterminate errors in chemical analyses most commonly distribute themselves in a manner which approaches a Gaussian distribution. For example, if the deviations from the mean of hundreds of repetitive pH measurements on a single sample were plotted against the frequency of occurrence of each deviation, a curve approximating that shown in Figure 4-2d would ordinarily result.

The frequent experimental observation of Gaussian behavior lends credibility to the idea that the indeterminate error observed in analytical measurements can be traced to the accumulation of a large number of small, independent, and uncontrolled uncertainties. Equally important, the Gaussian distribution of most analytical data permits the use of statistical techniques to estimate the limits of indeterminate error from the precision of such data.

CLASSICAL STATISTICS

Statistics permits a mathematical description of random processes such as the effect of indeterminate error on the results of a chemical analysis. It is

---

[2] In deriving the Gaussian curve, it is not necessary to assume, as we have, that the individual uncertainties have identical magnitudes.

important to realize, however, that the techniques of classical statistics apply exactly to an *infinite* number of observations only. When these techniques are applied to the two to five replicate analyses that the chemist can afford to make, conclusions as to the probable indeterminate error can be seriously incorrect and misleadingly optimistic. It is for this reason that modification of the classical techniques is necessary. Before considering these practical modifications, however, it is worthwhile to describe briefly some important relationships of classical statistics.

Properties of the Normal Error Curve.   The upper two curves of Figure 4-3 are normal error curves for two different analytical methods. The topmost curve represents data from the more precise of the two methods inasmuch as the results are distributed more closely about the central value.

As shown in Figure 4-3, normal error curves can be plotted in three different ways. In each the ordinate is the frequency of occurrence y for each value of the abscissa. The observed values x of the measurement are plotted as abscissa a; here the central value is the mean, which is symbolized by μ. Abscissa b consists of individual deviations from the mean, $x - \mu$; here the most frequently occurring deviation has a value of zero. We shall consider the third type of plot, shown by abscissa c, presently.

It is important to emphasize that the curves under discussion are idealized because they represent the theoretical distribution of experimental results to be expected as the number of analyses involved approaches infinity. For a physically realizable set of results, a discontinuous distribution such as shown in Figure 4-2d would be more likely. Classical statistics is based on curves such as those shown in Figure 4-3 rather than on curves such as Figure 4-2d.

The distribution data in Figure 4-3 can be described mathematically in terms of just three parameters, as shown by the expression

$$y = \frac{e^{-(x-\mu)^2/2\sigma^2}}{\sigma\sqrt{2\pi}} = \frac{e^{-z^2/2}}{\sigma\sqrt{2\pi}} \qquad (4\text{-}2)$$

In this equation x represents values of individual measurements, and μ is the arithmetic mean for an infinite number of such measurements. The quantity $(x - \mu)$ is thus the deviation from the mean; y is the frequency of occurrence for each value of $(x - \mu)$. The symbol π has its usual meaning, and e is the base for Napierian logarithms, 2.718 . . . . The parameter σ is called the *standard deviation* and is a constant that has a unique value for any set containing a large number of measurements. The breadth of the normal error curve is directly related to σ.

The exponential in Equation 4-2 can be simplified by introducing

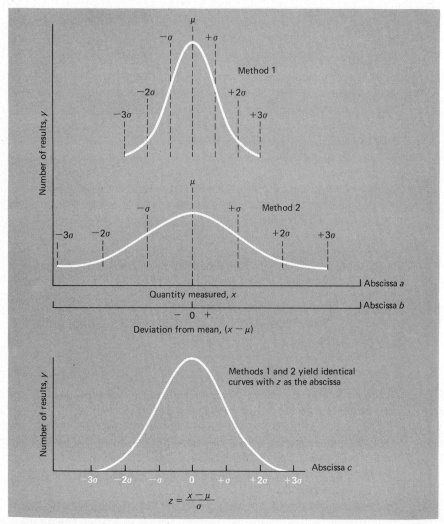

FIGURE 4-3 Normal error curves. The curves are for the measurement of the same quantity by two methods. Method 1 is more reliable; thus σ is smaller. Note the three types of abscissa. Abscissa a shows the measured quantity x with the maximum at μ. Abscissa b shows the deviation from the mean with the maximum at 0. Abscissa c shows z from Equation 4-2. Abscissa c reduces the two curves to a single one.

the variable

$$z = \frac{x - \mu}{\sigma} \qquad \textbf{(4-3)}$$

which then gives the deviation from the mean in units of standard deviations. As demonstrated by abscissa c in Figure 4-3, the substitution of z produces a single curve for all values of σ.

The Standard Deviation. Equation 4-2 indicates that a unique distribution curve exists for each value of the standard deviation. Regardless of the size of σ, however, it can be shown that 68.3% of the area beneath the curve lies within one standard deviation ($\pm 1\sigma$) of the mean, μ. Thus 68.3% of the values lie within these boundaries. Approximately 95.5% of all values will be within $\pm 2\sigma$; 99.7% will be within $\pm 3\sigma$. Values of $(x - \mu)$ corresponding to $\pm 1\sigma$, $\pm 2\sigma$, and $\pm 3\sigma$ are indicated by broken vertical lines in the upper curves of Figure 4-3. For the bottom curve the units for z shown on the abscissa are $\pm \sigma$.

These properties of the normal error curve are useful because they permit statements to be made about the probable magnitude of the indeterminate error in a given measurement *provided the standard deviation of the method of measurement is known.* Thus if σ were available, one could say that the chances are 68.3 out of 100 that the indeterminate error associated with any given single measurement is smaller than $\pm 1\sigma$, that the chances are 95.5 out of 100 that the error is less than $\pm 2\sigma$, and so forth. Clearly the standard deviation for a method of measurement is a useful quantity for estimating and reporting the probable size of indeterminate errors.

For a very large set of data the standard deviation is given by

$$\sigma = \sqrt{\frac{\sum_{i=1}^{N} (x_i - \mu)^2}{N}} \qquad (4\text{-}4)$$

Here the sum of the squares of the individual deviations from the mean $(x_i - \mu)$ is divided by the total number of measurements in the set, N. Extraction of the square root of this quotient gives σ.

Another precision term widely employed by statisticians is the *variance* which is equal to σ^2. Most experimental scientists prefer to employ σ rather than σ^2 because the units of the standard deviation are the same as those of the quantity measured.

APPLICATIONS OF STATISTICS TO SMALL SETS OF DATA

It has been found that direct application of classical statistics to a small number of replicate measurements (2 to 20 results) often leads to false conclusions regarding the probable magnitude of indeterminate error. Fortunately modifications of the relationships have been developed so that valid statements can be made about the random error associated with just two or three values.

Equations 4-2 and 4-4 are not directly applicable to a small number of replicate measurements because μ, the mean value of an infinitely large number of measurements (and the true value in the absence of determinate error), is never known. In its stead we are forced to employ $\bar{x}$, the mean of a small number of measurements. More often than not, $\bar{x}$ will differ somewhat from μ. This difference is, of course, the result of the indeterminate error whose probable magnitude we are trying to assess. It is important to note that *any error in $\bar{x}$ causes a corresponding error in σ* (Equation 4-4). Thus with a small set of data, not only is the mean $\bar{x}$ likely to differ from μ but, *equally important, the estimate of the standard deviation also may be misleading*. That is, we have *two* uncertainties to cope with, the one residing in the mean and the other in the standard deviation.

A decrease in the number of replicates in a set of data has two effects on the standard deviation. First, the influence of very high and very low values upon σ increases; that is, the reproducibility of σ becomes poorer. Second, the standard deviation as a measure of precision develops a negative *bias*. This bias is manifested by a greater frequency of low than high estimates for σ and a decrease in the average value of the standard deviation as the number of replicates becomes smaller.

The negative bias in σ for small sets of data is attributable to the fact that both a mean and a standard deviation must be extracted from the same small set. It can be shown that this bias can be largely eliminated by substituting the *number of degrees of freedom* $(N - 1)$ for N in Equation 4-4. That is, we define the standard deviation for a small number of measurements as

$$s = \sqrt{\frac{\sum_{i=1}^{N}(x_i - \bar{x})^2}{N - 1}} \tag{4-5}$$

Note that Equation 4-5 differs from Equation 4-4 in two regards. First, the denominator is now $(N - 1)$. Second, $\bar{x}$, the measured mean for the small set, replaces the true but unknown mean μ. To emphasize that the resulting standard deviation is but an approximation of the true value, it is given the symbol s rather than σ.

EXAMPLE

Calculate the standard deviation s for a subset consisting of the first five values in Table 4-1.

x_i	$|x_i - \bar{x}|$	$(x_i - \bar{x})^2$
9.990	8.0×10^{-3}	64.0×10^{-6}
9.985	3.0×10^{-3}	9.0×10^{-6}
9.973	9.0×10^{-3}	81.0×10^{-6}
9.980	2.0×10^{-3}	4.0×10^{-6}
9.982	0.0×10^{-3}	0.0×10^{-6}

$5\overline{)49.910}$
$9.9820 = \bar{x}$

$\sum_{i=1}^{5}(x_i - \bar{x})^2 = 158.0 \times 10^{-6}$

From Equation 4-4

$$s = \sqrt{\frac{158.0 \times 10^{-6}}{5-1}} = 6.3 \times 10^{-3} = \pm 0.006$$

Note that the data should not be rounded until the end.

The rationale for the use of $(N-1)$ in Equation 4-5 is as follows. When μ is unknown, we employ the set of replicate data to obtain two quantities, $\bar{x}$ and s. The need to establish the mean $\bar{x}$ from the data removes one degree of freedom. That is, if their signs are retained, the individual deviations from $\bar{x}$ must total zero; once $(N-1)$ deviations have been established, the final one is necessarily known as well. Thus only $(N-1)$ deviations provide independent measures of the precision for the set.

THE USES OF STATISTICS

Experimentalists employ statistical calculations to sharpen their judgments concerning the effects of indeterminate error. In the section that follows we shall examine a few of these applications,[3] namely:

1. The interval around the mean of a set within which the true mean can be expected to fall with a certain probability.
2. The number of times a measurement should be replicated in order that the experimental mean will fall, with a certain probability, within a predetermined interval around the true mean.
3. Whether or not an outlying value in a set of replicate results should be retained or rejected in calculating a mean for the set.

CONFIDENCE INTERVALS

The true mean value (μ) of a measurement is a constant that must always remain unknown. With the aid of statistical theory, however, limits may be set about the experimentally determined mean ($\bar{x}$) within which we may expect to find the true mean with a given degree of probability; the limits obtained in this manner are called *confidence limits*. The interval defined by these limits is known as the *confidence interval*.

Some of the properties of the confidence interval are worthy of mention. For a given set of data the size of the interval depends in part

---

[3] For other applications of statistical calculations, see W. J. Dixon and F. J. Massey, Jr., *Introduction to Statistical Analysis*, 3d ed. New York: McGraw-Hill, 1969; H. A. Laitinen and W. E. Harris, *Chemical Analysis*, 2d ed. New York: McGraw-Hill, 1975, Chapter 26; E. B. Wilson, Jr., *An Introduction to Scientific Research*. New York: McGraw-Hill, 1952, Chapters 9 and 10.

upon the odds for correctness desired. Clearly, for a prediction to be absolutely correct, we would have to choose an interval about the mean large enough to include all conceivable values that x_i might take. Such an interval, of course, has no predictive value. On the other hand, the interval does not need to be this large if we are willing to accept the probability of being correct 99 times in 100; it can be even smaller if 95% probability is acceptable. In short, as the probability for making a correct prediction is made less favorable, the interval included by the confidence limits becomes smaller.

The confidence interval, which is derived from the standard deviation s for the method of measurement, also depends in magnitude upon the certainty with which s is known. Often the chemist will have reason to believe that experimental value for s is an excellent approximation of σ. In other situations, however, a considerable uncertainty in s may exist. Under these circumstances the confidence interval will necessarily be larger.

Methods for Obtaining a Good Approximation of σ. Fluctuations in the calculated value for s decrease as the number of measurements N in Equation 4-5 increases; in fact it is proper to assume that s and σ are, for all practical purposes, identical when N is greater than about 20. It thus becomes feasible for the chemist to obtain a good approximation of s when the method of measurement is not too time-consuming and when an adequate amount of sample is available. For example, if it were necessary to measure the pH of numerous solutions in the course of an investigation, it might prove worthwhile to evaluate s in a series of preliminary experiments. This particular measurement is simple, requiring only that a pair of rinsed and dried electrodes be immersed in the test solution; the potential across the electrodes serves as a measure of the pH. To determine s, 20 to 30 portions of a solution of fixed pH could be measured, following exactly all steps of the procedure. Normally it would be safe to assume that the indeterminate error in this test would be the same as that in subsequent measurements and that the value of s calculated by means of Equation 4-5 would be a valid and accurate measure of the theoretical σ.

For analyses that are time-consuming, the foregoing procedure is not ordinarily practical. Here, however, precision data from a series of samples can often be pooled to provide an estimate of s which is superior to the value for any individual subset. Again, one must assume the same sources of indeterminate error among the samples. This assumption is usually valid provided the samples have similar compositions and each has been analyzed identically. To obtain a pooled estimate of s, deviations from the mean for each subset are squared; the squares for all of the subsets are then summed and divided by an appropriate number of degrees of freedom. The pooled s is obtained by extracting the square root of the quotient. One degree of freedom is lost for each subset. Thus the number of degrees of freedom for the pooled s is equal to the total number of measurements minus the number of subsets. An example of this calculation follows.

EXAMPLE

The mercury in samples of seven fish taken from Lake Erie was determined by a method based upon absorption of radiation by elemental mercury. The results are given in the accompanying table. Calculate a standard deviation for the method, based upon the pooled precision data.

SAMPLE NUMBER	NUMBER OF REPLICATIONS	RESULTS, Hg CONTENT, ppm	MEAN, ppm Hg	SUM OF SQUARES OF DEVIATIONS FROM MEAN
1	3	1.80, 1.58, 1.64	1.673	0.0259
2	4	0.96, 0.98, 1.02, 1.10	1.015	0.0115
3	2	3.13, 3.35	3.240	0.0242
4	6	2.06, 1.93, 2.12, 2.16, 1.89, 1.95	2.018	0.0611
5	4	0.57, 0.58, 0.64, 0.49	0.570	0.0114
6	5	2.35, 2.44, 2.70, 2.48, 2.44	2.482	0.0685
7	4	1.11, 1.15, 1.22, 1.04	1.130	0.0170
Number of measurements	28		Sum of squares =	0.2196

The value in column 5 for sample 1 was calculated as follows:

(x_i)	$\|(x_i - \bar{x}_1)\|$	$(x_i - \bar{x}_1)^2$
1.80	0.127	0.0161
1.58	0.093	0.0087
1.64	0.033	0.0011
3|5.02	Sum of squares =	0.0259
$\bar{x}_1 = 1.673$		

The other data in column 5 were obtained similarly. Then

$$s = \sqrt{\frac{0.0259 + 0.0115 + 0.0242 + 0.0611 + 0.0114 + 0.0685 + 0.0170}{28 - 7}}$$

$$= 0.10 \text{ ppm Hg}$$

Because the number of degrees of freedom is greater than 20, this estimate of s can be considered to be a good approximation of σ.

Confidence Interval When s Is a Good Approximation of σ. As indicated earlier (p. 61), the breadth of the normal error curve is determined by σ. It is also related to z in Equations 4-2 and 4-3. The area under a part of the error curve *relative* to the total area can be calculated for a desired value of z by means of Equation 4-2. This ratio (usually expressed as a percent) is called the *confidence level*, and it measures the probability for the absolute deviation $(x - \mu)$ being less than $z\sigma$. Thus the area under the curve encompassed by $z = \pm 1.96\sigma$ corresponds to 95% of the total area. Here the confidence level is 95%, and we may state that for a large number of measurements the calculated $(x - \mu)$ will be equal to or less than $\pm 1.96\sigma$ 95 times out of 100. Table 4-3 lists confidence intervals for various values of z.

The confidence limit for a single measurement can be obtained by rearranging Equation 4-3 and remembering that z can be either plus or minus in value. Thus

$$\text{confidence limit for } \mu = x \pm z\sigma \qquad (4\text{-}6)$$

The following example shows how Equation 4-6 is employed.

TABLE 4-3
Confidence Levels for Various Values of z

CONFIDENCE LEVEL, %	z
50	±0.67
68	±1.00
80	±1.29
90	±1.64
95	±1.96
96	±2.00
99	±2.58
99.7	±3.00
99.9	±3.29

EXAMPLE

Calculate the 50% and the 95% confidence limits for the first entry (1.80 ppm Hg) in the example on page 67.

Here we calculated that $s = 0.10$ ppm Hg and had sufficient data to assume $s \cong \sigma$. From Table 4-3 we see that $z = \pm 0.67$ and ± 1.96 for the two confidence levels. Thus from Equation 4-6

$$50\% \text{ confidence limit for } \mu = 1.80 \pm 0.67 \times 0.10$$
$$= 1.80 \pm 0.07$$
$$95\% \text{ confidence limit for } \mu = 1.80 \pm 1.96 \times 0.10$$
$$= 1.80 \pm 0.20$$

The chances are 50 in 100 that μ, the true mean (and in the absence of determinate error the true value), will be in the interval between 1.73 and 1.87 ppm Hg; there is a 95% chance that it will be in the interval between 1.60 and 2.00 ppm Hg.

Equation 4-6 applies to the result of a single measurement. It can be shown that the confidence interval is decreased by $\sqrt{N}$ for the average of N replicate measurements. Thus, a more general form of Equation 4-6 is

$$\text{confidence limit for } \mu = \bar{x} \pm \frac{z\sigma}{\sqrt{N}} \qquad \textbf{(4-7)}$$

EXAMPLE

Calculate the 50% and the 95% confidence limits for the mean value (1.67 ppm Hg) for sample 1 in the example on page 67. Again, $s \cong \sigma = 0.10$.

For the three measurements

$$50\% \text{ confidence limit} = 1.67 \pm \frac{0.67 \times 0.10}{\sqrt{3}} = 1.67 \pm 0.04$$

$$95\% \text{ confidence limit} = 1.67 \pm \frac{1.96 \times 0.10}{\sqrt{3}} = 1.67 \pm 0.11$$

Thus the chances are 50 in 100 that the true mean will lie in the interval of 1.63 to 1.71 ppm Hg and 95 in 100 that it will be between 1.56 and 1.78 ppm.

EXAMPLE

Calculate the number of replicate measurements needed to decrease the 95% confidence interval for the calibration of a 10-ml pipet to 0.005 ml, assuming that a procedure similar to the one for obtaining the data in Table 4-1 has been followed.

The standard deviation for the measurement is 0.0065 ml. Because s is based on 24 values, we may assume $s = \sigma = 0.0065$.

The confidence interval is given by

$$\text{confidence interval} = \pm \frac{z\sigma}{\sqrt{N}}$$

$$0.005 \text{ ml} = \frac{1.96 \times 0.0065}{\sqrt{N}}$$

$$N = 6.5$$

Thus, by employing the mean of seven measurements, we would have a somewhat better than 95% chance of knowing the true mean volume delivered by the pipet to ± 0.005 ml.

A consideration of Equation 4-7 indicates that the confidence interval for an analysis can be halved by employing the mean of four measurements. Sixteen measurements would be required to narrow the limit by another factor of 2. It is apparent that a point of diminishing return is rapidly reached in acquiring additional data. Thus the chemist ordinarily takes advantage of the relatively large gain afforded by averaging two to four measurements but can seldom afford the time required for further increases in confidence.

Confidence Limits When σ Is Unknown. Frequently a chemist must make use of an unfamiliar method. Furthermore, limitations in time or amount of available sample may preclude an accurate estimation of σ. Here a single set of replicate measurements must provide not only a mean value but also a precision estimate. As we have indicated earlier, s calculated from a limited set of data may be subject to considerable uncertainty; thus the confidence limits will be broader under these circumstances.

To account for the potential variability of s, use is made of the parameter t, which is defined as

$$t = \frac{x - \mu}{s} \qquad \text{(4-8)}$$

In contrast to z in Equation 4-3, t is dependent not only on the desired confidence level but also upon the number of degrees of freedom available in the calculation of s. Table 4-4 provides values for t under various circumstances. Note that the values for t become equal to those for z (Table 4-3) as the number of degrees of freedom becomes infinite.

The confidence limit can be derived from t by an equation analogous to Equation 4-7; that is,

$$\text{confidence limit for } \mu = \bar{x} \pm \frac{ts}{\sqrt{N}} \qquad \text{(4-9)}$$

TABLE 4-4
Values of *t* for Various Levels of Probability

DEGREES OF FREEDOM	FACTOR FOR CONFIDENCE INTERVAL, %				
	80	90	95	99	99.9
1	3.08	6.31	12.7	63.7	637
2	1.89	2.92	4.30	9.92	31.6
3	1.64	2.35	3.18	5.84	12.9
4	1.53	2.13	2.78	4.60	8.60
5	1.48	2.02	2.57	4.03	6.86
6	1.44	1.94	2.45	3.71	5.96
7	1.42	1.90	2.36	3.50	5.40
8	1.40	1.86	2.31	3.36	5.04
9	1.38	1.83	2.26	3.25	4.78
10	1.37	1.81	2.23	3.17	4.59
11	1.36	1.80	2.20	3.11	4.44
12	1.36	1.78	2.18	3.06	4.32
13	1.35	1.77	2.16	3.01	4.22
14	1.34	1.76	2.14	2.98	4.14
∞	1.29	1.64	1.96	2.58	3.29

EXAMPLE

A chemist obtained the following data for the alcohol content in a sample of blood: percent ethanol = 0.084, 0.089, and 0.079. Calculate the 95% confidence limit for the mean assuming (1) no additional knowledge about the precision of the method and (2) that on the basis of previous experiences $s \cong \sigma = 0.005\%$ ethanol.

1. $\bar{x} = \dfrac{(0.084 + 0.089 + 0.079)}{3} = 0.0840$

$$s = \sqrt{\frac{(0.00)^2 + (0.0050)^2 + (0.0050)^2}{3-1}} = 0.0050$$

Table 4-4 indicates that $t = \pm 4.30$ for two degrees of freedom and 95% confidence. Thus

$$95\% \text{ confidence limit} = 0.084 \pm \frac{4.3 \times 0.0050}{\sqrt{3}}$$

$$= 0.084 \pm 0.012$$

2. Because a good value of σ is available,

$$95\% \text{ confidence limit} = 0.084 \pm \frac{z\sigma}{\sqrt{N}}$$

$$= 0.084 \pm \frac{1.96 \times 0.0050}{\sqrt{3}}$$

$$= 0.084 \pm 0.006$$

Note that a sure knowledge of σ decreased the confidence interval by half.

REJECTION OF DATA

When a set of data contains an outlying result that appears to differ excessively from the average (or the median), the decision must be made to retain or to reject it. The choice of criterion for the rejection of a suspected result has its perils. If we set a stringent criterion that makes difficult the rejection of a questionable measurement, we run the risk of retaining results that are spurious and have an inordinate effect on the average of the data. On the other hand, if we set lenient limits on precision and make easy the rejection of a result, we are likely to discard measurements that rightfully belong in the set; we thus introduce a bias to the data. It is an unfortunate fact that no universal rule can be invoked to settle the question of retention or rejection.

TABLE 4-5
Critical Values for Rejection Quotient Q^a

NUMBER OF OBSERVATIONS	Q_{crit} (90% CONFIDENCE) REJECT IF $Q_{exp} >$
2	—
3	0.94
4	0.76
5	0.64
6	0.56
7	0.51
8	0.47
9	0.44
10	0.41

[a] Reproduced from R. B. Dean and W. J. Dixon, *Anal. Chem.*, **23**, 636 (1951). By permission of the American Chemical Society.

Of the numerous statistical criteria suggested to aid in deciding whether to retain or reject a measurement, the Q test[4] is to be preferred. Here the difference between the questionable result and its nearest neighbor is divided by the spread of the entire set. The resulting ratio, Q, is then compared with rejection values that are critical for a particular degree of confidence. Table 4-5 provides critical values of Q at the 90% confidence level.

EXAMPLE

The analysis of a calcite sample yielded CaO percentages of 55.95, 56.00, 56.04, 56.08, and 56.23, respectively. The last value appears anomalous; should it be retained or rejected?

The difference between 56.23 and 56.08 is 0.15%. The spread (56.23 − 55.95) is 0.28%. Thus

$$Q_{exp} = \frac{0.15}{0.28} = 0.54$$

For five measurements Q_{crit} is 0.64. Because $0.54 < 0.64$, retention is indicated.

Notwithstanding its superiority over other criteria, the Q test must be used with good judgment as well. For example, there will be situations in which the dispersion associated with the bulk of a set will be fortuitously small and the indiscriminate application of the Q test will result in rejection of a value that actually should be retained; indeed, in a three-number set containing a pair of identical values the experimental value for Q becomes indeterminately large. On the other hand, it has been pointed out[4] that the magnitudes of rejection quotients for small sets are likely to cause the retention of erroneous data.

The blind application of statistical tests to the decision for retention or rejection of a suspect measurement in a small set of data is not likely to be much more fruitful than an arbitrary decision. Indeed, the application of good judgment based upon an estimate of the precision to be expected may be a more sound approach, particularly if this estimate is based upon broad experience with the analytical method being employed. In the end, however, the only entirely valid reason for rejecting an experimental result from a small set is the sure knowledge that a mistake has been made in its acquisition. If one lacks this knowledge, *a cautious approach to the rejection of data is desirable.*

In summary, there are a number of recommendations for the treatment of a small set of results that contains a suspect value.

[4] R. B. Dean and W. J. Dixon, *Anal. Chem.*, **23**, 636 (1951).

1. Reexamine carefully all data relating to the suspected result to see if a gross error has affected its value. *A properly kept laboratory notebook containing careful notations of all observations is essential if this recommendation is to be helpful.*

2. If possible, estimate the precision that can be reasonably expected from the procedure to be sure that the outlying result actually is questionable.

3. Repeat the analysis if sufficient sample and time are available. Agreement of the newly acquired data with those that appear to be valid will lend weight to the notion that the outlying result should be rejected. Furthermore, the questionable result will have a smaller effect on the mean of the larger set of data if its retention is still indicated.

4. If more data cannot be secured, apply the Q test to the existing set to see if the doubtful result should be retained or rejected on statistical grounds.

5. If the Q test indicates retention, give consideration to reporting the median of the set rather than the mean. The median has the great virtue of allowing inclusion of all data in a set without undue influence from an outlying value. Moreover, it has been demonstrated that the median of a normally distributed set containing three measurements is more likely to provide a reliable estimate of the correct value than will the mean of the set after the outlying value has been arbitrarily discarded.[5]

Propagation of Errors in Computation

The scientist must frequently estimate the error in a result that has been computed from two or more data, each of which has an error associated with it. The way in which the individual errors accumulate depends upon the arithmetic relationship between the terms containing the errors.

PROPAGATION OF ERROR DURING ADDITION OR SUBTRACTION

Consider the summation

$$+0.50 \, (\pm 0.02)$$
$$+4.10 \, (\pm 0.03)$$
$$\underline{-1.97 \, (\pm 0.05)}$$
$$2.63 \, (\pm \ \ ? \ \ )$$

where the numbers in parentheses are the absolute indeterminate errors expressed as standard deviations. The uncertainty associated with the sum could be as large as ± 0.10 if the signs of the three individual standard

---

[5] National Bureau of Standards, *Technical News Bulletin* (July 1949); *J. Chem. Educ.*, **26**, 673 (1949).

deviations happened to be all positive or all negative. On the other hand, under fortuitous circumstances the three uncertainties could combine to give an accumulated error of zero. Neither of these is as probable as a combination leading to an uncertainty intermediate between the extremes. Statistics have shown that the most probable uncertainty in the case of sums or differences can be found by taking the square root of the sum of the individual *variances*. Thus in the present example

$$s_c^2 = s_1^2 + s_2^2 + s_3^2$$

where s_c is the standard deviation of the sum and s_1, s_2, and s_3 are the standard deviations of the three terms forming the sum. We see then that

$$s_c = \sqrt{(\pm 0.02)^2 + (\pm 0.03)^2 + (\pm 0.05)^2}$$

$$= \pm 0.06$$

and the sum could be reported as 2.63 (± 0.06).

Note that *the absolute error in an addition or a subtraction is determined from the absolute errors of the individual terms forming the sum or difference.* As we shall see, this relationship does not apply to products or quotients.

PROPAGATION OF ERROR IN MULTIPLICATION AND DIVISION

As an example, let us consider the following calculation:

$$\frac{4.10(\pm 0.02) \times 0.0050(\pm 0.0001)}{1.97(\pm 0.04)} = 0.0104 \, (\pm ?)$$

Note that the standard deviations of two of the numbers in this calculation are larger than the result itself. It is evident, then, that we cannot obtain the desired standard deviation by direct combination of the uncertainties as in an addition or a subtraction. Indeed, it can be shown that *for multiplication or division, the relative error of the product or quotient is determined by the relative errors of the numbers forming the computed result.* Thus, in this example we must first compute relative standard deviations.

$$(s_1)_r = \frac{(\pm 0.02) \times 100}{4.10} = \pm 0.49\%$$

$$(s_2)_r = \frac{(\pm 0.0001) \times 100}{0.0050} = \pm 2.0\%$$

$$(s_3)_r = \frac{(\pm 0.04) \times 100}{1.97} = \pm 2.0\%$$

Again, the *relative* variance of the result $(s_c)_r^2$ is equal to the sum of the individual *relative* variances.

$$(s_c)_r^2 = (s_1)_r^2 + (s_2)_r^2 + (s_3)_r^2$$

$$(s_c)_r = \sqrt{(\pm 0.49)^2 + (\pm 2.0)^2 + (\pm 2.0)^2} = \pm 2.9\%$$

In order to complete the calculation, we must find the *absolute* standard deviation of the result. Thus

$$s_c = 0.0104 \times 0.029 = 0.0003$$

and we can indicate the uncertainty of the answer as 0.0104 (± 0.0003).

The following example demonstrates the calculation of the standard deviation of the result for a more complex calculation.

EXAMPLE

Calculate the standard deviation of the result of the following computation:

$$\frac{[14.3(\pm 0.2) - 11.6(\pm 0.2)] \times 0.050(\pm 0.001)}{[820(\pm 10) + 1030(\pm 5)] \times 42.3(\pm 0.4)} = 1.725(\pm ?) \times 10^{-6}$$

First we must calculate the standard deviation of the sum and the difference. For the difference in the numerator,

$$s_1 = \sqrt{(\pm 0.2)^2 + (0.2)^2} = \pm 0.28$$

and for the sum in the denominator,

$$s_2 = \sqrt{(\pm 10)^2 + (\pm 5)^2} = \pm 11$$

We may then rewrite the equation as

$$\frac{2.7(\pm 0.28) \times 0.050(\pm 0.001)}{1850(\pm 11) \times 42.3(\pm 0.4)} = 1.725 \times 10^{-6}$$

The equation now contains only products and quotients; we thus compute the relative standard deviations of the individual quantities.

$$(s_1)_r = \frac{\pm 0.28}{2.7} \times 100 = \pm 10.4\%$$

$$(s_2)_r = \frac{\pm 0.001}{0.050} \times 100 = \pm 2.0\%$$

$$(s_3)_r = \frac{\pm 11}{1850} \times 100 = \pm 0.60\%$$

$$(s_4)_r = \frac{\pm 0.4}{42.3} \times 100 = \pm 0.95\%$$

and

$$(s_c)_r = \sqrt{(\pm 10.4)^2 + (\pm 2.0)^2 + (\pm 0.60)^2 + (\pm 0.95)^2} = \pm 10.6\%$$

The absolute standard deviation of the result is

$$s_c = (1.725 \times 10^{-6}) \times (\pm 0.106) = \pm 0.18 \times 10^{-6}$$

and our answer is thus written as

$$1.7(\pm 0.2) \times 10^{-6}$$

Note the amplification of error that results from the subtraction process in the numerator of the equation.

Significant Figure Convention

In reporting a measurement the experimenter should include not only what is considered to be its best value, be it a mean or a median, but also an estimate of its uncertainty. The latter is preferably reported as the standard deviation of the result. The deviation from the mean, the deviation from the median, or the spread may sometimes be encountered because these precision indicators are easier to calculate. Common practice also dictates that an experimental result should be rounded off so that it contains only the digits known with certainty plus the first uncertain one. This practice is called the *significant figure convention.*

For example, the average of the experimental quantities 61.60, 61.46, 61.55, and 61.61 is 61.555. The standard deviation of the sum is ± 0.069. Clearly the number in the second decimal place is subject to uncertainty. Such being the case, all numbers in succeeding decimal places are without meaning, and we are forced to round the average value accordingly. The question of taking 61.55 or 61.56 must be considered, 61.555 being equally spaced between them. A good guide to follow when rounding a 5 is always to round to the nearest even number; in this way, any tendency to round in a set direction is eliminated because there is an equal likelihood that the nearest even number will be the higher or the lower in any given situation. Thus we could report the foregoing results as 61.56 ± 0.07. On the other hand, if we had reason to doubt that ± 0.07 was a valid estimate of the precision, we might choose to present the result as 61.6 ± 0.1.

The significant figure convention is frequently used in lieu of a specific estimate for the precision of a result. Thus, by simply reporting 61.6 in this example, we would be saying in effect that we believe the first 6 and the 1 are certain digits but the value of the second 6 is in doubt. The disadvantage of this practice is obvious; all the reader can discern is the range of the uncertainty—here, greater than ± 0.05 and smaller than ± 0.5.

In employing the significant figure convention, it is important to appreciate that the zero not only functions as a number but also serves to locate decimal points in very small and very large numbers. A case in point is Avogadro's number. The first three digits, 6, 0, and 2, are known with certainty; the next is uncertain but is probably 3. Because the digits that follow the 6023 are not known, we substitute 19 zeros after the digit 3 to place the decimal point. Here the zeros indicate the order of magnitude of the number only and have no other meaning. It is clear that a distinction must be made between those figures that have physical significance (that is, *significant figures*) and those that are either unknown or meaningless owing to the inadequacies of measurement.

Zeros bounded by digits only on the left may or may not be significant. Thus the mass of a 20-mg weight known to carry no correction (to a tenth of a milligram) is known to three significant figures, 20.0 mg; when this is expressed as 0.0200 g, the number of significant figures does not change. If, on the other hand, we wish to express the volume of a 2-liter beaker as 2000 ml, the latter number will contain only one significant figure. The zeros simply indicate the order of magnitude. It can, of course, happen that the beaker in question has been found by experiment to contain 2.0 liters; here the zero following the decimal point is significant and implies that the volume is known to at least ±0.5 liter and might be known to ±0.05 liter. If this volume were to be expressed in milliliters, the zero following the 2 would still be significant, but the other two zeros would not. The use of exponential notation eliminates the difficulty. We could thus indicate the volume as 2.0×10^3 ml.

A certain amount of care is required in determining the number of significant figures to carry in the result of an arithmetic combination of two or more numbers. For addition and subtraction the number of significant figures can be seen by visual inspection. For example,

$$3.4 + 0.02 + 1.31 = 4.7$$

Clearly the second decimal place cannot be significant because an uncertainty in the first decimal place is introduced by the 3.4.

When data are being multiplied or divided, it is frequently assumed that the number of significant figures for the result is equal to that of the component quantity that contains the least number of significant figures. For example,

$$\frac{24 \times 0.452}{100.0} = 0.108 = 0.11$$

Here the 24 has two significant figures, and the result has therefore been rounded to agree. Unfortunately the rule does not always apply well. Thus the uncertainty in 24 in this calculation could be as small as 0.5 or as large as 5. The uncertainty in the quotient for these two limits is given in the accompanying table.

ASSUMED ABSOLUTE UNCERTAINTY IN 24	RELATIVE UNCERTAINTY	ABSOLUTE UNCERTAINTY IN 0.108	ROUND TO
>0.5	0.5/24 = 0.02	0.108 × 0.02 = 0.002	0.108
<5	5/24 = 0.2	0.108 × 0.2 = 0.02	0.11

Calculations such as this demonstrate the dilemma faced by the scientist who has only the significant figure convention to use as a guide to indeterminate errors.

Particular care is needed in rounding logarithms and antilogarithms. Consider, for example, the following computations:

$$\log 1.124 \times 10^{13} = 13.05077 = 13.0508$$

$$\log 1.125 \times 10^{13} = 13.05112 = 13.0511$$

$$\log 1.126 \times 10^{13} = 13.05154 = 13.0515$$

Note that a variation of ± 1 in the fourth digit of the number on the left results in a variation of about ± 4 in the sixth digit of the number on the right. Thus the log of 1.125×10^{13} should be reported as 13.0511 as shown.

This apparent gain in the number of significant figures is actually an artifact. The characteristic of the logarithm (13 in this case) serves only to locate the decimal point in the original number; all of the information about 1.125 is contained in the mantissa (0.05112). Thus the number of figures to the right of the decimal point in the logarithm should correspond to the number of significant figures in the original number. As shown by the example, *zeros immediately to the right of the decimal point are counted in rounding a mantissa.*

Two good general rules to follow in rounding logarithms and antilogarithms are the following:[6]

1. When converting numbers to logarithms, use as many *decimal places* in the mantissa as there are significant figures in the number.
2. When finding the antilogarithm, keep as many significant digits as there are decimal places in the mantissa. Thus antilogarithm 1.783 = 60.7. Here three *decimal* places in the mantissa lead to three significant digits in the number.

PROBLEMS

*1. To test a method for the analysis of barium, a solution was prepared by dissolving a weighed quantity of pure $BaCl_2 \cdot 2H_2O$ in water. After adding other components to simulate a solution of a typical sample, the mixture was diluted to exactly 500 ml. The proposed method was then applied to several 50.0-ml aliquots of the solution, each of which contained exactly 255 mg of Ba. The following results were obtained:

SAMPLE NUMBER	Ba FOUND, mg	SAMPLE NUMBER	Ba FOUND, mg
1	252	4	243
2	252	5	249
3	247	6	255

[6] See D. E. Jones, *J. Chem. Educ.*, **49**, 753 (1972).

For this set of data calculate (a) the mean, (b) the median, (c) the precision in terms of the spread, (d) the precision in terms of the average absolute and relative deviation from the mean, (e) the absolute and relative error of sample 1, (f) the absolute and relative error of the mean.

2. To test a method for the analysis of atmospheric SO_2, a standard sample was prepared by diluting measured quantities of SO_2 with appropriate volumes of air. Several 100-ml aliquots with an SO_2 concentration of 6.8 ppm were analyzed by the procedure. The following results were obtained:

SAMPLE NUMBER	SO_2, ppm	SAMPLE NUMBER	SO_2, ppm
1	7.0	5	7.3
2	7.1	6	6.8
3	6.7	7	6.4
4	6.3	8	7.4

For these data calculate (a) the mean, (b) the median, (c) the precision in terms of the spread, (d) the precision in terms of the average absolute and relative deviation (in parts per thousand) from the mean, (e) the absolute and relative error of sample 1, (f) the absolute and relative error of the mean.

*3. A standard sample known to contain 1.76% H_2O was analyzed by student A, who reported 1.73, 1.70, and 1.75% H_2O. Student B analyzed another standard with a composition of 7.49% H_2O and obtained 7.41, 7.44, and 7.46% H_2O. Compare (a) the absolute and relative deviations from the means of the two sets of data and (b) the absolute and relative errors associated with the means of the two sets.

4. Analysis for acetone in two standard samples yielded the following data:

ACETONE, %

Present	Found	Present	Found
	9.3		0.48
9.6	9.6	0.55	0.54
	9.7		0.50

(a) Compare the precision of the two analyses in terms of relative and absolute deviations from the means.
(b) Compare the errors associated with the two analyses in absolute and relative terms.

*5. A constant error of $+0.15$ mg of P is associated with a method for the analysis of P in organic compounds. Calculate the relative error (in parts per thousand) for the results of an analysis of a sample containing about 12% P if the following sample sizes are taken: (a) 10 mg, (b) 50 mg, (c) 100 mg, (d) 500 mg, (e) 1000 mg.

6. It was found that a solubility loss corresponding to 2.0 mg of Mo was associated with a gravimetric method for the determination of this element. Calculate the relative error in parts per thousand in an analysis of a sample that contained 14% Mo if the original sample weighed (a) 1.00 g, (b) 0.500 g, (c) 0.250 g, (d) 0.100 g.

*7. Calculate the absolute and relative standard deviations for the set of data in Problem 1.

8. Calculate the absolute and relative standard deviations for the set of data in Problem 2.

*9. Compare the absolute and relative standard deviations for the two sets of data in Problem 3.

10. Compare the absolute and relative standard deviations for the two sets of data in Problem 4.

*11. The following data were obtained for the weight-weight percent sulfur in fuel oils by absorption spectroscopy. Pool the data to obtain the standard deviation of the procedure.

SAMPLE NUMBER	S (w/w), %
1	0.62, 0.55, 0.56, 0.60
2	2.03, 1.97, 2.03
3	1.66, 1.60
4	1.09, 1.04, 1.05

12. Nine samples of illicit heroin preparations were analyzed in duplicate by a gas chromatographic technique. Pool the data to establish an absolute standard deviation for the procedure.

SAMPLE NUMBER	HEROIN, %	SAMPLE NUMBER	HEROIN, %	SAMPLE NUMBER	HEROIN, %
1	3.26, 3.30	4	9.6, 10.3	7	14.4, 14.8
2	1.27, 1.22	5	4.3, 4.1	8	20.9, 20.4
3	21.9, 21.8	6	7.8, 7.6	9	6.7, 6.2

*13. Calculate a pooled estimate of s from the following spectrophotometric analysis for NTA (nitrilotriacetic acid) in Ohio River water.

SAMPLE NUMBER	NTA, ppb
1	10, 14, 9, 16
2	29, 33, 34
3	17, 19, 14, 20, 22

14. The following data were obtained for the concentration of Ca in bovine serum. Pool the data to obtain a value of s for the procedure.

SAMPLE NUMBER	meq Ca/liter
1	3.172, 3.176, 3.174
2	3.946, 3.949
3	6.025, 6.042, 6.033, 6.020

*15. The pooling of 30 triplicate analyses for Cu in used jet engine oil yielded a standard deviation of $1.6\,\mu g/ml$ $(s \to \sigma)$. A sample of oil analyzed by this method was found to contain $16.2\,\mu g$ Cu/ml. Calculate the 90% and 99% confidence intervals for this analysis based upon (a) the result of a single analysis, (b) a mean of two analyses, and (c) the mean of four analyses.

16. The method described in Problem 15 was found to yield a pooled standard deviation for Pb of $s \to \sigma = 0.46\,\mu g$ Pb/ml. The analysis of an oil from a reciprocating aircraft engine showed a Pb content of $9.73\,\mu g$ Pb/ml. Calculate the 80% and 95% confidence intervals based upon (a) the result of a single analysis, (b) a mean of four analyses, (c) a mean of 16 analyses.

*17. How many replicate measurements would be needed to decrease the 90% and 99% confidence limits for the analysis described in Problem 15 to $\pm 1.0\,\mu g$ Cu/ml?

18. How many replicate measurements would be needed to decrease the 80% and 95% confidence limits for the analysis described in Problem 16 to $\pm 0.13\,\mu g$ Pb/ml?

*19. A chemist obtained the following data in a triplicate analysis of an insecticide preparation for its percent lindane: 6.38, 5.99, 6.25. Calculate the 90% confidence interval for the mean of the three data based upon (a) the precision of this analysis and (b) the knowledge—from previous experience—that $s \to \sigma = 0.28\%$ lindane.

20. An analyst obtained the following data in a duplicate analysis of air samples: ppm SO_2 = 8.6, 9.4. Calculate the 95% confidence interval and the 80% confidence interval for the mean of the data based upon (a) the precision of these two results and (b) the knowledge, from past experience with the method, that $s \to \sigma = 0.6$ ppm SO_2.

*21. A chemist obtained the following results for the determination of sulfur in a contaminated kerosene sample: 0.300%, 0.316%, 0.303%. Calculate the 95% confidence limits for the mean of this analysis.

22. Four replicate fluoride analyses on a sample of well water yielded the following data: 0.77, 0.85, 0.86, 0.75 ppm F^-. What are the 95% and 99% confidence limits for the mean of this analysis?

*23. A volumetric calcium analysis on triplicate samples of the blood serum of a patient believed to be suffering from a hyperparathyroid condition produced the following data: meq Ca/liter = 2.86, 2.95, 2.84. What is the 95% confidence interval for the mean of the data, assuming (a) no prior information about the precision of the analysis and (b) $s \to \sigma = 0.06$ meq Ca/liter?

24. A physiologist interested in the role of K^+ ion in the transmission of nerve signals developed a potentiometric analysis for the ion in sera. To evaluate the precision of the method he pooled the data from analyses performed over several weeks on samples that contained from 3 to 6 meq K^+/liter.

SAMPLE NUMBER	MEAN K^+ CONCN FOUND, meq/liter	NUMBER OF MEASUREMENTS	DEVIATION OF INDIVIDUAL RESULTS FROM THE MEAN, meq K^+/liter
1	5.55	5	0.08, 0.06, 0.25, 0.11, 0.00
2	4.97	2	0.15, 0.15
3	3.76	4	0.12, 0.11, 0.01, 0.00
4	6.64	9	0.03, 0.13, 0.09, 0.20, 0.27, 0.14, 0.05, 0.11, 0.02
5	3.88	5	0.12, 0.08, 0.01, 0.04, 0.17

(a) Calculate the standard deviation of each set of data.
(b) Calculate the standard deviation for the method by pooling the data.
(c) Would the value of s obtained in (b) be a good approximation of σ for the method?
(d) Calculate the 95% confidence interval of the mean for sample 4. Compare the results obtained with the standard deviation obtained in (a) and the pooled standard deviation from (b).
(e) Repeat the calculation in (d) for sample 2.

*25. An analytical chemist was interested in evaluating the indeterminate error in a gravimetric method for the determination of the hormone progesterone in oral tablets. Repeated use of the method yielded the following data:

SAMPLE NUMBER	NUMBER OF REPLICATE ANALYSES	MEAN PERCENT PROGESTERONE	INDIVIDUAL DEVIATIONS
1	5	3.66	0.02, 0.01, 0.06, 0.05, 0.04
2	3	3.45	0.06, 0.02, 0.04
3	8	3.55	0.00, 0.07, 0.04, 0.03, 0.02, 0.03, 0.02, 0.00
4	2	3.86	0.02, 0.02
5	3	3.12	0.02, 0.00, 0.02
6	6	3.97	0.06, 0.00, 0.04, 0.01, 0.01, 0.02

(a) Calculate the standard deviation for each set of data.
(b) Calculate a standard deviation for the method by pooling the data from the six samples.
(c) Would the value of s found in (b) be expected to be a good approximation of σ?
(d) Calculate the 95% confidence interval for sample 3, using s obtained in (a) and then the pooled s from (b).
(e) Repeat the calculations in (d) for sample 4.

26. The performance of a new photometer was tested by making 50 replicate measurements of the percent transmission of a solution. The standard deviation of these data was 0.12% T. How many replicate readings of the instrument should be taken for each subsequent measurement if the instrumental error associated with the mean is to be kept below
*(a) ±0.2% T with 99% certainty?
(b) ±0.1% T with 99% certainty?
*(c) ±0.2% T with 95% certainty?
(d) ±0.1% T with 95% certainty?

*27. What are the 90% and 95% confidence limits for a single measurement by the instrument described in Problem 26?

28. A standard method for the determination of tetraethyl lead (TEL) in gasoline is reported to have a standard deviation of 0.020 ml TEL/gal. If $s \rightarrow \sigma = 0.020$, how many replicate analyses should be made if the mean for the analysis of a sample is to be within
*(a) ±0.03 ml/gal of the true mean 99% of the time?
(b) ±0.03 ml/gal of the true mean 95% of the time?
(c) ±0.015 ml/gal of the true mean 90% of the time?

29. Calculate the 99.9%, the 99%, and the 80% confidence limits for an analysis by the method described in Problem 28 based upon
    (a) the result of a single measurement.
    (b) the mean of triplicate measurements.

*30. How many significant figures are there in
    (a) 0.00607?
    (b) 19931?
    (c) 0.00037?
    (d) 7357027?
    (e) 136.03?
    (f) 0.120?

31. How many significant figures are there in
    (a) 0.0336?
    (b) 737.37?
    (c) 3.3342?
    (d) 10.00?
    (e) 16?
    (f) 2.64×10^{-7}?

*32. Estimate the absolute and relative standard deviation for the results of the following calculations. Round the result to the proper number of significant figures. The numbers shown in parentheses are absolute standard deviations.
    (a) $y = 43.2(\pm 0.3) + 0.644(\pm 0.002) - 39.6(\pm 0.2) = 4.244$
    (b) $y = 1.29(\pm 0.04) + 0.007(\pm 0.002) + 3.52(\pm 0.06) = 4.817$
    (c) $y = 2.97(\pm 0.04) \times 10^{-6} \times 6.00(\pm 0.01) = 1.782 \times 10^{-5}$
    (d) $y = \dfrac{673(\pm 5) \times 20.4(\pm 0.3)}{397(\pm 3)} = 34.582$
    (e) $y = \dfrac{17.63(\pm 0.05) \times 10^{-6}}{343(\pm 1) \times 2.734(\pm 0.002) \times 10^{7}} = 1.8800 \times 10^{-15}$
    (f) $y = \dfrac{29.1(\pm 0.1) - 28.6(\pm 0.2)}{43.2(\pm 0.2)} = 1.157 \times 10^{-2}$
    (g) $y = [4.2(\pm 0.1) - 7.2(\pm 0.3)] \times 1.77(\pm 0.06) \times 10^{-3} = -5.310 \times 10^{-3}$
    (h) $y = \dfrac{3.33(\pm 0.02) \times 10^{-2} - 6.6(\pm 0.3) \times 10^{-3}}{14.1(\pm 0.1) - 1.96(\pm 0.02)} = 2.199 \times 10^{-3}$

33. Estimate the absolute and relative standard deviation for the results of the following calculations. Round the result to include only significant figures. The numbers shown in parentheses are absolute standard deviations.
    (a) $y = 1.76(\pm 0.03) \times 10^{-7} - 9.6(\pm 0.2) \times 10^{-8} = 8.000 \times 10^{-8}$
    (b) $y = 4.763(\pm 0.004) - 2.222(\pm 0.003) + 1.457(\pm 0.002) = 3.998$
    (c) $y = \dfrac{2.222(\pm 0.003)}{476,6(\pm 0.4)} = 4.6622 \times 10^{-3}$
    (d) $y = \dfrac{-10.02(\pm 0.04) \times 30.000(\pm 0.002)}{37425(\pm 5)} = -8.0321 \times 10^{-3}$
    (e) $y = \dfrac{1.70(\pm 0.04) \times 10^{-6} - 3.00(\pm 0.05) \times 10^{-5}}{2.743(\pm 0.006)} = -1.0317 \times 10^{-5}$
    (f) $y = \dfrac{10(\pm 1) - 6(\pm 2)}{0.0202(\pm 0.0002)} = 198.02$
    (g) $y = \dfrac{1.44(\pm 0.04) \times 10^{-6} - 6.34(\pm 0.02) \times 10^{-7}}{3.72(\pm 0.03) \times 10^{-18}} = 2.1667 \times 10^{11}$
    (h) $y = \dfrac{9.96(\pm 0.01) - 10.00(\pm 0.01)}{40.0(\pm 0.1)} = -1.000 \times 10^{-3}$

*34. Round the following answers to the correct number of significant figures.
   (a) $y = \log 2.97 \times 10^{16} = 16.4728$
   (b) $y = -\log 3 \times 10^{-3} = 2.5229$
   (c) $y = \log 12.3 = 1.0899$
   (d) $y = \text{antilog } 14.0 = 1.000 \times 10^{-14}$
   (e) $y = \text{antilog } (-12) = 1.000 \times 10^{-12}$
   (f) $y = \text{antilog } (-34.2) = 6.310 \times 10^{-35}$

35. Round the following answers to the correct number of significant figures.
   (a) $y = -\log 7.434 \times 10^{-6} = 5.12878$
   (b) $y = \log 432 = 2.6355$
   (c) $y = \log 2.1 \times 10^{17} = 17.32222$
   (d) $y = \text{antilog } 36.6 = 3.9811 \times 10^{36}$
   (e) $y = \text{antilog } (-10.08) = 8.3176 \times 10^{-11}$
   (f) $y = \text{antilog } 27 = 1.000 \times 10^{27}$

*36. Apply the Q test to the accompanying sets as a guide in determining whethe
   the outlying result should be retained or rejected.
   (a) 24.26, 24.50, 24.73, 24.63
   (b) 6.400, 6.416, 6.222, 6.408
   (c) 31.50, 31.68, 31.54, 31.82
   (d) 61.46, 61.38, 60.64, 61.50
   (e) 26.49, 26.56, 26.84

37. Apply the Q test to the accompanying sets as a guide in determining whether
   the outlying result should be retained or rejected.
   (a) 35.20, 35.72, 35.80
   (b) 65.30, 65.82, 65.85, 65.90
   (c) 29.03, 29.08, 28.97, 29.24
   (d) 16.88, 17.08, 17.02, 17.38
   (e) 29.03, 28.81, 29.08, 28.97

5

the solubility of precipitates

Reactions that yield products of limited solubility find wide application in three important analytical processes: (1) the separation of an analyte as a precipitate from soluble substances that would otherwise interfere with its ultimate measurement; (2) gravimetric analysis, in which a precipitate is formed whose weight is chemically related to the amount of analyte; and (3) volumetric analysis, based on determining the volume of a standard reagent required to precipitate the analyte essentially completely. The success of each of these applications requires that the solid produced have a relatively low solubility, be reasonably pure, and have a suitable particle size. In this chapter we consider the variables that influence the first of these three physical properties.

Examples illustrating the uses of solubility product constants to calculate the solubility of an ionic precipitate in water and in the presence of a common ion have been discussed in Chapter 3. The student should be thoroughly familiar with these principles before undertaking study of this chapter. Here we will be concerned with the way such variables as pH, concentration of complexing agents, and concentration of electrolytes affect the solubility of precipitates.

Effect of Competing Equilibria on the Solubility of Precipitates

The solubility of a precipitate increases in the presence of ionic or molecular species that react with the ions derived from the precipitate. This effect is illustrated by the examples shown in Table 5-1.

In the first example the solubility of barium sulfate is enhanced by the presence of a strong acid because sulfate ion, the conjugate base of the weak acid HSO_4^-, tends to react with hydronium ions. From the Le Châtelier principle it is evident that the added acid causes an increase in the hydrogen sulfate ion concentration. The consequent decrease in sulfate ion concentration is partially offset, however, by a shift of the first equilibrium to the right; the net result is an increase in solubility.

The second example shows that the solubility of silver bromide becomes greater in the presence of ammonia, which combines with silver ions to produce a silver ammine complex. Here ammonia molecules tend to decrease the silver ion concentration; a shift of the solubility equilibrium to the right occurs, resulting in an increase in solubility.

As a more general case, consider the sparingly soluble AB, which dissolves to give A and B ions:

$$
\begin{array}{ccc}
AB(s) \rightleftharpoons & A & + \quad B \\
 & + & + \\
 & C & D \\
 & \updownarrow & \updownarrow \\
 & AC & BD
\end{array}
$$

If A and B react with species C and D to form the soluble AC and BD,

TABLE 5-1
Increases in Solubility Brought About by Competing Equilibria

PRECIPITATE	SPECIES CAUSING SOLUBILITY INCREASES	EQUILIBRIA
$BaSO_4$	H_3O^+	$BaSO_4 \rightleftharpoons Ba^{2+} + SO_4^{2-}$ $+$ H_3O^+ $\updownarrow$ $HSO_4^- + H_2O$
$AgBr$	NH_3	$AgBr \rightleftharpoons Ag^+ + Br^-$ $+$ $2NH_3$ $\updownarrow$ $Ag(NH_3)_2^+$

introduction of either C or D into the solution will cause a shift in the solubility equilibrium in the direction that increases the solubility of AB.

Determination of the solubility of AB in a system such as this requires knowledge of the formal concentrations of the added C and D, as well as equilibrium constants for all three equilibria. Generally several algebraic expressions are needed to describe completely the concentration relationships in such a solution, and the solubility calculation requires the solution of multiple simultaneous equations. Frequently the solution of these algebraic equations is more formidable than the task of setting them up.

One point that should be constantly borne in mind when treating multiple equilibria is that *the validity and form of a particular equilibrium constant expression are in no way affected by the existence of additional competing equilibria in the solution.* Thus in the present example, the solubility product expression for AB describes the relationship between the equilibrium concentrations of A and B regardless of whether or not C and D are present in the solution. That is, at constant temperature the product [A][B] is a constant, provided only that some solid AB is present. To be sure, the *amount* of AB that dissolves is greater in the presence of C or D; the increase, however, is not because the ion product [A][B] has changed but rather because some of the precipitate has been converted to AC or BD.

A systematic approach by which any problem involving several equilibria can be attacked is presented in the following paragraphs. The approach will then be illustrated in the pages that follow.

SYSTEMATIC METHOD FOR SOLVING PROBLEMS INVOLVING SEVERAL EQUILIBRIA

1. Write chemical equations for all the reactions that appear to have any bearing on the problem.
2. State in terms of equilibrium concentrations what is being sought in the problem.
3. Write equilibrium constant expressions for all of the equilibria shown in step 1; find numerical values for the constants from appropriate tables.
4. Write mass-balance equations for the system. These are algebraic expressions relating the equilibrium concentrations of the various species to one another and to the formal concentrations of the substances present in the solution; they are derived by taking into account the way the solution was prepared.
5. Write a charge-balance equation. In any solution the concentrations of

the cations and anions must be such that the solution is electrically neutral. The charge-balance equation expresses this relationship.[1]

6. Count the number of unknown quantities in the equations developed in steps 3, 4, and 5, and compare this number with the number of independent equations. If the number of equations is equal to the number of unknown concentrations, the problem can be solved exactly by suitable algebraic manipulation. If there are fewer equations than unknowns, attempt to derive additional independent equations. If this cannot be done, it must be concluded that an exact solution to the problem is not possible; it may, however, be possible to arrive at an approximate solution.

7. Make suitable approximations to simplify the algebra or to decrease the number of unknowns so that the problem can be solved.

8. Solve the algebraic equations for the equilibrium concentrations that are necessary to give the answer as defined in step 2.

9. With the equilibrium concentrations obtained in step 8, check the approximations made in step 7 to be sure of their validity.

Step 6 in this scheme is particularly significant because it indicates whether an exact solution for the problem is theoretically feasible. If the number of independent equations is as great as the number of unknowns, the problem becomes purely algebraic, involving a solution to several simultaneous equations. On the other hand, if the number of equations is fewer than the number of unknowns, a search for other equations or approximations that will reduce the number of unknowns is essential. The student should never waste time in seeking a solution to a complex equilibrium problem without first establishing that sufficient data are available.

---

[1] As a simple example of a charge-balance equation, consider a solution prepared by dissolving NaCl in water. Such a solution has a net charge of zero although it contains both positive and negative ions. This neutrality is a direct consequence of the following relationship:

$$[Na^+] + [H_3O^+] = [Cl^-] + [OH^-]$$

Thus the solution is neutral by virtue of the fact that the sum of the concentrations of the positively charged species is equal to the sum of the concentrations of the negative species. Now consider an aqueous solution of $MgCl_2$. Here we must write

$$2[Mg^{2+}] + [H_3O^+] = [Cl^-] + [OH^-]$$

It is necessary to multiply the magnesium ion concentration by 2 in order to account for the two units of charge contributed by this ion; that is, charge balance is preserved because the chloride ion concentration is *twice* the magnesium ion concentration ($[Cl^-] = 2[Mg^{2+}]$). The concentration of a triply charged species, if present, would have to be multiplied by 3 for the same reason. Thus for a solution containing $Al_2(SO_4)_3$, $MgCl_2$, and water, the charge-balance equation would be

$$3[Al^{3+}] + 2[Mg^{2+}] + [H_3O^+] = 2[SO_4^{2-}] + [HSO_4^-] + [Cl^-] + [OH^-]$$

THE EFFECT OF pH ON SOLUBILITY

The solubilities of many precipitates are affected by the hydronium ion concentration of the solvent. Precipitates that exhibit this behavior contain an anion with basic properties, a cation with acidic properties, or both.

Solubility Product Calculations When the Hydronium Ion Concentration Is Fixed and Known. Analytical precipitations are frequently performed in solutions in which the hydronium ion concentration is fixed at some predetermined and known concentration. Calculation of solubility losses under these circumstances is a relatively straightforward process using the systematic approach, as illustrated by the following example.

EXAMPLE

Calculate the formal solubility of calcium oxalate in a solution that has a constant hydronium ion concentration of $1.00 \times 10^{-4} M$.

Step 1. Chemical equations.

$$CaC_2O_4(s) \rightleftharpoons Ca^{2+} + C_2O_4^{2-} \tag{5-1}$$

Because they are conjugate bases of weak acids, both oxalate ion and hydrogen oxalate ion will be involved in equilibria with the hydronium ion.

$$C_2O_4^{2-} + H_3O^+ \rightleftharpoons HC_2O_4^- + H_2O \tag{5-2}$$

$$HC_2O_4^- + H_3O^+ \rightleftharpoons H_2C_2O_4 + H_2O \tag{5-3}$$

Therefore this solution must contain other species capable of replacing the hydronium ions used up in establishing these equilibria.

Step 2. Definition of the unknown. What is sought? We wish to know the solubility of CaC_2O_4 in formula weights per liter. Because CaC_2O_4 is ionic, its formal solubility will be equal to the molar concentration of calcium ion; it will also be equal to the sum of the equilibrium concentrations of the oxalate species. That is,

$$solubility = [Ca^{2+}]$$

$$= [C_2O_4^{2-}] + [HC_2O_4^-] + [H_2C_2O_4]$$

Thus if we can calculate either of these quantities we shall have obtained a solution to the problem.

Step 3. Equilibrium constant expressions.

$$K_{sp} = [Ca^{2+}][C_2O_4^{2-}] = 2.3 \times 10^{-9} \tag{5-4}$$

Equation 5-2 is simply the reverse of the dissociation reaction for $HC_2O_4^-$.

We can thus use the value of K_2 for oxalic acid to provide a relationship between $[C_2O_4^{2-}]$ and $[HC_2O_4^-]$.[2]

$$K_2 = \frac{[H_3O^+][C_2O_4^{2-}]}{[HC_2O_4^-]} = 5.42 \times 10^{-5} \tag{5-5}$$

Similarly, for Equation 5-3

$$K_1 = \frac{[H_3O^+][HC_2O_4^-]}{[H_2C_2O_4]} = 5.36 \times 10^{-2} \tag{5-6}$$

Step 4. Mass-balance equations. Because the only source of Ca^{2+} and the various oxalate species is the dissolved CaC_2O_4, it follows that

$$[Ca^{2+}] = [C_2O_4^{2-}] + [HC_2O_4^-] + [H_2C_2O_4] \tag{5-7}$$

Furthermore, it is given that at equilibrium

$$[H_3O^+] = 1.0 \times 10^{-4} \tag{5-8}$$

Step 5. Charge-balance equations. A useful charge-balance equation cannot be written for this system because an amount of some unknown acid HX has been added to maintain $[H_3O^+]$ at 1.0×10^{-4}; an equation based on the electrical neutrality of the solution would require inclusion of the concentration of the anion $[X^-]$ associated with the unknown acid. As it turns out, an equation containing this additional unknown term is not needed.

Step 6. Comparison of equations and unknowns. We have four unknowns: $[Ca^{2+}]$, $[C_2O_4^{2-}]$, $[HC_2O_4^-]$, and $[H_2C_2O_4]$. We also have four independent algebraic relationships: Equations 5-4, 5-5, 5-6, and 5-7. Therefore an exact solution is possible, and the problem has now become one of algebra.

Step 7. Approximations. Because we have sufficient data, an exact solution to the problem can be obtained.

Step 8. Solution of the equations. A convenient way to solve the four equations is to make suitable substitutions into Equation 5-7 and thereby establish a relationship between $[Ca^{2+}]$ and $[C_2O_4^{2-}]$. We must first derive expressions for $[HC_2O_4^-]$ and $[H_2C_2O_4]$ in terms of $[C_2O_4^{2-}]$. Substitution of

---

[2] If desired, the equilibrium constant for the reaction in Equation 5-2 *as written* could be employed; that is,

$$K = \frac{[HC_2O_4^-]}{[H_3O^+][C_2O_4^{2-}]} = \frac{1}{K_2} = \frac{1}{5.42 \times 10^{-5}} = 1.85 \times 10^4$$

This expression could thus be employed in lieu of Equation 5-5. Nothing is gained by this added arithmetic operation, however.

1.00×10^{-4} for $[H_3O^+]$ in Equation 5-5 yields

$$\frac{(1.00 \times 10^{-4})[C_2O_4^{2-}]}{[HC_2O_4^-]} = 5.42 \times 10^{-5}$$

Thus

$$[HC_2O_4^-] = \frac{1.00 \times 10^{-4}}{5.42 \times 10^{-5}}[C_2O_4^{2-}] = 1.85[C_2O_4^{2-}]$$

Upon substituting this relationship and the hydronium ion concentration into Equation 5-6, we obtain

$$\frac{1.00 \times 10^{-4} \times 1.85[C_2O_4^{2-}]}{[H_2C_2O_4]} = 5.36 \times 10^{-2}$$

Thus

$$[H_2C_2O_4] = \frac{1.00 \times 10^{-4} \times 1.85[C_2O_4^{2-}]}{5.36 \times 10^{-2}} = 0.0034[C_2O_4^{2-}]$$

These values for $[H_2C_2O_4]$ and $[HC_2O_4^-]$ are substituted into Equation 5-7 to give

$$[Ca^{2+}] = [C_2O_4^{2-}] + 1.85[C_2O_4^{2-}] + 0.0034[C_2O_4^{2-}]$$

$$= 2.85[C_2O_4^{2-}]$$

or

$$[C_2O_4^{2-}] = \frac{[Ca^{2+}]}{2.85}$$

Substitution for $[C_2O_4^{2-}]$ in Equation 5-4 gives

$$[Ca^{2+}]\frac{[Ca^{2+}]}{2.85} = 2.3 \times 10^{-9}$$

$$[Ca^{2+}]^2 = 6.56 \times 10^{-9}$$

$$[Ca^{2+}] = 8.1 \times 10^{-5}$$

Thus from step 2 we conclude that

$$\text{solubility of } CaC_2O_4 = 8.1 \times 10^{-5} \text{ fw/liter}$$

Solubility Calculations Where the Hydronium Ion Concentration is Variable. Solutes containing basic anions (such as calcium oxalate) or acidic cations influence the hydronium ion concentration of their aqueous solutions. Thus if an auxiliary reagent is not employed to maintain a constant pH, the hydronium ion concentration becomes dependent upon the extent to which such solutes dissolve. For example, an aqueous solution of saturated calcium oxalate becomes basic as a consequence of the reactions

$$CaC_2O_4(s) \rightleftharpoons Ca^{2+} + C_2O_4^{2-}$$

$$C_2O_4^{2-} + H_2O \rightleftharpoons HC_2O_4^- + OH^-$$

$$HC_2O_4^- + H_2O \rightleftharpoons H_2C_2O_4 + OH^-$$

In contrast to the example just considered, the hydroxide ion concentration now becomes an unknown variable, and an additional algebraic equation must therefore be developed.

If the reaction between a precipitate and water is neglected, the calculated result is frequently in serious error. The magnitude of this error depends upon the solubility of the precipitate as well as the dissociation constant of the conjugate acid from which the anion of the precipitate is derived.

By employing the systematic method described earlier, sufficient algebraic equations can be readily derived to make computation of the solubility of a precipitate in pure water possible; exact solution of these equations is difficult, however, unless suitable approximations are made.[3]

Solubility of Metal Hydroxides in Water. In determining the solubility of metal hydroxides two equilibria may have to be considered. For example, with the divalent metal ion M^{2+} these are

$$M(OH)_2(s) \rightleftharpoons M^{2+} + 2OH^-$$

$$2H_2O \rightleftharpoons H_3O^+ + OH^-$$

Three algebraic equations are readily derived for this system, namely,

$$[M^{2+}][OH^-]^2 = K_{sp} \tag{5-9}$$

$$[H_3O^+][OH^-] = K_w \tag{5-10}$$

and from charge-balance considerations

$$2[M^{2+}] + [H_3O^+] = [OH^-] \tag{5-11}$$

If the hydroxide is reasonably soluble, the hydronium ion concentration will be small, and Equation 5-11 becomes

$$2[M^{2+}] \cong [OH^-]$$

Substitution of this expression into Equation 5-9 and rearrangement give

$$[M^{2+}] = \left(\frac{K_{sp}}{4}\right)^{1/3} = \text{solubility} \tag{5-12}$$

Note that this expression is identical with the one used in the example on page 30.

On the other hand, if the solubility of $M(OH)_2$ is very low, the situation is encountered in which $2[M^{2+}]$ is much smaller than $[H_3O^+]$. Equation 5-11 then becomes

$$[H_3O^+] \cong [OH^-] = 1.00 \times 10^{-7}$$

[3] See, for example, D. A. Skoog and D. M. West, *Fundamentals of Analytical Chemistry*, 3d ed. New York: Holt, Rinehart and Winston, 1976, pp. 94–99.

Again, substitution into Equation 5-9 and rearrangement yield

$$[M^{2+}] = \frac{K_{sp}}{[OH^-]^2} = \frac{K_{sp}}{1.00 \times 10^{-14}} = \text{solubility} \qquad \textbf{(5-13)}$$

EXAMPLE

Calculate the solubility of $Fe(OH)_3$ in water.

As a hypothesis let us assume that the charge-balance expression simplifies to

$$3[Fe^{3+}] + [H_3O^+] \cong 3[Fe^{3+}] = [OH^-]$$

Substitution for $[OH^-]$ into the solubility product expression gives

$$[Fe^{3+}][OH^-]^3 = [Fe^{3+}](3[Fe^{3+}])^3 = 4 \times 10^{-38}$$

$$[Fe^{3+}] = \left(\frac{4 \times 10^{-38}}{27}\right)^{1/4} = 2 \times 10^{-10}$$

and

$$\text{solubility} = 2 \times 10^{-10} \text{ fw/liter}$$

We have assumed, however, that

$$[OH^-] \cong 3[Fe^{3+}] = 3 \times 2 \times 10^{-10} = 6 \times 10^{-10}$$

which means that

$$[H_3O^+] = \frac{1.00 \times 10^{-14}}{6 \times 10^{-10}} = 1.7 \times 10^{-5}$$

Clearly $[H_3O^+]$ is not much smaller than $3[Fe^{3+}]$, as assumed; indeed, the reverse appears to be the case. That is,

$$3[Fe^{3+}] << [H_3O^+]$$

Thus the charge-balance equation reduces to

$$[H_3O^+] \cong [OH^-] = 1.00 \times 10^{-7}$$

Substitution for $[OH^-]$ in the solubility product expression yields

$$[Fe^{3+}] = \frac{4 \times 10^{-38}}{(1.00 \times 10^{-7})^3} = 4 \times 10^{-17}$$

$$\text{solubility} = 4 \times 10^{-17} \text{ fw/liter}$$

The assumption that $3[Fe^{3+}] << [H_3O^+]$ is clearly valid. Note the very large error in the first calculation where the faulty assumption was employed.

COMPLEX ION FORMATION AND SOLUBILITY

The solubility of a precipitate may be greatly altered by the presence of some species that will react with the anion or cation of the precipitate to form a stable complex. For example, the precipitation of aluminum with base is never complete in the presence of fluoride ion, even though aluminum hydroxide has an extremely low solubility; the fluoride complexes of aluminum(III) are sufficiently stable to prevent quantitative removal of the cation from solution. The equilibria involved can be represented by

$$Al(OH)_3(s) \rightleftharpoons Al^{3+} + 3OH^-$$
$$+$$
$$6F^-$$
$$\updownarrow$$
$$AlF_6^{3-}$$

Fluoride ions thus compete successfully with hydroxide ions for aluminum(III); as the fluoride concentration is increased, more and more of the precipitate is dissolved and converted to fluoroaluminate ions.

Quantitative Treatment of the Effect of Complex Formation on the Solubility of Precipitates. The solubility of a precipitate in the presence of a complexing reagent can be calculated, provided the equilibrium constant for the complex formation reaction is known. The techniques used are similar to those discussed in the preceding section.

EXAMPLE

Find the solubility of AgBr in a solution that is 0.020 F in NH_3.

Equilibria:

$$AgBr(s) \rightleftharpoons Ag^+ + Br^-$$

$$Ag^+ + NH_3 \rightleftharpoons AgNH_3^+$$

$$AgNH_3^+ + NH_3 \rightleftharpoons Ag(NH_3)_2^+$$

$$NH_3 + H_2O \rightleftharpoons NH_4^+ + OH^-$$

Definition of unknown:

$$\text{solubility of AgBr} = [Br^-]$$

$$= [Ag^+] + [AgNH_3^+] + [Ag(NH_3)_2^+]$$

Equilibrium constants:

$$[Ag^+][Br^-] = K_{sp} = 5.2 \times 10^{-13} \tag{5-14}$$

$$\frac{[AgNH_3^+]}{[Ag^+][NH_3]} = K_1 = 2.0 \times 10^3 \tag{5-15}$$

$$\frac{[Ag(NH_3)_2^+]}{[AgNH_3^+][NH_3]} = K_2 = 6.9 \times 10^3 \tag{5-16}$$

$$\frac{[NH_4^+][OH^-]}{[NH_3]} = K_b = 1.76 \times 10^{-5} \tag{5-17}$$

Mass-balance expressions:

$$[Br^-] = [Ag^+] + [AgNH_3^+] + [Ag(NH_3)_2^+] \qquad \textbf{(5-18)}$$

Because the NH_3 concentration was initially 0.020, we may also write

$$0.020 = [NH_3] + [AgNH_3^+] + 2[Ag(NH_3)_2^+] + [NH_4^+] \qquad \textbf{(5-19)}$$

Furthermore, the reaction of NH_3 with water produces one OH^- for each NH_4^+. Thus

$$[OH^-] \cong [NH_4^+] \qquad \textbf{(5-20)}$$

Charge-balance equation:

$$[NH_4^+] + [Ag^+] + [AgNH_3^+] + [Ag(NH_3)_2^+] = [Br^-] + [OH^-] \qquad \textbf{(5-21)}^4$$

Close examination of these eight equations reveals that there are only seven independent expressions inasmuch as Equation 5-21 is the sum of Equations 5-20 and 5-18. There are only seven unknowns, however, so a solution is possible.

Approximations:

1. $[NH_4^+]$ can be neglected in Equation 5-19. This assumption appears reasonable in light of the small value for K_b in Equation 5-17.
2. $[AgNH_3^+]$ and $2[Ag(NH_3)_2^+]$ are both significantly smaller than $[NH_3]$ in Equation 5-19. A reason for believing this assumption might be valid is the small value for K_{sp} (Equation 5-14). This suggests that little AgBr dissolves, and thus the concentration of the two complexes will be small. We cannot be certain that this assumption is proper at this stage, but it is worth trying.

Application of these approximations reduces Equation 5-19 to

$$[NH_3] = 0.020 \qquad \textbf{(5-22)}$$

and also decreases the number of unknowns to four: $[Ag^+]$, $[AgNH_3^+]$, $[Ag(NH_3)_2^+]$, and $[Br^-]$. Thus only four equations containing these unknowns are needed. Equations 5-14, 5-15, 5-16, and 5-18 serve.

Substituting Equation 5-22 into 5-15 and rearranging yield

$$\frac{[AgNH_3^+]}{[Ag^+]} = 2.0 \times 10^3 \times 0.020 = 40 \qquad \textbf{(5-23)}$$

Substitution of this relation and Equation 5-22 into Equation 5-16 gives, upon rearrangement,

$$\frac{[Ag(NH_3)_2^+]}{[Ag^+]} = 40 \times 6.9 \times 10^3 \times 0.020 = 5.52 \times 10^3 \qquad \textbf{(5-24)}$$

---

[4] We have neglected the $[H_3O^+]$ because its concentration will certainly be negligible in a 0.020-F solution of NH_3.

A relationship between $[Ag^+]$ and $[Br^-]$ can now be derived by substitution of Equations 5-24 and 5-23 into Equation 5-18. Thus

$$[Br^-] = [Ag^+] + 40[Ag^+] + 5.52 \times 10^3[Ag^+]$$

$$= 5.6 \times 10^3[Ag^+]$$

Thus the solubility product constant can be expressed as

$$\frac{[Br^-]}{5.6 \times 10^3} \times [Br^-] = 5.2 \times 10^{-13}$$

$$[Br^-] = 5.4 \times 10^{-5}$$

$$\text{solubility} = 5.4 \times 10^{-5} \text{ fw AgBr/liter}$$

To check the assumptions we first calculate $[Ag^+]$ with Equation 5-14:

$$[Ag^+] = \frac{5.2 \times 10^{-13}}{[Br^-]} = \frac{5.2 \times 10^{-13}}{5.4 \times 10^{-5}} = 9.6 \times 10^{-9}$$

Insertion of this value for $[Ag^+]$ in Equations 5-23 and 5-24 gives

$$[AgNH_3^+] = 40 \times 9.6 \times 10^{-9} = 3.8 \times 10^{-7}$$

$$[Ag(NH_3)_2^+] = 5.52 \times 10^3 \times 9.6 \times 10^{-9} = 5.3 \times 10^{-5}$$

Thus we see that assumption 2 is valid; that is, the concentration of the two complexes is small when compared with that of NH_3 (0.020 M).

To check assumption 1 we substitute Equations 5-22 and 5-20 into the dissociation constant for ammonia (Equation 5-17):

$$\frac{[NH_4^+]^2}{0.020} = 1.76 \times 10^{-5}$$

$$[NH_4^+] = 5.9 \times 10^{-4}$$

Again, we see that this concentration is small when compared with 0.020 M.

Complex Formation Involving a Common Ion of the Precipitate. Many precipitates tend to react with one of their constituent ions to form soluble complexes. For example, silver chloride forms chloro complexes believed to be of the composition $AgCl_2^-$, $AgCl_3^{2-}$, and so on. Such reactions cause increases in solubility at high concentrations of the common ion. This effect is illustrated by Figure 5-1, where the experimentally determined solubility of silver chloride is plotted against the logarithm of the potassium chloride concentration. At chloride concentrations less than 10^{-3} F, the experimental solubilities do not differ greatly from those calculated with the solubility product for silver chloride. At higher chloride ion concentrations, however, the calculated solubilities approach zero while the measured values rise precipitously; in about 0.3-F

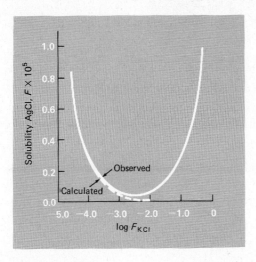

FIGURE 5-1 The solubility of silver chloride in potassium chloride solutions. The dashed curve is calculated from K_{sp}; the solid curve represents experimental values obtained by A. Pinkus and A. M. Timmermans in *Bull. Soc. Chim. Belges*, **46**, 46–73 (1937).

potassium chloride the solubility of silver chloride is the same as in pure water, and in 1-F solution it is approximately eight times larger. If complete information were available regarding the composition of the complexes and their formation constants, a quantitative description of these effects should be possible (see Problem 20 at the end of this chapter).

Solubility increases in the presence of large excesses of a common ion are by no means rare. Of particular interest are the amphoteric hydroxides such as those of aluminum and zinc, which form slightly soluble precipitates upon treatment with base; these redissolve in the presence of excess hydroxide ions to give the complex aluminate and zincate ions. For aluminum the equilibria can be represented as

$$Al^{3+} + 3OH^- \rightleftharpoons Al(OH)_3(s)$$

$$Al(OH)_3(s) + OH^- \rightleftharpoons Al(OH)_4^-$$

As with silver chloride, the solubilities of aluminum hydroxide and zinc hydroxide pass through minima and then increase rapidly with increasing concentrations of the common ion. The hydroxide ion concentration corresponding to the minimum solubility can be calculated readily if the equilibrium constants for the reactions are known.

Separation of Ions by Control of the Concentration of the Precipitating Reagent

When two ions react with a third to form precipitates of different solubilities, the less soluble species will be produced at a lower reagent concentration. If the solubilities are sufficiently different, quantitative removal of the first ion from solution may be achieved without precipitation of the second. Such separations require careful control of the precipitating reagent concentration at some suitable, predetermined level. Many important analytical separations, notably those involving sulfide ion, hydroxide ion, and organic reagents, are based on this method.

CALCULATION OF THE FEASIBILITY OF SEPARATIONS

An important application of solubility product calculations involves determining the feasibility and the optimum conditions for separations based on the control of reagent concentration. The following problem illustrates such an application.

EXAMPLE

Is the difference between the solubilities of their hydroxides sufficient to permit the quantitative separation of Fe^{3+} from Mg^{2+} in a solution that is $0.10\ F$ in each cation? If the separation is possible, what range of OH^- concentrations is permissible? Solubility product constants for the two hydroxides are

$$[Fe^{3+}][OH^-]^3 = 4 \times 10^{-38}$$

$$[Mg^{2+}][OH^-]^2 = 1.8 \times 10^{-11}$$

The K_{sp} for $Fe(OH)_3$ is so much smaller than that for $Mg(OH)_2$ as to suggest that the former will precipitate at a lower OH^- concentration.[5]

We can answer the questions posed in this problem by (1) calculating the OH^- concentration required to achieve the quantitative precipitation of Fe^{3+} from this solution and (2) determining the OH^- concentration at which $Mg(OH)_2$ will just begin to precipitate. If (1) is smaller than (2), a separation is feasible, and the range of OH^- concentrations to be used will be defined by the two values.

To determine (1), we must first decide what constitutes a quantitative removal of Fe^{3+} from the solution. Under no conditions can every iron(III) ion be precipitated; we must therefore *arbitrarily* set some limit below which, for all practical purposes, the further presence of this ion can be neglected. When its concentration has been decreased to $10^{-6}\ M$, only $\frac{1}{100,000}$ of the original quantity of Fe^{3+} will remain in the solution; for most purposes, removal of all but this fraction of an ion can be considered a quantitative separation.

---

[5] The reader should be aware, however, that it is only the enormous numerical difference between the two constants that permits this judgment. The units for K_{sp} for $Mg(OH)_2$ ($mole^3/liter^3$) are not the same as those for $Fe(OH)_3$ ($mole^4/liter^4$); strictly then, the two constants are not comparable.

We can readily calculate the OH^- concentration in equilibrium with 1.0×10^{-6} M Fe^{3+} by substituting directly into the solubility product expression:

$$(1.0 \times 10^{-6})[OH^-]^3 = 4 \times 10^{-38}$$

$$[OH^-] = 3.4 \times 10^{-11}$$

Thus if we maintain the OH^- concentration at 3.4×10^{-11} mole/liter, the Fe^{3+} concentration will be lowered to 1.0×10^{-6} mole/liter. Note that quantitative precipitation of $Fe(OH)_3$ is achieved in a distinctly acidic solution.

We must now consider question (2)—that is, what is the maximum OH^- concentration that can exist in solution without causing formation of $Mg(OH)_2$? Precipitation cannot occur until the Mg^{2+} concentration multiplied by the square of the OH^- concentration exceeds the solubility product, 1.8×10^{-11}. By substituting 0.1 (the molar Mg^{2+} concentration of the solution) into the solubility product expression, we can calculate the *maximum* OH^- concentration that can be tolerated:

$$0.10[OH^-]^2 = 1.8 \times 10^{-11}$$

$$[OH^-] = 1.3 \times 10^{-5}$$

When the OH^- concentration exceeds this level, the solution will be supersaturated with respect to $Mg(OH)_2$, and precipitation can begin.

From these calculations we conclude that quantitative separation of $Fe(OH)_3$ can be expected if the OH^- concentration is greater than 3.4×10^{-11} mole/liter and that $Mg(OH)_2$ will not precipitate until a concentration of 1.3×10^{-5} mole/liter is reached. Therefore it should be possible to separate Fe^{3+} from Mg^{2+} by maintaining the OH^- concentration between these levels.

SULFIDE SEPARATIONS

A number of important methods for the separation of metallic ions involve controlling the concentration of the precipitating anion by regulating the hydronium ion concentration of the solution. Such methods are particularly attractive because of the relative ease with which the hydronium ion concentration may be maintained at some predetermined level by the use of a suitable buffer.[6] Perhaps the best known of these methods makes use of hydrogen sulfide as the precipitating reagent. Hydrogen sulfide is a weak acid, dissociating as follows:

$$H_2S + H_2O \rightleftharpoons H_3O^+ + HS^- \qquad K_1 = \frac{[H_3O^+][HS^-]}{[H_2S]} = 5.7 \times 10^{-8}$$

[6] The preparation and properties of buffer solutions are considered in Chapter 9. An important property of a buffer is that it maintains the hydronium ion concentration at an approximately fixed and predetermined level.

$$HS^- + H_2O \rightleftharpoons H_3O^+ + S^{2-} \qquad K_2 = \frac{[H_3O^+][S^{2-}]}{[H_2S]} = 1.2 \times 10^{-15}$$

These equations may be combined to give an expression for the overall dissociation of hydrogen sulfide into sulfide ion:

$$H_2S + 2H_2O \rightleftharpoons 2H_3O^+ + S^{2-} \qquad K_1K_2 = \frac{[H_3O^+]^2[S^{2-}]}{[H_2S]} = 6.8 \times 10^{-23}$$

The constant for this reaction is simply the product of K_1 and K_2.

In sulfide separations the solutions are continuously kept saturated with hydrogen sulfide; thus the formal concentration of the reagent is essentially constant throughout the precipitation. Because it is such a weak acid, the actual molar concentration of hydrogen sulfide will correspond closely to its solubility in water, which is about 0.1 F. For practical purposes, then, we may assume that throughout any sulfide precipitation

$$[H_2S] \cong 0.10 \text{ mole/liter}$$

Substituting this value into the dissociation constant expression, we obtain

$$\frac{[H_3O^+]^2[S^{2-}]}{0.10} = 6.8 \times 10^{-23}$$

$$[S^{2-}] = \frac{6.8 \times 10^{-24}}{[H_3O^+]^2}$$

Thus the molar concentration of the sulfide ion varies inversely as the square of the hydronium ion concentration of the solution. This relationship is useful for calculating the optimum conditions for the separation of cations by sulfide precipitation.

EXAMPLE

Find the conditions under which Pb^{2+} and Tl^+ can, in theory, be separated quantitatively by H_2S precipitation from a solution that is 0.1 F in each cation.

The constants for the two solubility equilibria are

$$PbS(s) \rightleftharpoons Pb^{2+} + S^{2-} \qquad [Pb^{2+}][S^{2-}] = 7 \times 10^{-28}$$

$$Tl_2S(s) \rightleftharpoons 2Tl^+ + S^{2-} \qquad [Tl^+]^2[S^{2-}] = 1 \times 10^{-22}$$

PbS will precipitate at a lower S^{2-} concentration than the Tl_2S. Assuming again that lowering the Pb^{2+} concentration to 10^{-6} M or less constitutes quantitative removal, and substituting this value into the solubility product expression, we can evaluate the required sulfide ion concentration.

$$10^{-6}[S^{2-}] = 7 \times 10^{-28}$$

$$[S^{2-}] = 7 \times 10^{-22}$$

This value should then be compared with the S^{2-} concentration needed to initiate precipitation of Tl_2S from a 0.1-F solution:

$$(0.1)^2[S^{2-}] = 1 \times 10^{-22}$$

$$[S^{2-}] = 1 \times 10^{-20}$$

Thus to achieve a separation the S^{2-} concentration should be kept between 7×10^{-22} and 1×10^{-20} mole/liter. Now we must compute the H_3O^+ concentrations necessary to hold the S^{2-} concentration within these confines. Using the relationship derived previously,

$$[S^{2-}] = \frac{6.8 \times 10^{-24}}{[H_3O^+]^2}$$

and substituting the two limiting values for $[S^{2-}]$, we obtain

$$[H_3O^+]^2 = \frac{6.8 \times 10^{-24}}{7 \times 10^{-22}} = 0.97 \times 10^{-2}$$

$$[H_3O^+] = 0.098 \cong 0.1$$

and

$$[H_3O^+]^2 = \frac{6.8 \times 10^{-24}}{1 \times 10^{-20}}$$

$$[H_3O^+] = 0.026 \cong 0.03$$

By maintaining the H_3O^+ concentration between 0.03 and 0.1 M, it should, in theory, be possible to separate PbS without precipitation of Tl_2S. From a practical standpoint, however, it is doubtful that conditions could be controlled closely enough to give a clean separation.

Effect of Electrolyte Concentration on Solubility

It is found experimentally that precipitates tend to be more soluble in an electrolyte solution than in water, provided, of course, that the electrolyte contains no ions in common with the precipitate. The data plotted in Figure 5-2 demonstrate the magnitude of this effect for three precipitates. A twofold increase in the solubility of barium sulfate is observed when the potassium nitrate concentration is increased from 0 to 0.02 F. The same change in electrolyte concentration increases the solubility of barium iodate by a factor of only 1.25 and of silver chloride by 1.20.

The effect of an electrolyte upon solubility stems from the electrostatic attraction between the foreign ions and the ions of opposite charge in the precipitate. Such interactions shift the position of the equilibrium. It is important to realize that this effect is not peculiar to solubility equilibria but is observed with all other types as well. For example, the data in Table 5-2 show that the degree of dissociation of acetic acid

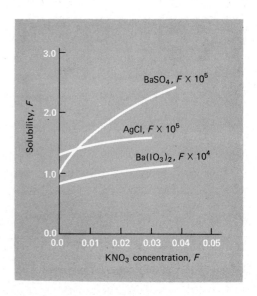

FIGURE 5-2 The effect of electrolyte concentration on the solubility of some salts.

increases significantly in the presence of sodium chloride. These experimental dissociation constants were obtained by measuring the equilibrium concentrations of hydronium and acetate ions in solutions containing the indicated salt concentrations. Again, the obvious shift in equilibrium can be attributed to the attraction of the ions of the electrolyte for the charged hydronium and acetate ions.

From data such as these, one must conclude that the equilibrium law, as we have presented it, is a *limiting law* in the sense that it applies

TABLE 5-2

Dissociation Constants for Acetic Acid in Solutions of Sodium Chloride at 25°C[a]

CONCENTRATION OF NaCl, F	APPARENT K'_a
0.00	1.75×10^{-5}
0.02	2.29×10^{-5}
0.11	2.85×10^{-5}
0.51	3.31×10^{-5}
1.01	3.16×10^{-5}

[a] From H. S. Harned and C. F. Hickey, *J. Amer. Chem. Soc.*, **59**, 1289 (1937). With permission of the American Chemical Society.

exactly only to very dilute solutions in which the electrolyte concentration is insignificant (that is, to ideal solutions only; see p. 26). We must now consider a more rigorous form of the law which can be applied to nonideal solutions.

SOME EMPIRICAL OBSERVATIONS

Extensive studies concerned with the influence of electrolyte concentration upon chemical equilibrium have led to a number of important generalizations. One is that the magnitude of the effect is highly dependent upon the charges of the species involved in the equilibrium. Where all are neutral particles, little variation in the equilibrium constant is observed. On the other hand, the effects become greater as the charges on the reactants or products increase. Thus, for example, of the two equilibria

$$AgCl(s) \rightleftharpoons Ag^+ + Cl^-$$

$$BaSO_4(s) \rightleftharpoons Ba^{2+} + SO_4^{2-}$$

the second is shifted farther to the right in the presence of moderate amounts of potassium nitrate than is the first (see Figure 5-2).

A second important generality is that over a considerable electrolyte concentration range the effects are essentially independent of the kind of electrolyte and dependent only upon a concentration parameter of the solution called the *ionic strength*. This quantity is defined by the equation

$$\text{ionic strength} = \mu = \tfrac{1}{2}(m_1 Z_1^2 + m_2 Z_2^2 + m_3 Z_3^2 + \cdots) \qquad \textbf{(5-25)}$$

where $m_1, m_2, m_3, \ldots$, represent the molar concentrations of the various ions in the solution, and $Z_1, Z_2, Z_3, \ldots$, are their respective charges.

EXAMPLE

Calculate the ionic strength of (1) a 0.1-F solution of KNO_3 and (2) a 0.1-F solution of Na_2SO_4.

1. For the KNO_3 solution, m_{K^+} and $m_{NO_3^-}$ are 0.1 and

$$\mu = \tfrac{1}{2}(0.1 \times 1^2 + 0.1 \times 1^2) = 0.1$$

2. For the Na_2SO_4 solution, $m_{Na^+} = 0.2$ and $m_{SO_4^{2-}} = 0.1$. Therefore

$$\mu = \tfrac{1}{2}(0.2 \times 1^2 + 0.1 \times 2^2) = 0.3$$

EXAMPLE

What is the ionic strength of a solution that is both 0.05 F in KNO_3 and 0.1 F in Na_2SO_4?

$$\mu = \tfrac{1}{2}(0.05 \times 1^2 + 0.05 \times 1^2 + 0.2 \times 1^2 + 0.1 \times 2^2)$$

$$= 0.35$$

From these examples it is apparent that the ionic strength of a strong electrolyte solution consisting solely of singly charged ions is identical with the total formal salt concentration. If the species carry multiple charges, however, the ionic strength is greater than the formal concentration.

For solutions with ionic strengths of 0.1 or less it is found that the electrolyte effect is independent of the *kind* of ions and dependent *only upon the ionic strength.* Thus the degree of dissociation of acetic acid is the same in the presence of sodium chloride, potassium nitrate, or barium iodide, provided the concentrations of these species are such that the ionic strength is fixed. It should be noted that this independence with respect to electrolyte species disappears at high ionic strength.

ACTIVITY AND ACTIVITY COEFFICIENTS

To describe the effect of ionic strength on equilibria in quantitative terms, chemists use a concentration parameter called the *activity,* which is defined as follows:

$$a_A = [A]f_A \qquad (5\text{-}26)$$

where a_A is the activity of the species A, $[A]$ is its molar concentration, and f_A is a dimensionless quantity called the *activity coefficient.* The activity coefficient (and thus the activity) of A varies with ionic strength such that employment of a_A instead of $[A]$ in an equilibrium constant expression frees the numerical value of the constant from dependence on the ionic strength. To illustrate, for the dissociation of acetic acid we write

$$K_a = \frac{a_{H_3O^+} \cdot a_{OAc^-}}{a_{HOAc}} = \frac{[H_3O^+][OAc^-]}{[HOAc]} \times \frac{f_{H_3O^+} \cdot f_{OAc^-}}{f_{HOAc}}$$

where $f_{H_3O^+}$, f_{OAc^-}, and f_{HOAc} vary with ionic strength to keep K_a numerically constant over a wide range of ionic strengths (in contrast to the *apparent* K_a' shown in Table 5-2).

Properties of Activity Coefficients. Activity coefficients have the following properties:

1. The activity coefficient of a species can be thought of as a measure of the effectiveness with which that species influences an equilibrium in which it is a participant. In very dilute solutions where the ionic strength is minimal, this effectiveness becomes constant, and the activity coefficient acquires a value of unity. Under such circumstances the activity and molar concentration become identical. As the ionic strength increases, however, an ion loses some of its effectiveness, and its activity coefficient decreases. We may summarize this behavior in terms of Equation 5-26. At moderate ionic strengths $f_A < 1$; as the solution approaches infinite dilution, however, $f_A \to 1$ and thus $a_A \to [A]$.

At high ionic strengths the activity coefficients for some species increase and may even become greater than one. Interpretation of the behavior of solutions in this region is difficult; we shall confine our discussion to regions of low or moderate ionic strengths (that is, where $\mu < 0.1$).

The variation of typical activity coefficients as a function of ionic strength is shown in Figure 5-3.

2. In solutions that are not too concentrated, the activity coefficient for a given species is independent of the specific nature of the electrolyte and dependent only upon the ionic strength.

3. For a given ionic strength, the activity coefficient of an ion departs farther from unity as the charge carried by the species increases. This effect is shown in Figure 5-3. The activity coefficient of an uncharged molecule is approximately unity, regardless of ionic strength.

4. For ions of the same charge, activity coefficients are approximately the same at any given ionic strength. The small variations that do exist can be correlated with the effective diameter of the hydrated ions.

5. The activity coefficient of a given ion describes its effective behavior in all equilibria in which it participates. For example, at a given ionic strength a single activity coefficient for cyanide ion describes the influence of that species upon any of the following equilibria:

$$HCN + H_2O \rightleftharpoons H_3O^+ + CN^-$$

$$AgCN(s) \rightleftharpoons Ag^+ + CN^-$$

$$Ni(CN)_4^{2-} \rightleftharpoons Ni^{2+} + 4CN^-$$

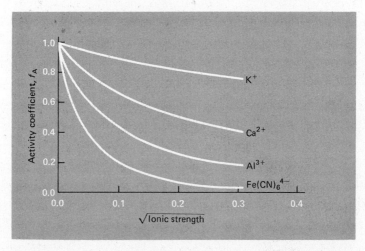

FIGURE 5-3   The effect of ionic strength on activity coefficients.

Evaluation of Activity Coefficients. In 1923 Debye and Hückel derived the following theoretical expression, which permits the calculation of activity coefficients of ions:[7]

$$-\log f_A = \frac{0.5085\, Z_A^2 \sqrt{\mu}}{1 + 0.3281\, a_A \sqrt{\mu}} \qquad \text{(5-27)}$$

where
$$-\log f_A = (0.5)\, Z_A^2 \sqrt{\mu}$$

f_A = activity coefficient of the species A
Z_A = charge on the species A
μ = ionic strength of the solution
a_A = the effective diameter of the hydrated ion A in ångström units (Å) (1 Å = 10^{-8} cm)

The constants 0.5085 and 0.3281 are applicable to solutions at 25°C; other values must be employed at different temperatures.

Unfortunately considerable uncertainty exists regarding the magnitude of a_A in Equation 5-27. Its value appears to be approximately 3 Å for most singly charged ions so that for these species, the denominator of the Debye–Hückel equation simplifies to approximately $(1 + \sqrt{\mu})$. For ions with higher charge a_A may become as large as 10 Å. It should be noted that the second term of the denominator becomes small with respect to the first when the ionic strength is less than 0.01; under these circumstances uncertainties in a_A are of little significance in calculating activity coefficients.

Kielland[8] has calculated values of a_A for numerous ions from a variety of experimental data. His "best values" for effective diameters are given in Table 5-3. Also presented are activity coefficients calculated from Equation 5-27 using these values for the size parameter.

Experimental verification of individual activity coefficients such as those shown in Table 5-3 is unfortunately impossible; all experimental methods give only a mean activity coefficient for the positively and negatively charged ions in a solution.[9] It should be pointed out, however, that *mean* coefficients calculated from the data in Table 5-3 agree satisfactorily with the experimental values.

[7] P. Debye and E. Hückel, *Physik. Z.*, **24**, 185 (1923).

[8] J. Kielland, *J. Amer. Chem. Soc.*, **59**, 1675 (1937).

[9] The mean activity of the electrolyte $A_m B_n$ is defined as follows:

$$f_\pm = \text{mean activity coefficient} = (f_A{}^m \cdot f_B{}^n)^{1/(m+n)}$$

The mean activity coefficient can be measured in any of several ways, but it is impossible experimentally to resolve this term into the individual activity coefficients for f_A and f_B. For example, if $A_m B_n$ is a precipitate, we can write

$$K_{sp} = [A]^m [B]^n \cdot f_A{}^m \cdot f_B{}^n = [A]^m [B]^n \cdot f_\pm{}^{(m+n)}$$

TABLE 5-3

Activity Coefficient for Ions at 25°C[a] $f_i = i = $ any ion.

ION	a_A EFFECTIVE DIAMETER, Å	ACTIVITY COEFFICIENT AT INDICATED IONIC STRENGTHS $= \mu$.				
		0.001	0.005	0.01	0.05	0.1
H_3O^+	9	0.967	0.933	0.914	0.86	0.83
Li^+, $C_6H_5COO^-$	6	0.965	0.929	0.907	0.84	0.80
Na^+, IO_3^-, HSO_3^-, HCO_3^-, $H_2PO_4^-$, $H_2AsO_4^-$, OAc^-	4–4.5	0.964	0.928	0.902	0.82	0.78
OH^-, F^-, SCN^-, HS^-, ClO_3^-, ClO_4^-, BrO_3^-, IO_4^-, MnO_4^-	3.5	0.964	0.926	0.900	0.81	0.76
K^+, Cl^-, Br^-, I^-, CN^-, NO_2^-, NO_3^-, $HCOO^-$	3	0.964	0.925	0.899	0.80	0.76
Rb^+, Cs^+, Tl^+, Ag^+, NH_4^+	2.5	0.964	0.924	0.898	0.80	0.75
Mg^{2+}, Be^{2+}	8	0.872	0.755	0.69	0.52	0.45
Ca^{2+}, Cu^{2+}, Zn^{2+}, Sn^{2+}, Mn^{2+}, Fe^{2+}, Ni^{2+}, Co^{2+}, $Phthalate^{2-}$	6	0.870	0.749	0.675	0.48	0.40
Sr^{2+}, Ba^{2+}, Cd^{2+}, Hg^{2+}, S^{2-}	5	0.868	0.744	0.67	0.46	0.38
Pb^{2+}, CO_3^{2-}, SO_3^{2-}, $C_2O_4^{2-}$	4.5	0.868	0.742	0.665	0.46	0.37
Hg_2^{2+}, SO_4^{2-}, $S_2O_3^{2-}$, CrO_4^{2-}, HPO_4^{2-}	4.0	0.867	0.740	0.660	0.44	0.36
Al^{3+}, Fe^{3+}, Cr^{3+}, La^{3+}, Ce^{3+}	9	0.738	0.54	0.44	0.24	0.18
PO_4^{3-}, $Fe(CN)_6^{3-}$	4	0.725	0.50	0.40	0.16	0.095
Th^{4+}, Zr^{4+}, Ce^{4+}, Sn^{4+}	11	0.588	0.35	0.255	0.10	0.065
$Fe(CN)_6^{4-}$	5	0.57	0.31	0.20	0.048	0.021

[a] From J. Kielland, *J. Amer. Chem. Soc.*, **59**, 1675 (1937). Reproduced with permission.

By measuring the solubility of A_mB_n in a solution in which the electrolyte concentration approaches zero (that is, where f_A and $f_B \to 1$), we could obtain K_{sp}. A second solubility measurement at some ionic strength μ_1 would give values for [A] and [B]. These data would then permit the calculation of $f_A{}^m \cdot f_B{}^n = f_\pm{}^{(m+n)}$ for ionic strength μ_1. It is important to understand that there are insufficient experimental data to permit the calculation of the *individual* quantities f_A and f_B, and that there appears to be no additional experimental information that would permit evaluation of these quantities. This situation is general; the *experimental* determination of individual activity coefficients appears to be impossible.

The Debye–Hückel relationship and the data in Table 5-3 give satisfactory activity coefficients for ionic strengths up to about 0.1. Beyond this value, however, the equation fails, and mean activity coefficients must be determined experimentally.

Solubility Calculations Employing Activity Coefficients. The use of activities rather than molar concentrations in equilibrium constant calculations yields more accurate information. Unless otherwise specified, values for K_{sp} found in tables are generally constants based upon activities (activity-based constants are sometimes called the *thermodynamic constants*). Thus for the precipitate $A_m B_n$ we may write

(handwritten: $KsP = a_A{}^m \, a_b{}^n = [A]^m[B]^n f_A{}^m f_B{}^n$)

$$K_{sp} = a_A{}^m \cdot a_B{}^n = [A]^m[B]^n \cdot f_A{}^m \cdot f_B{}^n$$

or

(handwritten: $[A]^m[B]^n = KsP = \dfrac{K'sP}{f_A{}^m f_B{}^n}$)

$$[A]^m[B]^n = \frac{K_{sp}}{f_A{}^m \cdot f_B{}^n}$$

$$= K'_{sp}$$

where the bracketed terms are *molar concentrations* of A and B. Division of (the thermodynamic constant, K_{sp},) by the product of the activity coefficients for A and B (or the mean activity coefficient) yields a *concentration constant* K'_{sp} that is applicable to a solution of a particular ionic strength. This constant can then be employed in the equilibrium calculations as discussed earlier. The following example will demonstrate the procedure.

EXAMPLE

(handwritten: $PbCl_2$)

Use activities to calculate the solubility of $Ba(IO_3)_2$ in a 0.033-F solution of $Mg(IO_3)_2$. The thermodynamic solubility product for $Ba(IO_3)_2$ has a value of 1.57×10^{-9} (Appendix 7). *(handwritten: $K'sp$)*

At the outset we may write

$$[Ba^{2+}][IO_3{}^-]^2 = \frac{1.57 \times 10^{-9}}{f_{Ba^{2+}}f_{IO_3{}^-}^2} = K'_{sp}$$

(handwritten: $Ksp = \dfrac{K'sp}{f_{Ba^{2+}} f^2{}_{IO_3^-}}$)

We must next calculate activity coefficients for Ba^{2+} and $IO_3{}^-$ ions from the ionic strength of the solution. Thus

(handwritten: $0.033 F \, Mg(IO3)_2$)

$$\mu = \tfrac{1}{2}[m_{Mg^{2+}} \times (2)^2 + m_{IO_3{}^-} \times (1)^2]$$

$$= \tfrac{1}{2}(0.033 \times 4 + 0.066 \times 1) = 0.099 \cong 0.1$$

In calculating μ, we have assumed that the Ba^{2+} and $IO_3{}^-$ ions from the precipitate do not significantly affect the ionic strength of the solution. This simplification seems justified, considering the low solubility of barium iodate. In situations where it is not possible to make the assumption, the concentrations of the two ions can be approximated by an ordinary calculation, assuming activities and concentrations to be identical. These concentrations can then be introduced to give a better value for μ.

Turning now to Table 5-3, we find that at an ionic strength of 0.1

$$f_{Ba^{2+}} = 0.38 \qquad f_{IO_3^-} = 0.78$$

If the calculated ionic strength did not match that of one of the columns in the table, $f_{Ba^{2+}}$ and $f_{IO_3^-}$ could be obtained from Equation 5-27.

We may thus write

$$K_{sp} = \frac{1.57 \times 10^{-9}}{(0.38)(0.78)^2} = 6.8 \times 10^{-9} = K'_{sp}$$

$$K_{sp} = [Ba^{2+}][IO_3^-]^2 = 6.8 \times 10^{-9}$$

Proceeding now as for an ordinary solubility calculation (Chapter 3),

$$\text{solubility} = s = [Ba^{2+}]$$

$$[IO_3^-] \cong 0.066$$

$$s(0.066)^2 = 6.8 \times 10^{-9}$$

$$s = 1.56 \times 10^{-6} \text{ fw/liter}$$

It is of interest to note that the calculated solubility, neglecting the effects of ionic strength, is 3.60×10^{-7} fw/liter.

Omission of Activity Coefficients in Equilibrium Calculations. We shall ordinarily neglect activity coefficients and simply use molar concentrations in applications of the equilibrium law. This recourse simplifies the calculations and greatly decreases the amount of data needed. For most purposes the errors introduced by the assumption of unity for the activity coefficient will not be large enough to lead to false conclusions. It should be apparent from the preceding example, however, that disregard of activity coefficients may introduce a significant numerical error in calculations of this kind; relative errors of 100% or more are not uncommon.

The student should be alert to the conditions under which the approximation of concentration for activity is likely to lead to the largest errors. Significant discrepancies will occur when the ionic strength is large (0.01 or larger) or when the ions involved have multiple charges (see Table 5-3). With dilute solutions (ionic strength <0.01) of nonelectrolytes or of singly charged ions, the use of concentrations in a mass-law calculation often provides reasonably accurate results.

It is also important to note that the decrease in solubility resulting from the presence of an ion common to the precipitate is in part counteracted by the larger electrolyte concentration associated with presence of the salt containing the common ion. This effect is illustrated by the sample calculation just completed.

Additional Variables That Affect the Solubility of Precipitates

The solubility of most precipitates is influenced by temperature and the presence of organic solvents. Heat is absorbed as most solids dissolve. Therefore the solubility of precipitates generally increases with rising temperatures; correspondingly, solubility product constants for most sparingly soluble compounds become larger at high temperatures.

The solubility of most inorganic substances is markedly less in mixtures of water and organic solvents than in pure water. The data for calcium sulfate in Table 5-4 are typical of this effect.

TABLE 5-4
Solubility of Calcium Sulfate in Aqueous Ethyl Alcohol Solution[a]

CONCENTRATION OF ETHYL ALCOHOL, %(w/w)	SOLUBILITY OF $CaSO_4$, g $CaSO_4$/100 g SOLVENT
0	0.208
6.2	0.100
13.6	0.044
23.8	0.014
33.0	0.0052
41.0	0.0029

[a] From T. Yamamoto, *Bull. Inst. Phys. Chem. Res.* (Tokyo), **9**, 352 (1930); W. C. Linke, *Seidell Solubilities of Inorganic and Metal-Organic Compounds*, 4th ed., vol. 1. Washington, D.C.: American Chemical Society, 1958, p. 685. With permission.

Rate of Precipitate Formation

It is important to stress that no conclusions can be drawn about the rate of a reaction from the magnitude of its equilibrium constant. Many reactions with highly favorable equilibrium constants approach equilibrium at an imperceptible rate.

Precipitation reactions are often slow, several minutes or even several hours being required for the attainment of equilibrium. Occasionally the chemist can take advantage of a slow rate to accomplish separations that would not be feasible if equilibrium were approached rapidly. For example, calcium can be separated from magnesium by precipitation as the oxalate, despite the fact that the latter ion also forms an oxalate of comparable solubility. The separation is possible because equilibrium for magnesium oxalate formation is approached at a much slower rate than that for calcium oxalate formation; if the calcium oxalate is filtered shortly after precipitation, a solid that is essentially free of contamination by magnesium is obtained. On the other hand, a precipitate that remains in contact with the liquid will become contaminated with magnesium oxalate.

PROBLEMS

See Chapter 3, Problems 1, 2, and 5 through 12 (pp. 43–44) for additional elementary solubility product problems.

Except where otherwise stated, assume that activity coefficients are all unity in the problems that follow.

*1. Calculate the grams of $Pb(SCN)_2$ ($K_{sp} = 2.0 \times 10^{-5}$) that will dissolve in 100 ml of
   (a) water.
   (b) 0.500-F $Pb(NO_3)_2$.
   (c) 0.500-F KSCN.

2. Calculate the milligrams of $Sr(IO_3)_2$ ($K_{sp} = 3.3 \times 10^{-7}$) that will dissolve in 250 ml of
   (a) water.
   (b) 0.200-F $NaIO_3$.
   (c) 0.200-F $Sr(NO_3)_2$.

*3. Calculate the solubility of the following hydroxides in water and in 0.0100-F KOH:
   (a) $Mn(OH)_2$.
   (b) $Pd(OH)_2$ ($K_{sp} = 1.2 \times 10^{-31}$).

4. Calculate the solubility of the following hydroxides in water and in 0.100-F NaOH:
   (a) $Ca(OH)_2$ ($K_{sp} = 5.5 \times 10^{-6}$).
   (b) $Cr(OH)_3$ ($K_{sp} = 6 \times 10^{-31}$).

*5. The solubility products for a series of iodides are:

$$
\begin{array}{ll}
\text{TlI} & K_{sp} = 6.5 \times 10^{-8} \\
\text{AgI} & K_{sp} = 8.3 \times 10^{-17} \\
\text{PbI}_2 & K_{sp} = 7.1 \times 10^{-9} \\
\text{BiI}_3 & K_{sp} = 8.1 \times 10^{-19}
\end{array}
$$

List the four compounds in order of decreasing formal solubility in
   (a) water.
   (b) 0.10-F NaI.
   (c) 0.10-F solution of the solute cation.
   (In each case assume that the cations do not react with water.)

6. The solubility products for a series of iodates are:

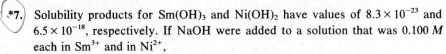

$$AgIO_3 \qquad K_{sp} = 3.0 \times 10^{-8}$$
$$Cu(IO_3)_2 \qquad K_{sp} = 7.4 \times 10^{-8}$$
$$Sm(IO_3)_3 \qquad K_{sp} = 4.7 \times 10^{-10}$$
$$Ce(IO_3)_4 \qquad K_{sp} = 4.7 \times 10^{-17}$$

List the four compounds in order of decreasing formal solubility in
(a) water.
(b) 0.20-F $NaIO_3$.
(c) 0.20-F solution of the solute cation.
(In each case assume that the cations do not react with water.)

*7. Solubility products for $Sm(OH)_3$ and $Ni(OH)_2$ have values of 8.3×10^{-23} and 6.5×10^{-18}, respectively. If NaOH were added to a solution that was $0.100\ M$ each in Sm^{3+} and in Ni^{2+},
(a) which ion would begin to precipitate first?
(b) what OH^- concentration would be required to lower the ion that precipitates first to a concentration of $1.0 \times 10^{-6}\ M$?
(c) would a quantitative separation of the two ions by control of OH^- be feasible (employ $1 \times 10^{-6}\ M$ as a criterion for quantitative separation)? If so, what range of OH^- concentration would permit the separation?

8. Determine whether it would be feasible in principle to perform the following separations from solutions in which the initial concentration of each species is 0.100. Use $1.0 \times 10^{-6}\ M$ as the criterion for quantitative removal. If a separation is feasible, specify the conditions that must be maintained.
(a) Ag^+ from Pb^{2+} with I^-
*(b) Cl^- from I^- with Ag^+
(c) Cu^+ from Ag^+ with SCN^-
*(d) Mg^{2+} from Pb^{2+} with OH^-
(e) Ca^{2+} from Ba^{2+} with SO_4^{2-}
*(f) Fe^{3+} from Mn^{2+} with OH^-

9. Determine which of the following separations is feasible by controlling the hydronium ion concentration of a saturated H_2S solution. Assume that the initial concentration of each ion is $0.100\ M$ and that lowering of a concentration to $1.0 \times 10^{-6}\ M$ constitutes quantitative removal. If a separation is possible, specify the range of H_3O^+ concentrations that could be employed.
*(a) Fe^{2+} and Zn^{2+}
(b) Ag^+ and Zn^{2+}
*(c) La^{3+} and Fe^{2+} $\qquad$ (K_{sp} for $La_2S_3 = 2.0 \times 10^{-13}$)
(d) Ce^{3+} and Mn^{2+} $\qquad$ (K_{sp} for $Ce_2S_3 = 6.10 \times 10^{-11}$)
*(e) Pb^{2+} and Zn^{2+}
(f) Fe^{2+} and Tl^+

*10. For a solution that is $0.050\ F$ in KNO_3 calculate K'_{sp} for
(a) $AgIO_3$.

    (b) $Zn(OH)_2$.

    (c) ZnS.

    (d) $Fe(OH)_3$.

11. For a solution that is $0.0333\ F$ in $Ba(NO_3)_2$ calculate K'_{sp} for

    (a) AgBr.

    (b) $PbSO_4$.

    (c) Tl_2S.

    (d) $Th(OH)_4$ (K_{sp} for $Th(OH)_4 = 4 \times 10^{-45}$).

*12. Calculate the solubilities of the following compounds in a 0.0333-F solution of $BaCl_2$, first employing activities and then neglecting them:

    (a) AgI.

    (b) $PbCl_2$.

    (c) $PbSO_4$ (assume that SO_4^{2-} does not react to form HSO_4^-).

    (d) $Cd_2Fe(CN)_6$ $[Cd_2Fe(CN)_6(s) \rightleftharpoons 2Cd^{2+} + Fe(CN)_6^{4-};\ K_{sp} = 3.2 \times 10^{-17}]$.

13. Calculate the solubilities of the following compounds in a 0.0167-F solution of $Ba(OH)_2$, first employing activities and then neglecting them:

    (a) AgSCN.

    (b) $Mn(OH)_2$.

    (c) $BaSO_4$ (assume that SO_4^{2-} does not react to form HSO_4^-).

    (d) $La(IO_3)_3$.

*14. Calculate the solubility of $PbSO_4$ in a solution that is (a) neutral and (b) $0.100\ F$ in HCl ($HSO_4^- + H_2O \rightleftharpoons H_3O^+ + SO_4^{2-};\ K_2 = 1.2 \times 10^{-2}$).

*15. Calculate the solubility of silver arsenate, Ag_3AsO_4, in a solution in which the H_3O^+ concentration is maintained at

    (a) $1.0 \times 10^{-6}\ M$.

    (b) $1.0 \times 10^{-4}\ M$.

    (c) $1.0 \times 10^{-2}\ M$.

16. Calculate the solubility of $Pb_3(PO_4)_2$ ($K_{sp} = 7.9 \times 10^{-45}$) in a solution that is maintained at a molar H_3O^+ concentration of

    (a) 1.0×10^{-3}.

    (b) 1.0×10^{-6}.

    (c) 1.0×10^{-9}.

*17. Calculate the solubility of Ag_2SO_3 ($K_{sp} = 1.5 \times 10^{-14}$) in a solution that is maintained at a molar H_3O^+ concentration of

    (a) 1.0×10^{-5}.

    (b) 1.0×10^{-8}.

    (c) 1.0×10^{-10}.

18. Calculate the solubility of PbOHBr $[PbOHBr(s) \rightleftharpoons Pb^{2+} + OH^- + Br^-;\ K_{sp} = 2.0 \times 10^{-15}]$ in a solution in which the H_3O^+ concentration is maintained at

    *(a) $1.0 \times 10^{-10}\ M$.

    (b) $1.0 \times 10^{-6}\ M$.

19. Calculate the solubility of the following sulfides in solutions in which $[H_3O^+]$ is first 0.010 and then $1.0 \times 10^{-4} \, M$:
    *(a) ZnS.
    (b) CdS.
    *(c) Tl$_2$S.
    (d) Ag$_2$S.

*20. The equilibrium constants for the reactions AgCl with Cl$^-$ are

$$AgCl(s) + Cl^- \rightarrow AgCl_2^- \qquad K_1 = \frac{[AgCl_2^-]}{[Cl^-]} = 2.0 \times 10^{-5}$$

$$AgCl_2^- + Cl^- \rightleftharpoons AgCl_3^{2-} \qquad K_2 = \frac{[AgCl_3^{2-}]}{[AgCl_2^-][Cl^-]} = 1$$

Calculate the solubility of AgCl in the following NaCl solutions:
(a) 1.0 F.
(b) 0.20 F.
(c) 0.010 F.
(d) 0.0010 F.

21. The equilibrium constant for formation of CuI_2^- is given by

$$Cu^+ + 2I^- \rightleftharpoons CuI_2^- \qquad K = \frac{[CuI_2^-]}{[Cu^+][I^-]^2} = 7.1 \times 10^8$$

What is the solubility of CuI in solutions having the following formal NaI concentrations:
(a) 2.0?
(b) 2.0×10^{-2}?
(c) 2.0×10^{-4}?
(d) 2.0×10^{-5}?

*22. Formation constants for the reactions Ag$^+$ with $S_2O_3^{2-}$ are

$$Ag^+ + S_2O_3^{2-} \rightleftharpoons AgS_2O_3^- \qquad K_1 = 6.6 \times 10^8$$

$$AgS_2O_3^- + S_2O_3^{2-} \rightleftharpoons Ag(S_2O_3)_2^{3-} \qquad K_2 = 4.4 \times 10^3$$

Calculate the solubility of AgI in 0.200-F Na$_2$S$_2$O$_3$ (assume that $S_2O_3^{2-}$ does not combine with H$_3$O$^+$).

gravimetric analysis

A gravimetric analysis is based upon the measurement of the weight of a substance that has a known composition and is chemically related to the analyte. Two types of gravimetric methods exist. In *precipitation methods* the species to be determined is caused to react chemically with a reagent to yield a product of limited solubility; after filtration and other suitable treatment a solid residue of known chemical composition is weighed. In *volatilization methods* the substance to be determined is separated as a gas from the remainder of the sample; here the analysis is based upon the weight of the substance volatilized or upon the weight of the nonvolatile residue. We shall be concerned principally with precipitation methods because these are more frequently encountered than methods involving volatilization.

Calculation of Results from Gravimetric Data

A gravimetric analysis requires two experimental measurements: the weight of sample taken and the weight of a product of known composition derived from the sample. Ordinarily these data are converted to a percentage of the analyte by a simple mathematical manipulation.

If A is the analyte, we may write

$$\% \, A = \frac{\text{weight of A}}{\text{weight of sample}} \times 100 \qquad \textbf{(6-1)}$$

Usually the weight of A is not measured directly. Instead, the species that is actually isolated and weighed either contains A or can be chemically related to A. In either case a *gravimetric factor* is needed to convert the weight of the precipitate to the corresponding weight of A. The properties of this factor are conveniently demonstrated with examples.

EXAMPLE

How many grams of Cl are contained in a precipitate of AgCl that weighs 0.204 g?

From the formula for AgCl we know that

$$\text{no. fw AgCl} = \text{no. fw Cl}$$

Because

$$\text{no. fw AgCl} = \frac{0.204}{\text{gfw AgCl}} = \text{no. fw Cl}$$

and also because

$$\text{wt Cl} = \text{no. fw Cl} \times \text{gfw Cl}$$

then

$$\text{wt Cl} = 0.204 \times \frac{\text{gfw Cl}}{\text{gfw AgCl}} = 0.204 \times \frac{35.45}{143.3}$$

$$= 0.204 \times 0.2474 = 0.0505 \text{ g}$$

EXAMPLE

To what weight of $AlCl_3$ would 0.204 g of AgCl correspond?

We know that each $AlCl_3$ yields three AgCl. Therefore

$$\text{no. fw AlCl}_3 = \frac{1}{3} \text{ no. fw AgCl} = \frac{1}{3} \times \frac{0.204}{\text{gfw AgCl}}$$

By the preceding arguments

$$\text{wt AlCl}_3 = 0.204 \times \frac{\text{gfw AlCl}_3}{3 \times \text{gfw AgCl}} = 0.204 \times \frac{133.3}{3 \times 143.3}$$

$$= 0.204 \times 0.310 = 0.0633 \text{ g}$$

Note how these two calculations resemble each other. In both the weight of one substance is given by a product involving the known weight of some other substance and a ratio that contains their respective gram formula weights. This ratio is the gravimetric factor. In the second example it was necessary to multiply the gram formula weight of silver chloride by 3 to balance the number of chlorides that appear in the numerator and denominator of the gravimetric factor.

EXAMPLE

What weight of Fe_2O_3 can be obtained from 1.63 g of Fe_3O_4? What is the gravimetric factor for this conversion?

Here it is necessary to assume that all Fe in the Fe_3O_4 is transformed into Fe_2O_3 and ample oxygen is available to accomplish this change. That is,

$$2Fe_3O_4 + [O] = 3Fe_2O_3$$

We see from this equation that $\frac{3}{2}$ fw of Fe_2O_3 are obtained from 1 fw of Fe_3O_4. Thus the number of formula weights of Fe_2O_3 is greater than the number of formula weights of Fe_3O_4 by a factor of $\frac{3}{2}$, or

$$\text{no. fw } Fe_2O_3 = \frac{3}{2} \times \text{no. fw } Fe_3O_4$$

$$\frac{\text{wt } Fe_2O_3}{\text{gfw } Fe_2O_3} = \frac{3}{2} \times \frac{\text{wt } Fe_3O_4}{\text{gfw } Fe_3O_4}$$

and

$$\text{wt } Fe_2O_3 = \frac{3}{2} \times \frac{\text{wt } Fe_3O_4}{\text{gfw } Fe_3O_4} \times \text{gfw } Fe_2O_3$$

Thus upon rearranging,

$$\text{wt } Fe_2O_3 = \text{wt } Fe_3O_4 \times \frac{3 \times \text{gfw } Fe_2O_3}{2 \times \text{gfw } Fe_3O_4}$$

Substitution of numerical values gives

$$\text{wt } Fe_2O_3 = 1.63 \times \frac{3 \times 159.7}{2 \times 231.5} = 1.687 = 1.69 \text{ g}$$

In this example

$$\text{gravimetric factor} = \frac{3 \times \text{gfw } Fe_2O_3}{2 \times \text{gfw } Fe_3O_4} = 1.035$$

A general definition for the gravimetric factor is:

$$\text{gravimetric factor} = \frac{a}{b} \times \frac{\text{gfw of substance sought}}{\text{gfw of substance weighed}}$$

where a and b are small integers that take such values as are necessary to make the number of formula weights in the numerator and denominator *chemically equivalent.*

Equation 6-1 can now be converted to the more useful form

$$\% A = \frac{\text{wt ppt} \times \left(\dfrac{a \times \text{gfw A}}{b \times \text{gfw ppt}} \right) \times 100}{\text{wt sample}} \qquad (6\text{-}2)$$

Additional examples of gravimetric factors are given in Table 6-1. Most chemical handbooks contain tabulations of these factors and their logarithms.

In all of the gravimetric factors considered thus far, chemical equivalence between numerator and denominator has been established simply by balancing the number of atoms of an element (other than oxygen) that is common to both. Occasionally this approach will be inadequate. Consider, for example, an indirect analysis for the iron in a sample of iron(III) sulfate that involves precipitation and weighing of barium sulfate. Here the gravimetric factor will contain no element common to numerator and denominator, and we must seek further for the means of establishing the chemical equivalence between these quantities. We note that

$$2 \text{ gfw Fe} \equiv 1 \text{ gfw Fe}_2(SO_4)_3 \equiv 3 \text{ gfw SO}_4 \equiv 3 \text{ gfw BaSO}_4$$

The gravimetric factor for calculating the percent Fe will therefore be

$$\text{gravimetric factor} = \frac{2 \times \text{gfw Fe}}{3 \times \text{gfw BaSO}_4}$$

Thus even though the species in the gravimetric factor are not directly related by a common element, their equivalence can be established through knowledge of the stoichiometry between them.

The examples that follow illustrate the use of the gravimetric factor in the calculation of the results from analyses.

TABLE 6-1
Typical Gravimetric Factors

SPECIES SOUGHT	SPECIES WEIGHED	GRAVIMETRIC FACTOR
In	In_2O_3	$\dfrac{2 \times \text{gfw In}}{\text{gfw In}_2O_3}$
HgO	$Hg_5(IO_6)_2$	$\dfrac{5 \times \text{gfw HgO}}{\text{gfw Hg}_5(IO_6)_2}$
I	$Hg_5(IO_6)_2$	$\dfrac{2 \times \text{gfw I}}{\text{gfw Hg}_5(IO_6)_2}$
K_3PO_4	K_2PtCl_6	$\dfrac{2 \times \text{gfw K}_3PO_4}{3 \times \text{gfw K}_2PtCl_6}$

EXAMPLE

A 0.703-g sample of a commercial detergent was ignited at red heat to destroy the organic matter. The residue was then taken up in hot HCl to convert the P to H_3PO_4. The phosphate was precipitated as $MgNH_4PO_4 \cdot 6H_2O$ by addition of Mg^{2+} followed by aqueous NH_3. After being filtered and washed, the precipitate was converted to $Mg_2P_2O_7$ by ignition at 1000°C. This residue weighed 0.432 g. Calculate the percent P in the sample.

$$\% \, P = \frac{0.432 \times \dfrac{2 \times \text{gfw P}}{\text{gfw Mg}_2\text{P}_2\text{O}_7} \times 100}{0.703}$$

$$= \frac{0.432 \times 0.2783 \times 100}{0.703} = 17.1$$

EXAMPLE

At elevated temperatures sodium oxalate is converted to sodium carbonate with evolution of carbon monoxide:

$$Na_2C_2O_4 \rightarrow Na_2CO_3 + CO$$

Ignition of a 1.3906-g sample of impure sodium oxalate yielded a residue weighing 1.1436 g. Calculate the percent purity of the sample.

Here it must be assumed that the difference between the initial and final weights represents the carbon monoxide evolved during the ignition; it is this weight loss that forms the basis for the analysis. From the equation for the process we see that

$$\text{no. fw CO} = \text{no. fw Na}_2\text{C}_2\text{O}_4$$

Thus

$$\% \, Na_2C_2O_4 = \frac{\text{wt CO} \times \dfrac{\text{gfw Na}_2\text{C}_2\text{O}_4}{\text{gfw CO}} \times 100}{\text{wt sample}}$$

$$= \frac{(1.3906 - 1.1436) \times 4.784 \times 100}{1.3906} = 84.97$$

EXAMPLE

A 0.2795-g sample of an insecticide containing only lindane ($C_6H_6Cl_6$; gfw = 290.8) and DDT ($C_{14}H_9Cl_5$; gfw = 354.5) was burned in a stream of oxygen in a quartz tube. The products ($CO_2, H_2O,$ and HCl) were passed through a solution of $NaHCO_3$. After acidification the chloride in this solution yielded 0.7161 g of AgCl. Calculate the percent lindane and DDT in the sample.

This problem contains two unknowns; we must therefore develop two independent equations that can be solved simultaneously.

One useful equation is

$$\text{wt } C_6H_6Cl_6 + \text{wt } C_{14}H_9Cl_5 = 0.2795 \text{ g}$$

A second equation is

$$\text{wt AgCl from } C_6H_6Cl_6 + \text{wt AgCl from } C_{14}H_9Cl_5 = 0.7161 \text{ g}$$

After the appropriate gravimetric factors have been inserted, the second equation becomes

$$\text{wt } C_6H_6Cl_6 \times \frac{6 \times \text{gfw AgCl}}{\text{gfw } C_6H_6Cl_6} + \text{wt } C_{14}H_9Cl_5 \times \frac{5 \times \text{gfw AgCl}}{\text{gfw } C_{14}H_9Cl_5} = 0.7161 \text{ g}$$

or

$$\text{wt } C_6H_6Cl_6 \times 2.957 + \text{wt } C_{14}H_9Cl_5 \times 2.021 = 0.7161$$

The first equation can be rearranged to give wt $C_{14}H_9Cl_5$; substitution into the second equation gives

$$2.957 \text{ wt } C_6H_6Cl_6 + 2.021 \,(0.2795 - \text{wt } C_6H_6Cl_6) = 0.7161$$

Thus

$$\text{wt } C_6H_6Cl_6 = 0.1616 \text{ g}$$

and

$$\% \, C_6H_6Cl_6 = \frac{0.1616}{0.2795} \times 100 = 57.82$$

$$\% \, C_{14}H_9Cl_5 = 100 - 57.82 = 42.18$$

Properties of Precipitates and Precipitating Reagents

The ideal precipitating reagent for a gravimetric analysis would react specifically with the analyte to produce a solid that would (1) have a sufficiently low solubility so that losses from that source would be negligible, (2) be readily filtered and washed free of contaminants, and (3) be unreactive and of known composition after drying or, if necessary, ignition. Few precipitates or reagents possess all these desirable properties; thus the chemist frequently finds it necessary to perform analyses using a product or a reaction that is far from ideal.

The variables that influence the solubility of precipitates were discussed in Chapter 5; we must now consider what can be done to achieve a pure and easily filtered solid.

FILTERABILITY AND PURITY OF PRECIPITATES

The ease with which a precipitate is isolated and purified depends on the particle size of the solid phase. The relationship between particle size and ease of filtration is straightforward. Coarse precipitates are readily retained by porous media and are thus rapidly filtered. Finely divided precipitates require dense filters; low filtration rates result. The effect of particle size upon purity of a precipitate is more complex. More often than not a decrease in soluble contaminants is found to accompany an increase in particle size.

In considering the purity of precipitates we shall use the term *coprecipitation*, which describes those processes by which *normally soluble* components of a solution are carried down during the formation of a precipitate. The student should clearly understand that contamination of a precipitate by a second substance whose solubility product has been exceeded *does not constitute coprecipitation*.

Factors That Determine the Particle Size of Precipitates. Particle size depends not only upon the chemical composition of a precipitate but also upon the conditions that exist at the time of its formation. Enormous variations are observed. At one extreme are *colloidal suspensions*, whose individual particles are so small as to be invisible to the naked eye (10^{-6} to 10^{-4} mm in diameter). These particles show no tendency to settle out from solution, nor are they retained upon common filtering media. At the other extreme are particles with dimensions on the order of several tenths of a millimeter. The temporary dispersion of such particles in the liquid phase is called a *crystalline suspension*. The particles of a crystalline suspension tend to settle out rapidly and are readily filtered.

No sharp discontinuities in physical properties occur as the dimensions of the particles in the solid phase increase from colloidal to those typical of crystals. Indeed some precipitates possess characteristics that are between these defined extremes. The majority, however, are easily recognizable as predominately colloidal or predominately crystalline. Thus although imperfect, this classification can be usefully applied to most solid phases.

The phenomenon of precipitation has long attracted the attention of chemists, but the mechanism of the process still is not fully understood. It is certain, however, that the particle size of the solid that forms is influenced in part by such experimental variables as the temperature, the solubility of the precipitate in the medium in which it is being formed, reactant concentrations, and the rate at which reactants are mixed. The effect of these variables can be accounted for, at least qualitatively, by assuming that the particle size is related to a single property of the system called its *relative supersaturation*,[1] where

---

[1] The name of P. P. von Weimarn is associated with the concept of relative supersaturation and its effect upon particle size. An account of von Weimarn's work is to be found in *Chem. Rev.*, **2**, 217 (1925). The von Weimarn viewpoint adequately suggests the general conditions that will lead to a satisfactory particle size for a precipitate. Other theories are superior in accounting for details of the precipitation process. See, for example, A. E. Nielsen, *The Kinetics of Precipitation*. New York: Macmillan, 1964.

$$\text{relative supersaturation} = \frac{Q - S}{S} \qquad\qquad \textbf{(6-3)}$$

Here Q is the concentration of the solute at any instant, and S is its equilibrium solubility.

During the formation of a sparingly soluble precipitate, each addition of precipitating reagent presumably causes the solution to be momentarily supersaturated (that is, $Q > S$). Under most circumstances this unstable condition is relieved, usually after a brief period, by precipitate formation. Experimental evidence suggests, however, that the particle size of the resulting precipitate varies inversely with the average degree of relative supersaturation that exists after each addition of reagent. Thus when $(Q - S)/S$ is large, the precipitate tends to be colloidal; when this parameter is low on the average, a crystalline solid results.

Mechanics of Precipitate Formation. The effect of relative supersaturation on particle size can be rationalized by postulating two precipitation mechanisms, *nucleation* and *particle growth*. The particle size of a freshly formed precipitate is governed by the extent to which one of these processes predominates over the other.

Nucleation is a process whereby some minimum number of ions or molecules (perhaps as few as four or five) unite to form a stable second phase. Further precipitation can occur either by the generation of additional nuclei or by the deposition of additional solid on the nuclei that have already been produced. If the former predominates, a precipitate containing a large number of small particles results; if growth predominates, a smaller number of larger particles will be produced.

The rate of nucleation is believed to increase exponentially with relative supersaturation, whereas the rate of particle growth bears an approximately linear relationship to this parameter. Thus when supersaturation is high, the nucleation rate far exceeds particle growth and is the predominant precipitation mechanism. At low relative supersaturations, on the other hand, the rate of particle growth may be the greater of the two. Under these circumstances deposition of solid on the particles already present may occur to the exclusion of further nucleation.

Experimental Control of Particle Size. Experimental variables that minimize supersaturation and thus lead to crystalline precipitates include elevated temperatures (to increase S), dilute solutions (to minimize Q), and slow addition of the precipitating agent with good stirring (also to lower the average value of Q).

The particle size of precipitates with solubilities that depend upon the acidity of the environment can often be enhanced by increasing S during precipitation. For example, large, easily filtered crystals of calcium oxalate can be obtained by forming the bulk of the precipitate in a somewhat acidic environment in which the salt is moderately soluble. The precipitation is then completed by slowly adding aqueous ammonia until the acidity is sufficiently low for quantitative removal of the calcium oxalate; the additional precipitate produced during this step forms on the solid.

A crystalline solid is much easier to manipulate than a colloidal suspension. Particle growth is thus preferable to further nucleation during the formation of a precipitate. If, however, the solubility S of a precipitate is very small, it is essentially impossible to avoid a momentarily large relative supersaturation as solutions are mixed; as a consequence, colloidal suspensions often cannot be avoided. For example, under conditions feasible for an analysis, the hydrous oxides of iron(III), aluminum, and chromium(III) and the sulfides of the most heavy metal ions can be formed only as colloids because of their very low solubilities. The same is true for the halide precipitates of silver ion.[2]

COLLOIDAL PRECIPITATES

Individual colloidal particles are so small that they are not retained on ordinary filtering media; furthermore, Brownian motion prevents their settling from the solution under the influence of gravity. Fortunately, however, the individual particles of most colloids can be caused to coagulate or agglomerate to give a filterable, noncrystalline mass that rapidly settles out of solution.

Coagulation of Colloids. Three experimental measures hasten the coagulation process: heating, stirring, and adding an electrolyte to the medium. To understand the effectiveness of these measures, we need to account for the stability of a colloidal suspension.

The individual particles in a typical colloidal suspension bear either a positive or a negative charge as a consequence of *adsorption* of cations or anions on their surfaces. The existence of this charge is readily demonstrated experimentally by observing the migration of the particles under the influence of an electric field.

Adsorption of ions upon an ionic solid has as its origin the normal bonding forces that are responsible for crystal growth. Thus a silver ion at the surface of a silver chloride particle has a partially unsatisfied bonding capacity by virtue of its surface location. Negative ions are attracted to this site by the same forces that hold chloride ions in the silver chloride lattice.

---

[2] Silver chloride illustrates that the relative supersaturation concept is imperfect. It ordinarily forms as a colloid, yet its formal solubility is not significantly different from other compounds such as $BaSO_4$ which generally form as crystals.

Chloride ions on the surface exert an analogous attraction for cations in the solvent.

The nature and magnitude of the charge on the particles of a colloidal suspension depend in a complex way on a number of variables. For colloidal suspensions of interest in analysis, however, the species adsorbed, and thus the charge on the particles, can be readily predicted from the empirical observation that lattice ions are generally more strongly adsorbed than any others. Thus a silver chloride particle will be positively charged in a solution containing an excess of silver ions, owing to the preferential adsorption of those ions. It will have a negative charge in the presence of excess chloride ions for the same reason. The silver chloride particles formed in a gravimetric chloride analysis initially carry a negative charge but become positive as an excess of the precipitating agent is added.

The extent of adsorption increases rapidly with increases in the concentration of the absorbed ion. Ultimately, however, the surface of each particle becomes saturated; under these circumstances further increases in concentration have little or no effect.

Figure 6-1 depicts schematically a colloidal silver chloride particle

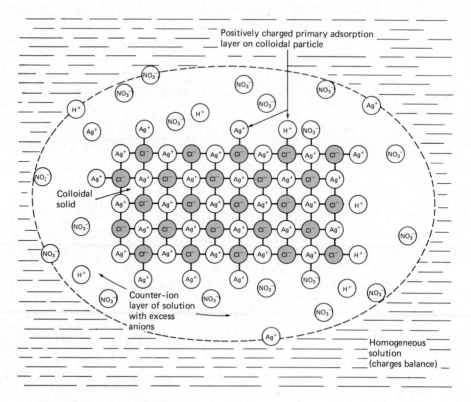

FIGURE 6-1 A colloidal AgCl particle suspended in a solution of $AgNO_3$.

in a solution containing an excess of silver ions. Attached directly to the solid surface are silver ions in the *primary adsorption layer*. Surrounding the charged particle is a *region of solution* called the *counter-ion layer*, within which there is an excess of negative ions sufficient to balance the charge of the adsorbed positive ions on the particle surface. The counter-ion layer forms as the result of electrostatic forces.

Considered together, the primarily adsorbed ions on the surface and their counter ions in the solution form an *electrical double layer* that exerts a repulsive force towards similarly constituted particles. This force is often sufficient to offset the normal cohesive forces that exist between small particles having the same chemical composition. Coagulation of a colloid therefore requires a reduction in the repulsive force of the double layer so that cohesion can occur.

The effect of the charged double layer on stabilization of a colloid is readily seen in the precipitation of chloride ion with silver ion. With the initial addition of silver nitrate, silver chloride is formed in an environment that has a high chloride ion concentration. The negative charge on the silver chloride particles is therefore high; the volume of the positive counter-ion layer surrounding each particle must also be relatively large to contain enough positive ions (hydronium or sodium ions, for example) to neutralize the negative charge of the primary layer. Coagulation does not occur under these circumstances. As more silver ions are added, the charge per particle decreases because the chloride ion concentration is decreased, and the number of particles is increased; the repulsive effect of the double layer thus becomes smaller. As chemical equivalence is approached, a sudden appearance of the coagulated colloid is observed. Here the number of adsorbed chloride ions per particle becomes small, and the double layer shrinks to a point where individual particles can approach one another closely enough to permit agglomeration. The agglomeration process can be reversed by adding a large excess of silver ions; here, of course, the charge of the double layer is reversed because the counter-ion layer is now negative.

Coagulation is often brought about by a short period of heating, especially if accompanied by stirring. The decrease in adsorption at elevated temperatures causes a corresponding lowering of net charge on each particle; in addition, the particles acquire kinetic energies sufficient to overcome the barrier to close approach.

An even more effective method of coagulation is to increase the electrolyte concentration of the solution by adding a suitable ionic compound. Under these circumstances the volume of solution that contains enough ions of opposite charge to neutralize the charge on the particle is lessened. Thus the introduction of an electrolyte has the effect of shrinking the counter-ion layer with the result that the surface charge on the particles is more completely neutralized. With their effective charge decreased, the particles can approach one another more closely.

Coprecipitation in Coagulated Colloids. Adsorption is the principal type of coprecipitation that affects coagulated colloids; other types are encountered with crystalline solids.

A coagulated colloid consists of irregularly arranged particles that form a loosely packed, porous mass. Within this mass large internal surface areas remain in contact with the solvent phase (see Figure 6-2). These surfaces will retain most of the primarily adsorbed ions that were on the uncoagulated particles. Even though the counter-ion layer surrounding the original colloidal particle is part of the solution, sufficient counter ions to impart electrical neutrality must accompany the particle (in the film of liquid surrounding the particle) through the processes of coagulation and filtration. *The net effect of surface adsorption is, therefore, the carrying down of an otherwise soluble species as a surface contaminant.*

Peptization of Colloids. Peptization refers to the process whereby a coagulated colloid reverts to its original dispersed state. Peptization frequently occurs when pure water is used to wash such a precipitate. Washing is not particularly effective in dislodging adsorbed contaminants; it does tend, however, to remove the electrolyte responsible for coagulation from the internal liquid in contact with the solid. As the electrolyte is removed, the counter-ion layers increase again in volume. The repulsive forces responsible for the original colloidal state are thus reestablished, and the particles detach themselves from the coagulated mass. The washings become cloudy as the freshly dispersed particles pass through the filter.

The chemist is thus faced with a dilemma in handling coagulated colloids. Although washing is needed to minimize contamination, there is also the risk of losses from peptization. This problem is commonly resolved by washing the agglomerated colloid with a solution containing a

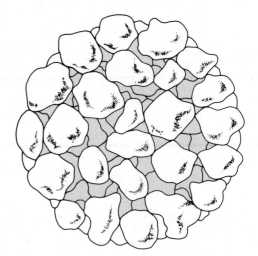

FIGURE 6-2 Coagulated colloidal particles.

volatile electrolyte that can subsequently be removed from the solid by heating. For example, silver chloride precipitates are ordinarily washed with dilute nitric acid, giving a product that is contaminated with the acid; no harm results, however, because the nitric acid is volatilized when the precipitate is dried at 110°C.

Practical Treatment of Colloidal Precipitates. Colloids are ordinarily precipitated from hot, stirred solutions to which sufficient electrolyte has been added to assure coagulation. Often the filterability of a coagulated colloid is improved by allowing it to stand for an hour or more in contact with the hot solution from which it was formed. During this process, which is known as *digestion*, weakly bound water appears to be lost from the precipitate; the result is a denser mass that is easier to filter.

A dilute solution of a volatile electrolyte is used to wash the filtered precipitate. Washing does not appreciably affect primarily adsorbed ions because the attraction between these species and the solid is too strong. Some exchange, however, may occur between the existing counter ions and one of the ions in the wash solution. Under any circumstances it must be expected that the precipitate will still be contaminated to some degree, even after extensive washing. The error introduced into the analysis from this source can range from 1 or 2 ppt (as in the coprecipitation of silver nitrate on silver chloride) to an intolerable level (as in the coprecipitation of heavy metal hydroxides upon the hydrous oxides of trivalent iron or aluminum).

A drastic way to minimize the effects of adsorption is *reprecipitation*. Here the filtered solid is redissolved and again precipitated. The first precipitate normally carries down only a fraction of the contaminant present in the original solvent. Thus the solution containing the redissolved precipitate will have a significantly lower contaminant concentration than the original. When precipitation is again carried out, less adsorption is to be expected. Reprecipitation adds substantially to the time required for an analysis; nevertheless, it is usually necessary for such precipitates as the hydrous oxides of iron(III) and aluminum, which possess extraordinary tendencies to adsorb the hydroxides of heavy metal cations such as zinc, cadmium, and manganese.

CRYSTALLINE PRECIPITATES

In general crystalline precipitates are more easily manipulated than coagulated colloids. The size of individual crystalline particles can be varied to a degree. As a consequence, the physical properties and purity of the solid are determined by experimental variables over which the chemist has a measure of control.

Methods of Improving Particle Size and Filterability. The particle size of crystalline solids can be improved by keeping the relative super-saturation low during the period in which the precipitate is formed. Equation 6-3 suggests that minimizing Q or maximizing S or both will accomplish this purpose.

The use of dilute solutions and the slow addition of precipitating agent with thorough mixing tends to minimize the momentary local super-saturation in the solution. Moreover, S can usually be increased by precipitation from hot solution. Significant improvement in particle size can be obtained with these simple measures.

Purity of Crystalline Precipitates. The specific surface area[3] of crystalline precipitates is relatively small; consequently coprecipitation by direct adsorption is negligible. However, other forms of coprecipitation, which involve the incorporation of contamination within the interior of crystals, may cause serious errors.

Two types of coprecipitation, *inclusion* and *occlusion*, are associated with crystalline precipitates. The two differ in the manner in which the contaminant is distributed throughout the interior of the solid. Inclusion involves the random distribution of foreign ions or molecules throughout the crystal. Occlusion, on the other hand, involves a nonhomogeneous distribution of ions or molecules of the contaminant within imperfections in the crystal lattice.

Inclusions can be either isomorphic or nonisomorphic. Isomorphic inclusion (also known as mixed crystal formation) occurs when the contaminant possesses dimensions and composition that permit its incorporation within the crystal structure with little or no strain of the lattice. A nonisomorphic inclusion appears to involve a solid solution of the contaminant in the precipitate. Both types tend to cause a homogeneous distribution of contaminant throughout the precipitate.

Occlusion occurs when whole droplets of the solution containing the impurities are trapped and surrounded by a rapidly growing crystal. Because the contaminants are located within the crystal, washing does little to decrease their amount. A lower precipitation rate may significantly lessen the extent of occlusion by providing time for the impurities to escape before they are trapped in the growing crystal. Digestion of the precipitate for as long as several hours is even more effective in eliminating occluded contaminants.

Digestion of Crystalline Precipitates. The heating of crystalline precipitates (without stirring) for some time after formation frequently yields a purer, more filterable product. The improvement in purity undoubtedly

---

[3] The specific surface is defined as the area exposed by a unit weight of solid; it ordinarily is expressed in terms of square centimeters per gram.

results from the solution and recrystallization that occur continuously and at an enhanced rate at elevated temperatures. During these processes many pockets of imperfection become exposed to the solution; the contaminant is thus able to escape from the solid, and more perfect crystals result.

Solution and recrystallization during digestion are probably responsible for the improvement in filterability as well. Bridging between adjacent particles yields larger crystalline aggregates that are more easily filtered. This view seems to be confirmed by the fact that little improvement in filtering characteristics occurs if the mixture is stirred during digestion.

DIRECTION OF COPRECIPITATION ERRORS

Coprecipitated impurities may cause the results of an analysis to be either too high or too low. If the contaminant is not a compound of the ion being determined, positive errors will always result. Thus a positive error will be observed when colloidal silver chloride adsorbs silver nitrate during a chloride analysis. On the other hand, when the contaminant contains the ion being determined, either positive or negative errors may be observed. In the determination of barium ions by precipitation as barium sulfate, for example, occlusion of barium salts occurs. If the occluded contaminant is barium nitrate, a positive error will be observed because this compound has a greater formula weight than the barium sulfate that would have formed had no coprecipitation occurred. If barium chloride were the contaminant, however, a negative error would arise because its formula weight is less than that of the sulfate salt.

PRECIPITATION FROM HOMOGENEOUS SOLUTION

In precipitation from homogeneous solution the precipitating agent is chemically generated in the solution. Local reagent excesses do not occur because the precipitating agent appears slowly and homogeneously throughout the entire solution; the relative supersaturation is thus kept low. In general homogeneously formed precipitates, both colloidal and crystalline, are better suited for analysis than solids formed by direct addition of a reagent.

Urea is often employed for the homogeneous generation of hydroxide ion. The reaction can be expressed by the equation

$$(H_2N)_2CO + 3H_2O = CO_2 + 2NH_4^+ + 2OH^-$$

This reaction proceeds slowly at temperatures just below boiling; typically 1 to 2 hr are needed to produce sufficient reagent for the completion of a precipitation. The method is particularly valuable for the precipitation of hydrous oxides or basic salts. For example, the hydrous oxides of iron(III)

TABLE 6-2
Methods for the Homogeneous Generation of Precipitants

PRECIPITANT	REAGENT	GENERATION REACTION	ELEMENTS PRECIPITATED
OH^-	Urea	$NH_2CO + 3H_2O = CO_2 + 2NH_4^+ + 2OH^-$	Al, Ga, Th, Bi, Fe, Sn
PO_4^{3-}	Trimethyl phosphate	$(CH_3O)_3PO + 3H_2O = 3CH_3OH + H_3PO_4$	Zr, Hf
$C_2O_4^{2-}$	Ethyl oxalate	$(C_2H_5)_2C_2O_4 + 2H_2O = 2C_2H_5OH + H_2C_2O_4$	Mg, Zn, Ca
SO_4^{2-}	Dimethyl sulfate	$(CH_3O)_2SO_2 + H_2O = 2CH_3OH + SO_4^{2-} + 2H_3O^+$	Ba, Ca, Sr, Pb
CO_3^{2-}	Trichloroacetic acid	$Cl_3CCOOH + 2OH^- = CHCl_3 + CO_3^{2-} + H_2O$	La, Ba, Ra
S^{2-}	Thioacetamide	$CH_3CNH_2 + H_2O = CH_3CNH_2 + H_2S$	Sb, Mo, Cu, Cd
8-Hydroxyquinoline	8-Acetoxyquinoline		Al, U, Mg, Zn

and aluminum are bulky, gelatinous masses that are heavily contaminated and difficult to filter when they are formed by the direct addition of base. In contrast, these same products are dense, readily filtered, and have considerably higher purity when they are produced by the homogeneous generation of hydroxide ion.

Homogeneous precipitation of crystalline precipitates also results in marked increases in crystal size as well as improvement in purity.

Representative analyses based upon precipitation by homogeneously generated reagents are given in Table 6-2.

DRYING AND IGNITION OF PRECIPITATES

After filtration, a gravimetric precipitate is heated until its weight becomes constant. Heating serves the purpose of removing the solvent and volatile electrolytes carried down with the precipitate; in addition, this treatment may induce chemical decomposition to give a product of known composition.

The temperature required to produce a suitable product varies, depending upon the precipitate. Figure 6-3 shows weight loss as a function of temperature for several common analytical precipitates. These data were obtained with an automatic thermobalance,[4] an instrument that measures the weight of a substance continuously as its temperature is increased at a constant rate in a furnace. The heating of three precipitates—silver chloride, barium sulfate, and aluminum oxide—serves simply to remove water and perhaps volatile electrolytes carried down during the precipitation process. Note the wide range in temperature required to produce an anhydrous precipitate of constant weight. Thus moisture is completely removed from silver chloride at a temperature above 110 to 120°C; dehydration of aluminum oxide, on the other hand, is not complete until a temperature greater than 1000°C is achieved. It is of interest to note that aluminum oxide, formed homogeneously with urea, can be completely dehydrated at a temperature of about 650°C.

The thermal curve for calcium oxalate is considerably more complex than the others shown in Figure 6-3. At temperatures below about 135°C, unbound water is eliminated to give the monohydrate $CaC_2O_4 \cdot H_2O$; a temperature of about 225°C converts this compound to the anhydrous oxalate. The abrupt change in weight at about 450°C signals the decomposition of calcium oxalate to calcium carbonate and carbon monoxide. The final step in the curve depicts the conversion of the carbonate to calcium oxide and carbon dioxide. It is evident that the weighing form employed for a calcium oxalate precipitate will depend upon the ignition conditions.

[4] For descriptions of thermobalances, see C. Duval, *Inorganic Thermogravimetric Analysis*, 2d ed. New York: Elsevier, 1962.

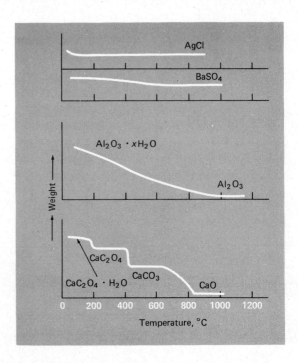

FIGURE 6-3 The effect of temperature on precipitate weights.

Applications of the Gravimetric Method

Gravimetric methods have been developed for most, if not all, inorganic anions and cations as well as for neutral species such as water, sulfur dioxide, carbon dioxide, and iodine. A variety of organic substances can also be readily determined gravimetrically. Examples include lactose in milk products, salicylates in drug preparations, phenolphthalein in laxatives, nicotine in pesticides, cholesterol in cereals, and benzaldehyde in almond extracts. Indeed gravimetric methods are among the most widely applicable of all chemical analyses.

INORGANIC PRECIPITATING AGENTS

Table 6-3 lists the common inorganic precipitating agents. These reagents typically cause formation of a slightly soluble salt or a hydrous oxide. The weighing form is either the salt itself or else an oxide. The lack of specificity of most inorganic reagents is clear from the many entries in the table.

Detailed procedures for the gravimetric determination of several inorganic species are given in Analyses 1-1 to 1-4, Chapter 20.

TABLE 6-3
Some Inorganic Precipitating Agents[a]

PRECIPITATING AGENT	ELEMENT PRECIPITATED[b]
$NH_3(aq)$	**Be** (BeO), **Al** (Al_2O_3), **Sc** (Sc_2O_3), Cr (Cr_2O_3),* **Fe** (Fe_2O_3), Ga (Ga_2O_3), Zr (ZrO_2), **In** (In_2O_3), Sn (SnO_2), U (U_3O_8)
H_2S	Cu (CuO),* **Zn** $(ZnO,$ or $ZnSO_4)$, **Ge** (GeO_2), As $(\underline{As_2O_3},$ or $As_2O_5)$, Mo (MoO_3), Sn (SnO_2),* Sb $(\underline{Sb_2O_3},$ or $Sb_2O_5)$, Bi (Bi_2S_3)
$(NH_4)_2S$	Hg $(\underline{HgS})$, Co (Co_3O_4)
$(NH_4)_2HPO_4$	**Mg** $(Mg_2P_2O_7)$, Al $(AlPO_4)$, Mn $(Mn_2P_2O_7)$, Zn $(Zn_2P_2O_7)$, Zr $(Zr_2P_2O_7)$, Cd $(Cd_2P_2O_7)$, Bi $(BiPO_4)$
H_2SO_4	Li, Mn, **Sr**, **Cd**, **Pb**, **Ba** (all as sulfates)
H_2PtCl_6	K $(K_2PtCl_6,$ or Pt), Rb $(\underline{Rb_2PtCl_6})$, Cs $(\underline{Cs_2PtCl_6})$
$H_2C_2O_4$	Ca (CaO), Sr (SrO), **Th** (ThO_2)
$(NH_4)_2MoO_4$	Cd $(CdMoO_4)$,* Pb $(\underline{PbMoO_4})$
HCl	**Ag** $(AgCl)$, Hg (Hg_2Cl_2), Na (as $NaCl$ from butyl alcohol), Si (SiO_2)
$AgNO_3$	**Cl** $(AgCl)$, Br $(\underline{AgBr})$, I$(\underline{AgI})$
$(NH_4)_2CO_3$	**Bi** (Bi_2O_3)
NH_4SCN	Cu $(Cu_2(SCN)_2)$
$NaHCO_3$	Ru, Os, Ir (precipitated as hydrous oxides; reduced with H_2 to metallic state)
HNO_3	Sn (SnO_2)
H_5IO_6	Hg $(Hg_5(IO_6)_2)$
$NaCl, Pb(NO_3)_2$	F $(PbClF)$
$BaCl_2$	SO_4^{2-} $(BaSO_4)$
$MgCl_2, NH_4Cl$	PO_4^{3-} $(Mg_2P_2O_7)$

[a] From W. F. Hillebrand, G. E. F. Lundell, H. A. Bright, and J. I. Hoffman, *Applied Inorganic Analysis*. New York: Wiley, 1953.

[b] Boldface type indicates that gravimetric analysis is the preferred method for the element or ion. The weighed form is indicated in parentheses. An asterisk indicates that the gravimetric method is seldom used. An underscored entry indicates the most reliable gravimetric method.

REDUCING REAGENTS

Table 6-4 lists several reagents that convert the analyte to its elemental form for weighing.

ORGANIC PRECIPITATING AGENTS

A number of organic reagents have been developed for the gravimetric analysis of inorganic species. Some of these yield products in which bonding between the reagent and the inorganic ion is primarily ionic. Others give rise to slightly soluble, nonionic complexes, or *coordination compounds*. Organic reagents tend to be more selective in their reactions than the majority of the inorganic reagents shown in Table 6-3. As a result, they are particularly useful for analysis.

We shall limit our discussion to three typical examples of organic reagents. A comprehensive treatment of this topic is found in the reference by Flagg.[5]

Sodium Tetraphenylboron. Sodium tetraphenylboron, $(C_6H_5)_4B^-Na^+$, is an important example of organic precipitating reagents that form saltlike precipitates. In cold mineral acid solutions it is a near-specific precipitating

TABLE 6-4
Some Reducing Reagents Employed in Gravimetric Methods

REDUCING AGENT	ANALYTE
SO_2	Se, Au
$SO_2 + H_2NOH$	Te
H_2NOH	Se
$H_2C_2O_4$	Au
H_2	Re, Ir
HCOOH	Pt
$NaNO_2$	Au
$TiCl_2$	Rh
$SnCl_2$	Hg
Electrolytic reduction	Co, Ni, Cu, Zn, Ag, In, Sn, Sb, Cd, Re, Bi

[5] J. F. Flagg, *Organic Reagents Used in Gravimetric and Volumetric Analysis.* New York: Interscience, 1948.

agent for potassium ion and for ammonium ion. The precipitates are stoichiometric, corresponding to the potassium or the ammonium salt, as the case may be; they are amenable to vacuum filtration and can be brought to constant weight at 105 to 120°C. Only mercury(II), rubidium, and cesium interfere and must be removed by prior treatment.

Dimethylglyoxime. An organic precipitating agent of unparalleled specificity is dimethylglyoxime. Its coordination compound with palladium is the only one that is sparingly soluble in acid solution. Similarly, only the nickel compound precipitates from a weakly alkaline environment. Nickel dimethylglyoxime is bright red and has the structure

The precipitate is bulky and has an exasperating tendency to creep as it is filtered and washed; these properties limit the amount of nickel that can be isolated conveniently with dimethylglyoxime.

Nickel dimethylglyoxime is readily dried at 110°C to give an anhydrous product having the exact composition shown by its formula.

8-Hydroxyquinoline. Approximately two dozen cations form sparingly soluble coordination compounds with 8-hydroxyquinoline, also known as *oxine*:

Typical of these is the product with magnesium:

TABLE 6-5
Gravimetric Methods for Organic Functional Groups

FUNCTIONAL GROUP	BASIS FOR METHOD	REACTION AND PRODUCT WEIGHED[a]
Carbonyl	Weight of precipitate with 2,4-dinitrophenylhydrazine	$RCHO + H_2NNHC_6H_3(NO_2)_2 \rightarrow \underline{RCH{=}NNHC_6H_3(NO_2)_2}(s) + H_2O$ (RCOR' reacts similarly)
Aromatic carbonyl	Weight of CO_2 formed at 230°C in quinoline; CO_2 distilled, absorbed, and weighed	$ArCHO \xrightarrow{230°C} Ar + \underline{CO_2}(g)$
Methoxyl and ethoxyl	Weight of AgI formed after distillation and decomposition of CH_3I or C_2H_5I	$\left.\begin{array}{l} ROCH_3 \quad + HI \rightarrow ROH \quad\; + CH_3I \\ RCOOCH_3 + HI \rightarrow RCOOH + CH_3I \\ ROC_2H_5 \quad + HI \rightarrow ROH \quad\; + C_2H_5I \end{array}\right\}$ $CH_3I + Ag^+ + H_2O \rightarrow \underline{AgI}(s) + CH_3OH$
Aromatic nitro	Weight loss of Sn	$RNO_2 + \tfrac{3}{2}\underline{Sn}(s) + 6H^+ \rightarrow RNH_2 + \tfrac{3}{2}Sn^{4+} + 2H_2O$
Azo	Weight loss of Cu	$RN{=}NR' + 2\underline{Cu}(s) + 4H^+ \rightarrow RNH_2 + R'NH_2 + 2Cu^{2+}$
Phosphate	Weight of Ba salt	$ROP(OH)_2 + Ba^{2+} \rightarrow \underline{ROPO_2Ba}(s) + 2H^+$
Sulfamic acid	Weight of $BaSO_4$ after reaction with HNO_2	$RNHSO_3H + HNO_2 + Ba^{2+} \rightarrow ROH + \underline{BaSO_4}(s) + N_2 + 2H^+$
Sulfinic acid	Weight of Fe_2O_3 after ignition of Fe^{3+} sulfinate	$3ROSOH + Fe^{3+} \rightarrow (ROSO)_3Fe + 3H^+$ $(ROSO)_3Fe \xrightarrow{O_2} CO_2 + H_2O + SO_2 + \underline{Fe_2O_3}(s)$

[a] The substance weighed is underlined.

The solubilities of metal oxinates vary from cation to cation. Moreover, because their formation involves the loss of hydrogen ions, their solubilities are pH dependent. In general the precipitates increase in solubility with increasing hydrogen ion concentration. Some are also soluble in strongly alkaline media. Thus control of the pH imparts considerable selectivity to 8-hydroxyquinoline.

GRAVIMETRIC ORGANIC FUNCTIONAL GROUP ANALYSIS

Several reagents react selectively with certain organic functional groups and thus permit the determination of most compounds containing these groups. A list of gravimetric functional group reagents is given in Table 6-5. Many of these reactions can also be used for volumetric and spectrophotometric methods. For the occasional analysis the gravimetric procedure will often be the method of choice because no calibration or standardization is required.

PROBLEMS

*1. Use chemical symbols to express the gravimetric factor for each of the following:

SOUGHT	WEIGHED		SOUGHT	WEIGHED
(a) CHF_3	CaF_2		(d) $(C_6H_5)_6Si_2$	$BaCO_3$
(b) $Fe_3Al_2Si_3O_{12}$	Fe_2O_3		(e) Na_3PO_4	$P_2O_5 \cdot 24MoO_3$
(c) $(C_6H_5)_6Si_2$	SiO_2		(f) Mo	$P_2O_5 \cdot 24MoO_3$

2. Use chemical symbols to express the gravimetric factor for each of the following:

SOUGHT	WEIGHED		SOUGHT	WEIGHED
(a) Mn_2O_3	Mn_3O_4		(d) $C_6H_3Cl_3$	$AgCl$
(b) $Na_2B_4O_7$	B_2O_3		(e) UO_2	U_3O_8
(c) Ni_3O_4	$Ni_2P_2O_7$		(f) K_2SO_4	$KB(C_6H_5)_4$

3. What weight of Ag_2SO_4 can be obtained from 0.676 g of

*(a) K_2SO_4? ***(c)** $Na[Ag(CN)_2]$?

(b) $AgNO_3$? **(d)** $Ag[Ag(CN)_2]$?

*4. A 7.24-g pesticide sample yielded 0.212 g of TlI after treatment with an excess of KI. Calculate the percentage of Tl_2SO_4 in the sample.

5. Calculate the percentage of copper in an ore specimen if a 0.663-g sample yielded 0.544 g of $Cu_2(SCN)_2$.

*6. The sulfur in three saccharine ($C_7H_5NO_3S$) tablets was oxidized to sulfate and then precipitated as $BaSO_4$. Calculate the average weight of saccharine in these tablets if 0.140 g of $BaSO_4$ was recovered.

7. The iodide ion in a 4.13-g sample that also contained chloride was converted to iodate by treatment with an excess of bromine:

$$3H_2O + 3Br_2 + I^- \rightarrow 6Br^- + IO_3^- + 6H^+$$

The unused bromine was eliminated by boiling. An excess of barium chloride was then added; 0.0972 g of $Ba(IO_3)_2$ was subsequently recovered. Calculate the percentage of potassium iodide originally present in the sample.

*8. What volume of a 0.50% (w/v) dimethylglyoxime ($C_4H_8N_2O_2$) solution should be used to precipitate the Ni^{2+} in a solution containing 80.0 mg of $NiCl_2$, assuming that a 4.0% excess of the reagent is needed to assure quantitative removal of the cation?

9. What volume of a 5.0% (w/v) solution of 8-hydroxyquinoline (C_9H_7NO) should be used to precipitate Al^{3+} from a solution containing 1.00 g of $Al_2(SO_4)_3$, assuming that a 10.0% excess of the reagent is needed to assure quantitative removal of the cation? (The formal ratio of reagent to Al^{3+} in the precipitate is 3 : 1.)

*10. If a 10.0% reagent excess is needed to minimize solubility losses, what volume of the following solutions will be needed to precipitate the sulfate from 0.500-g samples that contain 30.0% $Al_2(SO_4)_3$?
    (a) A solution that is 0.160 F in $BaCl_2 \cdot 2H_2O$.
    (b) A solution that is 4.20% (w/v) in $BaCl_2 \cdot 2H_2O$.

11. The cobalt in 0.50-g samples believed to contain about 2.5% Co_3O_4 is to be precipitated with α-nitroso-β-naphthol ($C_{10}H_7NO_2$). If a 10.0% excess of the reagent is needed to minimize solubility losses, what volume of a 0.45% (w/v) solution of the reagent in glacial acetic acid should be employed? (The formal ratio of reagent to Co in the precipitate is 3 : 1.)

*12. The bismuth oxycarbonate $(BiO)_2CO_3$ in a commercial drug tablet was analyzed by dissolving a 0.150-g sample in dilute HNO_3 and precipitating the Bi^{3+} as BiOCl by the addition of Cl^-. After drying at 100°C, this precipitate was found to weigh 0.0412 g. Calculate the percent $(BiO)_2CO_3$ in the sample.

13. After appropriate preliminary steps, the uranium in a 0.396-g specimen containing the mineral carnotite $[K_2(UO_3)_2(VO_4)_2 \cdot 3H_2O]$ was precipitated as the oxalate. Ignition of the precipitate yielded 0.147 g U_3O_8.
    (a) Calculate the percent carnotite in the sample.
    (b) How many pounds of vanadium are contained in 1 ton of the rock?

*14. A 0.513-g sample which contained epsomite ($MgSO_4 \cdot 7H_2O$) and inert material was dehydrated quantitatively by heating in a stream of dry air. The air was passed through a tube containing a desiccant. The original weight of the tube was 19.509 g; after water absorption its weight was 19.631 g. Calculate the percent $MgSO_4 \cdot 7H_2O$ in the sample, assuming that the mineral represented the only source of the water in the sample.

15. A 2.18-g sample containing the mineral hydrocerussite [$2PbCO_3 \cdot Pb(OH)_2$] was heated to a temperature sufficient to cause decomposition to PbO:

$$2PbCO_3 \cdot Pb(OH)_2(s) \rightarrow 3PbO(s) + 2CO_2 + H_2O$$

The ignited residue was found to weigh 1.667 g.
(a) Calculate the percent hydrocerussite in the sample.
(b) How many pounds of lead are contained in 1 ton of the material analyzed?

*16. The zinc in a 3.67-g sample of a foot powder was freed of organic matter by wet-ashing with a perchloric–nitric acid mixture and precipitated as $ZnNH_4PO_4$. Ignition of the filtered precipitate yielded 0.264 g of $Zn_2P_2O_7$. Calculate the percentage of zinc undecylenate, $Zn(C_{11}H_{19}O_2)_2$, in the sample.

17. Chloromycetin is an antibiotic with the formula $C_{11}H_{12}O_5N_2Cl_2$. A 0.962-g sample of an ophthalmic ointment was heated in a closed tube with metallic sodium to destroy the organic material and free the chloride; after dissolving the ignited mixture in water, the carbonaceous residue was removed by filtration. The chloride was then precipitated with $AgNO_3$ to give 0.0116 g of AgCl. Calculate the percentage of chloromycetin in the sample.

*18. Twenty tablets of a therapeutic iron preparation, with a total weight of 19.9 g, were ground and mixed thoroughly until homogeneous. A 2.87-g sample of the powdered material was then dissolved in HNO_3; the solution was heated to oxidize the Fe to the +3 state, and ammonia was added to precipitate $Fe_2O_3 \cdot xH_2O$. After ignition the Fe_2O_3 was found to weigh 0.313 g. On an average, how many milligrams of $FeSO_4 \cdot 7H_2O$ did each tablet contain?

19. To determine the phosphorous content of a fertilizer a 0.237-g sample was decomposed with hot concentrated nitric acid. The resulting solution was diluted, and then the PO_4^{3-} was precipitated as the quinoline salt of phosphomolybdic acid, $(C_9H_7N)_2H_3PO_4 \cdot 12MoO_3$. After filtration and drying the precipitate was found to weigh 0.777 g. Calculate the percent P_2O_5 in the sample.

*20. The nitrobenzene in a 0.643-g sample was determined by dilution with an HCl-methanol mixture followed by the introduction of 0.815 g of pure Sn. After being refluxed for 1 hr, the solution was cooled and filtered. The residual Sn was found to weigh 0.602 g. Calculate the percent $C_6H_5NO_2$ in the sample:

$$2C_6H_5NO_2 + 3Sn(s) + 12H^+ \rightarrow 2C_6H_5NH_2 + 4H_2O + 3Sn^{4+}$$

21. The KCNO content of a herbicide preparation was determined by dissolving a 3.62-g sample in water and adding an excess of semicarbazide hydrochloride. Reaction:

$$NH_2CONHNH_3^+ + CNO^- \rightarrow NH_2CONHNHCONH_2(s)$$

The resulting precipitate was filtered, washed, and dried; its weight was 0.383 g. Calculate the percent KCNO in the sample.

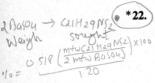

*22. The content of the hypnotic drug Captodiamine, $C_{21}H_{29}NS_2$ (gfw = 360), in a pharmaceutical preparation was established by means of a gravimetric sulfate analysis. Calculate the percentage of this active ingredient if an eight-tablet sample, weighing a total of 1.20 g, yielded 0.518 g of BaSO₄.

How many of these tablets should be used if the recommended oral dose is 100 mg of Captodiamine?

23. The label on a bottle of the tranquilizer Thoridazine, $C_{21}H_{26}N_2S_2$ (gfw = 371), was so damaged that the composition of the contents could no longer be read. A 12-pill sample was decomposed to convert the sulfur to sulfate. Calculate the average weight of Thoridazine in each tablet if the weight of BaSO₄ was 0.301 g.

This tranquilizer is to be administered to a nervous patient; 60 mg is the desired dosage. How many tablets should be prescribed for each dose?

*24. A 0.500-g sample containing only In_2O_3 and TeO_2 was analyzed by solution and precipitation with 8-hydroxyquinoline. The mixture of $In(C_9H_6ON)_3$ and $TeO(C_9H_6ON)_2$ was found to weigh 1.862 g. Calculate the percentage of each oxide present in the sample.

25. A 0.4482-g sample containing only $TlNO_3$ and $Pb(NO_3)_2$ was analyzed by solution and precipitation with NaI. The mixture of TlI and PbI_2 was found to weigh 0.5986 g after filtration, washing, and drying. Calculate the percentage of each nitrate present in the sample.

7
an introduction to
volumetric methods of analysis

A quantitative analysis based upon the measurement of volume is called a *volumetric* or *titrimetric method*. Volumetric methods are much more widely used than gravimetric methods because they are usually more rapid and convenient; in addition, they are often as accurate.

Definition of Some Terms

A titrimetric method employs one or more *standard solutions*, which are reagents whose concentrations are known exactly. Standard solutions are used to perform *titrations* in which the quantity of analyte in a solution is determined from the volume of standard solution that it consumes. Ordinarily a titration is performed by carefully adding the standard solution until reaction with the analyte is judged to be complete; the volume of standard reagent is then measured. Occasionally it is convenient or necessary to add an excess of the reagent and then determine the excess by *back-titration* with a second standard reagent.

The accuracy of a volumetric method is limited by the accuracy with which the concentration of the standard solution is known; for this reason much care is taken in the preparation of standard solutions. The concentration of a standard solution is established in one of two ways:

1. Directly, by dissolving a carefully weighed quantity of the pure reagent and diluting to an exactly known volume.
2. Indirectly, by titrating a solution containing a weighed quantity of a pure compound with the reagent solution.

In both methods a highly purified substance—called a *primary standard*—is required as the reference material. The process whereby the concentration of a standard solution is determined by titration against a primary standard is called a *standardization.*

The goal of every titration is the addition of standard solution in an amount that is chemically equivalent to the substance with which it reacts. This condition is achieved at the *equivalence point.* For example, the equivalence point in the titration of sodium chloride with silver nitrate is attained when exactly 1 fw of silver ion has been introduced for each formula weight of chloride ion in the sample. In the titration of sulfuric acid with sodium hydroxide the equivalence point occurs when 2 fw of the latter have been introduced for each formula weight of the former.

The equivalence point in a titration is a theoretical concept; in actual fact its location can be estimated only by observing physical changes associated with equivalence. These changes manifest themselves at the *end point* of the titration. It is to be hoped that any volume difference between the end point and equivalence point will be small. Differences do exist, however, owing to inadequacies in the physical changes and our ability to observe them; a *titration error* is the result.

A common method of end point detection in volumetric analysis involves the use of a supplementary chemical compound that exhibits a change in color as a result of concentration changes occurring near the equivalence point. Such a substance is called an *indicator.*

Reactions and Reagents Used in Volumetric Analysis

It is convenient to classify volumetric methods according to four reaction types, specifically, precipitation, complex formation, neutralization (acid-base), and oxidation-reduction. Each reaction type is unique in such matters as nature of equilibria involved, the indicators, reagents, and primary standards available, and the definition of equivalent weight.

PRIMARY STANDARDS

The accuracy of a volumetric analysis is critically dependent upon the primary standard used to establish, directly or indirectly, the concentration of the standard solution. Important requirements for a substance to serve as a good primary standard include the following:

1. Highest purity. Established methods should be available for confirming its purity.
2. Stability. It should not be attacked by constituents of the atmosphere.
3. Absence of hydrate water. If the substance were hygroscopic or efflorescent, drying and weighing would be difficult.
4. Availability at reasonable cost.
5. Reasonably high equivalent weight to minimize the relative error associated with the weighing operation.

Few substances meet or even approach these requirements. As a result, the number of primary standard substances available to the chemist is limited.

Occasionally it is necessary to use less pure substances in lieu of a primary standard. The assay (that is, the percent purity) of such a *secondary* standard must be established by careful analysis.[1]

STANDARD SOLUTIONS

An ideal standard solution for titrimetric analysis would have the following properties:

1. Its concentration should remain constant for months or years after preparation to eliminate the need for restandardization.
2. Its reaction with the analyte should be rapid in order to minimize the waiting period after each addition of reagent.
3. The reaction between the reagent and the analyte should be reasonably complete. As we shall presently show, satisfactory end points generally require this condition.
4. The reaction of the reagent with the analyte must be such that it can be described by a balanced chemical equation; otherwise the weight of the analyte cannot be calculated directly from the volumetric data. This requirement implies the absence of side reactions between the reagent and the unknown or with other constituents of the solution.
5. A method must exist for detecting the equivalence point between the reagent and the analyte; that is, a satisfactory end point is required.

Few volumetric reagents currently in use meet all of these requirements perfectly.

[1] Usage of the terms primary and secondary standard is a source of disagreement among chemists. Some reserve primary standard for the relatively few compounds that possess the properties listed at the start of this section (see, for example, H. A. Laitinen and W. B. Harris, *Chemical Analysis*, 2d ed. New York: McGraw-Hill, 1975, p. 101) and use secondary to denote species that have reliably established composition. Others prefer to consider all substances of known assay as primary standards and refer to solutions of known concentration as secondary standards. We prefer the former set of definitions.

End Points in Volumetric Methods

An end point is based upon a physical property which changes in a characteristic way at or near the equivalence point in the titration. The most common end point involves a color change due to the reagent, the analyte, or an indicator substance. Changes in other physical properties such as electrical potential, conductivity, temperature, and refractive index have also been employed to locate equivalence points.

CONCENTRATION CHANGES DURING TITRATION

End points are the result of marked changes in at least one reactant concentration during the titration. For most (but not all) end points, changes that occur in the region surrounding the equivalence point are particularly important. To illustrate, the second column of Table 7-1 contains data that show the changes in hydronium ion concentration that occur when 50.00 ml of 0.1000-F hydrochloric acid are titrated with standard 0.1000-F NaOH. To emphasize the *relative* changes in concentration that occur, the volumes selected for tabulation correspond to the increments needed to cause a tenfold decrease in concentration of the reacting species. Thus we see in column 2 of the table that 40.90 ml of base are needed to decrease the hydronium ion concentration from 0.100 M to 0.0100 M. An additional 8.1 ml are required to lower the concentration to 0.00100 M; 0.89 ml will cause another tenfold decrease and so on. Analogous increases in the hydroxide ion concentration occur simultaneously.

It is apparent from the data in Table 7-1 that the equivalence point region is characterized by the largest change in the *relative concentration* of the reacting species. Most end points depend upon these large changes.

Figure 7-1 consists of plots of the concentration data in Table 7-1. Note that the ordinate scale for these plots must be large to encompass the enormous concentration changes experienced by the reactants. As a result changes occurring in the region of most interest—that is, the equivalence point—are obscured.

To show more clearly the changes taking place in the equivalence point region it is useful to substitute p-values (see pp. 12 to 13) for one of the reactants in place of molar concentrations. Thus pH or pOH would be used for the titrations illustrated by Table 7-1; columns 3 and 4 list these values after various additions of reagent.

When the pH values in Table 7-1 are plotted against the volume of reagent added (Figure 7-2), a curve is obtained that gives a much clearer picture of the changes that occur in the solution near the equivalence point; it is evident from Figure 7-2 that a marked increase in pH occurs here. Also shown in Figure 7-2 is a curve in which pOH rather than pH is plotted as the ordinate for the same titration; here again a marked change in the p-function occurs in the region of equivalence.

TABLE 7-1

Concentration Changes during a Titration*

VOL 0.1000-F NaOH, ml	CONCENTRATION H$_3$O$^+$, MOLE/LITER	pH	pOH	VOL NaOH REQUIRED TO CAUSE A TENFOLD CHANGE IN H$_3$O$^+$ OR A UNIT CHANGE IN pH, ml
0.0	1.0×10^{-1}	1.00	13.00	
40.9	1.0×10^{-2}	2.00	12.00	40.9
49.01	1.0×10^{-3}	3.00	11.00	8.1
49.90	1.0×10^{-4}	4.00	10.00	0.89
49.990	1.0×10^{-5}	5.00	9.00	0.09
49.9990	1.0×10^{-6}	6.00	8.00	0.009
50.0000	1.0×10^{-7}	7.00	7.00	0.001
50.0010	1.0×10^{-8}	8.00	6.00	0.001
50.010	1.0×10^{-9}	9.00	5.00	0.009
50.10	1.0×10^{-10}	10.00	4.00	0.09
51.01	1.0×10^{-11}	11.00	3.00	0.91
61.1	1.0×10^{-12}	12.00	2.00	10.09

^{a} Although these data are useful for illustration purposes, it should be understood that their realization in the laboratory would not be possible because measurements of volumes and normalities to six significant figures are ordinarily impossible.

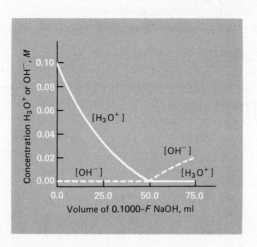

FIGURE 7-1 Changes in reactant concentrations during the titration of 50.00 ml of 0.1000-F HCl with 0.1000-F NaOH.

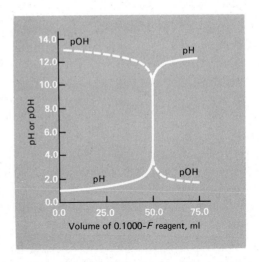

FIGURE 7-2 Titration curves for 50.00 ml of 0.1000-*F* HCl with 0.1000-*F* NaOH.

Curves such as either of those in Figure 7-2 are called *titration curves*. Plots of analogous data for titrations involving oxidation-reduction, precipitation, or complex formation reactions have the same general characteristics. Typical examples are derived and discussed in later chapters.

The shape of the titration curve, and especially the magnitude of the change in the equivalence point region, are of great interest to the analytical chemist because these determine how easily and how accurately the equivalence point can be located. The ease with which the physical change used to signal an end point can be detected is usually directly related to the magnitude and the rate of change in p-values. In general the most satisfactory end points occur where changes in p-functions are large and take place over a small range of reagent volumes.

Calculations Associated with Titrimetric Methods

In preparation for the discussion that follows, the student may find it helpful to review the material in Chapter 2 (pp. 15–18) on chemical units of weight and concentration.

Chemists frequently make use of the *equivalent weight* (or the *milliequivalent weight*) as the basis for volumetric calculations; *normality* is the corresponding unit of concentration. The manner in which these units are defined depends upon the type of reaction that serves as the basis for the analysis—that is, whether the titration reaction involves neutralization, oxidation-reduction, precipitation, or complex formation. Furthermore, the chemical behavior of the substance must be carefully specified if its equivalent weight is to be defined unambiguously. If the substance can react in more than one way, it will likewise have more than one equivalent or milliequivalent weight. *Thus the definition of the equivalent weight or the milliequivalent weight for a substance is always based on its behavior in a specific chemical reaction. Evaluation of the equivalent weight is impossible if the reaction involved is not specifically stated. Likewise, the concentration of a solution cannot be expressed in terms of normality without this information.*

EQUIVALENT OR MILLIEQUIVALENT WEIGHTS IN NEUTRALIZATION REACTIONS

The equivalent weight (eq wt) of a substance participating in a neutralization reaction is that weight which either contributes or reacts with 1 gfw of hydrogen ion *in that reaction*. The milliequivalent weight (meq wt) is $\frac{1}{1000}$ of the equivalent weight.

For strong acids or bases and for acids or bases containing only a single reactive hydrogen or hydroxide ion, the relationship between the equivalent weight and the formula weight can be readily determined. For example, the equivalent weight for potassium hydroxide and hydrochloric acid must be equal to their formula weights, respectively, because each has only a single reactive hydrogen ion or hydroxide ion. Similarly, we know that only one hydrogen in acetic acid, $HC_2H_3O_2$, is acidic; therefore the formula weight and equivalent weight for this acid must also be identical. Barium hydroxide, $Ba(OH)_2$, is a strong base containing two hydroxide ions that are indistinguishable. This base will necessarily react with two hydrogen ions in any acid-base reaction; thus its equivalent weight will be one-half its formula weight. For sulfuric acid, dissociation of the second hydrogen ion in water is not complete; the hydrogen sulfate ion, however, is a sufficiently strong acid so that both hydrogens participate in all aqueous neutralization reactions. The equivalent weight of H_2SO_4 as an acid is therefore always one-half its formula weight in aqueous solution.

The situation becomes more complex for a polyfunctional acid, which contains two or more hydrogen ions with differing tendencies to dissociate. For example, certain indicators undergo a color change when only the first of the three protons in phosphoric acid has been neutralized; that is,

$$H_3PO_4 + OH^- \rightarrow H_2PO_4^- + H_2O$$

Other indicators change color after two hydrogen ions have reacted:

$$H_3PO_4 + 2OH^- \rightarrow HPO_4^{2-} + 2H_2O$$

In the first titration the equivalent weight of phosphoric acid is equal to its formula weight; in the second, it is one-half its formula weight. (It is not practical to titrate the third proton; thus an equivalent weight for H_3PO_4

that is one-third the formula weight is not encountered in a neutralization titration.) *Without knowing which of these reactions is involved, an unambiguous definition of the equivalent weight for phosphoric acid is impossible.*

EQUIVALENT WEIGHT IN OXIDATION-REDUCTION REACTIONS

The equivalent weight of a participant in an oxidation-reduction reaction is that weight which directly or indirectly consumes or produces 1 mole of electrons. The numerical value for the equivalent weight is conveniently established by dividing the formula weight of the substance of interest by the change in oxidation number associated with its reaction. As an example, consider the oxidation of oxalate ion by permanganate:

$$5C_2O_4^{2-} + 2MnO_4^- + 16H^+ \rightarrow 10CO_2 + 2Mn^{2+} + 8H_2O$$

In this reaction the change in oxidation number for manganese is 5 because the element passes from the +7 to the +2 state; the equivalent weight for MnO_4^- or Mn^{2+} is therefore one-fifth of the corresponding formula weights. Each carbon atom in the oxalate ion is oxidized from the +3 to the +4 state. The equivalent weight of sodium oxalate is one-half its formula weight because it contains two carbon atoms. The equivalent weight of carbon dioxide, on the other hand, is equal to its formula weight because it contains but a single atom of carbon. These and other examples of equivalent weights are given in Table 7-2. Note that the equivalent weight for a substance is evaluated from the change in oxidation state that occurs in the titration and not from its oxidation state in the compound in question. Thus, for example, the oxidation state of manganese in Mn_2O_3 is +3. Before the reaction just considered could be used to determine the Mn_2O_3 content of a sample, the manganese would first have to be converted to permanganate ion by some reagent. Once in the +7 state, each manganese would be reduced to the +2 state in the titration step. The equivalent weight is thus gfw $Mn_2O_3/10$.

As in neutralization reactions, the equivalent weight for a given oxidizing or reducing agent is not invariant. Potassium permanganate, for example, reacts with reducing agents in at least four different ways, depending upon the conditions existing in the solution. The half-reactions are

$$MnO_4^- + e \rightarrow MnO_4^{2-}$$

$$MnO_4^- + 3e + 2H_2O \rightarrow MnO_2(s) + 4OH^-$$

$$MnO_4^- + 4e + 3H_2P_2O_7^{2-} + 8H^+ \rightarrow Mn(H_2P_2O_7)_3^{3-} + 4H_2O$$

$$MnO_4^- + 5e + 8H^+ \rightarrow Mn^{2+} + 4H_2O$$

The changes in oxidation number for manganese are 1, 3, 4, and 5. The equivalent weight of potassium permanganate would be equal to the gram

TABLE 7-2
Equivalent Weights of Some Manganese and Carbon Species Based on the Reaction $5C_2O_4^{2-} + 2MnO_4^- + 16H^+ \rightarrow 10CO_2 + 2Mn^{2+} + 8H_2O$

SUBSTANCE	EQUIVALENT WEIGHT
Mn	$\dfrac{\text{gfw Mn}}{5}$
$KMnO_4$	$\dfrac{\text{gfw } KMnO_4}{5}$.
$Ca(MnO_4)_2 \cdot 4H_2O$	$\dfrac{\text{gfw } Ca(MnO_4)_2 \cdot 4H_2O}{2 \times 5}$
Mn_2O_3	$\dfrac{\text{gfw } Mn_2O_3}{2 \times 5}$
CO_2	$\dfrac{\text{gfw } CO_2}{1}$
$C_2O_4^{2-}$	$\dfrac{\text{gfw } C_2O_4^{2-}}{2}$

formula weight for the first reaction and one-third, one-fourth, and one-fifth of the gram formula weight, respectively, for the others.

EQUIVALENT WEIGHTS IN PRECIPITATION AND COMPLEX FORMATION REACTIONS

It is awkward to devise a definition for equivalent weight that is entirely free of ambiguity for compounds involved in precipitation or complex formation reactions. As a result, many chemists prefer to avoid the use of the concept for reactions of this type and use formula weights exclusively instead. We are in sympathy with this practice. However, it is likely that the student will encounter situations where equivalent or milliequivalent weights are specified for substances involved in precipitation or complex formation; it is thus important to know how these quantities are defined.

    The equivalent weight of a participant in a precipitation or a complex formation reaction is that weight which reacts with or provides 1 gfw of the reacting cation if it is univalent, one-half of the gram formula weight if it is divalent, one-third of the gram formula weight if it is trivalent, and so on. Our earlier definitions of equivalent weight were based either on 1 mole of hydrogen ions or on 1 mole of electrons. Here 1 mole of a univalent cation or the equivalent thereof is used. *The cation referred to in this definition is always the cation directly involved in the reaction* and not necessarily the cation contained in the compound whose equivalent weight is being defined.

To illustrate the application of this definition, consider the reaction

$$Ag^+ + Cl^- \rightarrow AgCl(s)$$

Here the reacting cation is the univalent silver ion. The equivalent weights of some compounds that might be associated with this reaction are shown in Table 7-3.

The equivalent weight of the barium chloride dihydrate is one-half its formula weight, *not* because this weight contains one-half of the formula weight of divalent barium ions but rather because this is the weight that reacts with 1 fw of silver ions. A consideration of the last example will show why it is important to make such a fine distinction. Here we might be tempted to say that the equivalent weight of the AlOCl is one-third its formula weight based upon the presence of 1 mole of trivalent aluminum. This assignment would be correct if Al^{3+} were the cation involved in the reaction. In the example, however, silver ion is the reacting cation. Therefore the equivalent weight of AlOCl must be that weight which reacts with 1 fw of silver ions; that is, the formula weight and equivalent weight are identical.

As previously noted, a given compound may have more than one equivalent weight. For example, when silver is titrated with a solution of potassium cyanide, an end point can be detected for either of two reactions:

$$2CN^- + 2Ag^+ \rightarrow Ag[Ag(CN)_2](s)$$

or

$$2CN^- + Ag^+ \rightarrow Ag(CN)_2^-$$

TABLE 7-3

Equivalent Weights of Some Species That React To Form Precipitates Based on the Reaction $Ag^+ + Cl^- \rightarrow AgCl(s)$

SUBSTANCE	EQUIVALENT WEIGHT
Ag^+	gfw Ag^+
$AgNO_3$	gfw $AgNO_3$
Ag_2SO_4	$\dfrac{\text{gfw } Ag_2SO_4}{2}$
NaCl	gfw NaCl
$BaCl_2 \cdot 2H_2O$	$\dfrac{\text{gfw } BaCl_2 \cdot 2H_2O}{2}$
$AlCl_3$	$\dfrac{\text{gfw } AlCl_3}{3}$
AlOCl	gfw AlOCl

In the first reaction the equivalent weight of potassium cyanide would be identical to its formula weight; in the second it would be *twice* the formula weight. The latter is an example of an equivalent weight that is *greater* than a formula weight.

EQUIVALENT WEIGHTS OF COMPOUNDS NOT PARTICIPATING DIRECTLY IN A VOLUMETRIC REACTION

There is frequent need to define the equivalent weight of an analyte that is related only indirectly to the actual reactants of the titration. To illustrate, lead can be determined by an indirect method in which the cation is first precipitated as the chromate from an acetic acid solution. The precipitate is filtered, washed free of excess precipitant, and redissolved in dilute hydrochloric acid to give a solution of lead and dichromate ions. The latter may then be determined by an oxidation-reduction titration with a standard iron(II) solution. The reactions are

$$Pb^{2+} + CrO_4^{2-} \xrightarrow[\text{HOAc}]{\text{dil}} PbCrO_4(s) \qquad \text{(precipitate filtered and washed)}$$

$$2PbCrO_4(s) + 2H^+ \xrightarrow[\text{HCl}]{\text{dil}} 2Pb^{2+} + Cr_2O_7^{2-} + H_2O \qquad \text{(precipitate redissolved)}$$

$$Cr_2O_7^{2-} + 6Fe^{2+} + 14H^+ \longrightarrow 2Cr^{3+} + 6Fe^{3+} + 7H_2O \qquad \text{(titration)}$$

For the purpose of calculation we must ascribe an equivalent weight to lead. *Because the titration is an oxidation-reduction process, the equivalent weight of lead will have to be based on a change of oxidation number.* Clearly the lead exhibits no such change. It is, however, associated in a $1:1$ ratio with chromium, and this element changes from the $+6$ to the $+3$ state in the titration. Therefore we can say that a change in oxidation state of 3 is *associated* with each lead, and its equivalent weight in this sequence of reactions is one-third its atomic weight.

It is often helpful to make an inventory of the chemical relationships existing between the substance whose equivalent weight is sought and one of the participants in the titration. In this example the equations reveal that

$$2Pb^{2+} \equiv 2CrO_4^{2-} \equiv Cr_2O_7^{2-} \equiv 6Fe^{2+} \equiv 6e$$

Thus the quantity of each substance associated with the transfer of 1 mole of electrons is

$$\frac{2 \text{ gfw } Pb^{2+}}{6} \equiv \frac{2 \text{ gfw } CrO_4^{2-}}{6} \equiv \frac{1 \text{ gfw } Cr_2O_7^{2-}}{6} \equiv \frac{6 \text{ gfw } Fe^{2+}}{6} \equiv \frac{6 \text{ moles } e}{6}$$

Let us consider one more example. The nitrogen in the organic compound $C_9H_9N_3$ can be determined by quantitative conversion to ammonia, followed by titration with a standard solution of acid. The reactions are

$$C_9H_9N_3 + reagent \rightarrow 3NH_3 + products$$

$$NH_3 + H^+ \rightarrow NH_4^+$$

Here titration involves a neutralization process; therefore the equivalent weight for the analyte must be based on the consumption or production of hydrogen ions. Because each molecule of the analyte yields three ammonia molecules, it can be considered responsible for the consumption of three hydrogen ions. The equivalent weight, then, is the formula weight of $C_9H_9N_3$ divided by 3. By the same reasoning each nitrogen is converted to one ammonia molecule which reacts with one hydrogen ion; the equivalent weight of nitrogen, N, is thus equal to its formula weight.

CONCENTRATION UNITS USED IN VOLUMETRIC CALCULATIONS

Some of the ways by which chemists express the concentrations of solutions were discussed in Chapter 2; these included formal concentration, molar concentration, and various types of percentage composition. We now need two additional terms, *titer* and *normality*, which are commonly used to define the concentration of solutions used in volumetric analysis.

Titer. *The titer of a solution is the weight of a substance that is chemically equivalent to 1 ml of that solution.* Thus a silver nitrate solution having a titer of 1.00 mg of chloride would contain just enough silver nitrate in each milliliter to react completely with that weight of chloride ion. The titer might also be expressed in terms of milligrams or grams of potassium chloride, barium chloride, sodium iodide, or any other compound that reacts with silver nitrate. The concentration of a reagent to be used for the routine analysis of many samples is advantageously expressed in terms of its titer.

EXAMPLE

What is the barium chloride titer of a $0.125\text{-}F$ $AgNO_3$ solution?

$$titer = \frac{0.125 \text{ mfw } AgNO_3}{ml \ AgNO_3} \times \frac{1 \text{ mfw } BaCl_2}{2 \text{ mfw } AgNO_3} \times \frac{208 \text{ mg } BaCl_2}{mfw \ BaCl_2} = \frac{13.0 \text{ mg } BaCl_2}{ml \ AgNO_3}$$

EXAMPLE

The label on a bottle of dilute silver nitrate states that the solution has a titer of 8.78 mg NaCl. What is the formal concentration of this solution?

$$F_{AgNO_3} = \frac{8.78 \text{ mg NaCl}}{ml \ AgNO_3} \times \frac{1}{58.4 \text{ mg NaCl/mfw}} \times \frac{1 \text{ mfw } AgNO_3}{mfw \ NaCl} = \frac{0.150 \text{ mfw}}{ml}$$

Normality. *The normality, N, of a solution expresses the number of milliequivalents of solute contained in 1 ml of solution* or the number of equivalents contained in 1 liter. Thus a 0.20-N silver nitrate solution contains 0.20 milliequivalent (meq) of this solute in each milliliter of solution or 0.20 equivalent (eq) per liter.

SOME IMPORTANT WEIGHT-VOLUME RELATIONSHIPS

The raw data from a volumetric analysis are ordinarily expressed in units of milliliters, grams, and normality. Volumetric calculations involve conversion of such information into units of milliequivalents followed by reconversion into the metric weight of the desired chemical species. Two relationships, based on the foregoing definitions, are used for these transformations. The first involves converting the weight of a compound from units of grams to those of milliequivalents. This transformation is accomplished by dividing the weight of the substance by its milliequivalent weight; that is,

$$\text{no. meq A} = \frac{\text{wt A, g}}{\text{meq wt A, g/meq}}$$

EXAMPLE

Calculate the number of milliequivalents of chlordane, $C_{10}H_6Cl_8$ (gfw = 410), in 0.500 g of the pure insecticide, assuming that all of the chlorine present is ultimately titrated with Ag^+.

$$\text{no. meq } C_{10}H_6Cl_8 = \frac{\text{wt } C_{10}H_6Cl_8 \text{ (g)}}{\text{meq wt } C_{10}H_6Cl_8 \text{ (g/meq)}}$$

Because each formula weight of chlordane provides eight Cl^- which react with eight Ag^+, we may write

$$\text{no. meq } C_{10}H_6Cl_8 = \frac{0.500 \text{ g } C_{10}H_6Cl_8}{0.410 \text{ g/8 meq}} = 9.76$$

The second relationship permits calculation of the number of milliequivalents of solute contained in a given volume, V, provided the normality of the solution is known. Because by definition the normality is equal to the number of milliequivalents in each milliliter, it follows that

$$\text{no. meq A} = V_A(\text{ml}) \times N_A(\text{meq/ml})$$

EXAMPLE

The number of milliequivalents involved in a titration that required 27.3 ml of 0.200-N $KMnO_4$ is given by

$$\text{no. meq } KMnO_4 = 27.3 \text{ ml} \times 0.200 \text{ meq/ml}$$

$$= 5.46$$

Further applications of these relationships are illustrated in the following examples.

EXAMPLE

What weight of primary standard $K_2Cr_2O_7$ (gfw = 294.2) is needed to prepare exactly 2 liters of 0.1200-N reagent by the direct method? Titrations with dichromate involve the half-reaction

$$Cr_2O_7^{2-} + 14H^+ + 6e \rightleftharpoons 2Cr^{3+} + 7H_2O$$

The number of milliequivalents of $K_2Cr_2O_7$ required is first calculated:

$$\text{no. meq } K_2Cr_2O_7 = ml_{K_2Cr_2O_7} \times N_{K_2Cr_2O_7}$$

$$= 2000 \text{ ml} \times 0.1200 \text{ meq/ml}$$

$$= 240.0$$

Conversion of a weight in milliequivalents to a weight in grams involves multiplication by the milliequivalent weight.

$$\text{wt } K_2Cr_2O_7 = 240.0 \text{ meq} \times \frac{294.2 \text{ g/fw}}{6 \text{ eq/fw}} \times \frac{1}{1000 \text{ meq/eq}}$$

$$= 11.768 = 11.77 \text{ g}$$

EXAMPLE

What volume of 0.100-N HCl can be produced by diluting 150 ml of 1.24-N acid?

The number of milliequivalents of HCl must be the same in the two solutions. Therefore

no. meq HCl in diluted solution = no. meq HCl in concentrated solution

ml diluted solution $\times 0.100$ meq/ml $= 150$ ml $\times 1.24$ meq/ml

$$\text{ml diluted solution} = \frac{150 \times 1.24}{0.100} = 1860$$

$$= 1.86 \times 10^3$$

Thus a 0.100-N HCl solution would be obtained by diluting 150 ml of 1.24-N acid to 1.86×10^3 ml.

A Fundamental Relationship between Quantities of Reacting Substances. By definition 1 eq wt of an acid contributes 1 mole of hydrogen ions to a reaction. Also, 1 eq wt of a base consumes 1 mole of these ions. It then follows that at the equivalence point in a neutralization titration, the number of equivalents (or milliequivalents) of acid and of base will always be numerically equal. Similarly, at the equivalence point in an oxidation-reduction titration, the number of milliequivalents of oxidizing and reducing agent must also be equal. An identical relationship holds for precipitation and complex formation titrations. To generalize we may state that *at the equivalence point in any titration, the number of milliequivalents of standard is exactly equal to the number of milliequivalents of the substance with which it has reacted.* Nearly all volumetric calculations are based on this relationship.

CALCULATION OF CONCENTRATION OF STANDARD SOLUTIONS

The normality of a standard solution is computed either from a standardization titration or from the data related to its actual preparation.

EXAMPLE

A $Ba(OH)_2$ solution was standardized by titration against 0.1280-N HCl, 31.76 ml of the base being required to neutralize 46.25 ml of the acid. Calculate the normality of the $Ba(OH)_2$ solution.

Provided that the end point in the titration corresponds to the equivalence point,

$$\text{no. meq } Ba(OH)_2 = \text{no. meq HCl}$$

$$ml_{Ba(OH)_2} \times N_{Ba(OH)_2} = ml_{HCl} \times N_{HCl}$$

$$31.76 \text{ ml} \times N_{Ba(OH)_2} = 46.25 \text{ ml} \times 0.1280 \text{ meq/ml}$$

$$N_{Ba(OH)_2} = \frac{46.25 \times 0.1280}{31.76} = 0.1864 \text{ meq/ml}$$

EXAMPLE

Calculate the normality of an iodine solution if 37.34 ml were required to titrate a 0.2040-g sample of primary standard As_2O_3 (gfw = 197.8). The reaction is

$$I_2 + H_2AsO_3^- + H_2O \rightarrow 2I^- + H_2AsO_4^- + 2H^+$$

At the equivalence point

$$\text{no. meq } I_2 = \text{no. meq } As_2O_3$$

The number of milliequivalents of I_2 can be computed from the volume and normality; the number of milliequivalents of As_2O_3 can be calculated from the weight taken for the titration. Thus we may write

$$ml_{I_2} \times N_{I_2} = \frac{\text{wt } As_2O_3}{\text{meq wt } As_2O_3}$$

In its reaction with iodine, arsenic loses two electrons and is thus oxidized from the +3 to the +5 state. Therefore a total change of 4 is associated with each As_2O_3 molecule, making its equivalent weight one-fourth its formula weight. Thus

$$37.34 \text{ ml} \times N_{I_2} = \frac{0.2040 \text{ g}}{0.1978 \text{ g}/4 \text{ meq}}$$

$$N_{I_2} = \frac{0.2040}{37.34 \times 0.04945} = 0.1105$$

CALCULATION OF RESULTS FROM TITRATION DATA

EXAMPLE

The organic matter in a 3.77-g sample of a mercuric ointment was decomposed with HNO_3. After dilution the Hg^{2+} was titrated with a 0.114-N solution of NH_4SCN; exactly 21.3 ml of the reagent were required. Calculate the percent Hg (gfw = 201) and the percent $Hg(NO_3)_2$ (gfw = 325) in the ointment.

This titration involves the formation of a stable neutral complex, $Hg(SCN)_2$; that is,

$$Hg^{2+} + 2SCN^- \rightarrow Hg(SCN)_2$$

At the equivalence point

$$\text{no. meq } Hg^{2+} = \text{no. meq } NH_4SCN$$

$$= 21.3 \text{ ml} \times 0.114 \text{ meq/ml}$$

Because this is a complex formation reaction the milliequivalent weight of the reacting cation, Hg(II), is one-half its milliformula weight. Therefore

$$\text{wt Hg} = 21.3 \text{ ml} \times 0.114 \frac{\text{meq}}{\text{ml}} \times \frac{0.201 \text{ g}}{2 \text{ meq}}$$

and

$$\% \text{ Hg} = \frac{21.3 \times 0.114 \times 0.201/2}{3.77} \times 100 = 6.47$$

The percentage of $Hg(NO_3)_2$ is calculated identically:

$$\text{no. meq } Hg(NO_3)_2 = \text{no. meq } NH_4SCN$$

and

$$\% \text{ Hg(NO}_3)_2 = \frac{21.3 \times 0.114 \times 0.325/2}{3.77} \times 100 = 10.5$$

EXAMPLE

A 0.804-g sample of an iron ore was dissolved in acid. The iron was then reduced to the +2 state and titrated with 47.2 ml of a 0.112-N $KMnO_4$

solution. Calculate the results of the analysis in terms of percent Fe (gfw = 55.8) as well as percent Fe_2O_3 (gfw = 160).

The titration involves oxidation of Fe^{2+} to Fe^{3+}:

$$5Fe^{2+} + MnO_4^- + 8H^+ \rightarrow 5Fe^{3+} + Mn^{2+} + 4H_2O$$

At the equivalence point

$$\text{no. meq Fe} = \text{no. meq KMnO}_4$$

$$= 47.2 \text{ ml} \times 0.112 \text{ meq/ml}$$

Because Fe^{2+} loses one electron in this reaction, its milliequivalent weight is identical to its milliformula weight, and

$$g \text{ Fe} = 47.2 \text{ ml} \times 0.112 \frac{\text{meq}}{\text{ml}} \times \frac{55.8 \text{ g}}{1000 \text{ meq}}$$

Therefore

$$\% \text{ Fe} = \frac{47.2 \times 0.112 \times 55.8/1000}{0.804} \times 100 = 36.7$$

The percentage of Fe_2O_3 can be obtained in essentially the same way; thus at the equivalence point

$$\text{no. meq Fe}_2O_3 = \text{no. meq KMnO}_4$$

and by the same arguments

$$\% \text{ Fe}_2O_3 = \frac{47.2 \times 0.112 \times 160/2000}{0.804} \times 100$$

$$= 52.6$$

EXAMPLE

A 0.475-g sample containing $(NH_4)_2SO_4$ was dissolved in water and made alkaline with KOH. The liberated NH_3 was distilled into exactly 50.0 ml of 0.100-N HCl. The excess HCl was back-titrated with 11.1 ml of 0.121-N NaOH. Calculate the percent NH_3 (gfw = 17.0), as well as the percent $(NH_4)_2SO_4$ (gfw = 132) in the sample.

At the equivalence point the number of milliequivalents of acid and base are equal. In this titration, however, two bases are involved: NaOH and NH_3. Thus

$$\text{no. meq HCl} = \text{no. meq NH}_3 + \text{no. meq NaOH}$$

After rearranging,

$$\text{no. meq NH}_3 = \text{no. meq HCl} - \text{no. meq NaOH}$$

$$= (50.0 \times 0.100 - 11.1 \times 0.121)$$

Thus

$$\% \text{ NH}_3 = \frac{(50.0 \times 0.100 - 11.1 \times 0.121) \times 17.0/1000}{0.475} \times 100$$

$$= 13.1$$

The number of milliequivalents of $(NH_4)_2SO_4$ is the same as the number of milliequivalents of NH_3 by definition. Therefore

$$\% \text{ (NH}_4)_2\text{SO}_4 = \frac{(50.0 \times 0.100 - 11.1 \times 0.121) \times 132/2000}{0.475} \times 100$$

$$= 50.8$$

Here the milliequivalent weight of $(NH_4)_2SO_4$ is one-half the milliformula weight because

$$(NH_4)_2SO_4 \equiv 2NH_3 \equiv 2H^+$$

EXAMPLE

A standard 0.120-N $AgNO_3$ solution is to be used for the routine determination of salt in brines.

1. Express the titer of this solution as mg NaCl/ml.

$$\text{NaCl titer} = \frac{0.120 \text{ meq AgNO}_3}{\text{ml AgNO}_3} \times \frac{1.00 \text{ meq NaCl}}{1.00 \text{ meq AgNO}_3} \times \frac{58.4 \text{ mg NaCl}}{\text{meq NaCl}}$$

$$= \frac{7.01 \text{ mg NaCl}}{\text{ml AgNO}_3}$$

2. Calculate the milligrams of NaCl in each milliliter of brine if titration of a 50.0-ml sample required 21.4 ml of the $AgNO_3$ solution.

$$\frac{\text{mg NaCl}}{\text{ml brine}} = \frac{21.4 \text{ ml AgNO}_3}{50.0 \text{ ml brine}} \times \frac{7.01 \text{ mg NaCl}}{\text{ml AgNO}_3} = 3.00$$

PROBLEMS

1. Calculate the p-functions for each ion in
   *(a) a solution that is 0.0200 F in KI.
   (b) a solution that is 0.0200 F in $BaCl_2$.
   *(c) a solution that is 2.6×10^{-3} F in $Ba(OH)_2$.
   (d) a solution that is 0.060 F in HCl and 0.030 F in $BaCl_2$.
   *(e) a solution that is 4.6×10^{-4} F in $CaCl_2$ and 5.3×10^{-4} F in $Cd(NO_3)_2$.

2. Calculate the pPb and/or pIO_3 for a solution that is
   *(a) saturated with $Pb(IO_3)_2$ ($K_{sp} = 3.2 \times 10^{-13}$).
   (b) 0.0325 F in $Mg(IO_3)_2$.

$\checkmark$*(c) formed by mixing 25.0 ml of 0.200-F Pb^{2+} with 25.0 ml of 0.300-F $NaIO_3$.

(d) formed by mixing 25.0 ml of 0.200-F Pb^{2+} with 25.0 ml of 0.300-F $Mg(IO_3)_2$.

*(e) 1.2 F in $Pb(NO_3)_2$.

(f) 3.2 F in $LiIO_3$.

3. Convert the following p-functions into molar concentrations.

   *(a) pH = 8.32   *(e) pLi = -0.267

   (b) pOH = 0.116   (f) pNO_3 = 8.77

   *(c) pBr = 0.027   *(g) pMn = 0.0036

   (d) pCa = 13.18   (h) pCl = 1.190

*4. Classify the following reactions as to type, and indicate the equivalent weight for each of the substances listed on the right as a fraction or multiple of its gram formula weight.

   (a) $Fe^{3+} + 6Br^- \rightleftharpoons FeBr_6^{3-}$ $Fe(OH)_2Cl$, Fe_2O_3, $NaBr$, $C_6H_2Br_4$

   (b) $I_2 + H_2S \rightleftharpoons S + 2I^- + 2H^+$ I_2, H_2S, As_2S_5, KI

   (c) $Mg(OH)_2(s) + 2H_3O^+ \rightleftharpoons Mg^{2+} + 4H_2O$ H_2SO_4, $Mg(OH)_2$, $Mg_2P_2O_7$, MgO

   (d) $2Ce^{4+} + H_2AsO_3^- + H_2O \rightleftharpoons$ Ce^{3+}, NaH_2AsO_3, As_2O_5, $Ce_2(SO_4)_3$
       $2Ce^{3+} + H_2AsO_4^- + 2H^+$

5. Classify the following reactions as to type, and indicate the equivalent weight for each of the substances listed on the right as a fraction or multiple of its gram formula weight.

   (a) $Ag^+ + 2S_2O_3^{2-} \rightleftharpoons Ag(S_2O_3)_2^{3-}$ $Na_2S_2O_3$, $Ag_2(CN)_2$, $AgNO_3$, S

   (b) $C_6H_5NH_3^+ + OH^- \rightleftharpoons C_6H_5NH_2 + H_2O$ $Ba(OH)_2$, $C_6H_5NH_3Cl$, N_2,
                                                     $C_6H_5NH_2$

   (c) $Hg^{2+} + 2SCN^- \rightleftharpoons Hg(SCN)_2(aq)$ $Hg(NO_3)_2$, $BiOSCN$, $Ba(SCN)_2$, S

   (d) $Br_2 + C_2O_4^{2-} \rightleftharpoons 2CO_2 + 2Br^-$ Br_2, Br_3^-, $Na_2C_2O_4$, CO_2

*6. Classify the following reactions as to type, and indicate the equivalent weight for each of the substances listed on the right as a fraction or multiple of its gram formula weight.

   (a) $Ce^{3+} + 3IO_3^- \rightleftharpoons Ce(IO_3)_3(s)$ $Ce_2(SO_4)_3$, $NaIO_3$, I_2O_5, $AlOH(IO_3)_2$

   (b) $H_3AsO_4 + 2OH^- \rightleftharpoons HAsO_4^{2-} + 2H_2O$ As_2O_3, H_3AsO_4, KOH, As

   (c) $Cr_2O_7^{2-} + 6Fe^{2+} + 14H^+ \rightleftharpoons$ $K_2Cr_2O_7$, Fe_3O_4, Na_2CrO_4, Fe
       $2Cr^{3+} + 6Fe^{3+} + 7H_2O$

   (d) $Ag^+ + 2NH_3 \rightleftharpoons Ag(NH_3)_2^+$ NH_3, Ag_2SO_4, $(NH_4)_2SO_4$, N_2

7. Classify the following reactions as to type, and indicate the equivalent weight for each of the substances listed on the right as a fraction or multiple of its gram formula weight.

   (a) $Ni^{2+} + 4CN^- \rightleftharpoons Ni(CN)_4^{2-}$ KCN, $Ba(CN)_2$, $NiCl_2$, Ni_3O_4

   (b) $H_3AsO_4 + OH^- \rightleftharpoons H_2AsO_4^- + H_2O$ As_2O_3, H_3AsO_4, KOH, As

   (c) $IO_3^- + 3H_2SeO_3 \rightleftharpoons I^- + 3H_2SeO_4$ Se, $Al_2(SeO_3)_3$, KIO_3, I_2O_5

   (d) $BiO^+ + Cl^- \rightleftharpoons BiOCl(s)$ BiO^+, Bi^{3+}, $BaCl_2 \cdot 2H_2O$, HCl

*8. A solution contains 1.76 g of H_3AsO_4 in 500 ml. Calculate the concentration of this solution in terms of its

(a) formality. $0.0248 F$

(b) normality in an acid-base titration in which one proton reacts. $0.0249 N$

(c) normality in an acid-base titration in which two protons react. $0.0496 N$

(d) normality as an oxidizing agent for I^-. $0.640 N$

$$H_3AsO_4 + 3I^- + 2H^+ \rightarrow H_3AsO_3 + I_3^- + H_2O$$

(e) normality for the titration of Ag^+ where the reaction product is $Ag_3AsO_4(s)$. $0.0744 N$

(f) Ag^+ titer, based upon formation of $Ag_3AsO_4(s)$. 8.14 mg/ml

9. A solution contains 3.02 g of KCN in 600 ml. Calculate

(a) its formal concentration.

(b) its normality as a base.

$$CN^- + H_3O^+ \rightarrow HCN + H_2O$$

(c) its normality for the determination of Cd^{2+}.

$$Cd^{2+} + 4CN^- \rightarrow Cd(CN)_4^{2-}$$

(d) its normality for the determination of Ag^+.

$$Ag^+ + CN^- \rightleftharpoons AgCN(s)$$

(e) its normality in a titration with basic $KMnO_4$.

$$CN^- + 2MnO_4^- + 2OH^- \rightarrow 2MnO_4^{2-} + CNO^- + H_2O$$

(f) its Cd^{2+} titer [see (c)].

*10. A solution is 0.160 F with respect to permanganate. What is its normality as an oxidizing agent

(a) in a strongly alkaline environment where MnO_4^{2-} is the reduction product? $0.160 N$

(b) in a strongly acidic solution where Mn^{2+} is produced? $0.800 N$

(c) in a neutral solution where $MnO_2(s)$ is produced? $0.480 N$

(d) in the presence of pyrophosphate ion ($H_2P_2O_7^{2-}$) where the product is $0.640 N$ $Mn(H_2P_2O_7)_3^{3-}$? (Mn is in the +3 oxidation state.)

*11. A solution contains 1.63 g of $K_4Fe(CN)_6 \cdot 3H_2O$ in 400 ml of solution. Calculate

(a) its formal concentration. $9.66 \times 10^{-3} F$

(b) its normality in the standardization of Ce^{4+}. $9.66 \times 10^{-3} N$

$$Ce^{4+} + Fe(CN)_6^{4-} \rightleftharpoons Ce^{3+} + Fe(CN)_6^{3-}$$

(c) its normality for the determination of Zn^{2+}. $2.90 \times 10^{-2} N$

$$2Fe(CN)_6^{4-} + 3Zn^{2+} + 2K^+ \rightarrow K_2Zn_3[Fe(CN)_6]_2(s)$$

(d) its $Zn_2P_2O_7$ titer [for reaction see (c)]. 2.21 mg $Zn_2P_2O_7$/ml

12. Potassium hydrogen iodate, $KH(IO_3)_2$, is available in high purity and can be employed for the standardization of several reagents. A solution containing

1.02 g of $KH(IO_3)_2$/250 ml was prepared. Calculate

(a) its formal concentration. 0.25

(b) its normality as an acid.

(c) its normality as an oxidizing reagent where I^- is the reaction product.

(d) its normality as a reducing agent where IO_4^- is the reaction product.

(e) its normality as an oxidant in strong HCl solution where its reaction product is ICl (here I is in the +1 state).

(f) its KSCN titer.

$$2SCN^- + 3IO_3^- + 3Cl^- + 4H^+ \rightarrow 2HCN + 2SO_4^{2-} + 3ICl + H_2O$$

*13. How many milliequivalents of solute are contained in 20.0 ml of

(a) 0.0125-N $Ba(OH)_2$? 0.250

(b) 0.0125-F $Ba(OH)_2$? 0.500

(c) 0.0125-F KOH? 0.250

(d) 0.0125-N KCN 0.250 $[Ni^{2+} + 4CN^- \rightleftharpoons Ni(CN)_4^{2-}]$?

(e) 0.0125-F KCN 0.125 $[Ni^{2+} + 4CN^- \rightleftharpoons Ni(CN)_4^{2-}]$?

(f) 0.0125-F $K_2Cr_2O_7$ 1.50 $(Cr_2O_7^{2-} + 14H^+ + 6e \rightleftharpoons 2Cr^{3+} + 7H_2O)$?

14. How many milliformula weights of solute are contained in 10.0 ml of

(a) 0.160-F H_2SO_4?

(b) 0.160-N H_2SO_4?

(c) 0.160-N NaOH?

(d) 0.160-N $Na_2S_2O_3$ $[Cd^{2+} + 3S_2O_3^{2-} \rightarrow Cd(S_2O_3)_3^{4-}]$?

(e) 0.160-N $Na_2S_2O_3$ $(2S_2O_3^{2-} + I_2 \rightleftharpoons S_4O_6^{2-} + 2I^-)$?

(f) a solution of $AgNO_3$ having a titer of 4.64 mg $AlCl_3$?

*15. How many grams of solute are contained in

(a) 10.6 ml of 0.150-N KIO_3 (reaction product I_2)? $0.0680\ g$

(b) 2.50 liters of 0.0750-N H_2SO_4? 9.2

(c) 27.2 ml of 0.200-N $Hg(NO_3)_2$ 0.484 $[Hg^{2+} + 2Br^- \rightleftharpoons HgBr_2(aq)]$? $1\,mol = 2\,e.q.$

(d) 3.50 liters of 0.100-N $KBrO_3$ 9.74 $(BrO_3^- + 3H_3AsO_3 \rightleftharpoons Br^- + 3H_3AsO_4)$? $1\,mol = 6\,e.q.$

(e) 300 ml of 0.0150-N $Ba(OH)_2$? 0.385

(f) 300 ml of 0.0500-F $Ba(OH)_2$? 0.770

16. How many milligrams of solute are contained in

(a) 2.00 liters of 0.150-N $KBrO_3$ (reaction product Br^-)?

(b) 37.5 ml of 0.0800-N H_3PO_4 (reaction product HPO_4^{2-})?

(c) 25.0 ml of 0.120-N KBr (reaction product Br_2)?

(d) 25.0 ml of 0.120-N KBr (reaction product $HgBr_4^{2-}$)?

(e) 25.0 ml of 0.120-N KBr (reaction product AgBr)?

*17. Describe the preparation of 2.00 liters of 0.0400-N H_2SO_4 from

(a) 3.00-F H_2SO_4. $13.3\,ml$ $N_1V_1 = N_2V_2$

(b) 11.0% (w/w) H_2SO_4 solution. $71.4\,g$

(c) 0.222-N H_2SO_4. $360\,ml$

(d) the concentrated reagent [95% H_2SO_4 (w/w), $d = 1.84$ g/ml]. $2.2\,ml$

18. Describe the preparation of 750 ml of 0.0800-N KOH from
    (a) a 6.00-F solution.
    (b) a 8.92% (w/w) KOH solution.
    (c) a concentrated solution [50.0% KOH (w/w), $d = 1.505$ g/ml].
    (d) a 0.313-N solution.

*19. Describe how 500.0 ml of the following solutions should be prepared for the analytical application shown.

7.29 g (a) 0.150-N KSCN from the primary standard reagent $1 e.q \ KSCN = 1 \ mol \ KSCN$
     97 g KSCN = 1 mol KSCN.
$$2SCN^- + Hg^{2+} \rightarrow Hg(SCN)_2(aq)$$

3.71 g (b) 0.150-N H$_3$AsO$_3$ from primary standard grade As$_2$O$_3$ 1 mol H$_3$AsO$_3$ = 3 e.q.
     g fw = 126
$$H_3AsO_3 + I_2 + H_2O \rightarrow H_3AsO_4 + 2I^- + 2H^+$$

18.1 g (c) 0.200-N HCl from constant-boiling HCl containing 20.2 g HCl per 100 g solution
$$H^+ + OH^- \rightarrow H_2O \quad 1 mol = 1 e.q. \ HCl$$

     g fw = 407   g fw = 56
5.49 g (d) a K$_2$Cr$_2$O$_7$ solution having an iron titer of 12.50 mg/ml from the primary standard reagent
$$Cr_2O_7^{2-} + 6Fe^{2+} + 14H^+ \rightarrow 2Cr^{3+} + 7H_2O + 6Fe^{2+}$$

3.75 ml (e) 0.135-N K$_2$Cr$_2$O$_7$ from a 3.00-F solution [see (d) for reaction]   1 mol K$_2$Cr$_2$O$_7$ 6 e.q

20. Describe how 2.00 liters of the following solutions should be prepared for the analytical applications shown.
    (a) 0.0750-N KCN from the solid
$$2CN^- + Ag^+ \rightarrow Ag(CN)_2^-$$
    (b) 0.0800-N H$_3$PO$_4$ from the concentrated reagent that has a density of 1.69 g/ml and is 85.0% H$_3$PO$_4$ by weight
$$H_3PO_4 + OH^- \rightarrow H_2PO_4^- + H_2O$$
    (c) 0.0825-N Na$_2$B$_4$O$_7$ from pure Na$_2$B$_4$O$_7 \cdot 10H_2O$
$$B_4O_7^{2-} + 2H^+ + 5H_2O \rightarrow 4H_3BO_3$$
    (d) a KBrO$_3$ solution with a hydrazine titer of 1.20 mg/ml, from pure KBrO$_3$
$$3N_2H_4 + 2BrO_3^- \rightarrow 3N_2 + 2Br^- + 6H_2O$$

0.0810 N *21. A solution of perchloric acid was standardized by dissolving 0.235 g of primary standard grade HgO in a solution containing an excess of KBr.
$$HgO(s) + 4Br^- + H_2O \rightarrow HgBr_4^{2-} + 2OH^-$$
Calculate the normality of the HClO$_4$ if the liberated OH$^-$ required 26.8 ml of the acid.

22. A 0.412-g sample of primary standard grade sodium carbonate required 39.2 ml

of a sulfuric acid solution to reach the end point for the reaction

$$CO_3^{2-} + 2H^+ \rightarrow H_2O + CO_2(g)$$

What is the normality of the H_2SO_4? What is its formality?

*23. What is the normality of an $AgNO_3$ solution that has a titer of 4.64 mg $AlCl_3$/ml?

24. What is the normality of a solution of $KMnO_4$ that has a titer of 7.25 mg Fe_3O_4/ml? Reaction:

$$MnO_4^- + 5Fe^{2+} + 8H^+ \rightarrow Mn^{2+} + 5Fe^{3+} + 4H_2O$$

*25. A 0.612-g sample of primary standard Na_2CO_3 was decomposed with a 50.0-ml aliquot of dilute sulfuric acid. The resulting solution was boiled to remove CO_2, and the excess H_2SO_4 was back-titrated with 9.65 ml of a NaOH solution. In a separate experiment 25.0 ml of the NaOH were found to neutralize 26.3 ml of H_2SO_4. Calculate the normality of the H_2SO_4 and the NaOH.

26. A 0.339-g sample of sodium sulfate, which had an assay of 95.5% Na_2SO_4, was titrated with a solution of $BaCl_2$ [$Ba^{2+} + SO_4^{2-} \rightarrow BaSO_4(s)$]. What was the normality of the $BaCl_2$ solution if the end point was observed when 33.7 ml of the reagent were added? What was the SO_4^{2-} titer of the $BaCl_2$ solution?

*27. A 0.261-g sample of a sulfur-containing organic compound was burned in a stream of O_2, and the resulting SO_2 was absorbed in a solution of H_2O_2 ($H_2O_2 + SO_2 \rightarrow H_2SO_4$). The acid produced was titrated with 24.6 ml of 0.137-N KOH. Calculate the percent S in the compound.

28. A 0.362-g specimen containing the mineral hausmannite, Mn_3O_4, was analyzed by treatment that converted the Mn quantitatively to MnO_4^-. The resulting solution required 36.3 ml of a 0.113-N Fe^{2+} solution to reduce the MnO_4^- to Mn^{2+}. Calculate the percent Mn_3O_4 in the sample.

*29. The CO concentration of a gas was determined by passing a 2.60-liter sample through a heated tube containing I_2O_5.

$$5CO + I_2O_5 \rightarrow 5CO_2 + I_2$$

The I_2 formed was sublimed into 26.0 ml of 0.0101-N $Na_2S_2O_3$.

$$I_2 + 2S_2O_3^{2-} \rightarrow 2I^- + S_4O_6^{2-}$$

The excess $Na_2S_2O_3$ required 6.72 ml of 0.0122-N I_2 solution. Calculate the milligrams of CO per liter of sample.

30. The arsenic in a 1.76-g sample of a pesticide was converted to H_3AsO_4 by suitable treatment. The acid was then neutralized, and exactly 50.0 ml of 0.125-N $AgNO_3$ were added to precipitate the As quantitatively as Ag_3AsO_4. The excess Ag^+ in the filtrate and washings from the precipitate was titrated with 9.65 ml of 0.100-N KSCN. Calculate the percent As_2O_3 in the sample.

*31. The ethyl acetate concentration in an alcoholic solution of the ester was determined after diluting a 15.0-ml sample to exactly 200 ml. A 25.0-ml portion of the diluted solution was refluxed with 30.0 ml of 0.0575-N KOH.

$$CH_3COOC_2H_5 + OH^- \rightarrow CH_3OO^- + C_2H_5OH$$

After cooling, the excess OH$^-$ was back-titrated with 5.62 ml of 0.0404-N HCl. Calculate the grams of ethyl acetate per 100 ml of the original sample.

32. The thiourea in a 1.25-g organic sample was extracted into a dilute H_2SO_4 solution and titrated with 33.1 ml of 0.00862-N Hg^{2+}.

$$4(NH_2)_2CS + Hg^{2+} \rightarrow [(NH_2)_2CS]_4Hg^{2+}$$

Calculate the percent $(NH_2)_2CS$ in the sample.

precipitation titrations

Volumetric methods based upon the formation of sparingly soluble silver salts are among the oldest known; these procedures were and still are routinely employed for the analysis for silver, as well as for the determination of such ions as chloride, bromide, iodide, and thiocyanate. Volumetric precipitation methods that do not involve silver as one of the reactants are relatively limited in scope.

Titration Curves for Precipitation Reactions

Titration curves are useful for deducing the properties required of an indicator as well as the titration error that its use is likely to cause. The following example demonstrates how precipitation titration curves are derived from solubility product data. The student may find it useful to review the material in Chapter 5 on solubility of precipitates before studying this example in detail.

EXAMPLE

Derive a curve for the titration of 50.00 ml of 0.00500-F NaBr with 0.01000-F AgNO₃.

We shall derive data for both pBr and pAg, although only one of the two (the one that affects the indicator behavior) is needed.

Initial point. At the outset the solution is 0.00500 F in Br⁻ and 0.000 F in Ag⁺. Thus pBr = $-\log(5.00 \times 10^{-3}) = 2.301 \cong 2.30$. The pAg is indeterminate.

After addition of 5.00 ml reagent. Here the bromide ion concentration will have been decreased by precipitation as well as by dilution. Thus

$$F_{NaBr} = \frac{50.00 \times 0.00500 - 5.00 \times 0.01000}{50.00 + 5.00} = 3.64 \times 10^{-3}$$

The first term in the numerator is the number of milliformula weights of NaBr present originally; the second term represents the number of milliformula weights of AgNO₃ added and hence the number of milliformula weights of AgBr produced. Account is taken of the volume change in the denominator.

Bromide ions in the solution arise from both the unreacted NaBr and the slight solubility of AgBr. Thus the *molar* concentration of Br⁻ is larger than the residual concentration of NaBr by an amount equal to the formal solubility of the precipitate. That is,

$$[Br^-] = 3.64 \times 10^{-3} + [Ag^+]$$

The second term on the right accounts for the contribution of AgBr to the molar bromide concentration because one Ag⁺ ion is formed for each Br⁻ ion from this source. Unless the concentration of NaBr is very small, this second term can be neglected. That is,

$$[Ag^+] << 3.64 \times 10^{-3}$$

and

$$[Br^-] \cong 3.64 \times 10^{-3}$$

$$pBr = -\log(3.64 \times 10^{-3}) = 2.439 = 2.44$$

A convenient way to find pAg is to take the negative logarithm of the solubility product expression for AgBr. That is,

$$-\log([Ag^+][Br^-]) = -\log K_{sp} = -\log(5.2 \times 10^{-13})$$

$$-\log[Ag^+] - \log[Br^-] = -\log K_{sp} = 12.28$$

or

$$pAg + pBr = pK_{sp} = 12.28$$

$$pAg = 12.28 - 2.44 = 9.84$$

Other points up to chemical equivalence can be derived in this same way.

Equivalence point. Here neither NaBr nor $AgNO_3$ is in excess; necessarily, then,

$$[Ag^+] = [Br^-]$$

Substitution into the solubility product expression yields

$$[Ag^+] = [Br^-] = \sqrt{5.2 \times 10^{-13}} = 7.21 \times 10^{-7}$$

$$pAg = pBr = -\log(7.21 \times 10^{-7}) = 6.14$$

After addition of 25.10 ml reagent. The solution now contains an excess of $AgNO_3$; thus

$$F_{AgNO_3} = \frac{25.10 \times 0.01000 - 50.00 \times 0.00500}{75.10} = 1.33 \times 10^{-5}$$

and the molar concentration of silver ion is

$$[Ag^+] = 1.33 \times 10^{-5} + [Br^-] \cong 1.33 \times 10^{-5}$$

Here the second term in the right-hand side of the equation accounts for Ag^+ ions resulting from the slight solubility of AgBr; it can ordinarily be neglected.

$$pAg = -\log(1.33 \times 10^{-5}) = 4.876 = 4.88$$

$$pBr = 12.28 - 4.88 = 7.40$$

Additional points defining the titration curve beyond the equivalence point can be obtained in an analogous way.

SIGNIFICANT FIGURES IN TITRATION-CURVE CALCULATIONS

In deriving titration curves, concentration data associated with the equivalence point region are often of low precision because they are based upon small differences between large numbers. For example, in the calculation for F_{AgNO_3}, following the introduction of 25.10 ml of 0.01000-F $AgNO_3$, the numerator $(0.2510 - 0.2500 = 0.0010)$ contains only two significant figures; at best then, F_{AgNO_3} is known to two significant figures as well. To minimize the rounding error, however, we retained three digits (1.33×10^{-5}) in this calculation and rounded after calculating pAg.

In rounding the calculated value for pAg it is important to recall (p. 79) that it is the *mantissa only of a logarithm (that is, the number to the right of the decimal point) that should be rounded to the proper number of significant figures* because the characteristic serves merely to locate the decimal point. Thus in the example under consideration we rounded pAg to two figures to the right of the decimal point; that is, pAg = 4.88.

Note that we would have been justified in carrying an additional significant figure for the points well away from the equivalence point. For example, the initial pBr could have been reported as 2.301. Nothing is gained, however, because it is the equivalence point region that is of primary interest. Thus here, and in later derivations of titration curves, we will round p-functions to two places to the right of the decimal. This limited precision suffices because the changes in p-functions are generally great enough so that they are not obscured by the uncertainty in the equivalence point data.

FACTORS INFLUENCING THE SHARPNESS OF END POINTS

A sharp and easily located end point is observed when small additions of reagent cause large changes in p-function. It is therefore of interest to examine those variables that influence the magnitude of such changes during a titration.

Reagent Concentration. Table 8-1 lists pAg and pBr data, computed by the methods illustrated, for three concentrations of bromide and silver ion. The pAg data are plotted in Figure 8-1 and clearly show the effect of concentration on titration curves. It is apparent that an increase in analyte

TABLE 8-1
Changes in pAg and pBr during Titration with Solutions of Different Concentration

VOLUME AgNO₃, ml	50.0 ml OF 0.0500-F Br⁻ WITH 0.1000-F AgNO₃		50.0 ml OF 0.00500-F Br⁻ WITH 0.01000-F AgNO₃		50.0 ml OF 0.000500-F Br⁻ WITH 0.001000-F AgNO₃	
	pAg	pBr	pAg	pBr	pAg	pBr
0.00	—	1.30	—	2.30	—	3.30
10.00	10.68	1.60	9.68	2.60	8.68	3.60
20.00	10.13	2.15	9.13	3.15	8.13	4.15
23.00	9.72	2.56	8.72	3.56	7.72	4.56
24.90	8.41	3.87	7.41	4.87	6.50	5.78[a]
24.95	8.10	4.18	7.10	5.18	6.33	5.95[a]
25.00	6.14	6.14	6.14	6.14	6.14	6.14
25.05	4.18	8.10	5.18	7.10	5.95	6.33[b]
25.10	3.88	8.40	4.88	7.40	5.78	6.50[b]
27.00	2.58	9.70	3.58	8.70	4.58	7.70
30.00	2.20	10.08	3.20	9.08	4.20	8.08

[a] Here the approximation that $[Ag^+] << F_{NaBr}$ could not be used.

[b] Here the approximation that $[Br^-] << F_{AgNO_3}$ could not be used.

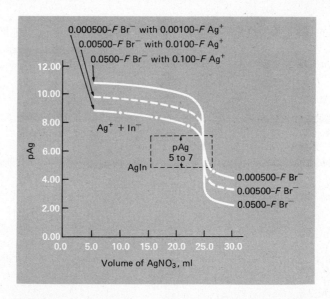

FIGURE 8-1   The effect of reagent concentration on titration
curves. 50.00 ml of NaBr are titrated for each curve.

and reagent concentration enhances the change in pAg in the equivalence
point region; an analogous effect is observed when pBr is plotted rather
than pAg.

These effects have practical significance for the titration of
bromide ion. If the analyte concentration is sufficient to permit the use of a
silver nitrate solution that is $0.1\ F$ or stronger, easily detected end points
are observed, and the titration error is minimal. On the other hand, with
solutions that are $0.001\ F$ or less the change in pAg or pBr is so small that
end point detection becomes difficult; a large titration error is thus to be
expected.

As will be shown later, the influence of reagent concentration upon
end point sharpness applies not only to precipitation titrations but to
titrations based upon other reaction types as well.

Completeness of Reaction.   Figure 8-2 shows how the solubility of the
species produced influences curves for titrations in which $0.1\text{-}F$ silver
nitrate serves as the reagent. Clearly the greatest change in pAg occurs in
the titration of iodide ion which, of all the anions considered, forms the
least soluble silver salt and hence reacts most completely with silver ion.
The smallest change in pAg is observed for the reaction that is least
favorable—that is, in the titration of bromate ion. Reactions that produce
silver salts with solubilities intermediate between these extremes yield
titration curves with end point changes that are also intermediate in

FIGURE 8-2 The effect of completeness of reaction on titration curves. 50.00 ml of a 0.0500-F solution of the anion are titrated with 0.1000-F AgNO$_3$ for each curve.

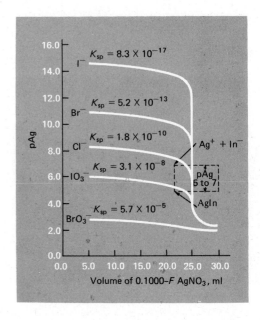

magnitude. Again, we shall see that this effect is common to all reaction types.

TITRATION CURVES FOR MIXTURES

The methods developed in the previous section can be extended to mixtures that form precipitates of differing solubilities. To illustrate, consider the titration of a 50.00-ml solution that is 0.0800 F in iodide ion and 0.1000 F in chloride with 0.2000-F silver nitrate.

Because silver iodide has a much smaller solubility than silver chloride, the initial additions of the reagent will result in formation of the iodide exclusively. Here the titration curve should be similar to that for iodide shown in Figure 8-2. It is of interest, then, to determine the extent to which iodide is precipitated before appreciable formation of silver chloride takes place.

With the first appearance of silver chloride the solubility product expressions for both precipitates are satisfied; division of one by the other provides a useful relationship.

$$\frac{\cancel{[Ag^+]}[I^-]}{\cancel{[Ag^+]}[Cl^-]} = \frac{8.3 \times 10^{-17}}{1.82 \times 10^{-10}} = 4.56 \times 10^{-7}$$

or

$$[I^-] = 4.56 \times 10^{-7} \, [Cl^-]$$

It is apparent that the iodide concentration will have been reduced to a minuscule fraction of the chloride ion concentration prior to the onset of precipitation by silver chloride. That is, formation of silver chloride will not occur until very near the equivalence point for iodide or after the addition of, for all practical purposes, 20 ml of reagent in this titration. Here the chloride ion concentration, because of dilution, will be approximately

$$F_{Cl^-} \cong [Cl^-] = \frac{50.00 \times 0.1000}{50.00 + 20.00} = 0.0714$$

and

$$[I^-] = 4.56 \times 10^{-7} \times 0.0714 = 3.26 \times 10^{-8}$$

The percentage of iodide unprecipitated at this point can be calculated as follows:

$$\text{original no. mfw } I^- = 50.00 \times 0.0800 = 4.00$$

$$\% \text{ } I^- \text{ unprecipitated} = \frac{3.26 \times 10^{-8} \times 70.00 \times 100}{4.00} = 5.7 \times 10^{-5}$$

Thus, to within about 6×10^{-5} percent of the equivalence point for iodide, no silver chloride should form, and the titration curve should be indistinguishable from that for the iodide alone. The first half of the titration curve shown by the solid line in Figure 8-3 was derived on this basis.

When chloride ion begins to precipitate, the rapid decrease in pAg is terminated abruptly. The pAg is most conveniently calculated from the

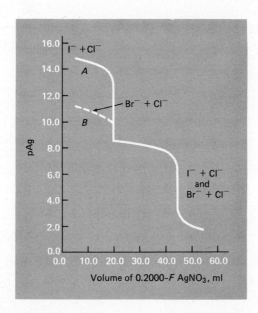

FIGURE 8-3 Curves for the titration of halide mixtures with 0.2000-F AgNO$_3$. Curve A: 50.00 ml of a solution that is 0.0800 F in I$^-$ and 0.1000 F in Cl$^-$. Curve B: 50.00 of a solution that is 0.0800 F in Br$^-$ and 0.1000 F in Cl$^-$.

solubility product constant for silver chloride:

$$[Cl^-] \cong 0.0714$$

$$[Ag^+] = \frac{1.82 \times 10^{-10}}{0.0714} = 2.55 \times 10^{-9}$$

$$pAg = -\log(2.55 \times 10^{-9}) = 8.59$$

Further additions of silver nitrate decrease the chloride ion concentration, and the curve then becomes that for the titration of chloride by itself. For example, after 25.00 ml of reagent have been added,

$$F_{Cl^-} \cong [Cl^-] = \frac{50.00 \times 0.1000 + 50.00 \times 0.0800 - 25.00 \times 0.2000}{75.00}$$

Here the first two terms in the numerator give the number of milliformula weights of chloride and iodide, respectively, and the third is the number of milliformula weights of titrant added. Thus

$$[Cl^-] = 0.0533$$

$$[Ag^+] = \frac{1.82 \times 10^{-10}}{0.0533} = 3.41 \times 10^{-9}$$

$$pAg = 8.47$$

The remainder of the curve can be derived in the same way as a curve for chloride by itself.

Curve A in Figure 8-3 is the titration curve for the chloride-iodide mixture just considered; it is a composite of the individual curves for the two anionic species. Two equivalence points are evident. It is to be expected that the change associated with the first equivalence point will become less distinct as the solubilities of the two precipitates approach one another. Curve B demonstrates this effect for a mixture of bromide and chloride ions. In this titration the initial pAg values are lower because solubility of silver bromide is greater than that of silver iodide. Beyond the first equivalence point, however, the two titration curves become identical.

It is possible to obtain experimental curves similar to those shown in Figure 8-3 by measuring the potential of a silver electrode immersed in the solution; this technique, which is discussed in Chapter 14, permits the analysis of the individual components in mixtures.

CHEMICAL INDICATORS FOR PRECIPITATION TITRATIONS

A chemical indicator produces a visually detectable change—usually of color or turbidity—in the solution. The indicator functions by reacting competitively with one of the reactants or products of the titration. Thus in the titration of species A with a reagent B and in the presence of an indicator (In) that can react with B, chemical description of the solution throughout the titration will be

$$A + B \rightleftharpoons AB$$

$$In + B \rightleftharpoons InB$$

For indicator action it is necessary, of course, that InB impart a significantly different appearance to the solution than does In. In addition, the amount of InB required for an observable change should be so small that no appreciable consumption of B occurs when InB is formed. Finally, the equilibrium constant for the indicator reaction must be such that the ratio [InB]/[In] is shifted from a small to a large value as a consequence of the change in [B] (or pB) that occurs in the equivalence point region. This last condition is most likely to be realized where the change in pB is large.

To illustrate, consider the application of a typical indicator which exhibits a full color change as pAg varies from 7 to 5, to the three bromide titrations described in Table 8-1 and Figure 8-1. It is clear that each titration requires a different volume of titrant to encompass this range. Thus the data in the second column of Table 8-1 indicate that less than 0.10 ml of 0.1-F $AgNO_3$ is needed; that is, the color change will begin beyond 24.95 ml and be complete before 25.05 ml of silver have been added. An abrupt color change and a minimal titration error can be expected. In contrast, with 0.001-F $AgNO_3$ the color change will commence at about 24.5 ml and will be complete at 25.8 ml. Location of the end point under these circumstances is impossible. The end point change for a titration with 0.01-F reagent would require somewhat less than 0.2 ml; here the indicator would be usable, but the uncertainty with respect to the equivalence point would be substantial.

Consider now the applicability and the effectiveness of this same indicator with regard to the titrations represented by the curves in Figure 8-2. During the titration of bromate and iodate ions the indicator will exist largely as AgIn throughout; thus no color change is observed after the first addition of silver nitrate. The solubilities of silver bromide and silver iodide are small enough to prevent formation of significant amounts of AgIn until the equivalence point region is reached; thus from Figure 8-2 it is seen that pAg remains above 7 until just before the equivalence point for the bromide titration and just beyond the equivalence point for the iodide. In both titrations the excess silver ion required to complete the color change corresponds to less than 0.01 ml of reagent; the titration error would thus be negligible.

An indicator with a pAg range of 5 to 7 would not be satisfactory for a chloride titration because formation of appreciable amounts of AgIn would begin approximately 1 ml short of the equivalence point and extend over a range of about 1 ml; exact location of the end point would be impossible. On the other hand, an indicator with a pAg range of 4 to 6 would be perfectly satisfactory. No satisfactory chemical indicator exists for the iodate and bromate titrations because the pAg changes in the equivalence point region are too small.

Examples of indicators employed for precipitation titrations with silver ions are described in the paragraphs that follow.

The Formation of a Second Precipitate; the Mohr Method. The formation of a second, highly colored precipitate is the basis for end point detection with the *Mohr method*. This procedure has been widely applied to the titration of chloride ion and bromide ion with standard silver nitrate. Chromate ion is the indicator, the end point being signaled by the appearance of brick-red silver chromate, Ag_2CrO_4.

The formal solubility of silver chromate is several times greater than that for silver chloride. Thus the latter precipitate tends to form first in the titration mixture. By adjusting the chromate concentration to a suitable level, formation of silver chromate can be retarded until the silver ion concentration in the mixture is equal to the theoretical equivalence point for the titration of chloride. As we have shown (p. 169), at equivalence we may write

$$[Ag^+] = [Cl^-] = \sqrt{K_{sp}}$$
$$[Ag^+] = \sqrt{1.82 \times 10^{-10}} = 1.35 \times 10^{-5}$$

The chromate concentration required to initiate precipitation of silver chromate under these conditions can be computed with the solubility product constant for silver chromate.

$$[CrO_4^{2-}] = \frac{K_{sp}}{[Ag^+]^2} = \frac{1.1 \times 10^{-12}}{(1.35 \times 10^{-5})^2} = 6 \times 10^{-3} \text{ mole/liter}$$

In principle, then, an amount of chromate ion necessary to give this concentration could be added, and the red color of silver chromate would signal the appearance of the first excess of silver ion over its equivalence concentration. In practice, however, end point detection is difficult if not impossible in solutions with chromate-ion concentrations as large as $6 \times 10^{-3} M$ because the yellow color imparted by this concentration of indicator masks the color of the silver chromate. As a result, indicator concentrations smaller than $6 \times 10^{-3} M$ must be employed (the optimum concentration appears to be about $2.5 \times 10^{-3} M$). This decrease in the chromate concentration requires a silver concentration greater than $1.35 \times 10^{-5} M$ to produce a red precipitate. In addition, a finite amount of silver nitrate must be added to produce a detectable quantity of the precipitate. Both factors cause an overconsumption of reagent. The problem is serious where dilute solutions are involved, but the error is small with 0.1-F solutions. A correction may be made by determining an *indicator blank*— that is, by determining the silver ion consumption for a suspension of chloride-free calcium carbonate in about the same volume and with the same quantity of indicator as the sample. The blank titration mixture serves as a convenient color standard for subsequent titrations. An alternative that largely eliminates the indicator error is to use the Mohr method to standardize the silver nitrate solution against pure sodium chloride. The "working normality" obtained for the solution will compensate not only for the overconsumption of reagent but also for the acuity of the analyst in detecting the color change.

Attention must be paid to the acidity of the medium because the equilibrium

$$2CrO_4^{2-} + 2H^+ \rightleftharpoons Cr_2O_7^{2-} + H_2O$$

is displaced to the right as the hydrogen ion concentration is increased. Because silver dichromate is considerably more soluble than the chromate, the indicator reaction in acid solution requires substantially larger silver ion concentrations, if indeed it occurs at all. If the medium is made strongly alkaline, there is danger that silver will precipitate as its oxide:

$$2Ag^+ + 2OH^- \rightleftharpoons 2AgOH(s) \rightleftharpoons Ag_2O(s) + H_2O$$

Thus the determination of chloride by the Mohr method must be carried out in a medium that is neutral or nearly so (pH 7 to 10). The presence of either sodium hydrogen carbonate or borax in the solution tends to maintain the hydrogen ion concentration within suitable limits.

Formation of a Colored Complex; the Volhard Method. A standard thiocyanate solution may be used to titrate silver ion by the *Volhard method*:

$$Ag^+ + SCN^- \rightleftharpoons AgSCN(s)$$

Iron(III) ion serves as the indicator, imparting a red coloration to the solution with the first slight excess of thiocyanate:

$$Fe^{3+} + SCN^- \rightleftharpoons \underset{red}{Fe(SCN)^{2+}}$$

The titration must be carried out in acid solution to prevent precipitation of iron(III) as a hydrous oxide. The titration error in the Volhard method is small because the indicator is highly sensitive to thiocyanate ions. Thus 1 or 2 ml of a saturated iron(III) ammonium sulfate solution (about 40%) will impart a faint orange color to 100 ml of solution that also contains about 0.01 ml of 0.10-F thiocyanate. To avoid a premature end point in the titration, however, the solution must be shaken vigorously and the titration continued until the indicator color is permanent. Otherwise reaction between the silver ions adsorbed on the surface of the precipitate and the titrant is prohibitively slow.

The most important application of the Volhard method is for the indirect determination of halide ions. A measured excess of standard silver nitrate solution is added to the sample, and the excess silver ion is determined by back-titration with a standard thiocyanate solution. The requirement of a strongly acidic environment represents a distinct advantage for the Volhard technique over other methods for halide analysis because such ions as carbonate, oxalate, and arsenate (which form slightly soluble silver salts in neutral but not acidic media) do not interfere.

In contrast to the other silver halides, silver chloride is more soluble than silver thiocyanate. As a consequence, the reaction

$$AgCl(s) + SCN^- \rightleftharpoons AgSCN(s) + Cl^-$$

occurs to a significant extent near the end point during back-titration of the excess silver ion. The result is an end point that fades and an overconsumption of thiocyanate ion, which in turn leads to low values for the chloride analysis.

We can readily estimate the magnitude of the error arising from the reaction by first evaluating the equilibrium constant for the process; to obtain this information the ratio of the solubility products of silver chloride to silver thiocyanate is computed. That is,

$$K = \frac{[Ag^+][Cl^-]}{[Ag^+][SCN^-]} = \frac{[Cl^-]}{[SCN^-]} = \frac{1.82 \times 10^{-10}}{1.1 \times 10^{-12}} = 1.7 \times 10^2$$

As mentioned earlier, about 0.01 ml of 0.10-F thiocyanate is required to produce a color in 100 ml of solution when the usual amount of indicator is employed. Thus the thiocyanate concentration must be approximately 10^{-5} F to give a detectable end point. The equilibrium chloride ion concentration of the solution at this point is then

$$[Cl^-] = K[SCN^-] = 1.7 \times 10^2 \times 10^{-5}$$

$$= 1.7 \times 10^{-3} \text{ mole/liter}$$

Insofar as this chloride concentration results primarily from reaction between the thiocyanate reagent and the silver chloride, we can readily calculate the volume of the overtitration that results. Thus with a total volume of 100 ml at the end point the excess of 0.1-F thiocyanate will amount to

$$\frac{1.7 \times 10^{-3} \times 100}{0.1} = 1.7 \text{ ml}$$

In actual practice the overconsumption of reagent is often greater.

Two general methods are employed to avoid the error resulting from the reaction between thiocyanate and silver chloride. The first involves the use of the maximum allowable indicator concentration [about 0.2-M iron(III) ion].[1] The more popular way involves isolation of the

[1] E. H. Swift, G. M. Arcand, R. Lutwack, and D. J. Meier, *Anal. Chem.*, **22**, 306 (1950).

precipitated silver chloride before back-titration with the thiocyanate. Filtration, followed by titration of an aliquot of the filtrate, yields excellent results provided the precipitated silver chloride is first briefly digested. The time required for filtration is, of course, a disadvantage. Probably the most widely employed modification is that of Caldwell and Moyer,[2] which consists of coating the silver chloride with nitrobenzene, thereby substantially removing it from contact with the solution. The coating is accomplished by shaking the titration mixture with a few milliliters of the organic liquid prior to back-titration.

Adsorption Indicators.   An adsorption indicator is an organic compound that is adsorbed on or desorbed from the surface of the solid formed during a precipitation titration. Ideally the adsorption or desorption occurs near the equivalence point and results not only in a color change but also a transfer of color from the solution to the solid (or the reverse). Analytical procedures based upon adsorption indicators are sometimes called *Fajans methods* in honor of the scientist who was active in their development.

A typical adsorption indicator is the organic dye *fluorescein*, which is useful for the titration of chloride ion with silver nitrate. In aqueous solution fluorescein partially dissociates into hydrogen ions and negatively charged fluoresceinate ions that impart a yellowish green color to the medium. The fluoresceinate ion forms a highly colored silver salt of limited solubility. In its application as an indicator, however, *the concentration of the dye is never large enough to exceed the solubility product for silver fluoresceinate.*

In the early stages of a Fajans titration for chloride with silver ions the dye anion is not appreciably adsorbed by the precipitate; it is in fact repelled from the surface by the negative charge resulting from adsorbed chloride ions. When the equivalence point is passed, however, the precipitate particles become positively charged by virtue of the strong adsorption by excess silver ions; under these conditions retention of fluoresceinate ions *in the counter-ion layer* occurs. The net result is the appearance of the red color of silver fluoresceinate *on the surface of the precipitate.* It is important to emphasize that the color change is an *adsorption* (not a precipitation) process inasmuch as the solubility product of the silver fluoresceinate is never exceeded. The adsorption is reversible, the dye being desorbed upon back-titration with chloride ion.

The successful application of an adsorption indicator requires that the precipitate and the indicator have the following properties:

1. The particles of the precipitate must be of colloidal dimensions to maximize the quantity of indicator adsorbed.

[2] J. R. Caldwell and H. V. Moyer, *Ind. Eng. Chem., Anal. Ed.,* **7,** 38 (1935).

2. The precipitate must strongly adsorb its own ions. We have seen (Chapter 6) that this property is characteristic of colloidal precipitates.

3. The indicator dye must be strongly held in the counter-ion layer by the primarily adsorbed ion. In general this type of adsorption correlates with a low solubility of the salt formed between the dye and the lattice ion. At the same time this species must have sufficient solubility so that it does not precipitate.

4. The pH of the solution must be such as to assure that the ionic form of the indicator predominates. The active form of most adsorption indicators is an ion that is the conjugate acid or base of the dye molecule; it is thus capable of combining with hydrogen or hydroxide ions to form the inactive parent molecule.

Titrations involving adsorption indicators are rapid, accurate, and reliable. Their application, however, is limited to a relatively few precipitation reactions in which a colloidal precipitate is rapidly formed. In the presence of high electrolyte concentrations end points with adsorption indicators tend to be less satisfactory, owing to coagulation of the precipitate and the consequent decrease in the surface on which adsorption can occur.

Most adsorption indicators are weak acids. Their use is thus confined to neutral or slightly acidic solutions where the indicator exists predominantly as the anion. A few cationic adsorption indicators are known; these are suitable for titrations in strongly acidic solutions. For such indicators adsorption of the dye and coloration of the precipitate occur in the presence of an excess of the anion of the precipitate (that is, when the precipitate particles possess a negative charge).

Finally, many adsorption indicators sensitize silver-containing precipitates toward photodecomposition, which may cause difficulties.

OTHER METHODS FOR END POINT DETECTION

Electroanalytical methods, which can be applied for detection of end points for some precipitation reactions, are described in Chapters 14 and 15.

Applications of Precipitation Titrations

Most applications of precipitation titrations are based upon the use of a standard silver nitrate solution and are sometimes called *argentometric* methods as a consequence. Table 8-2 lists some typical applications of argentometry. Note that many of these analyses are based upon precipitation of the analyte with a measured excess of silver nitrate followed by a Volhard titration with standard potassium thiocyanate. Both of these

TABLE 8-2

Typical Argentometric Precipitation Methods

SUBSTANCE DETERMINED	END POINT	REMARKS
AsO_4^{3-}, Br^-, I^-, CNO^-, SCN^-	Volhard	Removal of silver salt not required
CO_3^{2-}, CrO_4^{2-}, CN^-, Cl^-, $C_2O_4^{2-}$, PO_4^{3-}, S^{2-}	Volhard	Removal of silver salt required before back-titration of excess Ag^+
BH_4^-	Modified Volhard	Titration of excess Ag^+ following: $BH_4^- + 8Ag^+ + 8OH^- \rightleftharpoons 8Ag(s) + H_2BO_3^- + 5H_2O$
Epoxide	Volhard	Titration of excess Cl^- following hydro-halogenation
K^+	Modified Volhard	Precipitation of K^+ with known excess of $B(C_6H_5)_4^-$, addition of excess Ag^+ which precipitates $AgB(C_6H_5)_4$, and back-titration of the excess
Br^-, Cl^-	Mohr	
Br^-, Cl^-, I^-, SeO_3^{2-}	Adsorption indicator	
$V(OH)_4^+$, fatty acids, mercaptans	Electroanalytical	Direct titration with Ag^+
Zn^{2+}	Modified Volhard	Precipitate as $ZnHg(SCN)_4$; filter, dissolve in acid, add excess Ag^+; back-titrate excess Ag^+
F^-	Modified Volhard	Precipitate as PbClF; filter, dissolve in acid, add excess Ag^+; back-titrate excess Ag^+

TABLE 8-3

Miscellaneous Volumetric Precipitation Methods

REAGENT	ION DETERMINED	REACTION PRODUCT	END POINT
$K_4Fe(CN)_6$	Zn^{2+}	$K_2Zn_3[Fe(CN)_6]_2$	Diphenylamine
$Pb(NO_3)_2$	SO_4^{2-}	$PbSO_4$	Erythrosin B
	MoO_4^{2-}	$PbMoO_4$	Eosin A
$Pb(OAc)_2$	PO_4^{3-}	$Pb_3(PO_4)_2$	Dibromofluorescein
	$C_2O_4^{2-}$	PbC_2O_4	Fluorescein
$Th(NO_3)_4$	F^-	ThF_4	Alizarin red
$Hg_2(NO_3)_2$	Cl^-, Br^-	Hg_2X_2	Bromophenol blue
$NaCl$	Hg_2^{2+}	Hg_2X_2	Bromophenol blue

reagents are obtainable in primary standard quality; however, potassium thiocyanate is somewhat hygroscopic and is ordinarily standardized against silver nitrate. Both silver nitrate and potassium thiocyanate solutions are stable indefinitely.

Table 8-3 lists some miscellaneous volumetric methods based on reagents other than silver nitrate.

Specific directions for argentometric titrations are found in section 2 of Chapter 20.

PROBLEMS

*1. A solution contains 2.83 g of silver nitrate in a total volume of 150 ml.
   (a) What is the formal concentration of this solution?
   (b) What is its normality as a titrant for bromide ion?
   (c) What is its $CrOHCl_2$ titer?
   (d) What is its CHI_3 titer, based on the reaction

$$CHI_3 + 3Ag^+ + H_2O = 3AgI + CO + 3H^+$$

2. A silver nitrate solution has a titer of 1.28 mg KSCN/ml.
   (a) What is the formal concentration of this solution?
   (b) How many milligrams of Ag^+ are contained in each milliliter of this solution?
   (c) With what volume of 0.0820-F KSCN will 40.0 ml of this solution react?
   (d) How many milliliters of this solution will be needed to react with 25.0 ml of 0.0150-F $Ba(SCN)_2$?

*3. Calculate the normality of a silver nitrate solution if 24.0 ml are required to react with
   (a) 0.185 g of pure NaCl.
   (b) 18.5 ml of a 0.0860-F KSCN solution.
   (c) 18.5 ml of a 0.0860-F Ba(SCN)$_2$ solution.
   (d) 18.5 ml of a KSCN solution that has a titer of 1.48 mg AgNO$_3$/ml.

4. What minimum volume of 0.0916-N AgNO$_3$ will be needed to assure a 5.0% excess of silver ion in the precipitation of AgCl from
   (a) 0.247 g of impure KCl?
   (b) 16.8 ml of 0.119-F MgCl$_2$?
   (c) 16.8 ml of 0.119-N MgCl$_2$?
   (d) 0.360 g of a sample that assays 67.4% MgCl$_2$.

*5. Describe a method of preparation for each of the following standard solutions; these are to be employed for the analyses shown in Table 8-3.
   (a) 1.50 liters of 0.0500-N Pb(NO$_3$)$_2$ from the pure compound
   (b) 500 ml of 0.0450-N Hg$_2$(NO$_3$)$_2$ from a sample of Hg$_2$(NO$_3$)$_2$ that is 98.3% pure
   (c) 800 ml of 0.0600-N K$_4$Fe(CN)$_6$ from the pure compound
   (d) 2.00 liters of 0.0150-N Th(NO$_3$)$_4$ from a sample of Th(NO$_3$)$_4$ that is 92.6% pure

6. A solution of NH$_4$SCN was standardized against a 0.206-g sample of primary standard grade AgNO$_3$ with iron(III) as indicator. Exactly 36.7 ml of the NH$_4$SCN were used.
   (a) Calculate the normality of the NH$_4$SCN.
   (b) What is the Hg titer of the solution in terms of the reaction

$$Hg^{2+} + 2SCN^- \rightarrow Hg(SCN)_2$$

*7. A 0.317-g sample of pure zinc was dissolved in acid, neutralized, and titrated with 45.1 ml of a sodium sulfide solution (product ZnS).
   (a) Calculate the normality of the Na$_2$S solution.
   (b) What is its Zn$_2$P$_2$O$_7$ titer?

8. Calculate the percentage of BaCl$_2 \cdot$2H$_2$O if a 1.36-g sample required 46.7 ml of 0.105-N AgNO$_3$ in a Mohr titration.

*9. A 0.632-g sample of a pesticide was decomposed by fusion with sodium carbonate, following which the residue was leached with hot water. The fluoride in the sample was then precipitated as PbClF by addition of HCl and Pb(NO$_3$)$_2$. The precipitate was filtered, washed, and dissolved in 5% HNO$_3$. The liberated Cl$^-$ was precipitated with 40.0 ml of 0.300-N AgNO$_3$. After the AgCl had been coated with nitrobenzene, the excess Ag$^+$ was back-titrated with 11.1 ml of 0.199-N NH$_4$SCN. Calculate the percent F and the percent Na$_2$SeF$_6$ in the sample.

10. A 0.888-g sample of a fertilizer was decomposed and treated in such a way as to yield an aqueous solution of HPO$_4^{2-}$. The addition of 50.0 ml of 0.187-N

$AgNO_3$ resulted in quantitative precipitation of Ag_3PO_4. The excess Ag^+ in the filtrate and washings from this precipitate required 12.1 ml of 0.105-N KSCN. Calculate the percent P_2O_5 in the sample.

*11. The formaldehyde in a 2.73-g sample of a seed disinfectant was steam distilled, and the aqueous distillate was collected in a 500-ml volumetric flask. After dilution to volume a 50.0-ml aliquot was treated with 25.0 ml of 0.136-F KCN solution to convert the formaldehyde to potassium cyanohydrin.

$$K^+ + CH_2O + CN^- \rightarrow KOCH_2CN$$

The excess KCN was then removed by addition of 50.0 ml of 0.100-N $AgNO_3$.

$$2CN^- + 2Ag^+ \rightarrow Ag_2(CN)_2(s)$$

The excess Ag^+ in the filtrate and washings required a 22.6-ml titration with 0.141-N NH_4SCN. Calculate the percent CH_2O in the sample.

12. Chloroisodiamine chloride is an antihypertensive agent with the formula $C_{14}H_{18}Cl_6N_2$ (gfw = 427). A 1.71-g sample of a medication containing this drug was heated in a closed tube with metallic sodium to destroy the organic matter and to free the chloride. The ignited mixture was taken up with water, acidified, and treated with 25.0 ml of 0.0500-F $AgNO_3$. The excess silver ion required a 5.32-ml titration with 0.0720-F KSCN. Calculate the percentage of active ingredient, assuming that it was the only source of chloride in the sample.

*13. The Association of Official Analytical Chemists recommends a Volhard titration for analysis of the insecticide heptachlor, $C_{10}H_5Cl_7$. The percentage of heptachlor is given by the calculation

$$\% \text{ heptachlor} = \frac{(ml_{Ag} \times N_{Ag} - ml_{SCN} \times N_{SCN}) \times 37.32}{\text{wt of sample}}$$

What does this calculation reveal concerning the stoichiometry of the titration?

14. Monochloroacetic acid, a preservative in fruit juices, reacts quantitatively with $AgNO_3$:

$$ClCH_2COOH + Ag^+ + H_2O = HOCH_2COOH + AgCl + H^+$$

A 200-ml juice sample was acidified with H_2SO_4, following which the $ClCH_2COOH$ was isolated by extraction with ethyl ether. The analyte was returned to aqueous solution by extraction with 1-F NaOH; the ether layer was then discarded. The aqueous phase was acidified, and 50.0 ml of a silver nitrate solution were added to assure complete precipitation of the AgCl. The excess $AgNO_3$ in the filtrate and washings required a 14.1-ml titration with 0.0601-N NH_4SCN. A blank carried through the entire procedure required 47.0 ml of the NH_4SCN solution. Calculate the milligrams $ClCH_2COOH$ per 100 ml of sample.

*15. The residual fluoride on 48.6-g of lettuce was converted to calcium fluoride by drying and igniting the sample in the presence of CaO. The resulting CaF_2 was decomposed with acid in the presence of SiO_2, and the resulting SiF_4 was distilled. The F^- in the distillate was converted to $ThF_4(s)$ by titration with 6.75 ml of 0.00806-N $Th(NO_3)_4$; alizarin red served as an indicator. Calculate the parts per million F^- in the sample.

16. A 0.177-g sample of an alloy was dissolved in acid and treated with 50.0 ml of 0.0982-N $K_4Fe(CN)_6$. After the formation of the solid $K_2Zn_3[Fe(CN)_6]_3$ was complete, the excess $K_4Fe(CN)_6$ was back-titrated with 7.97 ml of 0.116-N Zn^{2+} employing diphenylamine and a small amount of $K_3Fe(CN)_6$ as indicator. Calculate the percent Zn in the alloy.

*17. The elemental Se dispersed in a detergent for dandruff control was determined by suspending a 10.0-ml sample in a warm, ammoniacal solution that contained an excess of $AgNO_3$:

$$6Ag^+ + 3Se(s) + 6NH_3 + 3H_2O \rightarrow 2Ag_2Se(s) + Ag_2SeO_3(s) + 6NH_4^+$$

The mixed precipitate was filtered and washed free of excess Ag^+. Treatment with excess nitric acid dissolved the Ag_2SeO_3 but not the Ag_2Se. The Ag^+ from the Ag_2SeO_3 consumed 19.4 ml of 0.0166-N KSCN in a Volhard titration. How many milligrams of Se were contained per milliliter of sample?

18. A 9.63-g sample containing $BaCl_2$ and BaI_2 as well as inert materials was dissolved and diluted to 500 ml in a volumetric flask. A 50.0-ml aliquot was titrated with 0.0807-N $AgNO_3$. Bromophenol blue served as an adsorption indicator, which gave a color change when both the I^- and Cl^- were precipitated quantitatively. A 100-ml aliquot was then titrated with the $AgNO_3$ employing eosin, which is adsorbed after only the I^- is titrated. The first titration required 37.4 ml of the $AgNO_3$ and the second 27.6 ml. Calculate the percent $BaCl_2$ and BaI_2 in the original sample.

*19. A 0.2330-g sample that contained only $BaCl_2$ and KBr required 20.77 ml of 0.1000-N $AgNO_3$ to reach the end point for a Mohr titration. Calculate the percentage of each compound in the sample.

20. For each of the following precipitation titrations calculate the concentration of the cation and the anion at reagent volumes corresponding to the equivalence point as well as ± 10.0 ml, ± 1.00 ml, and ± 0.10 ml of equivalence. Construct a titration curve from the data plotting the p-function of the cation versus reagent volume.

*(a) 25.00 ml of 0.05000-F $AgNO_3$ with 0.02500-F NH_4SCN

(b) 25.00 ml of 0.05000-F KI with 0.02500-F $AgNO_3$

*(c) 50.00 ml of 0.1000-F Na_2SeO_3 with 0.2000-F $AgNO_3$
  (K_{sp} $Ag_2SeO_3 = 9.7 \times 10^{-16}$)

(d) 50.00 ml of 0.1500-F Na_2SO_4 with 0.3000-F $Pb(NO_3)_2$

*(e) 40.00 ml of 0.04000-F $Cd(NO_3)$ with 0.02000-F $K_4Fe(CN)_6$
  (K_{sp} $Cd_2Fe(CN)_6 = 3.2 \times 10^{-17}$)

(f) 40.00 ml of 0.05000-F $Pb(NO_3)_2$ with 0.1000-F $NaIO_3$
  (K_{sp} $Pb(IO_3)_2 = 3.2 \times 10^{-13}$)

9
theory of neutralization titrations

The standard solutions employed in neutralization titrations are always prepared from strong acids or strong bases because these compounds react more completely than their weaker counterparts; as with precipitation titrations, end points improve as reactions become more complete. Other factors that influence the magnitude of the equivalence point pH or pOH change, and thus the quality of end points, include the nature of the analyte as well as the concentrations of analyte and titrant. The judicious selection of an acid-base indicator requires an understanding of the ways these several factors affect a particular titration; it is for this reason that we need to know how to derive curves for neutralization titrations.

The student may find it helpful to review the material in Chapter 2 on acids and bases, as well as the introductory treatment of equilibria involving these species in Chapter 3, before reading the discussion that follows.

Indicators for Acid-Base Titrations

Many compounds, both synthetic and naturally occurring, differ in color depending upon the pH of the solutions in which they are dissolved. Some of these substances have been used for thousands of years to indicate the alkalinity or acidity of water. They are also of importance to the modern chemist who employs them to estimate the pH of solutions and signal the end point in acid-base titrations.

THEORY OF INDICATOR BEHAVIOR

Acid-base indicators are generally weak organic acids or bases which, upon dissociation or association, undergo internal structural changes that give rise to alterations in color. We can symbolize the typical reaction of an acid-base indicator as follows:

$$H_2O \ + \ \underset{\text{(acid color)}}{HIn} \ \rightleftharpoons \ H_3O^+ \ + \ \underset{\text{(base color)}}{In^-}$$

or

$$\underset{\text{(base color)}}{In} \ + \ H_2O \ \rightleftharpoons \ \underset{\text{(acid color)}}{InH^+} \ + \ OH^-$$

For the first indicator HIn will be the major constituent in strongly acid solutions and will be responsible for the "acid color" of the indicator, whereas In$^-$ will represent its "basic color." For the second indicator the species In will predominate in basic solutions and thus be responsible for the "basic color," whereas InH$^+$ will constitute the "acid color."

Equilibrium expressions for these processes are

$$\frac{[H_3O^+][In^-]}{[HIn]} = K_a$$

and

$$\frac{[InH^+][OH^-]}{[In]} = K_b$$

These expressions can be rearranged to give

$$\frac{[In^-]}{[HIn]} = \frac{K_a}{[H_3O^+]}$$

and

$$\frac{[InH^+]}{[In]} = \frac{K_b}{[OH^-]} = \frac{K_b[H_3O^+]}{K_w}$$

A solution containing an indicator will show continuous changes in color with variations in pH. The human eye is not very sensitive to these changes, however. Typically a five- to tenfold excess of one form is required before the color of that species appears predominant to the observer; further increases in the ratio have no detectable effect. It is only in the region where the ratio varies from a five- to a tenfold excess of one form to a similar excess of the other that the color of the solution appears to change. Thus the subjective "color change" involves a major alteration in the position of the indicator equilibrium. Using HIn as an example, we may write that the indicator exhibits its pure acid color to the average observer when

$$\frac{[In^-]}{[HIn]} \leq \frac{1}{10}$$

and its basic color when

$$\frac{[In^-]}{[HIn]} \geq \frac{10}{1}$$

The color appears to be intermediate for ratios between these two values. These numerical estimates, of course, represent average behavior only; some indicators require smaller ratio changes and others larger. Furthermore, considerable variation in the ability to judge colors exists among observers; indeed, a color-blind person may be unable to discern any change.

If the two concentration ratios are substituted into the dissociation constant expression for the indicator, the range of hydronium ion concentrations needed to effect the indicator color change can be evaluated. Thus for the full acid color

$$\frac{[H_3O^+][In^-]}{[HIn]} = \frac{[H_3O^+]1}{10} = K_a$$

$$[H_3O^+] = 10K_a$$

and similarly for the full basic color

$$\frac{[H_3O^+]10}{1} = K_a$$

$$[H_3O^+] = \frac{1}{10}K_a$$

To obtain the indicator range we take the negative logarithms of the two expressions. That is,

$$\text{indicator pH range} = -\log 10K_a \text{ to } -\log \frac{K_a}{10}$$

$$= -1 + pK_a \text{ to } -(-1) + pK_a$$

$$= pK_a \pm 1$$

Thus an indicator with an acid dissociation constant of 1×10^{-5} will show a complete color change when the pH of the solution in which it is dissolved changes from 4 to 6. A similar relationship is easily derived for an indicator of the basic type.

TYPES OF ACID-BASE INDICATORS

A list of compounds possessing acid-base indicator properties is large and includes a variety of organic structures. An indicator covering almost any

TABLE 9-1
Some Important Acid-Base Indicators[a]

COMMON NAME	TRANSITION RANGE (pH)	COLOR CHANGE		INDICATOR TYPE[b]
		ACID	BASE	
Thymol blue	1.2–2.8	Red	Yellow	1
	8.0–9.6	Yellow	Blue	
Methyl yellow	2.9–4.0	Red	Yellow	2
Methyl orange	3.1–4.4	Red	Orange-yellow	2
Bromocresol green	3.8–5.4	Yellow	Blue	1
Methyl red	4.4–6.2	Red	Yellow	2
Chlorophenol red	4.8–6.4	Yellow	Red	1
Bromothymol blue	6.0–7.6	Yellow	Blue	1
Phenol red	6.4–8.2	Yellow	Red	1
Neutral red	6.8–8.0	Red	Yellow	2
Cresol purple	7.6–9.2	Yellow	Purple-red	1
Phenolphthalein	8.0–9.8	Colorless	Red-violet	1
Thymolphthalein	9.3–10.5	Colorless	Blue	1
Alizarin yellow	10.1–11.1	Yellow	Lilac	2

[a] Data taken from I. M. Kolthoff and C. Rosenblum, *Acid-Base Indicators*. New York: Macmillan, 1937, p. 386.
[b] (1) Acid type: $HIn + H_2O \rightleftarrows H_3O^+ + In^-$
(2) Base type: $In + H_2O \rightleftarrows InH^+ + OH^-$

desired pH range can ordinarily be found. A few common indicators are given in Table 9-1.

TITRATION ERRORS WITH ACID-BASE INDICATORS

Two types of titration errors can be distinguished. The first is a determinate error that occurs when the pH at which the indicator changes color differs from the pH at chemical equivalence. This type of error can be minimized through judicious indicator selection; a blank will frequently provide an appropriate correction if such is necessary.

The second type is an indeterminate error arising from the limited ability of the eye to distinguish reproducibly the point at which a color change occurs. The magnitude of this random error will depend upon the change in pH per milliliter of reagent at the equivalence point, the concentration of the indicator, and the sensitivity of the eye to the two indicator colors. On the average the visual uncertainty with an acid-base indicator corresponds to about ± 0.5 pH unit. By matching the color of the solution being titrated with that of a reference standard containing a similar amount of indicator at the appropriate pH, the uncertainty can often be reduced to ± 0.1 pH unit or less. It must be understood that these uncertainties are approximations that will vary considerably from indicator to indicator as well as from person to person.

VARIABLES THAT INFLUENCE THE BEHAVIOR OF INDICATORS

The pH interval over which a given indicator exhibits a color change is influenced by the temperature, the ionic strength of the medium, the presence of organic solvents, and the presence of colloidal particles. Some of these effects, particularly the last two, can cause the transition range to shift by one or more pH units.[1]

Titration Curves for Strong Acids or Strong Bases

When both reagent and analyte are strong, the net neutralization reaction can be expressed as

$$H_3O^+ + OH^- = 2H_2O$$

and derivation of a curve for such a titration is analogous to that for a precipitation titration.

TITRATION OF A STRONG ACID WITH A STRONG BASE

The hydronium ions in an aqueous solution of a strong acid come from two sources: (1) the reaction of the solute with water and (2) the dissociation of water itself. In any but the most dilute solutions, however, the contribution from the solute far exceeds that from the solvent. Thus for a solution of HCl having a concentration greater than about $1 \times 10^{-6} F$ we may write

$$[H_3O^+] = F_{HCl}$$

For a solution of a strong base such as sodium hydroxide an analogous situation exists. That is,

$$[OH^-] = F_{NaOH}$$

Clearly, for the completely dissociated $Ba(OH)_2$

$$[OH^-] = 2 \times F_{Ba(OH)_2}$$

[1] For a discussion of these effects see H. A. Laitinen, *Chemical Analysis*. New York: McGraw-Hill, 1960, pp. 50–55.

A useful relationship for calculating the pH of basic solutions can be obtained by taking the negative logarithm of each side of the ion product constant expression for water. That is,

$$-\log K_w = -\log([H_3O^+][OH^-]) = -\log[H_3O^+] - \log[OH^-]$$

$$pK_w = pH + pOH$$

At 25°C pK_w has a value of 14.00.

As will be seen from the following example, the derivation of a curve for the titration of a strong acid with a strong base is simple because the hydronium ion or the hydroxide ion concentration is obtained directly from the formal concentration of the acid or base that is present in excess.

EXAMPLE

Derive a curve for the titration of 50.00 ml of 0.0500-F HCl with 0.1000-F NaOH. Round pH data to two places to the right of the decimal point.

Initial pH. The solution is 5.00×10^{-2} F in HCl. Because HCl is completely dissociated,

$$[H_3O^+] = 5.00 \times 10^{-2}$$

$$pH = -\log(5.00 \times 10^{-2}) = -\log 5.00 - \log 10^{-2}$$

$$= -0.699 + 2 = 1.301 = 1.30$$

pH after addition of 10.00 ml NaOH. The volume of the solution is now 60.00 ml and part of the HCl has been neutralized. Thus

$$[H_3O^+] = \frac{50.00 \times 0.0500 - 10.00 \times 0.1000}{60.00} = 2.50 \times 10^{-2}$$

$$pH = 2 - \log 2.50 = 1.60$$

Additional data, which define the curve short of the equivalence point, are derived in the same way. The results of such calculations are given in column 2 of Table 9-2.

pH after addition of 25.00 ml NaOH. Here the solution contains neither an excess of HCl nor of NaOH; thus the pH is governed by the dissociation of water, which now is the only source of hydronium and hydroxide ions.

$$[H_3O^+] = [OH^-] = \sqrt{K_w} = 1.00 \times 10^{-7}$$

$$pH = 7.00$$

pH after addition of 25.10 ml NaOH. Here

$$F_{NaOH} = \frac{25.10 \times 0.1000 - 50.00 \times 0.0500}{75.10} = 1.33 \times 10^{-4}$$

Provided the concentration of OH^- from the dissociation of water is negligible with respect to F_{NaOH},

$$[OH^-] = 1.33 \times 10^{-4}$$

$$pOH = 3.88$$

$$pH = 14.00 - 3.88 = 10.12$$

Additional data for this titration, calculated in the same way, are given in column 2 of Table 9-2.

Effect of Concentration. The effects of reagent and analyte concentrations on neutralization titration curves are shown by the two sets of data in Table 9-2; these data are plotted in Figure 9-1.

For the titration with 0.1-F NaOH (curve A) the change in pH in the equivalence point region is large. With 0.001-F NaOH the change is markedly less but still pronounced.

Figure 9-1 shows that the selection of an indicator is not critical when the reagent concentration is approximately 0.1 F. Here the volume differences among titrations with the three indicators are of the same magnitude as the uncertainties associated with the reading of the buret and are thus negligible. On the other hand, bromocresol green would clearly be unsuited for a titration with the 0.001-F reagent. Not only would the color change occur continuously over a substantial range of titrant volumes but the transition to the alkaline form would be essentially complete before the

TABLE 9-2

Changes in pH during the Titration of a Strong Acid with a Strong Base

VOLUME NaOH, ml	50.0 ml OF 0.0500-F HCl WITH 0.1000-F NaOH	50.0 ml OF 0.000500-F HCl WITH 0.001000-F NaOH
	pH	pH
0.00	1.30	3.30
10.00	1.60	3.60
20.00	2.15	4.15
24.00	2.87	4.87
24.90	3.87	5.87
25.00	7.00	7.00
25.10	10.12	8.12
26.00	11.12	9.12
30.00	11.80	9.80

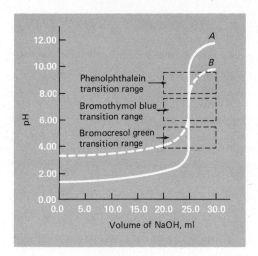

FIGURE 9-1 Curve for the titration of HCl with standard NaOH. Curve *A*: 50.00 ml 0.0500-*F* HCl with 0.1000-*F* NaOH. Curve *B*: 50.00 ml 0.000500-*F* HCl with 0.001000-*F* NaOH.

equivalence point was reached; a significant determinate error would result. The use of phenolphthalein would be subject to similar objections. Of the three indicators, only bromothymol blue would provide a satisfactory end point with a minimal determinate titration error.

TITRATION OF A STRONG BASE WITH A STRONG ACID

An analogous situation exists for the titration of strong bases with strong acids. In the region short of the equivalence point the solution is highly alkaline, the hydroxide ion concentration being numerically equal to the normality of the base. The solution is neutral at the equivalence point for precisely the same reason noted previously. Finally, the solution becomes acidic in the region beyond the equivalence point; here the pH is computed from the excess of strong acid that has been introduced. A curve for the titration of a strong base with 0.1-*F* hydrochloric acid is shown in Figure 9-5 (p. 206). The choice of indicator is subject to the same considerations as noted for the titration of a strong acid with a strong base.

Titration Curves for Weak Acids or Weak Bases

In generating a curve for the titration of a weak acid it is necessary to evaluate the pH for a solution of the acid itself, its conjugate base, and mixtures of these two solutes; derivation of a titration curve for a weak base involves similar calculations. The first two were considered on pages 32 to 39. Calculation of the pH for solutions containing appreciable quantities of conjugate acid-base pairs is treated in the section that follows.

THE pH OF SOLUTIONS CONTAINING CONJUGATE ACID-BASE PAIRS

A solution containing a conjugate acid-base pair may be acidic, basic, or neutral depending upon the position of two competitive equilibria. For a solution of the weak acid HA and its sodium salt NaA, these equilibria are

$$HA + H_2O \rightleftharpoons H_3O^+ + A^- \qquad K_a = \frac{[H_3O^+][A^-]}{[HA]} \qquad \text{(9-1)}$$

and

$$A^- + H_2O \rightleftharpoons OH^- + HA \qquad K_b = \frac{K_w}{K_a} = \frac{[OH^-][HA]}{[A^-]} \qquad \text{(9-2)}$$

If the first equilibrium lies farther to the right than the second, the solution will be acidic. Conversely, if the second equilibrium is more favorable, the solution will be basic. From the two equilibrium constant expressions it is evident that the relative concentrations of the hydronium and hydroxide ions will depend not only upon the relative magnitudes of K_a and K_b but also upon the ratio between the concentrations of the acid and its conjugate base in the solution.

A similar situation prevails for a weak base and its conjugate acid. Thus for a solution containing both ammonia and ammonium chloride the equilibria are

$$NH_3 + H_2O \rightleftharpoons NH_4^+ + OH^-$$

$$K_b = \frac{[NH_4^+][OH^-]}{[NH_3]} = 1.76 \times 10^{-5}$$

and

$$NH_4^+ + H_2O \rightleftharpoons NH_3 + H_3O^+$$

$$K_a = \frac{[H_3O^+][NH_3]}{[NH_4^+]} = \frac{K_w}{K_b}$$

$$= \frac{1.00 \times 10^{-14}}{1.76 \times 10^{-5}} = 5.68 \times 10^{-10}$$

We see that K_b is much larger than K_a; thus the first equilibrium ordinarily predominates, and the solutions are basic. When the ratio of $[NH_4^+]$ to $[NH_3]$ becomes greater than about 200, however, the difference in equilibrium constants is offset, and such solutions then become acidic.

Calculation of the pH of a Solution Containing a Weak Acid and Its Conjugate Base. Consider a solution that has a formal concentration of acid F_{HA} and a formal concentration of conjugate base F_{NaA}. The important equilibria and equilibrium constant expressions are given in Equations 9-1 and 9-2.

The reaction shown in Equation 9-1 tends to decrease the concentration of HA by an amount equal to $[H_3O^+]$. Reaction 9-2, conversely, increases the concentration of HA by an amount equal to $[OH^-]$. Thus the *molar* concentration of HA is given by

$$[HA] = F_{HA} - [H_3O^+] + [OH^-] \qquad \textbf{(9-3)}$$

Similarly, the reaction in Equation 9-1 increases $[A^-]$ by an amount equal to $[H_3O^+]$, whereas reaction 9-2 decreases $[A^-]$ by an amount corresponding to $[OH^-]$. Thus

$$[A^-] = F_{NaA} + [H_3O^+] - [OH^-] \qquad \textbf{(9-4)}$$

Equations 9-3 and 9-4 are also readily derived from mass- and charge-balance considerations.[2]

In most of the problems we shall encounter the difference between $[H_3O^+]$ and $[OH^-]$ will be much smaller than the two formal concentrations; under these circumstances Equations 9-3 and 9-4 simplify to

$$[HA] \cong F_{HA} \qquad \textbf{(9-5)}$$

$$[A^-] \cong F_{NaA} \qquad \textbf{(9-6)}$$

Substitution of these quantities into the dissociation constant expression and rearrangement yield

$$[H_3O^+] = K_a \frac{F_{HA}}{F_{NaA}} \qquad \textbf{(9-7)}[3]$$

The assumption leading to Equations 9-5 and 9-6 sometimes breaks down with acids or bases that have dissociation constants greater than about 10^{-3} or where the formal concentration of either the acid or its conjugate base (or both) is very small. As in earlier calculations, the assumptions should be made. The provisional value for $[H_3O^+]$ and $[OH^-]$ should then be used to test the assumptions.[4]

---

[2] See D. A. Skoog and D. M. West, *Fundamentals of Analytical Chemistry*, 3d ed. New York: Holt, Rinehart and Winston, 1976, p. 198.

[3] An alternative form of Equation 9-7 is frequently encountered in biological and biochemical texts. It is obtained by expressing each term in the form of its negative logarithm and inverting the concentration ratio to keep all signs positive. Thus

$$-\log [H_3O^+] = -\log K_a - \log \frac{F_{HA}}{F_{NaA}}$$

Therefore

$$pH = pK_a + \log \frac{F_{NaA}}{F_{HA}} \qquad \textbf{(9-8)}$$

This expression is known as the *Henderson–Hasselbalch* equation.

[4] Ordinarily when $[H_3O^+]$ is significant, the solution will be acidic, and $[OH^-]$ can be neglected. Similarly, if the solution is basic, only $[OH^-]$ will be important, and $[H_3O^+]$ is of no consequence. Finally, in a neutral or nearly neutral solution the difference between $[H_3O^+]$ and $[OH^-]$ approaches zero.

Within the limits imposed by the assumptions made in its derivation, Equation 9-7 states that the hydronium ion concentration of a solution containing a weak acid and its conjugate base is dependent only upon the *ratio* between the formal concentrations of these two solutes. Furthermore, this ratio remains *independent of the dilution* because the concentration of each component changes in a proportionate manner upon a change in volume. Thus *the hydronium ion concentration of a solution containing appreciable quantities of a weak acid and its conjugate base tends to be independent of dilution and depends only upon the formal ratio of concentrations between the two solutes.* This independence of pH from dilution is one manifestation of the *buffering* properties of such solutions.

EXAMPLE

What is the pH of a solution that is $0.400\ F$ in formic acid and $1.00\ F$ in sodium formate?

The equilibrium governing the hydronium ion concentration in this solution is

$$H_2O + HCOOH \rightleftharpoons H_3O^+ + HCOO^-$$

for which

$$K_a = \frac{[H_3O^+][HCOO^-]}{[HCOOH]} = 1.77 \times 10^{-4}$$

$$[HCOO^-] \cong F_{HCOO^-} = 1.00$$

$$[HCOOH] \cong F_{HCOOH} = 0.400$$

Substitution into Equation 9-7 gives

$$[H_3O^+] = 1.77 \times 10^{-4} \times \frac{0.400}{1.00}$$

$$= 7.08 \times 10^{-5}$$

Thus

$$pH = -\log(7.08 \times 10^{-5}) = 4.15$$

If we compare the pH of this solution with that of a solution that is $0.400\ F$ in formic acid only (Equation 3-15, p. 34), we find that the addition of sodium formate has had the effect of raising the pH from 2.07 to 4.15.

Calculation of the pH of a Solution of a Weak Base and Its Conjugate Acid. The calculation of pH for a solution of a weak base and its conjugate acid is completely analogous to that discussed in the preceding section.

EXAMPLE

Calculate the pH of a solution that is $0.28\ F$ in NH_4Cl and $0.070\ F$ in NH_3.

The equilibrium of interest is

$$NH_3 + H_2O \rightleftharpoons NH_4^+ + OH^-$$

for which $K_b = 1.76 \times 10^{-5}$.

Normally we may assume that

$$[NH_3] \cong F_{NH_3} = 0.070$$

$$[NH_4^+] \cong F_{NH_4^+} = 0.28$$

Substituting these values into the equilibrium constant expression,

$$\frac{0.28\,[OH^-]}{0.070} = 1.76 \times 10^{-5}$$

$$[OH^-] = 4.40 \times 10^{-6}$$

$$pOH = -\log(4.40 \times 10^{-6}) = 5.36$$

$$pH = 14.00 - 5.36 = 8.64$$

BUFFER SOLUTIONS

A *buffer solution* is defined as a solution that resists changes in pH as a result of either dilution or small additions of acids or bases. The most effective buffer solution contains large and approximately equal concentrations of a conjugate acid-base pair.

Effect of Dilution. The pH of a buffer solution remains essentially independent of dilution until its concentrations are decreased to the point where the approximations used to develop Equations 9-5 and 9-6 become invalid. The following example illustrates the behavior of a typical buffer during dilution.

EXAMPLE

Calculate the pH of the buffer described in the example on page 196 (original pH 4.15) upon dilution by a factor of (1) 50 and (2) 10,000.

1. Upon the dilution by a factor of 50

$$F_{HCOOH} = \frac{0.400}{50} = 8.00 \times 10^{-3}$$

$$F_{HCOONa} = \frac{1.00}{50} = 2.00 \times 10^{-2}$$

If we assume, as before, that $([H_3O^+] - [OH^-])$ is small with respect to the two formalities (p. 195), we obtain

$$\frac{[H_3O^+] \times 2.00 \times 10^{-2}}{8.00 \times 10^{-3}} = 1.77 \times 10^{-4}$$

$$[H_3O^+] = 7.08 \times 10^{-5}$$

$$pH = 4.15$$

Here

$$[OH^-] = \frac{1.00 \times 10^{-14}}{7.08 \times 10^{-5}} = 1.41 \times 10^{-10}$$

Thus the assumption $(7.08 \times 10^{-5} - 1.41 \times 10^{-10}) \ll 8.00 \times 10^{-3}$ and 2.00×10^{-2} is reasonably good.

2. Upon dilution by a factor of 10,000

$$F_{HCOOH} = 4.00 \times 10^{-5}$$

$$F_{HCOONa} = 1.00 \times 10^{-4}$$

Here the more exact statements given by Equations 9-3 and 9-4 must be used. That is,

$$[HCOOH] = 4.00 \times 10^{-5} - [H_3O^+] + [OH^-]$$

$$[HCOO^-] = 1.00 \times 10^{-4} + [H_3O^+] - [OH^-]$$

The solution is acidic; thus it seems probable that $[OH^-] \ll [H_3O^+]$ (see footnote 4, p. 195). Therefore

$$[HCOOH] \cong 4.00 \times 10^{-5} - [H_3O^+]$$

$$[HCOO^-] = 1.00 \times 10^{-4} + [H_3O^+]$$

Substitution of these quantities into the dissociation constant expression gives

$$\frac{[H_3O^+](1.00 \times 10^{-4} + [H_3O^+])}{4.00 \times 10^{-5} - [H_3O^+]} = 1.77 \times 10^{-4}$$

This equation rearranges to the quadratic form

$$[H_3O^+]^2 + 2.77 \times 10^{-4}[H_3O^+] - 7.08 \times 10^{-9} = 0$$

The solution is (see Appendix 3)

$$[H_3O^+] = 2.36 \times 10^{-5}$$

$$pH = 4.63$$

Thus a 10,000-fold dilution caused the pH to increase from 4.15 to 4.63, whereas a 50-fold dilution had essentially no effect.

Figure 9-2 contrasts the behavior of buffered and unbuffered solutions with dilution. For each the initial solute concentration is 1.00 F. The resistance of the buffered solution to pH changes is clear.

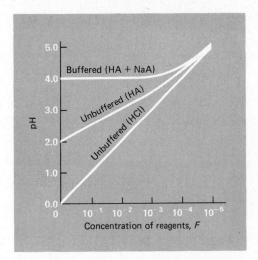

FIGURE 9-2 The effect of dilution on the pH of buffered and unbuffered solutions. The dissociation constant for HA is assumed to be 1.00×10^{-4}. Initial solute concentrations are $1.00\ F$.

Addition of Acids and Bases to Buffers. The resistance of buffer mixtures to pH changes from added acids or bases is conveniently illustrated with examples.

EXAMPLE

Calculate the pH change that takes place when 100 ml of (1) 0.0500-F NaOH and (2) 0.0500-F HCl are added to 400 ml of a buffer solution that is $0.200\ F$ in NH_3 and $0.300\ F$ in NH_4Cl.

The initial pH of the solution is obtained by assuming that

$$[NH_3] \cong F_{NH_3} = 0.200$$

$$[NH_4^+] \cong F_{NH_4Cl} = 0.300$$

and substituting into the dissociation constant expression for NH_3

$$\frac{0.300\,[OH^-]}{0.200} = 1.76 \times 10^{-5}$$

$$[OH^-] = 1.17 \times 10^{-5}$$

$$pH = 14.00 - (-\log 1.17 \times 10^{-5}) = 9.07$$

1. Addition of NaOH converts part of the NH_4^+ in the buffer to NH_3:

$$NH_4^+ + OH^- \rightarrow NH_3 + H_2O$$

The formal concentrations of NH_3 and NH_4Cl then become

$$F_{NH_3} = \frac{400 \times 0.200 + 100 \times 0.0500}{500} = \frac{85.0}{500} = 0.170$$

$$F_{NH_4Cl} = \frac{400 \times 0.300 - 100 \times 0.0500}{500} = \frac{115}{500} = 0.230$$

When substituted into the dissociation constant expression, these values yield

$$[OH^-] = \frac{1.76 \times 10^{-5} \times 0.170}{0.230} = 1.30 \times 10^{-5}$$

$$pH = 14.00 - (-\log 1.30 \times 10^{-5}) = 9.11$$

and

$$\Delta pH = 9.11 - 9.07 = 0.04$$

2. Addition of HCl converts part of the NH_3 to NH_4^+; thus

$$F_{NH_3} = \frac{400 \times 0.200 - 100 \times 0.0500}{500} = \frac{75}{500} = 0.150$$

$$F_{NH_4^+} = \frac{400 \times 0.300 + 100 \times 0.0500}{500} = \frac{125}{500} = 0.250$$

$$[OH^-] = 1.76 \times 10^{-5} \times \frac{0.150}{0.250} = 1.06 \times 10^{-5}$$

$$pH = 14.00 - (-\log 1.06 \times 10^{-5}) = 9.02$$

$$\Delta pH = 9.02 - 9.07 = -0.05$$

EXAMPLE

Contrast the behavior of 400 ml of an unbuffered solution with a pH 9.07 ($[OH^-] = 1.17 \times 10^{-5}$) when treated in the same way as the buffered solution described in the previous example.

1. After addition of NaOH

$$[OH^-] = \frac{400 \times 1.17 \times 10^{-5} + 100 \times 0.0500}{500} \cong \frac{5.00}{500} = 1.00 \times 10^{-2}$$

$$pH = 14.00 - (-\log 1.00 \times 10^{-2}) = 12.00$$

$$\Delta pH = 12.00 - 9.07 = 2.93$$

2. After addition of HCl

$$[H_3O^+] = \frac{100 \times 0.0500 - 400 \times 1.17 \times 10^{-5}}{500} \cong \frac{5.00}{500} = 1.00 \times 10^{-2}$$

$$pH = -\log 1.00 \times 10^{-2} = 2.00$$

$$\Delta pH = 2.00 - 9.07 = -7.07$$

Buffer Capacity. The foregoing examples demonstrate that a solution containing a conjugate acid-base pair shows remarkable resistance to changes in pH. The ability of a buffer to prevent a significant change in pH is directly related to the total concentration of the buffering species as well as their concentration ratios. For example, 400-ml portions of a solution formed by diluting the buffer in the example we have been considering by a factor of 10 would exhibit changes of about 0.4 to 0.5 pH unit when treated with the same amounts of sodium hydroxide or hydrochloric acid; the change for the more concentrated buffer was only about 0.04 to 0.05.

The *buffer capacity* of a solution is defined as the number of equivalents of strong acid or base needed to cause 1.00 liter of the buffer to undergo a 1.00-unit change in pH. As we have seen, the buffer capacity is dependent upon the concentration of the conjugate acid-base pair. It is also dependent upon their concentration ratio and is at a maximum when this ratio is unity.[5]

Preparation of Buffers. In principle a buffer solution of any desired pH can be prepared by combining calculated quantities of a suitable conjugate acid-base pair. In practice, however, buffers prepared in this way are found to differ somewhat in pH from the predicted values. These differences arise in part from the uncertainties that exist in the numerical values for many dissociation constants and the simplifications used in calculations. Moreover, the ionic strength of a buffer is usually so high that good values for the activity coefficients of the ions in the solution cannot be obtained from the Debye–Hückel relationship. Thus in the NH_3/NH_4^+ buffer considered in the example (p. 199) the ionic strength is about 0.30; the concentration equilibrium constant K'_a (see p. 104) would therefore be significantly larger than 1.76×10^{-5} and quite uncertain (about 4×10^{-5}).

Empirically derived recipes for preparation of buffer solutions of known pH are available in chemical handbooks and reference works.[6] Because of their widespread employment, two buffer systems deserve specific mention. McIlvaine buffers cover a pH range from about 2 to 8 and are prepared by mixing solutions of citric acid with disodium hydrogen phosphate. Clark and Lubs buffers, which encompass a pH range from 2 to 10, make use of three systems: phthalic acid, potassium hydrogen phthalate; potassium dihydrogen phosphate, dipotassium hydrogen phosphate; and boric acid, sodium borate.

TITRATION CURVES FOR WEAK ACIDS

Derivation of a curve for the titration of a weak acid (or weak base) requires a different calculation for each of four distinct situations. (1) At

[5] See D. A. Skoog and D. M. West, *Fundamentals of Analytical Chemistry*, 3d ed. New York: Holt, Rinehart and Winston, 1976, p. 204.

[6] See, for example, L. Meites, ed., *Handbook of Analytical Chemistry*. New York: McGraw-Hill, 1963, pp. 5-12 and 11-5 to 11-7.

the outset the solution contains only a weak acid or a weak base, and the pH is calculated from the formal concentration of that solute. (2) After various increments of reagent (in quantities up to, but not including, an equivalent amount) have been added, the solution consists of a series of buffers; the pH of each can be calculated from the formal concentrations of the product and the residual weak acid or base. (3) At the equivalence point the solution contains the conjugate of the weak acid or base being titrated, and the pH is calculated from the formal concentration of this product. (4) Finally, beyond the equivalence point the excess of strong acid or base that is used as titrant represses the basic or acidic character of the product to such an extent that the pH is governed largely by the concentration of the excess reagent.

Derivation. To illustrate these calculations, consider derivation of a curve for the titration of 50.00 ml of 0.1000-F acetic acid ($K_a = 1.75 \times 10^{-5}$) with 0.1000-$F$ sodium hydroxide.

EXAMPLE

Initial pH. Here we must calculate the pH of a 0.1000-F solution of HOAc; using the method shown on page 33, a value of 2.88 is obtained.

pH after addition of 10.00 ml of reagent. A buffer solution consisting of NaOAc and HOAc has now been produced. The formal concentrations of the two constituents are given by

$$F_{HOAc} = \frac{50.00 \times 0.1000 - 10.00 \times 0.1000}{60.00} = \frac{4.000}{60.00}$$

$$F_{NaOAc} = \frac{10.00 \times 0.1000}{60.00} = \frac{1.000}{60.00}$$

Upon substituting these concentrations into the dissociation constant expression for acetic acid, we obtain

$$\frac{[H_3O^+]\ 1.000/\cancel{60.00}}{4.000/\cancel{60.00}} = K_a = 1.75 \times 10^{-5}$$

$$[H_3O^+] = 7.00 \times 10^{-5}$$

$$pH = 4.16$$

Calculations similar to this will delineate the curve throughout the buffer region. Data from such calculations are given in column 2 of Table 9-3. When the acid has been 50% neutralized (in this particular titration, after an addition of exactly 25.00 ml of base), note that the formal concentrations of acid and conjugate base are identical; within the limits of the usual approximations, so also are their molar concentrations. Thus these terms cancel one another in the equilibrium constant expression, and the

TABLE 9-3

Changes in pH during the Titration of a Weak Acid with a Strong Base

VOLUME NaOH, ml	50.00 ml OF 0.1000-F HOAc WITH 0.1000-F NaOH	50.00 ml OF 0.001000-F HOAc WITH 0.001000-F NaOH
	pH	pH
0.00	2.88	3.91
10.00	4.16	4.30
25.00	4.76	4.80
40.00	5.36	5.38
49.00	6.45	6.46
49.90	7.46	7.47
50.00	8.73	7.73
50.10	10.00	8.09
51.00	11.00	9.00
60.00	11.96	9.96
75.00	12.30	10.30

hydronium ion concentration is numerically equal to the dissociation constant; that is, the pH is equal to the pK_a. Likewise, in the titration of a weak base the hydroxide ion concentration is numerically equal to the dissociation constant of the base at the midpoint.

EXAMPLE

Equivalence point pH. At the equivalence point in the titration the acetic acid has been converted to sodium acetate. The solution is therefore similar to one formed by dissolving that base in water. After evaluating the formal acetate concentration, the pH calculation is identical to that described on page 39 for a weak base. In the present example the NaOAc concentration is 0.0500 F. Thus

$$[OH^-] \cong \sqrt{\frac{K_w}{K_a} \times 0.0500} = \sqrt{\frac{1.00 \times 10^{-14} \times 0.0500}{1.75 \times 10^{-5}}}$$

$$[OH^-] = 5.34 \times 10^{-6}$$

$$pH = 14.00 - (-\log 5.34 \times 10^{-6})$$

$$= 8.73$$

pH after addition of 50.10 ml of base. After 50.10 ml of base have been added, hydroxide ions arise from both the excess of sodium hydroxide and the equilibrium between acetate ion and water. The contribution of the latter is small enough to be neglected, however, because the excess of strong base will tend to repress this reaction. This fact becomes evident when we consider that the hydroxide ion concentration was only $5.34 \times 10^{-6} M$ at the equivalence point of the titration; once an excess of strong base has been added, the contribution from the reaction of the acetate will be even smaller. Thus

$$[OH^-] \cong F_{NaOH} = \frac{50.10 \times 0.1000 - 50.00 \times 0.1000}{100.1} = 1.0 \times 10^{-4}$$

$$pOH = 14.00 - (-\log 1.0 \times 10^{-4}) = 10.00$$

Note that titration curves for a weak acid and a strong base become identical with those for a strong acid with a strong base in the region slightly beyond the equivalence point.

Effect of Concentration. The second and third columns of Table 9-3 contain pH data for the titration of 0.1000-F and of 0.001000-F acetic acid, respectively, with sodium hydroxide solutions of the same strengths. Substantial portions of the curve for the dilute solution required the use of more exact expressions for the concentration relationships; an example of this calculation appears on page 198.

Figure 9-3 is a plot of the data in Table 9-3. Note that the initial pH values are higher and the equivalence point pH is lower for the more dilute solution than for the 0.1000-F solution. At intermediate titrant volumes, however, the pH values differ only slightly. Acetic acid–sodium acetate

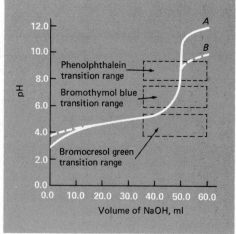

FIGURE 9-3 Curves for the titration of acetic acid with NaOH. Curve *A*: 0.1000-F acid with 0.1000-F base. Curve *B*: 0.001000-F acid with 0.001000-F base.

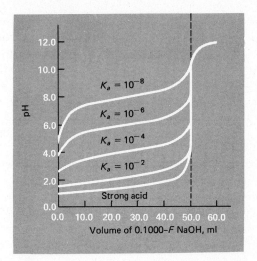

FIGURE 9-4 Influence of acid strength upon titration curves. Each curve represents the titration of 50.00 ml of 0.1000-F acid with 0.1000-F NaOH.

buffers exist within this region; Figure 9-3 is graphical confirmation of the fact that the pH of buffers is largely independent of dilution.

Titration curves for 0.1000-F solutions of acids with differing strengths are shown in Figure 9-4. Note that the change in pH in the equivalence point region becomes smaller as the acid becomes weaker; that is, as the reaction between the acid and the base becomes less complete. The relation between completeness of reaction and reagent concentrations illustrated by Figures 9-3 and 9-4 is analogous to these effects on precipitation titration curves (pp. 170 to 172).

Indicator Choice; Feasibility of Titration. Figures 9-3 and 9-4 clearly indicate that the choice of indicator for the titration of a weak acid is more limited than that for a strong acid. For example, bromocresol green is totally unsuited for titration of 0.1000-F acetic acid; nor would bromothymol blue be satisfactory because its full color change would occur over a range between about 47 and 50 ml of 0.100-F base. It might still be possible to employ bromothymol blue by titrating to the full basic color of the indicator rather than to the change in color; this technique would require a comparison standard containing the same concentration of indicator as the solution being titrated. An indicator such as phenolphthalein, which changes color in the basic region, would be most satisfactory for this titration.

The end point pH change associated with the titration of 0.00100-F acetic acid (curve B, Figure 9-3) is so small that a significant titration error is likely to be introduced. By employing an indicator with a transition range between that of phenolphthalein and bromothymol blue and by using a color comparison standard, it should be possible to establish the end point with a reproducibility of a few percent.

Figure 9-4 illustrates that similar problems exist as the strength of the acid being titrated decreases. Precision on the order of ± 2 ppt can be achieved in the titration of a 0.100-F acid solution with a dissociation constant of 10^{-8} provided a suitable color comparison standard is available. With more concentrated solutions somewhat weaker acids can be titrated with reasonable precision.

Titration Curves for Weak Bases

Figure 9-5 shows theoretical titration curves for a series of weak bases of differing strengths. These curves were derived by methods analogous to those employed for derivation of the titration curves for acetic acid. Clearly, indicators with *acidic* transition ranges must be employed for weaker bases.

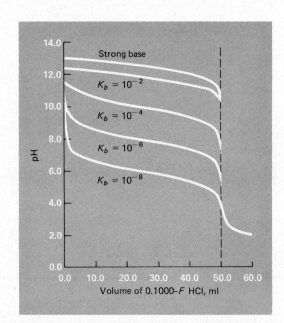

FIGURE 9-5 Influence of base strength upon titration curves. Each curve represents the titration of 50.00 ml of 0.1000-F base with 0.1000-F HCl.

Titration Curves for a Mixture of a Weak and Strong Acid (or a Weak and Strong Base)

The hydronium ion concentration of a solution containing both a weak and a strong acid can be expressed as

$$[H_3O^+] = F_{HS} + [A^-] \qquad (9\text{-}9)$$

The contribution of the strong acid is given by its formality F_{HS}, whereas the acidity due to the weak acid is equal to the anion concentration $[A^-]$ (we have assumed that the dissociation of water makes a vanishingly small contribution to the hydronium ion concentration and can thus be neglected).

Two other algebraic equations that can be employed in conjunction with Equation 9-9 to provide an exact solution for the hydronium ion concentration of a mixture of acids are the dissociation constant expression for HA and the mass-balance expression

$$F_{HA} = [A^-] + [HA] \tag{9-10}$$

If the acid HA is sufficiently weak ($K_a < 10^{-4}$), its dissociation will be small enough so that $[A^-]$ in Equation 9-9 is ordinarily negligible with respect to F_{HS}. Under these circumstances the pH of the mixture is simply governed by the formality of the strong acid. The initial points in the titration curve for a mixture of this type will therefore be essentially identical to that of the strong acid alone. At the equivalence point for the strong acid the solution will contain the product of the strong acid titration, which will not directly contribute to the pH, plus the as-yet untitrated weak acid, which most assuredly will. The remainder of the titration will thus be described by a curve that is typical for the weak acid. The shape of the curve, and hence the information obtainable from it, will depend in large measure upon the strength of the weak acid. Figure 9-6 depicts the titration of mixtures containing hydrochloric acid and several weak acids. It may be seen that a break in the titration curve at the first equivalence point will be essentially nonexistent if the weak acid has a relatively large dissociation constant (curve A); for such mixtures only the total acidity (that is, HCl + HA) can be obtained from the titration. Conversely, where the weak acid has a very small dissociation constant, only the strong acid content

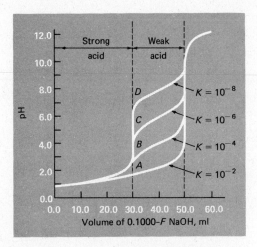

FIGURE 9-6 Curves for the titration of strong acid–weak acid mixtures with 0.1000-F NaOH. Each titration involves 25.00 ml of a solution that is 0.1200 F with respect to HCl and 0.0800 F with respect to HA.

can be determined. For weak acids of intermediate strength two useful end points may exist.

Titration Curves for Polyfunctional Acid-Base Systems

An examination of Appendix 8 reveals that many weak acids of interest to chemists contain two or more ionizable protons. For example, three equilibria are established in an aqueous solution of phosphoric acid:

$$H_3PO_4 + H_2O \rightleftharpoons H_3O^+ + H_2PO_4^- \qquad K_1 = \frac{[H_3O^+][H_2PO_4^-]}{[H_3PO_4]}$$

$$= 7.11 \times 10^{-3}$$

$$H_2PO_4^- + H_2O \rightleftharpoons H_3O^+ + HPO_4^{2-} \qquad K_2 = \frac{[H_3O^+][HPO_4^{2-}]}{[H_2PO_4^-]}$$

$$= 6.34 \times 10^{-8}$$

$$HPO_4^{2-} + H_2O \rightleftharpoons H_3O^+ + PO_4^{3-} \qquad K_3 = \frac{[H_3O^+][PO_4^{3-}]}{[HPO_4^{2-}]}$$

$$= 4.2 \times 10^{-13}$$

With this acid, as with most others, each successive dissociation becomes less favorable; thus $K_3 < K_2 < K_1$.

Polyfunctional bases are also common, an example being Na_2CO_3. Carbonate ion, the conjugate base of the hydrogen carbonate ion, is involved in the stepwise equilibria:

$$CO_3^{2-} + H_2O \rightleftharpoons HCO_3^- + OH^- \qquad (K_b)_1 = \frac{[HCO_3^-][OH^-]}{[CO_3^{2-}]}$$

$$= 2.13 \times 10^{-4}$$

$$HCO_3^- + H_2O \rightleftharpoons H_2CO_3 + OH^- \qquad (K_b)_2 = \frac{[H_2CO_3][OH^-]}{[HCO_3^-]}$$

$$= 2.25 \times 10^{-8}$$

Numerical values for the basic dissociation constants for Na_2CO_3 are related to the stepwise acid dissociation constants of the corresponding carbonic acids (p. 33); that is,

$$(K_b)_1 = \frac{K_w}{(K_a)_2}$$

and

$$(K_b)_2 = \frac{K_w}{(K_a)_1}$$

where $(K_a)_1$ and $(K_a)_2$ are the first and second dissociation constants, respectively, of H_2CO_3.

The pH of solutions containing a polyfunctional acid or a buffer mixture involving such an acid can be evaluated rigorously through the use of the systematic approach to the multiequilibrium problem introduced earlier (p. 89). Solution of the several simultaneous equations that result from this treatment, however, is usually difficult and time-consuming without the aid of a computer. However, for acids or bases with successive dissociation constants that differ by a factor of about 10^3 (or more) a simplifying assumption can be applied which decreases the algebraic labor of these calculations. In most of the discussion that follows we shall consider examples to which this approximation is applicable. As a consequence, only one new type of pH calculation, that for a solution of a salt of the type NaHA, will be required.

SOLUTIONS OF SALTS OF THE TYPE NaHA

A solute of the type NaHA is both an acid and a base. Thus its acid dissociation reaction can be formulated as

$$HA^- + H_2O \rightleftharpoons A^{2-} + H_3O^+$$

It is also, however, the conjugate base of H_2A; that is,

$$HA^- + H_2O \rightleftharpoons H_2A + OH^-$$

One of these reactions produces hydronium ions and the other hydroxide ions. Whether an HA^- solution is acidic or basic will be determined by the relative magnitude of the equilibrium constants for these processes:

$$K_2 = \frac{[H_3O^+][A^{2-}]}{[HA^-]} \tag{9-11}$$

$$K_b = \frac{K_w}{K_1} = \frac{[H_2A][OH^-]}{[HA^-]} \tag{9-12}$$

If K_b is greater than K_2, the solution will be basic; otherwise it will be acidic. A substance such as NaHA that can behave both as an acid and as a base is said to be *amphiprotic*.

A solution of NaHA can be described in terms of material balance:

$$F_{NaHA} = [HA^-] + [H_2A] + [A^{2-}] \tag{9-13}$$

It can also be described in terms of charge balance:

$$[Na^+] + [H_3O^+] = [HA^-] + 2[A^{2-}] + [OH^-]$$

Because the sodium ion concentration is equal to the formal concentration of the salt, the last equation can be rewritten as

$$F_{NaHA} = [HA^-] + 2[A^{2-}] + [OH^-] - [H_3O^+] \tag{9-14}$$

Frequently neither the hydroxide nor the hydronium ion concentration can be neglected in solutions of this type; therefore a fifth equation is needed to take care of the five unknowns. The ion product constant for water will serve:

$$K_w = [H_3O^+][OH^-]$$

The derivation of a rigorous expression for the hydronium ion concentration from these five equations is difficult. However, a reasonable approximation, applicable to solutions of most acid salts, can be obtained. Subtracting Equation 9-13 from 9-14 yields, upon rearrangement,

$$[H_2A] = [A^{2-}] + [OH^-] - [H_3O^+]$$

With the aid of Equations 9-11 and 9-12 we can express $[H_2A]$ and $[A^{2-}]$ in terms of $[HA^-]$; that is,

$$\frac{K_w[HA^-]}{K_1[OH^-]} = \frac{K_2[HA^-]}{[H_3O^+]} + [OH^-] - [H_3O^+]$$

Replacement of $[OH^-]$ by the equivalent expression $K_w/[H_3O^+]$ gives

$$\frac{[H_3O^+][HA^-]}{K_1} = \frac{K_2[HA^-]}{[H_3O^+]} + \frac{K_w}{[H_3O^+]} - [H_3O^+]$$

Multiplication of both sides by $[H_3O^+]$ and rearrangement yield

$$[H_3O^+]^2\left(\frac{[HA^-]}{K_1} + 1\right) = K_2[HA^-] + K_w$$

Finally,

$$[H_3O^+] = \sqrt{\frac{K_1K_2[HA^-] + K_1K_w}{[HA^-] + K_1}} \qquad (9\text{-}15)$$

Under most circumstances it can be *assumed* that

$$[HA^-] \cong F_{NaHA} \qquad (9\text{-}16)$$

and Equation 9-15 takes the form

$$[H_3O^+] = \sqrt{\frac{K_1K_2F_{NaHA} + K_1K_w}{F_{NaHA} + K_1}} \qquad (9\text{-}17)$$

Equation 9-16 is a reasonable approximation provided the two equilibrium constants that contain $[HA^-]$ are small and the formal concentration of NaHA is not too low.

Frequently K_1 will be so much smaller than F_{NaHA} that it can be neglected in the denominator, thus permitting further simplification of Equation 9-17. Moreover, K_1K_w will often be much smaller than $K_1K_2F_{NaHA}$. Thus provided $K_1 \ll F_{NaHA}$ and $K_1K_w \ll K_1K_2F_{NaHA}$, Equation 9-17 simplifies to

$$[H_3O^+] \cong \sqrt{K_1K_2} \qquad (9\text{-}18)$$

Note that solutions of this type tend to have a constant pH over a substantial range of concentrations.

EXAMPLE

Calculate the hydronium ion concentration of a 0.100-F NaHCO$_3$ solution.

We must first examine the assumptions required for Equation 9-18. The dissociation constants, K_1 and K_2, for H$_2$CO$_3$ are 4.45×10^{-7} and 4.7×10^{-11}, respectively. Obviously K_1 is much smaller than F_{NaHA}; further, $K_1 K_w$ is equal to 4.45×10^{-21}, whereas $K_1 K_2 F_{NaHA}$ is about 2×10^{-18}. We can therefore use the simplified equation. Thus

$$[H_3O^+] = \sqrt{4.45 \times 10^{-7} \times 4.7 \times 10^{-11}}$$

$$[H_3O^+] = 4.6 \times 10^{-9}$$

EXAMPLE

Calculate the hydronium ion concentration of a 1.0×10^{-3} F solution of Na$_2$HPO$_4$.

Here the pertinent dissociation constants (those containing the principal solute species [HPO$_4^{2-}$]) are K_2 and K_3 for H$_3$PO$_4$, 6.34×10^{-8} and 4.2×10^{-13}, respectively. Considering again the assumptions implicit in Equation 9-18, we find that 6.34×10^{-8} is indeed much smaller than the formal concentration of Na$_2$HPO$_4$; the denominator can again be simplified. On the other hand, $K_2 K_w$ is by no means much smaller than $K_2 K_3 F_{Na_2HPO_4}$, the former being about 6×10^{-22} and the latter about 3×10^{-23}. We should therefore use the partially simplified version of Equation 9-17:

$$[H_3O^+] = \sqrt{\frac{6.34 \times 10^{-8} \times 4.2 \times 10^{-13} \times 1.0 \times 10^{-3} + 6.34 \times 10^{-22}}{1.0 \times 10^{-3}}}$$

$$= 8.1 \times 10^{-10}$$

Use of Equation 9-18 would have yielded a value of 1.6×10^{-10} M.

EXAMPLE

Find the hydronium ion concentration of a 0.0100-F NaH$_2$PO$_4$ solution.

Here the two dissociation constants of importance (those containing the principal solute species [H$_2$PO$_4^-$]) are K_1 and K_2 for H$_3$PO$_4$; these have values of 7.11×10^{-3} and 6.34×10^{-8}. Because K_1 is not much smaller than $F_{NaH_2PO_4}$, the denominator of Equation 9-17 cannot be simplified. On the other hand, the numerator can be simplified to $K_1 K_2 F_{NaH_2PO_4}$. Thus Equation 9-17 becomes

$$[H_3O^+] = \sqrt{\frac{7.11 \times 10^{-3} \times 6.34 \times 10^{-8} \times 1.0 \times 10^{-2}}{0.010 + 7.11 \times 10^{-3}}}$$

$$= 1.6 \times 10^{-5}$$

Use of Equation 9-18 would have yielded a value of 2.1×10^{-5}.

DERIVATION OF TITRATION CURVES FOR
POLYFUNCTIONAL ACIDS

Titration curves for solutions of polyfunctional acids or bases are ordinarily more complex than those that we have so far considered. Multiple inflection points are observed when the component acids (or bases) differ appreciably in strength, and more than one useful equivalence point may exist if the change in pH is sufficiently abrupt in the region of these inflections.

The techniques described in the first part of this chapter enable us to derive, with reasonable accuracy, theoretical titration curves for polybasic acids, provided the ratio between the dissociation constants of the two acids is somewhat greater than 10^3. If the ratio is smaller than this, the error becomes prohibitive, particularly in the region of the first equivalence point, and a more rigorous treatment of the equilibrium relationships is required.

In the example that follows we shall derive points on the titration curve of maleic acid, a bifunctional weak organic acid with the empirical formula $H_2C_4H_2O_4$. Employing H_2M to symbolize the free acid, we may write two dissociation equilibria as follows:

$$H_2M + H_2O \rightleftharpoons H_3O^+ + HM^- \qquad K_1 = 1.20 \times 10^{-2}$$

$$HM^- + H_2O \rightleftharpoons H_3O^+ + M^{2-} \qquad K_2 = 5.96 \times 10^{-7}$$

We see that $K_1/K_2 = (1.20 \times 10^{-2})/(5.96 \times 10^{-7}) = 2.0 \times 10^4$. With a ratio this large it is feasible to neglect the second dissociation step when deriving points in the early part of the curve. That is, we assume that the concentration of H_3O^+ resulting from dissociation of HM^- is negligible with respect to the contribution from the dissociation of H_2M. This assumption is tantamount to saying that $[M^{2-}] << (HM^-)$ and $[H_2M]$. It can be shown that to within a few tenths of a milliliter of the first equivalence point this assumption does not lead to serious error. Shortly beyond the first equivalence point the second equilibrium is assumed to be sufficiently dominant so that the basic reaction of HM^-,

$$HM^- + H_2O \rightleftharpoons H_2M + OH^-$$

does not significantly influence the pH. Here we are assuming that $[H_2M] << [HM^-]$ and $[M^{2-}]$.

EXAMPLE

Derive a curve for the titration of 25.00 ml of 0.1000-F maleic acid with 0.1000-F NaOH.

Initial pH. Insofar as the initial pH is determined by the first dissociation equilibrium

$$H_2M + H_2O \rightleftharpoons H_3O^+ + HM^-$$

we may write that

$$[H_3O^+] = [HM^-]$$

$$[H_2M] \cong 0.1000 - [HM^-] = 0.1000 - [H_3O^+]$$

Because K_1 is large, the approximation that $0.1000 \gg [H_3O^+]$ will introduce a serious error into the calculated value for $[H_3O^+]$. Thus

$$1.20 \times 10^{-2} = \frac{[H_3O^+]^2}{1.00 \times 10^{-1} - [H_3O^+]}$$

$$[H_3O^+]^2 + 1.20 \times 10^{-2}[H_3O^+] - 1.20 \times 10^{-3} = 0$$

$$[H_3O^+] = \frac{-1.20 \times 10^{-2} + \sqrt{1.44 \times 10^{-4} + 4.80 \times 10^{-3}}}{2}$$

$$= 2.92 \times 10^{-2}$$

$$pH = -\log 2.92 \times 10^{-2} = 1.54$$

pH after addition of 5.00 ml base. With the addition of base a buffer is formed consisting of the weak acid H_2M and its conjugate base HM^-. Again, we can neglect the dissociation of HM^- to give M^{2-} and treat the solution as a simple buffer system. Then

$$F_{H_2M} = \frac{(25.00 \times 0.1000) - (5.00 \times 0.1000)}{30.00} = 6.67 \times 10^{-2}$$

$$F_{NaHM} = \frac{5.00 \times 0.1000}{30.00} = 1.67 \times 10^{-2}$$

Then from Equations 9-5 and 9-6 we write

$$[H_2M] \cong 6.67 \times 10^{-2}$$

$$[HM^-] \cong 1.67 \times 10^{-2}$$

Substituting these values into the equilibrium constant expression, we obtain a tentative value of 4.8×10^{-2} mole/liter for $[H_3O^+]$. It is clear the approximation that $[H_3O^+]$ is much smaller than the two formal concentrations is not valid. Thus we must employ Equations 9-3 and 9-4.

$$[HM^-] = 1.67 \times 10^{-2} + [H_3O^+] - [OH^-] \cong 1.67 \times 10^{-2} + [H_3O^+]$$

$$[H_2M] = 6.67 \times 10^{-2} - [H_3O^+] + [OH^-] \cong 6.67 \times 10^{-2} - [H_3O^+]$$

Because the solution is quite acidic, the approximation that $[OH^-]$ is very small is surely valid. Substitution of these expressions into the dissociation constant relationship gives

$$\frac{[H_3O^+](1.67 \times 10^{-2} + [H_3O^+])}{6.67 \times 10^{-2} - [H_3O^+]} = 1.20 \times 10^{-2}$$

or

$$[H_3O^+]^2 + 2.87 \times 10^{-2}[H_3O^+] - 8.00 \times 10^{-4} = 0$$

and

$$[H_3O^+] = 1.74 \times 10^{-2}$$

$$pH = 1.76$$

Additional points in the first buffer region can be computed in a similar way.

First equivalence point. At the first equivalence point

$$[HM^-] \cong \frac{2.50}{50.0} = 5.00 \times 10^{-2}$$

Simplification of the numerator in Equation 9-17 is clearly justified. On the other hand, the concentration of HM^- is relatively close to the value for K_1. Hence

$$[H_3O^+] \cong \sqrt{\frac{K_1 K_2 F_{NaHM}}{K_1 + F_{NaHM}}} = \sqrt{\frac{(1.20 \times 10^{-2})(5.96 \times 10^{-7})(5.00 \times 10^{-2})}{(1.20 \times 10^{-2} + 5.00 \times 10^{-2})}}$$

$$= 7.60 \times 10^{-5}$$

$$pH = -\log 7.60 \times 10^{-5} = 4.12$$

Second buffer region. Further additions of base to the solution form a new buffer system consisting of HM^- and M^{2-}. When enough base has been added so that the reaction of HM^- with water to give OH^- may be neglected (a few tenths of a milliliter), the pH of the mixture is readily obtained from K_2. With the introduction of 25.50 ml of NaOH, for example, the number of milliformula weights of Na_2M is equal to the number of milliformula weights of base in excess of that required to produce NaHM, namely, $(25.50 - 25.00)0.1000$. Thus

$$F_{Na_2M} = \frac{(25.50 - 25.00)0.1000}{50.50} = \frac{0.050}{50.50}$$

The number of milliformula weights of NaHM remaining will be equal to the amount present at the first equivalence point (25.00×0.1000) minus the number of milliformula weights of Na_2M produced. Here

$$F_{NaHM} = \frac{(25.00 \times 0.1000) - (25.50 - 25.00)0.1000}{50.50}$$

$$= \frac{2.450}{50.50}$$

Now we may write

$$\frac{[H_3O^+](0.050/\cancel{50.50})}{(2.45/\cancel{50.50})} = 5.96 \times 10^{-7}$$

$$[H_3O^+] = 2.92 \times 10^{-5}$$

The assumption that $[H_3O^+]$ is small with respect to the two formal concentrations is valid and

$$pH = 4.54$$

Second equivalence point. After the addition of 50.00 ml of sodium hydroxide the solution is 0.0333 F in Na_2M. Reaction of the base M^{2-} with water is the predominant equilibrium in the system and the only one that must be taken into account. Thus

$$M^{2-} + H_2O \rightleftharpoons HM^- + OH^-$$

$$K_b = \frac{K_w}{K_2} = \frac{1.00 \times 10^{-14}}{5.96 \times 10^{-7}} = \frac{[OH^-][HM^-]}{[M^{2-}]}$$

$$[OH^-] \cong [HM^-]$$

$$[M^{2-}] = 0.0333 - [OH^-] \cong 0.0333$$

$$\frac{[OH^-]^2}{0.0333} = \frac{1.00 \times 10^{-14}}{5.96 \times 10^{-7}}$$

$$[OH^-] = 2.36 \times 10^{-5}$$

$$pOH = 4.63$$

$$pH = 14.00 - 4.63 = 9.37$$

pH beyond the second equivalence point. Further additions of sodium hydroxide repress the basic reaction of M^{2-}. The pH is calculated from the concentration of NaOH added in excess of that required for the complete neutralization of H_2M. An example of this calculation is given on page 191.

Figure 9-7 depicts the titration of 0.1000-F maleic acid with 0.1000-F NaOH. This curve was derived by use of the calculations shown in the preceding example. Two end points are apparent; in principle the judicious choice of indicator would permit either the first or both acidic hydrogens of maleic acid to be titrated. The second end point is clearly the better one to use, however, inasmuch as the pH change is more pronounced.

Figure 9-8 shows titration curves for three other dibasic acids. These curves illustrate that a well-defined end point corresponding to the first equivalence point is observed only where the degree of dissociation of the two acids is sufficiently different.

The ratio of K_1 to K_2 for oxalic acid (curve B) is approximately 1000. The curve for this titration shows an inflection corresponding to the first equivalence point. However, the magnitude of the pH change is too small to permit the location of this equivalence point by means of a chemical indicator; thus only the second equivalence point can be used for accurate analyses.

Curve A illustrates the theoretical titration curve for the tribasic phosphoric acid. Here the ratio of K_1 to K_2 is approximately 10^5, a figure that is about 100 times greater than that for oxalic acid; two well-defined

FIGURE 9-7 Titration curve for 25.00 ml of 0.1000-F maleic acid, H_2M.

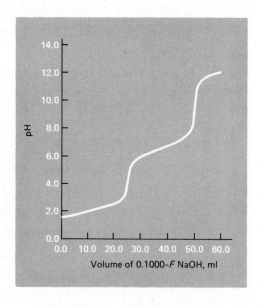

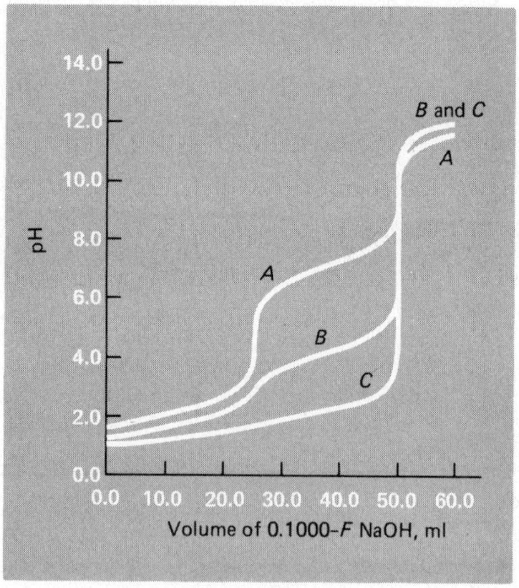

FIGURE 9-8 Curves for the titration of polybasic acids. A 0.1000-F NaOH solution is used to titrate 25.00 ml of 0.1000-$F H_3PO_4$ (curve A), 0.1000-F oxalic acid (curve B), and 0.1000-$F H_2SO_4$ (curve C).

end points are observed, either of which is satisfactory for analytical purposes. If the indicator with an acid transition range is used, one equivalent of base will be consumed per mole of acid. With an indicator exhibiting a color change in the basic region, two equivalents of base will be used. The third hydrogen of phosphoric acid is so slightly dissociated ($K_3 = 4.2 \times 10^{-13}$) that it does not yield an end point of any practical value.

The buffering effect of the third dissociation is noticeable, however, causing the pH for curve A to be lower than that for the other two in the region beyond their second equivalence points.

In general the titration of polyfunctional acids or bases yields individual end points that are of practical value only where the ratio of the two dissociation constants is at least 10^4. If the ratio is much less than this, the pH change at the first equivalence point will be too small for accurate detection—only the second end point will prove satisfactory for analysis.

Titration Curves for Polyfunctional Bases

The derivation of a titration curve for a polyfunctional base involves no new principles. To illustrate, consider the titration of a sodium carbonate solution with standard hydrochloric acid. The important equilibrium constants are

$$CO_3^{2-} + H_2O \rightleftharpoons OH^- + HCO_3^- \qquad (K_b)_1 = \frac{K_w}{(K_a)_2} = 2.13 \times 10^{-4}$$

$$HCO_3^- + H_2O \rightleftharpoons OH^- + H_2CO_3 \qquad (K_b)_2 = \frac{K_w}{(K_a)_1} = 2.25 \times 10^{-8}$$

The reaction of carbonate ion with water governs the initial pH of the solution; the method shown on page 215 is used in the pH calculation. With the first additions of acid, a carbonate–hydrogen carbonate buffer is established; the hydroxide ion concentration is calculated from $(K_b)_1$ [or the hydronium ion concentration from $(K_a)_2$]. Sodium hydrogen carbonate is the principal solute species at the first equivalence point; Equation 9-18 will give the hydronium ion concentration of this solution. With further introductions of acid a hydrogen carbonate–carbonic acid buffer governs the pH; here the hydroxide ion concentration is calculated from $(K_b)_2$ [or the hydronium ion concentration from $(K_a)_1$]. At the second equivalence point the solution consists of carbonic acid and sodium chloride; the hydronium ion concentration is estimated in the usual way for a simple weak acid, using $(K_a)_1$. Finally, when excess hydrochloric acid has been introduced, the dissociation of the weak acid is repressed to a point where the hydronium ion concentration is essentially that of the formal concentration of the strong acid.

Figure 9-9 illustrates a titration curve for a solution of sodium carbonate. Two end points are observed, the second being appreciably sharper than the first. It is apparent that mixtures of sodium carbonate and sodium hydrogen carbonate can be analyzed by neutralization methods. Thus titration to a phenolphthalein end point would yield the number of milliformula weights of carbonate present, whereas titration to a bromo-cresol green color change would require an amount of acid equal to twice

FIGURE 9-9 Curve for the titration of 25.00 ml of 0.1000-F Na_2CO_3 with 0.1000-F HCl.

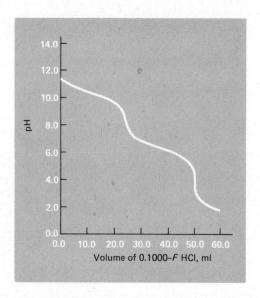

Volume of 0.1000-F HCl, ml

the number of milliformula weights of carbonate plus the number of milliformula weights of bicarbonate in the sample.

Composition of Solutions of a Polybasic Acid as a Function of pH

To understand clearly the compositional changes that occur in the course of a titration involving a polybasic acid, it is instructive to plot the *relative* amount of the free acid as well as each of its anions as a function of the pH of the solution. For this purpose we shall define so-called α-values for each of the anion-containing species. An α-value is the fraction of the total weak acid concentration represented by a particular species. Turning again to the maleic acid system, if we define C_T as the sum of the concentrations of the maleate-containing species, then mass balance requires that

$$C_T = [H_2M] + [HM^-] + [M^{2-}] \tag{9-19}$$

and by definition

$$\alpha_0 = \frac{[H_2M]}{C_T}$$

$$\alpha_1 = \frac{[HM^-]}{C_T}$$

$$\alpha_2 = \frac{[M^{2-}]}{C_T}$$

The sum of the α-values for a system must equal unity; that is,

$$\alpha_0 + \alpha_1 + \alpha_2 = 1$$

We can readily express α_0, α_1, and α_2 in terms of $[H_3O^+]$, K_1, and K_2. Rearrangement of the dissociation constant expressions gives

$$[HM^-] = \frac{K_1[H_2M]}{[H_3O^+]}$$

$$[M^{2-}] = \frac{K_1K_2[H_2M]}{[H_3O^+]^2}$$

After substituting these quantities, the mass-balance equation becomes

$$C_T = [H_2M] + \frac{K_1[H_2M]}{[H_3O^+]} + \frac{K_1K_2[H_2M]}{[H_3O^+]^2}$$

which can be converted to

$$[H_2M] = \frac{C_T[H_3O^+]^2}{[H_3O^+]^2 + K_1[H_3O^+] + K_1K_2}$$

Substituting this value for $[H_2M]$ into the equation defining α_0 gives

$$\alpha_0 = \frac{[H_2M]}{C_T} = \frac{[H_3O^+]^2}{[H_3O^+]^2 + K_1[H_3O^+] + K_1K_2} \qquad (9\text{-}20)$$

By similar manipulation it is easily shown that

$$\alpha_1 = \frac{[HM^-]}{C_T} = \frac{K_1[H_3O^+]}{[H_3O^+]^2 + K_1[H_3O^+] + K_1K_2} \qquad (9\text{-}21)$$

$$\alpha_2 = \frac{[M^{2-}]}{C_T} = \frac{K_1K_2}{[H_3O^+]^2 + K_1[H_3O^+] + K_1K_2} \qquad (9\text{-}22)$$

Note that the denominator is the same for each expression; calculation of α-values at any desired pH is thus relatively simple. Furthermore, the

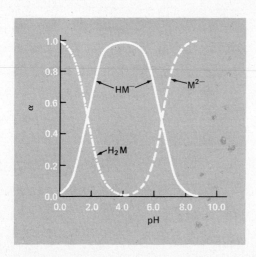

FIGURE 9-10 Composition of H_2M solutions as a function of pH.

equations illustrate that the fractional amount of each species at any fixed pH is independent of the total concentration, C_T.

The three curves plotted in Figure 9-10 show the α-values for each maleate-containing species as a function of pH. A consideration of these curves in conjunction with the titration curve for maleic acid (Figure 9-7) gives a clear picture of all concentration changes that occur during the course of the titration. For example, Figure 9-7 reveals that before the addition of any base the pH of the solution is 1.5. Referring to Figure 9-10, we see that at this pH, α_0 for H_2M is roughly 0.7, whereas α_1 for HM^- is approximately 0.3. For all practical purposes α_2 is zero. Thus approximately 70% of the maleic acid exists in the undissociated form and 30% exists as HM^-. With addition of base the pH rises, as does the fraction of HM^-. At the first equivalence point (pH = 4.12) essentially all of the maleate is present as HM^- ($\alpha_1 \rightarrow 1$). Beyond the first equivalence point HM^- decreases and M^{2-} increases. At the second equivalence point (pH = 9.37) it is evident that essentially all of the maleate exists as M^{2-}.

PROBLEMS

*1. Calculate the pH of a 0.100-F solution of
   (a) glycolic acid.
   (b) hydrazoic acid.
   (c) formic acid.
   (d) iodic acid.
   (e) picric acid.
   (f) ammonium chloride.

2. Calculate the pH of a 0.0800-F solution of
   (a) hydrogen cyanide.
   (b) lactic acid.
   (c) nitrous acid.
   (d) sulfamic acid.
   (e) trichloroacetic acid.
   (f) hydroxylamine hydrochloride.

*3. Calculate the pH of a 0.0100-F solution of
   (a) ammonia.
   (b) ethanolamine.
   (c) pyridine.
   (d) aniline.
   (e) sodium cyanide.
   (f) sodium lactate.

4. Calculate the pH of a 0.0800-F solution of
   (a) methylamine.
   (b) trimethylamine.
   (c) hydroxylamine.
   (d) sodium phenolate.
   (e) piperidine.
   (f) sodium hypochlorite.

*5. Calculate the pH of a solution produced by mixing 30.0 ml of 0.0200-F $HClO_4$ with 30.0 ml of
   (a) water. 2
   (b) 0.0100-F NaOH. 2·30
   (c) 0.0200-F NaOH. 7·00
   (d) 0.0300-F NaOH. 11·7
   (e) 0.0200-F $Ba(OH)_2$. 12·0
   (f) 0.0200-F hypochlorous acid HOCl. 2·00

6. Calculate the pH of a solution prepared by mixing 25.0 ml of 0.0300-F NaOH with 25.0 ml of
   (a) water.
   (b) 0.0200-F Ba(OH)$_2$.
   (c) 0.0300-F HCl.
   (d) 0.0200-F HCl.
   (e) 0.0400-F HCl.
   (f) 0.0200-F NH$_3$.

*7. Calculate the pH of a solution prepared by mixing 40.0 ml of 0.0300-F sodium hypochlorite, NaOCl, with 40.0 ml of
   (a) water. 9.84
   (b) 0.0100-F NaOH. 11.70
   (c) 0.0200-F HCl. 7.22
   (d) 0.0300-F HCl. 4.67
   (e) 0.0400-F HCl. 2.30
   (f) 0.0300-F HOCl. 7.52

8. Calculate the pH of a solution prepared by mixing 25.0 ml of 0.0500-F NH$_3$ with 25.0 ml of
   (a) water.
   (b) 0.0100-F NaOH.
   (c) 0.0300-F HCl.
   (d) 0.0500-F HCl.
   (e) 0.0700-F HCl.
   (f) 0.0250-F NH$_4$Cl.

*9. Calculate the pH of a solution prepared by mixing 50.0 ml of 0.200-F ethylamine hydrochloride, C$_2$H$_5$NH$_3$Cl, with 50.0 ml of
   (a) water.
   (b) 0.0400-F HCl.
   (c) 0.160-F NaOH.
   (d) 0.200-F NaOH.
   (e) 0.260-F NaOH.
   (f) 0.120-F ethylamine.

10. Calculate the pH of a solution prepared by mixing 100 ml of 0.0250-F propanoic acid, CH$_3$CH$_2$COOH, with 100 ml of
   (a) water.
   (b) 0.0100-F HClO$_4$.
   (c) 0.0150-F NaOH.
   (d) 0.0250-F NaOH.
   (e) 0.0350-F NaOH.
   (f) 0.0350-F sodium propanoate.

*11. Calculate the pH of a 0.0500-F solution of
   (a) carbonic acid.
   (b) arsenic acid.
   (c) o-phthalic acid.
   (d) malic acid.
   (e) phosphoric acid.
   (f) arsenious acid.

12. Calculate the pH of a 0.0800-F solution of
   (a) sodium oxalate.
   (b) sodium carbonate.
   (c) sodium malonate.
   (d) sodium sulfite.
   (e) sodium malate.
   (f) sodium phosphate.

*13. Calculate the pH of a 0.0600-F solution of
   (a) sodium phosphite.
   (b) sodium sulfide.
   (c) sodium phthalate.
   (d) sodium arsenate.
   (e) sodium tartrate.
   (f) sodium periodate.

14. Calculate the pH of a 0.0300-F solution of
   (a) malonic acid.
   (b) phosphorous acid.
   (c) periodic acid.
   (d) sulfurous acid.
   (e) oxalic acid.
   (f) hydrogen sulfide.

*15. Calculate the pH of a 0.0200-F solution of
   (a) NaH_2AsO_4.
   (b) Na_2HAsO_4.
   (c) $NaHC_2O_4$.
   (d) Na_2HPO_4.
   (e) NaH_2AsO_3.
   (f) potassium hydrogen phthalate.

16. Calculate the pH of a 0.100-F solution of
   (a) sodium hydrogen malate.
   (b) sodium hydrogen tartrate.
   (c) potassium hydrogen sulfite.
   (d) sodium hydrogen malonate.

*17. Calculate the pH of a buffer that is
   (a) 0.150 F in NH_3 and 0.100 F in $(NH_4)_2SO_4$.
   (b) 0.0200 F in phenol and 0.0300 F in sodium phenolate.
   (c) 0.0120 F in sulfamic acid and 0.0200 F in sodium sulfamate.
   (d) 0.0100 F in piperidine and 0.0200 F in piperidine hydrochloride.

18. Calculate the pH of a buffer that is
   (a) 0.0500 F in aniline and 0.100 F in aniline hydrochloride.
   (b) 1.20 F in sodium hypochlorite and 0.100 F in hypochlorous acid.
   (c) 0.200 F in trichloroacetic acid and 0.100 F in sodium trichloroacetate.
   (d) 0.150 F in hydroxylamine hydrochloride and 0.100 F in hydroxylamine.

*19. What is the pH of a buffer prepared by
   (a) dissolving 5.00 g of acetic acid and 8.00 g of sodium acetate and diluting to 250 ml?
   (b) dissolving 9.00 g of sodium acetate in water, adding 10.0 ml of 6.00-F HCl, and diluting to 500 ml?
   (c) dissolving 40.0 g of acetic acid in water, adding 100 ml of 2.17-F KOH, and diluting to 2.00 liters?

20. What is the pH of a buffer prepared by
   (a) dissolving 16.2 g of salicylic acid in water, adding 100 ml of 0.50-F NaOH, and diluting to 250 ml?
   (b) dissolving 10.0 g of salicylic acid and 10.0 g of sodium salicylate in water and diluting to 500 ml?
   (c) dissolving 9.50 g of sodium salicylate, adding 25.0 ml of 1.00-F HCl, and diluting to 250 ml?

*21. Calculate the molar ratio of conjugate acid to base or conjugate base to acid in a buffer that has a pH of 8.90 and contains
   (a) NH_3 and NH_4Cl.
   (b) CH_3NH_2 and CH_3NH_3Cl.
   (c) HCN and NaCN.
   (d) HOCl and NaOCl.

22. Calculate the molar ratio of conjugate acid to base or conjugate base to acid in a buffer that has a pH of 5.16 and contains
   (a) HNO_2 and $NaNO_2$.
   (b) CH_3CH_2COOH and CH_3CH_2COONa.
   (c) C_5H_5N and C_5H_5NHCl.
   (d) $C_6H_5NH_2$ and $C_6H_5NH_3Cl$.

*23. Calculate the pH change that occurs when the following solutions are diluted by a factor of 100.
  (a) 0.0400-F salicylic acid
  (b) 0.0200-F sodium salicylate and 0.0400-F salicylic acid
  (c) 0.200-F sodium salicylate and 0.400-F salicylic acid

24. Calculate the pH change that occurs when the following solutions undergo a 50-fold dilution.
  (a) 0.0400-F HNO_2
  (b) 0.0500-F $NaNO_2$ and 0.0400-F HNO_2
  (c) 0.500-F $NaNO_2$ and 0.400-F HNO_2

*25. Calculate the change in pH that occurs when 50.0 ml of 0.0100-F HCl are added to 50.0 ml of each of the undiluted solutions in Problem 23.

26. Calculate the change in pH that occurs when 50.0 ml of 0.0200-F HCl are added to 50.0 ml of each of the undiluted solutions in Problem 24.

*27. Calculate the change in pH that occurs when 50.0 ml of 0.0100-F NaOH are added to 50.0 ml of each of the undiluted solutions in Problem 23.

28. Calculate the change in pH that occurs when 50.0 ml of 0.0200-F NaOH are added to 50.0 ml of each of the undiluted solutions in Problem 24.

*29. How many grams of NH_4Cl must be added to 1.00 liter of 0.125-F NH_3 to give a buffer of pH 10.16 (assume no volume change)?

30. How many grams of sodium acetate must be added to 5.00 liters of 0.176-F acetic acid to produce a buffer of pH 4.96 (assume no volume change)?

*31. What volume of 0.500-F HCl must be added to 1.00 liter of 0.225-F NH_3 to produce a buffer with a pH of 9.40?

32. What volume of 0.400-F NaOH must be added to 800 ml of 0.180-F acetic acid to produce a buffer with a pH of 4.40?

*33. Calculate the pH of the solution that results when 40.0 ml of 0.200-F NaOH are mixed with 30.0 ml of
  (a) 0.300-F H_3PO_4.   (c) 0.200-F H_3PO_4.   (e) 0.100-F H_3PO_4.
  (b) 0.300-F NaH_2PO_4.   (d) 0.200-F NaH_2PO_4.

34. Calculate the pH of the solution that results when 20.0 ml of 0.400-F HCl are mixed with 30.0 ml of
  (a) 0.300-F Na_3PO_4.   (c) 0.200-F Na_3PO_4.   (e) 0.100-F Na_3PO_4.
  (b) 0.300-F Na_2HPO_4.   (d) 0.200-F Na_2HPO_4.

*35. Calculate the pH of the solution that results when 40.0 ml of 0.200-F H_3PO_4 are mixed with 30.0 ml of
  (a) 0.200-F NaOH.   (c) 0.400-F NaOH.   (e) 0.400-F Na_2HPO_4.
  (b) 0.200-F Na_2HPO_4. (d) 0.400-F NaH_2PO_4. (f) 0.400-F Na_3PO_4.

36. Calculate the pH of the solution that results when 20.0 ml of 0.400-F Na_3PO_4 are mixed with 30.0 ml of
- **(a)** 0.200-F HCl.
- **(c)** 0.400-F HCl.
- **(e)** 0.400-F NaH_2PO_4.
- **(b)** 0.200-F Na_2HPO_4.
- **(d)** 0.400-F Na_2HPO_4.
- **(f)** 0.400-F H_3PO_4.

*37. What volume of 1.20-F NaOH should be added to 250 ml of 0.800-F H_3PO_4 to yield a buffer with a pH of **(a)** 3.00? **(b)** 7.00?

38. What volume of 1.50-F HCl should be added to 250 ml of 0.800-F Na_3PO_4 to yield a buffer with a pH of **(a)** 11.70? **(b)** 6.40?

39. Calculate the pH of the solution that results from the addition of 0.00, 10.00, 20.00, 35.00, 39.00, 40.00, 41.00, 45.00, and 50.00 ml of 0.100-F HCl in the titration of
- ***(a)** 50.00 ml of 0.0800-F methylamine.
- **(b)** 50.00 ml of 0.0800-F hydroxylamine.
- ***(c)** 50.00 ml of 0.0800-F pyridine.
- **(d)** 50.00 ml of 0.0800-F sodium cyanide.

40. Calculate the pH of the solution that results from the addition of 0.00, 5.00, 12.50, 20.00, 24.00, 25.00, 26.00, and 30.00 ml of 0.200-F NaOH in the titration of
- ***(a)** 50.00 ml of 0.1000-F benzoic acid.
- **(b)** 50.00 ml of 0.1000-F hypochlorous acid.
- **(c)** 50.00 ml of 0.1000-F formic acid.
- ***(d)** 50.00 ml of 0.1000-F aniline hydrochloride.

*41. Calculate the pH of the following carbonate-containing solutions:

FORMAL CONCENTRATION

	H_2CO_3	$NaHCO_3$	Na_2CO_3
(a)	0.0500	0.000	0.000
(b)	0.000	0.200	0.000
(c)	0.000	0.000	0.150
(d)	0.0500	0.150	0.000
(e)	0.000	0.150	0.0500

42. Calculate the pH of the following phthalate-containing solutions where H_2A symbolizes o-phthalic acid, $C_6H_4(COOH)_2$:

FORMAL CONCENTRATION

	H_2A	NaHA	Na_2A
(a)	0.0250	0.000	0.000
(b)	0.000	0.0300	0.000
(c)	0.000	0.000	0.0750
(d)	0.0400	0.0600	0.000
(e)	0.000	0.0600	0.0400

*43. Calculate the pH of the following arsenate-containing solutions:

FORMAL CONCENTRATION

	H_3AsO_4	NaH_2AsO_4	Na_2HAsO_4
(a)	0.0250	0.000	0.000
(b)	0.000	0.0800	0.000
(c)	0.000	0.000	0.150
(d)	0.115	0.220	0.000
(e)	0.000	0.220	0.115

44. Calculate the pH of the following oxalate-containing solutions:

FORMAL CONCENTRATION

	H_2Ox	NaHOx	Na_2Ox
(a)	0.0660	0.000	0.000
(b)	0.000	0.400	0.000
(c)	0.000	0.000	0.0250
(d)	0.325	0.210	0.000
(e)	0.000	0.210	0.325

*45. Calculate the pH of a solution that is
  (a) 0.0200 F in HCl and 0.150 F in succinic acid.
  (b) 0.0200 F in HCl and 0.150 F in oxalic acid.

46. Calculate the pH of a solution that is
  (a) 0.0100 F in HCl and 0.125 F in nitrous acid.
  (b) 0.0100 F in HCl and 0.125 F in chloroacetic acid.

*47. What is the pH of a buffer prepared by mixing 500 ml of 1.000-F H_3PO_4 with 500 ml of (use approximate method)
  (a) 0.600-F NaOH?　　　　　(d) 0.800-F Na_2HPO_4?
  (b) 1.600-F NaOH?　　　　　(e) 1.400-F Na_2HPO_4?
  (c) 0.800-F NaH_2PO_4?

48. What is the pH of a buffer prepared by mixing 400 ml of 0.250-F sodium phthalate (Na_2A) with 600 ml of (use approximate method)
  (a) 0.100-F HCl?　　　　　(d) 0.250-F H_2A?
  (b) 0.200-F HCl?　　　　　(e) 0.100-F H_2A?
  (c) 0.250-F NaHA?

49. Derive a curve for the titration of 50.00 ml of a 0.05000-F solution of compound A with a 0.1000-F solution of compound B. Calculate the pH after the following additions: 0.00, 10.00, 20.00, 24.00, 25.00, 26.00, 35.00, 45.00, 49.00, 50.00, 51.00, and 60.00.

	A	B
*(a)	H_2SO_4	NaOH
*(b)	Na_2CO_3	HCl
(c)	H_3AsO_4	NaOH

50. For pH values of 1.00, 3.00, 5.00, 7.00, 9.00, 11.00, and 13.00, calculate α-values for each species in an aqueous solution of

   *(a) oxalic acid.   (c) fumaric acid.

   *(b) hydrogen sulfide.   (d) succinic acid.

*51. Use α-values to identify the major (that is, $\geq 5\%$)

   (a) phosphate-containing species in a solution with a pH of 8.00.

   (b) oxalate-containing species in a solution with a pH of 4.00.

   (c) sulfite-containing species in a solution with a pH of 6.00.

   (d) phosphate-containing species in a solution with a pH of 3.00.

*52. Use the information in Problem 51 to calculate the molar concentration of

   (a) HPO_4^{2-} in a solution that is $0.0600\ F$ in phosphate-containing species and is buffered to a pH of 8.00.

   (b) $H_2C_2O_4$ in a solution that is $0.200\ F$ in oxalate-containing species and is buffered to a pH of 4.00.

   (c) SO_3^{2-} in a solution prepared by dissolving 0.500 fw of Na_2SO_3 in sufficient dilute HCl to give 1.00 liter with a pH of 6.00.

   (d) H_3PO_4 in a solution prepared by dissolving 0.100 fw of $Na_2HPO_4 \cdot 2H_2O$ in sufficient dilute HCl to give 400 ml with a pH of 3.00.

53. Use α-values to identify the major (that is, $\geq 5\%$)

   (a) arsenate-containing species in a solution with a pH of 9.00.

   (b) carbonate-containing species in a solution with a pH of 5.00.

   (c) phosphite-containing species in a solution with a pH of 7.00.

54. Use the information in Problem 53 to calculate the molar concentration of

   (a) $H_2AsO_4^{-}$ in a solution that is $0.100\ F$ in arsenate-containing species and is buffered to a pH of 9.00.

   (b) CO_3^{2-} in a solution prepared by dissolving 0.600 fw of $NaHCO_3$ in sufficient dilute NaOH to give 900 ml with a pH of 5.00.

   (c) $H_2PO_3^{-}$ in a solution prepared by dissolving 0.500 fw of Na_2HPO_3 in sufficient dilute HCl to give 800 ml with a pH of 7.00.

10

applications of neutralization titrations

Volumetric methods based upon neutralization involve the titration of hydronium or hydroxide ions produced directly or indirectly from the sample. Neutralization methods are extensively employed in chemical analysis. For most applications water serves conveniently as the solvent; it should be recognized, however, that the acidic or basic character of a solute is determined in part by the nature of the solvent in which it is dissolved. Consequently the substitution of some other solvent for water may permit a titration that cannot be successfully performed in an aqueous environment.

Reagents for Neutralization Reactions

In Chapter 9 we noted that the most pronounced pH changes in the equivalence point region occur when strong acids and strong bases are involved in the titration. It is for this reason that standard solutions for neutralization titrations are prepared from such acids and bases.

PREPARATION OF STANDARD ACID SOLUTIONS

Hydrochloric acid is the most commonly used standard acid for volumetric analysis. Dilute solutions of the reagent are indefinitely stable and can be used in the presence of most cations without complicating precipitation reactions. It is reported that 0.1-N solutions can be boiled for as long as 1 hr without loss of acid, provided that water lost by evaporation is periodically replaced; 0.5-N solutions can be boiled for at least 10 min without significant loss.

Solutions of perchloric acid and sulfuric acid are also stable and can serve as standard reagents in titrations where the presence of chloride ion would cause precipitation difficulties. Standard solutions of nitric acid are seldom used because of their oxidizing properties.

A standard acid solution is ordinarily prepared by diluting an approximate volume of the concentrated reagent and standardizing it against a primary standard base. Less frequently the composition of the concentrated acid is established through careful density measurement, following which a weighed quantity is diluted to an exact volume. (Tables relating density of reagents to composition are found in most chemistry or chemical engineering handbooks.) A stock solution with an exactly known hydrochloric acid concentration can also be prepared by distillation of the concentrated reagent; under controlled conditions the final quarter of the distillate, which is known as *constant boiling* HCl, has a fixed and known composition, its acid content being dependent only upon the atmospheric pressure. For a pressure, P, between 670 and 780 mm Hg, the weight in air of the distillate that contains exactly one equivalent of acid is given by[1]

$$\frac{\text{g constant boiling HCl}}{\text{equivalent}} = 164.673 + 0.02039\,P \qquad (10\text{-}1)$$

Standard solutions can be prepared by diluting a weighed quantity of this acid to an exactly known volume.

STANDARDIZATION OF ACIDS

Sodium Carbonate. Sodium carbonate is frequently used as a standard for acid solutions. Primary standard-grade sodium carbonate is available commercially; it can also be prepared by heating purified sodium hydrogen carbonate at 270 to 300°C for 1 hr:

$$2NaHCO_3 \rightarrow Na_2CO_3 + H_2O + CO_2(g)$$

As shown in Figure 9-9, two end points are observed in the titration of

---

[1] Association of Official Analytical Chemists, *Official Methods of Analysis*, 12th ed. Washington, D.C.: Association of Official Analytical Chemists, 1975, p. 944.

sodium carbonate. The first, corresponding to conversion of carbonate to hydrogen carbonate, occurs at a pH of about 8.4. The second, involving the formation of carbonic acid, is observed at about pH 4.0; the second end point is always used for standardization because the change in pH is greater. An even sharper end point change can be achieved by boiling the solution briefly to eliminate the reaction product, carbonic acid. The sample is titrated to the first appearance of the acid color of the indicator (such as bromocresol green or methyl orange). At this point the solution contains a large amount of carbonic acid and a small amount of unreacted hydrogen carbonate. Boiling effectively destroys this buffer by eliminating the carbonic acid:

$$H_2CO_3 \rightarrow CO_2(g) + H_2O$$

As a result, the solution again acquires an alkaline pH owing to the residual hydrogen carbonate ion. The titration is completed after the solution has cooled. Now, however, considerably larger changes in pH attend the final additions of acid; a sharper indicator transition is thus observed.

As an alternative, the acid can be introduced in an amount slightly in excess of that needed to convert the sodium carbonate to carbonic acid. The solution is boiled, as before, to remove carbon dioxide; after cooling, the excess acid is back-titrated with a dilute solution of base. Any indicator suitable for a strong acid–strong base titration can be employed. The volume ratio between the acid and the base must be established by an independent titration.

Directions for the standardization of hydrochloric acid solutions against sodium carbonate are to be found in Section 3, Chapter 20.

Other Primary Standards for Acids. *Tris*-(hydroxymethyl)amino-methane $(HOCH_2)_3CNH_2$, known also as TRIS or THAM, is available in primary standard purity from commercial sources. It possesses the advantage of a substantially larger equivalent weight than sodium carbonate.

Other standards include sodium tetraborate, mercury(II) oxide, and calcium oxalate; details concerning their use can be found in standard reference works.[2]

PREPARATION OF STANDARD SOLUTIONS OF BASE

Sodium hydroxide is the most common basic reagent, although potassium hydroxide and barium hydroxide are also encountered. None of these is obtainable in primary standard purity; after preparation to approximate strength, their solutions must be standardized.

Standard base solutions are reasonably stable so long as they are

---

[2] See, for example, I. M. Kolthoff and V. A. Stenger, *Volumetric Analysis*, vol. 2. New York: Interscience, 1947, pp. 73–93; L. Meites, *Handbook of Analytical Chemistry*. New York: McGraw-Hill, 1963, p. 3-34.

protected from prolonged exposure to glass and contact with the atmosphere. Sodium hydroxide reacts slowly with glass to form silicates; a standard solution that is to be employed for longer than a week or two should be stored in a polyethylene bottle or a glass bottle that has been coated with paraffin.

Effect of Carbon Dioxide upon Standard Base Solutions. In solution as well as the solid state the hydroxides of sodium, potassium, and barium avidly react with atmospheric carbon dioxide to produce the corresponding carbonates:

$$CO_2 + 2OH^- \rightarrow CO_3^{2-} + H_2O$$

The absorption of carbon dioxide by a standardized solution of a base does not necessarily alter its acid titer. For example, if potassium or sodium hydroxide solutions are employed where circumstances permit the use of an indicator with an acid transition range (bromocresol green, for example), each carbonate ion in the reagent will have reacted with two hydronium ions of the analyte (see Figure 9-9). That is,

$$CO_3^{2-} + 2H_3O^+ \rightarrow H_2CO_3 + 2H_2O$$

This consumption is chemically equivalent to the amount of base used to form the carbonate, and no error will be incurred.

Unfortunately most applications of standard base require the use of an indicator with a basic transition range (phenolphthalein, for example). Here each carbonate ion will have consumed only one hydronium ion when the color change of the indicator is observed:

$$CO_3^{2-} + H_3O^+ \rightarrow HCO_3^- + H_2O$$

The effective normality of the base is thus diminished and a determinate error (called a *carbonate error*) will result.

When carbon dioxide is absorbed by standard barium hydroxide, precipitation of barium carbonate occurs:

$$CO_2 + Ba^{2+} + 2OH^- \rightarrow BaCO_3(s) + H_2O$$

The acid titer is thus decreased regardless of the indicator employed in the titration; a carbonate error is the inevitable consequence.

The solid reagents used to prepare standard solutions of base represent a further source of carbonate ion. The extent of contamination is frequently great, owing to the absorption of atmospheric carbon dioxide by the solid. As a result, even freshly prepared solutions of base are likely to contain significant quantities of carbonate. Its presence will not cause a carbonate error provided the analysis is performed with the same indicator that was used for standardization; this restriction, however, causes the reagent to lose much of its versatility.

Several methods exist for the preparation of carbonate-free

hydroxide solutions. Barium hydroxide may be employed as the reagent; the sparingly soluble carbonate can be further diminished by the addition of a neutral barium salt such as the chloride or the nitrate. A barium salt can also be used to eliminate the carbonate from a potassium or sodium hydroxide solution. The presence of barium ion is frequently undesirable, however, owing to its tendency to form slightly soluble salts with anions that may be present in the sample.

Carbonate-free solutions of the alkali-metal hydroxides may be prepared by direct solution of the freshly cleaned metals. Most chemists think that the possibility of fire and explosion during the solution process represents an unacceptable risk.

The preferred method for preparing sodium hydroxide solutions takes advantage of the very low solubility of sodium carbonate in concentrated solutions of the alkali. An approximately 50% aqueous solution of sodium hydroxide is prepared (or purchased from commercial sources); after the sodium carbonate has settled, a portion of the supernatant liquid is decanted and diluted to give the desired concentration. Details for this procedure are given in Section 3 of Chapter 20. Alternatively, the concentrated sodium hydroxide solution can be filtered to eliminate the sodium carbonate.

A carbonate-free base solution must be prepared from water that contains no carbon dioxide. Distilled water, which is frequently supersaturated with respect to carbon dioxide, should be boiled briefly to eliminate the gas; the water is allowed to cool to room temperature before the introduction of base because hot alkali solutions rapidly absorb carbon dioxide.

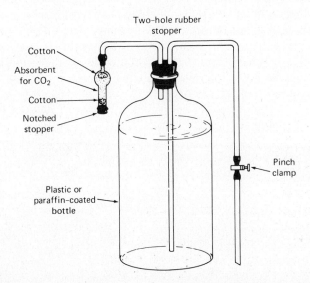

Two-hole rubber
stopper

Cotton

Absorbent
for CO_2

Cotton

Notched
stopper

Pinch
clamp

Plastic or
paraffin-coated
bottle

FIGURE 10-1 Arrangement for the storage of standard base solutions.

Figure 10-1 shows an arrangement for preventing the uptake of atmospheric carbon dioxide by basic solutions during storage. Air entering the vessel is passed over a solid absorbent for CO_2 such as soda lime or Ascarite.[3] The contamination that occurs as the solution is transferred from this storage bottle to the buret is negligible. Absorption during a titration can be minimized by covering the open end of the buret with a small test tube or beaker.

If a standard solution of base is to be used for no longer than a week or so, storage in a tightly stoppered polyethylene bottle will usually provide sufficient protection against the uptake of atmospheric carbon dioxide. Care should be taken to keep the bottle stoppered except during the brief periods when the contents are being transferred to a buret.

The parts of any ground-glass fitting will freeze upon prolonged exposure to alkaline solutions. For this reason glass-stoppered containers should not be used for the storage of strong bases; similarly, burets equipped with glass stopcocks should be promptly drained and thoroughly cleaned after being used to dispense these reagents. A better alternative is the employment of burets equipped with Teflon stopcocks.

STANDARDIZATION OF BASES

Several excellent primary standards are available for the standardization of bases. Most are weak organic acids that require the use of an indicator with a basic transition range.

Potassium Hydrogen Phthalate, $KHC_8H_4O_4$. Potassium hydrogen phthalate possesses many qualities that are desirable in a primary standard. It is a nonhygroscopic crystalline solid with a high equivalent weight. For most purposes, the commercial analytical-grade salt can be used without the need for further purification. Potassium hydrogen phthalate of certificated purity is available from the National Bureau of Standards for the most exacting work.

Directions for the standardization of sodium hydroxide solutions are given in Section 3, Chapter 20.

Other Primary Standards for Bases. Benzoic acid is obtainable in primary standard purity and can be used for the standardization of bases. Because its solubility in water is limited, the reagent is ordinarily dissolved in ethanol prior to dilution and titration.

Potassium hydrogen iodate, $KH(IO_3)_2$, is an excellent primary standard with a high equivalent weight. It is also a strong acid; as a result,

---

[3] Arthur H. Thomas Company, Philadelphia, Pa. Ascarite consists of sodium hydroxide deposited on asbestos.

virtually any indicator with a pH transition range between 4 and 10 can be used.

Typical Applications of Neutralization Titrations

The most obvious application of neutralization methods is to determine the innumerable inorganic, organic, and biological species that possess inherent acidic or basic properties. Equally important, however, are the many applications that involve conversion of the analyte to an acid or a base by suitable chemical treatment followed by titration with a standard strong base or acid.

Two types of end points are commonly used for neutralization titrations. The first, based upon the color change of an indicator, was discussed in Chapter 9. The second involves the direct measurement of pH throughout the titration by means of a glass-calomel electrode system; here the potential of the glass electrode is directly proportional to pH. Appropriate plots of the data permit establishment of the end point. The potentiometric procedure is considered in Chapter 14.

ELEMENTAL ANALYSIS

Several important elements that occur in organic and biological systems are conveniently determined by methods that involve an acid-base titration as the final step. Generally the elements susceptible to this type of analysis are nonmetallic. Principal among these are carbon, nitrogen, chlorine, bromine, sulfur, phosphorus, and fluorine; in addition, similar methods exist for several less commonly encountered species. In each instance the element is converted to an inorganic acid or base that can then be titrated. A few examples follow.

Nitrogen. Nitrogen occurs in many important materials such as proteins, synthetic drugs, fertilizers, explosives, and potable water supplies. The analysis for nitrogen is thus of singular importance to research and to industry.

Two principal methods exist for the determination of organic nitrogen. The *Dumas* method is suitable for the analysis of virtually all organic nitrogen compounds. It involves mixing the sample with powdered copper(II) oxide and igniting the mixture in a combustion tube. At elevated temperatures the sample is oxidized to carbon dioxide, water, elemental nitrogen, and perhaps nitrogen oxides. The last are reduced to elemental nitrogen as the gases pass through a packing of hot copper. The ignition products are swept by a stream of carbon dioxide into a gas buret filled with highly concentrated potassium hydroxide. This solution completely absorbs the carbon dioxide, water, and any sulfur dioxide or hydrogen halide that may be present. Elemental nitrogen is not absorbed and partially displaces the liquid from the buret; its volume is thus measured directly.

The *Kjeldahl method*, which was first described in 1883, is one of the most widely used of all chemical analyses. It requires no special equipment and is readily adapted to the routine analysis of large numbers of samples. The Kjeldahl method (or one of its modifications) is the standard means for determining protein nitrogen in grains, meats, and other biological materials.

In essence the sample is oxidized in hot, concentrated sulfuric acid during which the bound nitrogen is converted to ammonium ion. The solution is then treated with an excess of strong base; following distillation the liberated ammonia is titrated.

The critical step in the Kjeldahl method is the oxidation with sulfuric acid. The carbon and hydrogen in the sample are converted to carbon dioxide and water, respectively. The fate of nitrogen, however, depends upon its state of combination in the original sample. If it existed as an amine or an amide, as in proteinaceous matter, conversion to ammonium ion is nearly always quantitative. On the other hand, nitrogen in higher oxidation states, such as nitro, azo, and azoxy groups, will be converted to the elemental state or to nitrogen oxides during the oxidation step and will not be retained in the sulfuric acid. Low results due to the existence of nitrogen in these oxidation states can be eliminated by pretreatment of the sample with a reducing agent; complete conversion to ammonium ion during the digestion step is thus assured. One prereduction scheme calls for the addition of salicylic acid and sodium thiosulfate to the concentrated sulfuric acid solution containing the sample; the digestion is then performed in the usual way.

Certain aromatic heterocyclic compounds such as pyridine and its derivatives are particularly resistant to complete oxidation by sulfuric acid. Thus unless special precautions are followed, low results attend the analysis for nitrogen in samples containing these species (see Figure 4-1, p. 50).

The oxidation process is the most time-consuming step in the Kjeldahl method; an hour or more may be needed for refractory samples. Of the many modifications aimed at improving the kinetics of the process, the one proposed by Gunning is now almost universally employed. Here a neutral salt such as potassium sulfate is added to increase the boiling point of the sulfuric acid solution and thus the temperature at which the oxidation occurs. Care is needed, however, because oxidation of the ammonium ion can occur if the salt concentration is too great. This problem is enhanced if evaporation of the sulfuric acid is excessive during digestion.

Attempts to hasten the Kjeldahl oxidation by introducing such stronger oxidizing agents as perchloric acid, potassium permanganate, and hydrogen peroxide fail because ammonium ions are partially oxidized to volatile nitrogen oxides.

Many substances catalyze the oxidation step. Mercury, copper, and selenium, either combined or in the elemental state, are effective. Mercury(II), if present, must be precipitated with hydrogen sulfide prior to the distillation step; otherwise some ammonia will be retained as an ammine complex.

Figure 10-2 illustrates a typical distillation arrangement for a Kjeldahl analysis. The long-necked container, which is used for both oxidation and distillation, is called a *Kjeldahl flask*. After the oxidation is judged complete, the contents of the flask are cooled, diluted with water, and then made basic to liberate the ammonia:

$$NH_4^+ + OH^- \rightarrow NH_3(g) + H_2O$$

To avoid the loss of ammonia during neutralization a concentrated sodium hydroxide solution, more dense than the diluted oxidation mixture, is carefully poured down the side of the flask to form a second, lower layer. The flask is quickly joined to the distillation apparatus; only then are the two layers mixed by gently swirling the flask.

In addition to the Kjeldahl flask, the apparatus shown in Figure 10-2 includes a spray trap that prevents droplets of the strongly alkaline

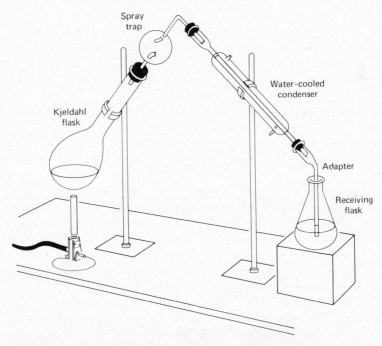

FIGURE 10-2 Apparatus for the distillation and collection of ammonia.

solution from being carried over in the vapor stream. A water-cooled condenser is provided. During distillation the end of the adapter tube extends below the surface of an acidic solution in the receiving flask.

Two methods exist for titration of the collected ammonia. In one the receiver contains a known quantity of standard acid; after distillation is complete, the excess acid is back-titrated with standard base. An indicator with an acidic transition interval is required, owing to the presence of ammonium ions at equivalence. A convenient alternative that requires only one standard solution involves use of an unmeasured excess of boric acid in the receiving flask; its reaction with ammonia is

$$HBO_2 + NH_3 \rightleftharpoons NH_4^+ + BO_2^-$$

The borate ion produced is a reasonably strong base and can be titrated with a standard solution of hydrochloric acid:

$$BO_2^- + H_3O^+ \rightleftharpoons HBO_2 + H_2O$$

At the equivalence point the solution contains boric acid and ammonium ions; an indicator with an acidic transition interval is thus required.

Details of the Kjeldahl method are found in Analysis 3-8, Chapter 20.

Sulfur. Sulfur in organic and biological materials is conveniently determined by burning the sample in a stream of oxygen. The sulfur dioxide (and sulfur trioxide) formed during the oxidation is collected in a dilute solution of hydrogen peroxide:

$$SO_2(g) + H_2O_2 \rightarrow H_2SO_4$$

The sulfuric acid is then titrated with standard base.

Other Elements. Table 10-1 lists other elements that can be determined by neutralization methods.

DETERMINATION OF INORGANIC SUBSTANCES

Numerous inorganic species can be determined by titration with strong acids or bases. A few examples follow.

Ammonium Salts. Ammonium salts can be conveniently determined by conversion to ammonia with strong base and distillation in the Kjeldahl apparatus shown in Figure 10-2. The ammonia is then collected and titrated as in the Kjeldahl method.

Nitrates and Nitrites. The method just considered can also be applied to the determination of inorganic nitrate or nitrite by reducing these species to ammonium ion. Devarda's alloy (50% Cu, 45% Al, 5% Zn) is commonly

TABLE 10-1
Elemental Analysis Based on Neutralization Titrations

ELE-MENT	CONVERTED TO	ABSORPTION OR PRECIPITATION PRODUCTS	TITRATION
N	NH_3	$NH_3(g) + H_3O^+ \rightarrow NH_4^+ + H_2O$	Excess HCl with NaOH
S	SO_2	$SO_2(g) + H_2O_2 \rightarrow H_2SO_4$	NaOH
C	CO_2	$CO_2(g) + Ba(OH)_2 \rightarrow$ $BaCO_3(s) + H_2O$	Excess $Ba(OH)_2$ with HCl
Cl(Br)	HCl	$HCl(g) + H_2O \rightarrow Cl^- + H_3O^+$	NaOH
F	SiF_4	$SiF_4(g) + H_2O \rightarrow H_2SiF_6$	NaOH
P	H_3PO_4	$12H_2MoO_4 + 3NH_4^+ + H_3PO_4 \rightarrow$ $(NH_4)_3PO_4 \cdot 12MoO_3(s) +$ $12H_2O + 3H^+$	
		$(NH_4)_3PO_4 \cdot 12MoO_3(s) + 26OH^- \rightarrow$ $HPO_4^{2-} + 12MoO_4^{2-} + 14H_2O +$ $3NH_3(g)$	Excess NaOH with HCl

used as the reducing agent. Granules of the alloy are introduced to a strongly alkaline solution of the sample in a Kjeldahl flask. The ammonia is distilled after reaction is complete. Arnd's alloy (60% Cu, 40% Mg) can also be used as the reducing agent.

Carbonate and Carbonate Mixtures. The qualitative and quantitative determination of constituents in a solution containing sodium carbonate, sodium hydrogen carbonate, and sodium hydroxide, alone or admixed, provides interesting examples of neutralization titrations. No more than two of these three constituents can exist in appreciable amount in any solution because reaction will eliminate the third. Thus the mixing of sodium hydroxide with sodium hydrogen carbonate results in the formation of sodium carbonate until one or the other of the original reactants is exhausted. If the sodium hydroxide is used up, the solution will contain sodium carbonate and sodium hydrogen carbonate; if sodium hydrogen carbonate is used up, sodium carbonate and sodium hydroxide will remain. Finally, if equiformal amounts are mixed, the principal solute species will be sodium carbonate.

The analysis of such mixtures requires two titrations with standard acid. An indicator with a transition in the vicinity of pH 8 to 9 is used for one; an acid-range indicator is used for the other. The composition of the solution can be deduced from the relative volumes of acid needed to titrate

TABLE 10-2

Volume Relationship in the Analysis of Mixtures Containing Carbonate, Hydrogen Carbonate, and Hydroxide Ions

CONSTITUENTS PRESENT	RELATIONSHIP BETWEEN VOLUME OF ACID NEEDED TO REACH A PHENOLPHTHALEIN END POINT, V_{ph}, AND A BROMOCRESOL GREEN END POINT, V_{bg}
NaOH	$V_{ph} = V_{bg}$
Na_2CO_3	$V_{ph} = \frac{1}{2}V_{bg}$
$NaHCO_3$	$V_{ph} = 0,\ V_{bg} > 0$
NaOH, Na_2CO_3	$V_{ph} > \frac{1}{2}V_{bg}$
Na_2CO_3, $NaHCO_3$	$V_{ph} < \frac{1}{2}V_{bg}$

equal volumes of the sample (see Table 10-2 and Figure 9-9). Once the composition has been established, the volume data can be used to determine the concentration of each component in the sample.

EXAMPLE

A solution to be analyzed contained $NaHCO_3$, Na_2CO_3, NaOH, alone or in permissible combination. Titration of a 50.0-ml portion to a phenolphthalein end point required 22.1 ml of 0.100-F HCl. A second 50.0-ml aliquot required 48.4 ml of the HCl when titrated to a bromocresol green end point. Calculate the formal composition of the original solution.

Had the solution contained only NaOH, the volume of acid required would have been the same, regardless of indicator (that is, $V_{ph} = V_{bg}$). In fact, however, the second titration required a total of 48.4 ml. Because less than half of this amount was involved in the first titration, the solution must have contained some $NaHCO_3$ in addition to Na_2CO_3. We can now calculate the concentration of the two constituents. When the phenolphthalein end point was reached, the CO_3^{2-} originally present was converted to HCO_3^-. Thus

$$\text{no. mfw } Na_2CO_3 = 22.1 \times 0.100 = 2.21$$

The titration from the phenolphthalein to the bromocresol green end point $(48.4 - 22.1 = 26.3\ \text{ml})$ involved both the hydrogen carbonate originally present as well as that formed by titration of the carbonate. Thus

$$\text{no. mfw } NaHCO_3 + \text{no. mfw } Na_2CO_3 = 26.3 \times 0.100$$

Hence

$$\text{no. mfw } NaHCO_3 = 2.63 - 2.21 = 0.42$$

The formal concentrations are readily calculated from these data:

$$F_{Na_2CO_3} = \frac{2.21}{50.0} = 0.0442 \text{ mfw/ml}$$

$$F_{NaHCO_3} = \frac{0.42}{50.0} = 0.0084 \text{ mfw/ml}$$

In practice the titration used in this example is not entirely satisfactory because the pH change corresponding to the hydrogen carbonate equivalence point is not sufficient to give a sharp color change with a chemical indicator (see Figure 9-9). Titration to a color match with a solution containing an approximately equivalent amount of sodium hydrogen carbonate is helpful; nevertheless, errors of 1% or more must be expected.

The limited solubility of barium carbonate can be used to improve the titration of carbonate-hydroxide or carbonate–hydrogen carbonate mixtures. The *Winkler* method for the determination of carbonate-hydroxide mixtures involves titration of both components in an aliquot with an acid-range indicator. An excess of neutral barium chloride is then added to a second aliquot to precipitate the carbonate ion, following which the hydroxide ion is titrated to a phenolphthalein end point. Provided the concentration of the excess barium ion is about 0.1 *F*, the solubility of barium carbonate is too low to interfere with the titration.

An accurate analysis of a carbonate–hydrogen carbonate mixture can be achieved by establishing the total equivalents through titration of an aliquot to the acidic end point with an indicator such as bromocresol green. The hydrogen carbonate in a second aliquot is converted to carbonate by the addition of a known excess of standard base. After a large excess of barium chloride has been introduced, the excess base is titrated with standard acid to a phenolphthalein end point.

The presence of solid barium carbonate does not hamper end point detection in either of these methods.

Directions for the determination of sodium carbonate are given in Analysis 3-4, Chapter 20.

DETERMINATION OF ORGANIC FUNCTIONAL GROUPS

Neutralization titrations are convenient for the direct or indirect determination of several organic functional groups. Brief descriptions of methods for the more common groups follow.

Carboxylic and Sulfonic Acid Groups. Carboxylic and sulfonic acid groups are the two most common structures that impart acidity to organic compounds. Most carboxylic acids have dissociation constants that range between 10^{-4} and 10^{-6} and are thus readily titrated. An indicator that changes color in the basic range is required; phenolphthalein is widely used for this purpose.

Many carboxylic acids are not sufficiently soluble in water to permit a direct titration in this medium. Where this problem exists, the acid can be dissolved in ethanol and titrated with aqueous base. Alternatively, the acid may be dissolved in an excess of standard base; the unreacted base is then back-titrated with standard acid.

Sulfonic acids are generally strong acids and readily dissolve in water. Their titration with a base therefore is straightforward.

Neutralization titrations are often employed to determine equivalent weights of purified organic acids; the data serve as an aid in qualitative identification.

Amine Groups. Aliphatic amines generally have basic dissociation constants on the order of 10^{-5} and can thus be titrated directly with a solution of a strong acid. Aromatic amines such as aniline and its derivatives, on the other hand, are usually too weak for titration in aqueous medium ($K_b \sim 10^{-10}$); the same is true for cyclic amines with aromatic character such as pyridine and its derivatives. Saturated cyclic amines such as piperidine, on the other hand, are often similar to aliphatic amines.

Many amines that are not susceptible to neutralization titration in aqueous media are readily determined in a nonaqueous solvent, such as anhydrous acetic acid, that enhances their basicity.

Ester Groups. Esters are commonly determined by saponification with a measured quantity of standard base:

$$R_1COOR_2 + OH^- \rightarrow R_1COO^- + HOR_2$$

The excess base is then titrated with standard acid.

Esters vary widely in their reactivity toward saponification. Some require several hours of heating with a base to complete the saponification. A few react rapidly enough to permit direct titration with standard base. Typically the ester is refluxed with standard 0.5-N base for 1 to 2 hr. After cooling the excess base is determined with standard acid.

PROBLEMS

1. Identify the acid on the left side of each equation, and express its equivalent weight as a fraction or multiple of its gram formula weight.
   *(a) $H_3PO_4 + 2NaOH \rightarrow Na_2HPO_4 + 2H_2O$
   (b) $Ba(OH)_2 + 2KHC_8H_4O_4 \rightarrow BaC_8H_4O_4 + K_2C_8H_4O_4 + 2H_2O$
   *(c) $Na_2B_4O_7 + 2HCl + 5H_2O \rightarrow 4H_3BO_3 + 2NaCl$
   (d) $C_2H_4(NH_2)_2 + 2HCl \rightarrow C_2H_4(NH_3Cl)_2$
   *(e) $C_6H_4(COOCH_3)_2 + 2NaOH \rightarrow 2C_6H_4(COONa)_2 + 2CH_3OH$
   (f) $HgO + 2HCl \rightarrow 2H_2O + HgCl_2$

*2. Express the equivalent weight of each base on the left side of each equation in Problem 1 in terms of a fraction or multiple of its gram formula weight.

*3. Describe the preparation of
(a) 2.0 liters of 0.15-N NaOH from the solid.
(b) 5.00 liters of 0.125-N HCl from constant-boiling HCl (the HCl was distilled when the barometer read 759 mm Hg).
(c) 2.5 liters of 0.075-N H_3PO_4 (see Problem 1a for reaction) from the concentrated reagent which had a density of 1.26 and contains 40.0% (w/w) H_3PO_4.
(d) 500 ml of 0.0300-N $KHC_8H_4O_4$ from the pure compound (see Problem 1b for reaction).

4. Describe the preparation of
(a) 1.5 liters of 0.020-N $Ba(OH)_2$ from the solid.
(b) 4.0 liters of 1.5-N H_2SO_4 from a concentrated reagent that had a density of 1.48 and contained 58% (w/w) H_2SO_4.
(c) 750 ml of 0.0590-N $Na_2B_4O_7$ from pure $Na_2B_4O_7 \cdot 10H_2O$ (see Problem 1c for reaction).
(d) 1500 ml of 0.200-N HCl from constant-boiling HCl that was distilled at a pressure of 745 mm Hg.

*5. The following data were obtained for the standardization of HCl against samples of pure HgO that were dissolved in a solution containing excess KBr.

$$HgO(s) + 2Br^- + H_2O \rightarrow HgBr_2(aq) + 2OH^-$$

The OH^- ions were then titrated.

g HgO	ml HCl
0.2734	38.99
0.3016	42.96
0.2465	35.10
0.2512	35.84

(a) Calculate the mean normality for the acid.
(b) Calculate the relative standard deviation for the standardization.
(c) Calculate the 95% confidence interval for the mean, based on these data.

6. The following data were obtained for the standardization of HCl against pure Na_2CO_3. An excess of acid was added; the solution was boiled to remove CO_2, and the excess acid was back-titrated with a solution of base. The volume ratio between the reagents was then obtained and found to be 1.540 ml HCl per ml of NaOH.

g Na_2CO_3	ml HCl	ml NaOH
0.2367	40.00	5.92
0.3130	45.00	2.64
0.2888	40.00	1.44
0.2464	43.50	7.38

(a) Calculate the mean normality for the acid.

(b) Calculate the relative standard deviation for the standardization.

(c) Calculate the 95% confidence interval for the mean based on these data.

7. Suggest a range of sample weights for the indicated primary standard if it is desired to use between 35 and 45 ml of titrant.

*(a) 0.20-N KOH versus $KH(IO_3)_2$

*(b) 0.025-N $Ba(OH)_2$ versus benzoic acid (C_6H_5COOH)

*(c) 0.35-N $HClO_4$ versus $NaHCO_3$ where a weighed quantity is first ignited to give Na_2CO_3, which is then titrated

$$2NaHCO_3 \rightarrow Na_2CO_3 + H_2O + CO_2$$

(d) 0.050-N HCl versus TRIS, $(HOCH_3)_3CNH_2$

$$(HOCH_3)_3CNH_2 + H_3O^+ \rightarrow (HOCH_3)_3CNH_3^+ + H_2O$$

(e) 0.15-N NaOH versus $Na_2B_4O_7 \cdot 10H_2O$ (see Problem 1c for reaction)

(f) 0.075-N $HClO_4$ versus HgO (see Problem 1f for reaction)

*8. A 50.0-ml sample of a white dinner wine required 33.2 ml of 0.0267-N NaOH to achieve a phenolphthalein end point. Express the acidity of the wine in terms of grams tartaric acid ($H_2C_4H_4O_6$, gfw = 150) per 100 ml (assume that two hydrogens of the acid are titrated).

9. A 50.0-ml sample of a household cleaning solution was diluted to exactly 250 ml in a volumetric flask. A 25.0-ml aliquot of this solution required 36.7 ml of 0.221-N HCl to reach a bromocresol green end point. Calculate the weight-volume percentage of NH_3 in the sample (assuming that all of its alkalinity results from that constituent).

*10. A 1.26-g sample of a recrystallized organic acid required a 31.7-ml titration with 0.112-N NaOH to reach a phenolphthalein end point. What is the equivalent weight of the acid?

11. To establish the identity of the cation in a pure carbonate a 0.521-g sample was dissolved in 50.0 ml of 0.106-N HCl and boiled to remove CO_2. The excess HCl was back-titrated with 17.3 ml of 0.0812-N NaOH. Identify the carbonate.

*12. The active ingredient in Antabuse, a drug used for treatment of chronic alcoholism, is tetraethylthiuram disulfide:

$$\overset{\displaystyle S}{\overset{\|}{}} \quad \overset{\displaystyle S}{\overset{\|}{}}$$
$$(C_2H_5)_2NCSSCN(C_2H_5)_2$$

(gfw = 296). The sulfur in a 0.341-g sample of an Antabuse preparation was oxidized to SO_2, which was absorbed in H_2O_2 to give H_2SO_4. The acid was titrated with 25.1 ml of 0.0286-N base. Calculate the percent active ingredient in the preparation.

13. Neohetramine, $C_{16}H_{22}ON_4$ (gfw = 286), is a common antihistamine. A 0.221-g

sample containing this compound was analyzed by the Kjeldahl method. The ammonia produced was collected in HBO_2; the resulting BO_2^- was titrated with 37.3 ml of 0.0213-N HCl. Calculate the percent neohetramine in the sample.

*14. To obtain the percent protein in a wheat product the percent nitrogen present is generally multiplied by 5.70. A 1.863-g sample of a wheat flour was analyzed by the Kjeldahl method. The ammonia formed was distilled into 50.0 ml of 0.0476-N HCl; a 6.41-ml back-titration with 0.0501-N base was required. Calculate the percent protein in the flour.

15. The formaldehyde content of a pesticide preparation was determined by weighing 0.723 g of the liquid sample into a flask containing 25.0 ml of 0.501-N NaOH and 50 ml of 3% H_2O_2. Upon heating, the following reaction took place:

$$OH^- + HCHO + H_2O_2 \rightarrow HCOO^- + 2H_2O$$

After cooling, the excess base was titrated with 12.1 ml of 0.496-N H_2SO_4. Calculate the percent HCHO in the sample.

*16. A 4.00-liter sample of urban air (at 22°C and 1.00 atm) was bubbled through a solution containing 25.0 ml of 0.0207-N $Ba(OH)_2$; $BaCO_3$ precipitated. The excess base was back-titrated to a phenolphthalein end point with 17.2 ml of 0.0135-N HCl. Calculate the parts per million CO_2 in the air (that is, the ml $CO_2/10^6$ ml air).

17. Air (at 16°C and 1.00 atm) was bubbled at a rate of 3.00 liters/min for 10.0 min through a trap containing 75 ml of 1% H_2O_2 ($H_2O_2 + SO_2 \rightarrow H_2SO_4$); the H_2SO_4 was titrated with 8.75 ml of 0.00333-N NaOH. Calculate the parts per million SO_2 (that is, ml $SO_2/10^6$ ml air).

*18. What should be the normality of a $Ba(OH)_2$ solution if its titer is to be 1.50 mg $HClO_4$/ml?

19. What is the H_3AsO_4 titer of a 0.0512-N NaOH solution if phenolphthalein is to serve as the indicator?

*20. A 1.19-g sample containing $(NH_4)_2SO_4$, NH_4NO_3, and nonreactive substances was dissolved and diluted to 100 ml in a volumetric flask. A 25.0-ml aliquot was made basic with strong alkali and the liberated NH_3 was distilled into 50.0 ml of 0.0617-N HCl. The excess HCl required 4.23 ml of 0.0814-N NaOH.

A 10.0-ml aliquot of the sample was made alkaline after the addition of Devarda's alloy (p. 236). The NH_3 from both NH_4^+ and NO_3^- was then distilled into 50.0 ml of the standard acid and back-titrated with 17.3 ml of the base.

Calculate the percent $(NH_4)_2SO_4$ and NH_4NO_3 in the sample.

21. A 0.117-g sample of a phosphorus-containing compound was digested in a mixture of HNO_3 and H_2SO_4, which resulted in formation of CO_2, H_2O, and

H_3PO_4. Addition of ammonium molybdate yielded a solid having the composition $(NH_4)_3PO_4 \cdot 12MoO_3$. The precipitate was filtered, washed, and dissolved in 40.0 ml of 0.225-F NaOH.

$$(NH_4)_3PO_4 \cdot 12MoO_3(s) + 26OH^- \rightarrow HPO_4^{2-} + 12MoO_4^{2-} + 14H_2O + 3NH_3(g)$$

After boiling the solution to remove the NH_3, the excess NaOH was titrated with 11.2 ml of 0.166-N HCl. Calculate the percent P in the sample.

*22. A 0.284-g specimen containing sodium azide was treated with an unmeasured excess of sodium nitrite and 50.0 ml of 0.100-N perchloric acid solution. Upon completion of the reaction

$$2H_3O^+ + N_3^- + NO_2^- \rightarrow 3H_2O + N_2O(g) + N_2(g)$$

the excess acid was titrated with 13.0 ml of standard 0.0800-N NaOH. Calculate the percentage of sodium azide in the sample.

23. A 1.88-g sample of impure mercury(II) oxide was brought into solution, following which an excess of potassium iodide was introduced. Reaction:

$$HgO + H_2O + 4I^- \rightarrow HgI_4^{2-} + 2OH^-$$

Calculate the percentage of HgO in the sample if 41.8 ml of 0.114-N HCl were needed to titrate the liberated base.

*24. Triphenyltetrazolium chloride, $C_{19}H_{15}ClN_4$, is used to establish the viability of seeds. A 0.307-g sample containing this substance was analyzed by the Kjeldahl method. The ammonia produced was distilled into 50.0 ml of 0.0600-N HCl; back-titration of the excess acid required 3.47 ml of 0.0718-N NaOH.
(a) Calculate the percentage of nitrogen in the sample.
(b) Calculate the percentage of triphenyltetrazolium chloride (gfw = 335) if this substance was the only source of nitrogen in the sample.

25. A random 30-tablet sample, weighing a total of 42.0 g, was analyzed for its riboflavin ($C_{17}H_{20}N_4O_6$) content by the Kjeldahl method. The ammonia produced was collected in an unmeasured excess of 4% H_3BO_3; titration of the BO_2^- produced required 10.45 ml of 0.0508-N HCl.
(a) Calculate the average weight (in milligrams) of riboflavin (gfw = 376.4) in each tablet based upon this analysis.
(b) How many of these tablets should be taken daily in order to assure an intake of at least 3 mg of riboflavin?

*26. A 0.802-g sample containing $NaHC_2O_4$, $H_2C_2O_4 \cdot 2H_2O$, and unreactive species was titrated to a phenolphthalein end point with 39.6 ml of 0.101-N NaOH. The resulting solution was then evaporated to dryness and ignited to decompose the sodium oxalate.

$$Na_2C_2O_4 \rightarrow Na_2CO_3 + CO$$

The residue was boiled with 50.0 ml of 0.123-N HCl; after cooling, the excess acid was back-titrated with 2.76 ml of the base. Calculate the percent $NaHC_2O_4$ and $H_2C_2O_4 \cdot 2H_2O$ in the original sample.

*27. A 1.09-g sample of commercial KOH, which was contaminated by K_2CO_3, was dissolved in water, and the resulting solution was diluted to 500 ml. A 50.0-ml aliquot of this solution was treated with 25.0 ml of 0.0823-N HCl and boiled to remove CO_2. The excess acid consumed 5.55 ml of 0.0464-N NaOH (phenolphthalein indicator). An excess of neutral $BaCl_2$ was added to another 50.0-ml aliquot to precipitate the carbonate as $BaCO_3$. The solution was then titrated with 18.6 ml of the acid to a phenolphthalein end point. Calculate the percent KOH, K_2CO_3, and water in the sample, assuming that these are the only compounds present.

28. A 0.445-g sample containing $NaHCO_3$, Na_2CO_3, and H_2O was dissolved and diluted to exactly 250.0 ml. A 25.0-ml aliquot of the diluted sample was then boiled with 50.00 ml of 0.0134-N HCl. After cooling, the excess acid in the solution required 4.15 ml of 0.0111-N NaOH when titrated to a phenolphthalein end point. A second 25.0-ml aliquot was then treated with an excess of $BaCl_2$ and 25.0 ml of the base; precipitation of all of the carbonate resulted, and 8.42 ml of the HCl were required to titrate the excess base. Calculate the composition of the mixture.

*29. Calculate the volume of 0.0545-N HCl needed to titrate
   (a) 25.0 ml of 0.0400-F Na_3PO_4 to a thymolphthalein end point.
   (b) 25.0 ml of 0.0400-F Na_3PO_4 to a bromocresol green end point.
   (c) 50.0 ml of a solution that is 0.0222 F in Na_3PO_4 and 0.0361 F in Na_2HPO_4 to a bromocresol green end point.
   (d) 30.0 ml of a solution that is 0.0222 F in Na_3PO_4 and 0.0361 F in NaOH to a thymolphthalein end point.

30. Calculate the volume of 0.0646-N NaOH needed to titrate
   (a) 25.0 ml of a solution that is 0.0250 F in HCl and 0.0150 F in H_3PO_4 to a bromocresol green end point.
   (b) the solution in (a) to a thymolphthalein end point.
   (c) 25.0 ml of a solution that is 0.0712 F in NaH_2PO_4 to a thymolphthalein end point.
   (d) 30.0 ml of a solution that is 0.0225 F in H_3PO_4 and 0.0275 F in NaH_2PO_4 to a thymolphthalein end point.

*31. Solutions containing NaOH, Na_2CO_3, and $NaHCO_3$, alone or in compatible combination, were titrated with standard 0.126-N HCl. Tabulated below are the acid volumes needed to titrate 25.0-ml portions of each solution to (1) a phenolphthalein end point and (2) a bromocresol green end point. Use this information to deduce the composition of the solutions. In addition, calculate the number of milligrams of each solute per milliliter of solution.

	(1)	(2)
(a)	24.12	24.13
(b)	14.63	43.25
(c)	31.34	38.83
(d)	15.11	30.21
(e)	0.00	26.72

32. Solutions containing NaOH, Na_3AsO_4, and Na_2HAsO_4, alone or in compatible combination, were titrated with standard 0.0860-N HCl. Tabulated below are the acid volumes needed to titrate 25.0-ml portions of each solution to (1) a phenolphthalein end point and (2) a bromocresol green end point. Use this information to deduce the composition of the solutions. In addition, calculate the number of milligrams of each solute per milliliter of solution.

	(1)	(2)
(a)	0.00	16.76
(b)	20.83	30.12
(c)	21.62	43.22
(d)	19.93	19.95
(e)	14.12	40.32

*33. A series of solutions can contain HCl, H_3PO_4, or NaH_2PO_4, alone or in any compatible combination of these solutes. Tabulated below are the volumes of 0.117-N NaOH needed to titrate 25.0-ml portions of each solution to (1) a bromocresol green end point and (2) a thymolphthalein end point. Use this information to deduce the composition of the solutions. In addition, calculate the number of milligrams of each solute per milliliter of solution.

	(1)	(2)
(a)	14.12	14.11
(b)	21.67	26.32
(c)	18.16	47.32
(d)	0.00	12.25
(e)	19.96	19.98

34. A series of solutions can contain $HClO_4$, maleic acid, and sodium hydrogen maleate, alone or in any compatible combination of these solutes. Tabulated below are the volumes of 0.0994-N NaOH needed to titrate 25.0-ml portions of each solution to (1) a bromocresol green end point and (2) a thymolphthalein end point. Use this information to deduce the composition of the solutions. In addition, calculate the number of milligrams of each solute per milliliter of solution.

	(1)	(2)
(a)	0.00	17.77
(b)	11.60	30.12
(c)	24.03	24.03
(d)	10.50	21.01
(e)	20.24	31.63

11 complex formation

titrations

Many metal ions react with electron pair donors to form coordination compounds or complex ions. The donor species, or *ligand*, must have at least one pair of unshared electrons available for bond formation. The water molecule, ammonia, and halide ions are common ligands.

Although exceptions exist, a specific cation ordinarily forms bonds with a maximum of two, four, or six ligands, its *coordination number* being a statement of this maximum. The species formed as a result of coordination can be electrically positive, neutral, or negative. Thus, for example, copper(II) (with a coordination number of four) forms a cationic ammine complex $Cu(NH_3)_4^{2+}$, a neutral complex with glycine, $Cu(NH_2CH_2COO)_2$, and an anionic complex with chloride ion, $CuCl_4^{2-}$.

Titrimetric methods based upon complex formation (sometimes called *complexometric methods*) have been used for at least a century. The truly remarkable growth in their analytical applications is of more recent origin, however, and is based upon a particular class of coordination compounds called *chelates*. A chelate is produced when a metal ion coordinates with two (or more) donor groups of a single ligand. The copper complex of glycine mentioned earlier is an example. Here copper is bonded to both the oxygen of the carboxylate groups and the nitrogen of the amine groups. A chelating agent that has two donor groups available for coordination bonding is called *bidentate*, whereas one with three groups is called *terdentate*. Quadri-, quinque-, and sexadentate chelating agents are known.

From the standpoint of volumetric analysis the great virtue of a reagent that yields a chelate over one that simply forms a complex with the metal ion lies in the fact that chelation is essentially a single-step process, whereas complex formation may involve the production of one or more intermediate species. Consider, for example, the equilibrium that exists between the metal ion M with a coordination number of four and the quadridentate ligand D.[1]

$$M + D \rightleftharpoons MD$$

The equilibrium expression for this process is

$$K_f = \frac{[MD]}{[M][D]}$$

where K_f is the *formation constant*.

Similarly, the equilibrium between M and the bidentate ligand B can be represented by

$$M + 2B \rightleftharpoons MB_2$$

This equation, however, is the summation of a two-step process that involves formation of the intermediate MB

$$M + B \rightleftharpoons MB$$

$$MB + B \rightleftharpoons MB_2$$

for which

$$K_1 = \frac{[MB]}{[M][B]} \quad \text{and} \quad K_2 = \frac{[MB_2]}{[MB][B]}$$

The product of K_1 and K_2 yields the equilibrium constant for the overall process[2]

$$\beta_2 = K_1 K_2 = \frac{[MB]}{[M][B]} \times \frac{[MB_2]}{[MB][B]} = \frac{[MB_2]}{[M][B]^2}$$

In a like manner, the reaction between M and the unidentate ligand A involves the overall equilibrium

$$M + 4A \rightleftharpoons MA_4$$

and the equilibrium constant β_4 for the formation of MA_4 from M and A is

---

[1] The electrostatic charges associated with M and D determine the charge of the product but are not important in terms of the present argument.

[2] It is customary to use the symbol β_i to indicate an overall formation constant. Thus, for example, $\beta_2 = K_1 K_2$, $\beta_3 = K_1 K_2 K_3$, $\beta_4 = K_1 K_2 K_3 K_4$, and so forth.

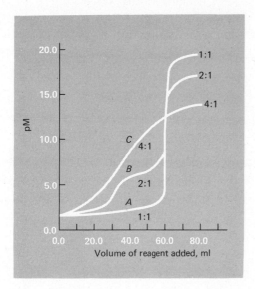

FIGURE 11-1 Curves for complex formation titrations. Titration of 60.0 ml of 0.020-F M with a 0.020-F solution of the quadridentate ligand D to give MD as the product (curve A), with a 0.040-F solution of the bidentate ligand B to give MB_2 (curve B), and with a 0.080-F solution of the unidentate ligand A to give MA_4 (curve C). The overall formation constant for each product is 1.0×10^{20}.

numerically equal to the product of the equilibrium constants for the four constituent processes.

Each of the titration curves depicted in Figure 11-1 is based upon a reaction that has an overall equilibrium constant of 10^{20}. Curve A is derived for the formation of MD in a single step. Curve B involves the formation of MB_2 in a two-step process for which K_1 is 10^{12} and K_2 is 10^8. Curve C represents formation of MA_4 for which the equilibrium constants for the four individual steps are 10^8, 10^6, 10^4, and 10^2, respectively. These curves demonstrate the clear superiority of a ligand that combines with a metal ion in a 1:1 ratio because the change in pM in the equivalence point region is largest with such a system. Thus polydentate ligands, which generally combine in lower ratios with metal ions, are ordinarily superior reagents for complex formation titrations.

Regardless of reaction type, the error associated with a titration decreases with increasing completeness of the reaction upon which the titration is based. In this respect polydentate ligands usually offer a distinct advantage over unidentate titrants; formation constants associated with reactions between metal ions and the former tend to be larger than those involving the latter.

Titrations with Inorganic Complexing Reagents

Complexometric titrations are among the oldest of volumetric methods.[3]

---

[3] For further information see I. M. Kolthoff and V. A. Stenger, *Volumetric Analysis*, vol. 2. New York: Interscience, 1947, pp. 282, 331.

TABLE 11-1

Typical Inorganic Complex Formation Titrations[a]

TITRANT	ANALYTE	REMARKS
$Hg(NO_3)_2$	Br^-, Cl^-, SCN^-, CN^-, thiourea	Products are neutral mercury(II) complexes; various indicators used
$AgNO_3$	CN^-	Product is $Ag(CN)_2^-$; indicator is I^-; titrate to first turbidity of AgI
$NiSO_4$	CN^-	Product is $Ni(CN)_4^{2-}$; indicator is AgI; titrate to first turbidity of AgI
KCN	Cu^{2+}, Hg^{2+}, Ni^{2+}	Products $Cu(CN)_4^{2-}$, $Hg(CN)_2$, $Ni(CN)_4^{2-}$; various indicators used

[a] For further applications and selected references see L. Meites, *Handbook of Analytical Chemistry*. New York: McGraw-Hill, 1963, p. 3-226.

For example, the titration of iodide ion with mercury(II)

$$Hg^{2+} + 4I^- \rightleftharpoons HgI_4^{2-}$$

was first reported in 1834; the determination of cyanide based upon formation of the dicyanoargentate(I) ion, $Ag(CN)_2^-$, was described by Liebig in 1851. Table 11-1 lists typical nonchelating complexing reagents as well as some of the uses to which they have been put.

Titrations with Aminopolycarboxylic Acids

Numerous tertiary amines that also contain carboxylic acid groups form remarkably stable chelates with many metal ions. Their potential as analytical reagents was first recognized by Schwarzenbach in 1945; since that time they have been extensively investigated. The revival of interest in volumetric complex formation methods can be traced to these reagents.[4]

The discussion that follows is based on ethylenediaminetetraacetic acid, a titrant that is used extensively in chemical analysis; titrations with other chelating reagents can be treated analogously.

[4] These reagents are the subject of several excellent monographs. See, for example, G. Schwarzenbach and H. Flashka, *Complexometric Titrations*, trans. H. M. N. H. Irving, 2d ed. London: Methuen & Co., 1969; F. J. Welcher, *The Analytical Uses of Ethylenediaminetetraacetic Acid*. Princeton, N.J.: Van Nostrand, 1958; A. Ringbom, *Complexation in Analytical Chemistry*. New York: Interscience, 1963. For annotated summaries of applications see L. Meites, *Handbook of Analytical Chemistry*. New York: McGraw-Hill, 1963, pp. 3-167 to 3-234.

ETHYLENEDIAMINETETRAACETIC ACID

Ethylenediaminetetraacetic acid (also called ethylenedinitrilotetraacetate and often abbreviated EDTA) has the structure

$$
\begin{array}{c}
\text{HOOC—CH}_2 \\
\hspace{2em} \diagdown \\
\hspace{3em} \text{N—CH}_2\text{—CH}_2\text{—N} \\
\hspace{2em} \diagup \hspace{6em} \diagdown \\
\text{HOOC—CH}_2
\end{array}
\begin{array}{c}
\text{CH}_2\text{—COOH} \\
\\
\\
\text{CH}_2\text{—COOH}
\end{array}
$$

EDTA is the most widely used of the polyaminocarboxylic acids. It is a weak acid for which $K_1 = 1.0 \times 10^{-2}$, $K_2 = 2.1 \times 10^{-3}$, $K_3 = 6.9 \times 10^{-7}$, and $K_4 = 5.5 \times 10^{-11}$. These values indicate that the first two protons are lost much more readily than the remaining two. In addition to the four acidic hydrogens, each nitrogen atom has an unshared pair of electrons; the molecule thus has six potential sites for bonding with a metal ion and is a sexadentate ligand.

The abbreviations H_4Y, H_3Y^-, H_2Y^{2-}, HY^{3-}, and Y^{4-} are often employed in referring to EDTA and its ions.

The free acid H_4Y as well as the dihydrate of the sodium salt $Na_2H_2Y \cdot 2H_2O$ are available in reagent quality. The former can serve as a primary standard after drying for several hours at 130 to 145°C. It is then dissolved in the minimum amount of base required for complete solution.

Under normal atmospheric conditions the dihydrate $Na_2H_2Y \cdot 2H_2O$ contains 0.3% moisture in excess of the stoichiometric amount. For all but the most exacting work this excess is sufficiently reproducible to permit use of a corrected weight for the salt in the direct preparation of a standard solution. If necessary, the pure dihydrate can be prepared by drying at 80°C in an atmosphere of 50% relative humidity for several days.

Composition of EDTA Solutions as a Function of pH. Five EDTA-containing species can exist in an aqueous solution; the relative amount of each species in a given solution depends upon the pH. A convenient way of showing this relationship involves plotting α-values (see p. 218) for the various species as a function of pH, where

$$
\alpha_0 = \frac{[H_4Y]}{C_T}
$$

$$
\alpha_1 = \frac{[H_3Y^-]}{C_T}
$$

and so forth. It will be recalled that C_T is the sum of the equilibrium concentrations of all species. Thus α-values give the mole fractions of any particular form. Figure 11-2 illustrates how the several α-values for EDTA depend upon pH. It is apparent that H_2Y^{2-} is the predominant species in moderately acid media (pH 3 to 6). In the range between pH 6 and 10, HY^{3-} is the major constituent. Only at pH values greater than 10 does Y^{4-} become a major component of the solution.

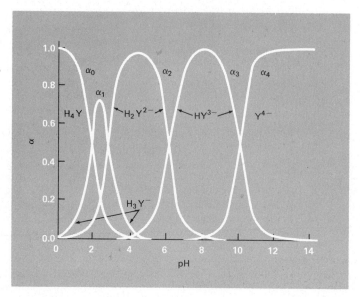

FIGURE 11-2   Composition of EDTA solutions as a function of pH.

These relationships strongly influence the equilibria that exist in solutions containing EDTA and various cations.

COMPLEXES OF EDTA AND METAL IONS

A particularly valuable property of EDTA as a titrant is that it combines with metal ions in a 1:1 ratio regardless of the charge on the cation. In moderately acid solution these reactions can be formulated as[5]

$$M^+ + H_2Y^{2-} \rightleftharpoons MHY^{2-} + H^+$$

$$M^{2+} + H_2Y^{2-} \rightleftharpoons MY^{2-} + 2H^+$$

$$M^{3+} + H_2Y^{2-} \rightleftharpoons MY^- + 2H^+$$

$$M^{4+} + H_2Y^{2-} \rightleftharpoons MY + 2H^+$$

Figure 11-2 indicates that reactions performed in neutral to moderately basic solutions are best written as

$$M^{n+} + HY^{3-} \rightleftharpoons MY^{(n-4)+} + H^+$$

EDTA is a remarkable reagent, not only because it forms chelates with all cations but also because most of these chelates are sufficiently

[5] In this chapter and those that follow we shall revert to the use of H^+ as a convenient shorthand notation for H_3O^+; from time to time we shall refer to the species represented by H^+ as the hydrogen ion.

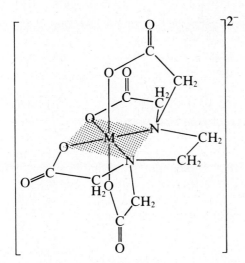

FIGURE 11-3 Structure of a metal-EDTA chelate. (Bond lengths are not to scale.)

stable to form the basis for a volumetric analysis. This great stability undoubtedly results from the several complexing sites within the molecule that give rise to structures that effectively surround and isolate the cation. One form of the complex is depicted in Figure 11-3. Note that all six ligand groups in the EDTA are involved in bonding the divalent metal ion.

Table 11-2 lists formation constants K_{MY} for common EDTA

TABLE 11-2
Formation Constants for EDTA Complexes[a]

CATION	K_{MY}	$\log K_{MY}$	CATION	K_{MY}	$\log K_{MY}$
Ag^+	2.1×10^7	7.32	Cu^{2+}	6.3×10^{18}	18.80
Mg^{2+}	4.9×10^8	8.69	Zn^{2+}	3.2×10^{16}	16.50
Ca^{2+}	5.0×10^{10}	10.70	Cd^{2+}	2.9×10^{16}	16.46
Sr^{2+}	4.3×10^8	8.63	Hg^{2+}	6.3×10^{21}	21.80
Ba^{2+}	5.8×10^7	7.76	Pb^{2+}	1.1×10^{18}	18.04
Mn^{2+}	6.2×10^{13}	13.79	Al^{3+}	1.3×10^{16}	16.13
Fe^{2+}	2.1×10^{14}	14.33	Fe^{3+}	1.3×10^{25}	25.1
Co^{2+}	2.0×10^{16}	16.31	V^{3+}	7.9×10^{25}	25.9
Ni^{2+}	4.2×10^{18}	18.62	Th^{4+}	1.6×10^{23}	23.2

[a] Data from G. Schwarzenbach, *Complexometric Titrations*. New York: Interscience (London: Chapman & Hall Ltd.), 1957, p. 8. With permission. (Constants valid at 20°C and an ionic strength of 0.1.)

complexes. Note that the constant refers to the equilibrium involving the species Y^{4-} with the metal ion. That is,

$$M^{n+} + Y^{4-} \rightleftharpoons MY^{(n-4)+} \qquad K_{MY} = \frac{[MY^{(n-4)+}]}{[M^{n+}][Y^{4-}]} \qquad (11\text{-}1)$$

DERIVATION OF EDTA TITRATION CURVES

The general approach to derivation of a titration curve for a metal-ion EDTA reaction does not differ fundamentally from that used for precipitation or neutralization titrations. Here, however, it is ordinarily necessary to account for more than one equilibrium; as a result, the computational detail is somewhat greater than in the earlier examples.

Effect of pH. Inspection of metal-ion EDTA equilibria (see p. 252) reveals that the extent of complex formation will depend upon the pH of the environment. Thus an alkaline medium is needed for titrations involving cations such as calcium and magnesium, which form relatively weak complexes. On the other hand, titrations involving cations that form more stable complexes such as zinc and nickel can be performed successfully in moderately acidic solutions.

Because of this pH dependence, EDTA titrations are generally carried out in solutions that are buffered to a constant and predetermined pH. The restriction of a constant pH permits a considerable simplification in the computation of data for titration curves.

In the derivation of EDTA titration curves for buffered solutions, it is convenient to employ the term α_4, defined as

$$\alpha_4 = \frac{[Y^{4-}]}{C_T} \qquad (11\text{-}2)$$

In this expression C_T is the total concentration of *uncomplexed* EDTA species. That is,

$$C_T = [Y^{4-}] + [HY^{3-}] + [H_2Y^{2-}] + [H_3Y^-] + [H_4Y]$$

Thus α_4 represents the fraction of the total uncomplexed reagent that exists as Y^{4-}.

A derivation similar to that shown on page 218 reveals that α_4 depends only upon the pH and the four acid dissociation constants of EDTA, K_1, K_2, K_3, and K_4. Specifically,

$$\alpha_4 = \frac{K_1 K_2 K_3 K_4}{[H^+]^4 + K_1[H^+]^3 + K_1 K_2[H^+]^2 + K_1 K_2 K_3[H^+] + K_1 K_2 K_3 K_4} \qquad (11\text{-}3)$$

Table 11-3 lists values of α_4 for selected pH levels; these were calculated with Equation 11-3. Substitution of $\alpha_4 C_T$ for $[Y^{4-}]$ (Equation 11-2) into the formation constant expression (Equation 11-1) and rear-

TABLE 11-3
**Values for α_4 for EDTA in Solutions of
Various pH**

pH	α_4	pH	α_4
2.0	3.7×10^{-14}	7.0	4.8×10^{-4}
3.0	2.5×10^{-11}	8.0	5.4×10^{-3}
4.0	3.6×10^{-9}	9.0	5.2×10^{-2}
5.0	3.5×10^{-7}	10.0	3.5×10^{-1}
6.0	2.2×10^{-5}	11.0	8.5×10^{-1}
		12.0	9.8×10^{-1}

rangement yield

$$K'_{MY} = \alpha_4 K_{MY} = \frac{[MY^{(n-4)+}]}{[M^{n+}]C_T} \tag{11-4}$$

where K'_{MY} is a *conditional* or *effective* formation constant that describes equilibrium conditions *only at the pH for which α_4 is applicable.*

Conditional constants provide a simple means by which the equilibrium concentrations of the metal ion and the complex can be calculated at any point in a titration curve. Note that the expression for the conditional constant differs from that of the formation constant only in that the term C_T replaces the equilibrium concentration of the completely dissociated anion $[Y^{4-}]$. This difference is significant, however, because C_T is more readily determined from the stoichiometry of the reaction than is $[Y^{4-}]$.

EXAMPLE

Derive a curve for the titration of 50.0 ml of 0.0100-F Ca^{2+} with 0.0100-F EDTA in a solution buffered to a constant pH of 10.0.

Calculation of conditional constant. The conditional formation constant for the calcium-EDTA complex at pH 10 can be obtained from the formation constant of the complex (Table 11-2) and the α_4 value for EDTA at pH 10 (Table 11-3). Thus

$$K'_{CaY} = \alpha_4 K_{CaY} = (3.5 \times 10^{-1})(5.0 \times 10^{10})$$
$$= 1.75 \times 10^{10}$$

Preequivalence point values for pCa. Before the equivalence point has been reached, the molar concentration of Ca^{2+} will be equal to the sum of the contributions from the untitrated excess of the ion and from dissociation of the complex, the latter being numerically equal to C_T. It is ordinarily reasonable to assume that C_T is small with respect to the formal concentration of the uncomplexed calcium ion. Thus, for example, after the addition of 25.0 ml of reagent

$$[Ca^{2+}] = \frac{50.0 \times 0.0100 - 25.0 \times 0.0100}{75.0} + C_T \cong 3.33 \times 10^{-3}$$

and

$$pCa = -\log 3.33 \times 10^{-3} = 2.48$$

Equivalence point pCa. Here the solution is $0.00500\ F$ in CaY^{2-}; the only source of Ca^{2+} ions will be from dissociation of this complex. It also follows that the calcium ion concentration must be identical to the sum of the concentrations of the uncomplexed EDTA ions, C_T. Thus

$$[Ca^{2+}] = C_T$$

$$[CaY^{2-}] = 0.00500 - [Ca^{2+}] \cong 0.00500$$

The conditional formation constant for CaY^{2-} at pH 10.0 is

$$\frac{[CaY^{2-}]}{[Ca^{2+}]C_T} = 1.75 \times 10^{10}$$

Thus upon substitution

$$\frac{0.00500}{[Ca^{2+}]^2} = 1.75 \times 10^{10}$$

$$[Ca^{2+}] = 5.35 \times 10^{-7}$$

$$pCa = 6.27$$

Postequivalence point pCa. Beyond the equivalence point formal concentrations of CaY^{2-} and EDTA are directly obtained. For example, after 60.0 ml of reagent have been added,

$$F_{CaY^{2-}} = \frac{50.0 \times 0.0100}{110} = 4.55 \times 10^{-3}$$

$$F_{EDTA} = \frac{10.0 \times 0.0100}{110} = 9.1 \times 10^{-4}$$

As an approximation we may write

$$[CaY^{2-}] = 4.55 \times 10^{-3} - [Ca^{2+}] \cong 4.55 \times 10^{-3}$$

$$C_T = 9.1 \times 10^{-4} + [Ca^{2+}] \cong 9.1 \times 10^{-4}$$

and

$$\frac{4.55 \times 10^{-3}}{[Ca^{2+}] \times 9.1 \times 10^{-4}} = 1.75 \times 10^{10} = K'_{CaY}$$

$$[Ca^{2+}] = 2.86 \times 10^{-10}$$

$$pCa = 9.54$$

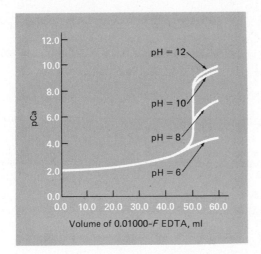

FIGURE 11-4 Influence of pH upon the titration of 0.0100-F Ca^{2+} with 0.01000-F EDTA.

Figure 11-4 shows titration curves for calcium ion in solutions buffered to various pH levels. It is apparent that appreciable changes in pCa can be achieved only if the pH of the solution is maintained at about 8 or greater. As shown in Figure 11-5, however, cations with larger formation constants provide good end points even in acidic solutions. Figure 11-6 shows the minimum permissible pH for satisfactory end points in the titration of various metal ions in the absence of competing complexing agents. Note that a moderately acidic environment is satisfactory for many divalent heavy-metal cations, and that a strongly acidic medium can be tolerated in the titration of iron(III).

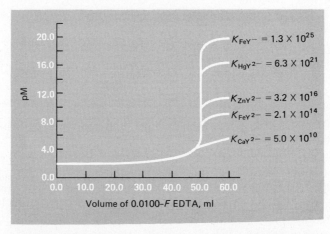

FIGURE 11-5 Titration curves for 50.0 ml of 0.0100-F cation solutions at pH 6.0.

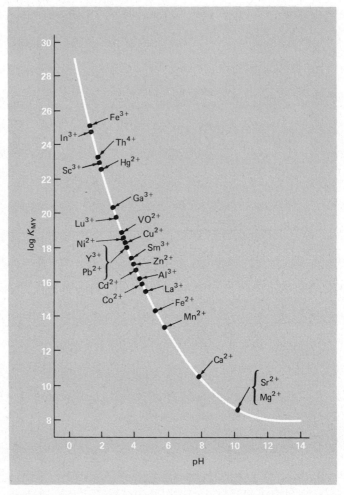

FIGURE 11-6 Minimum pH needed for satisfactory titration of various cations with EDTA. [From C. N. Reilley and R. W. Schmid, *Anal. Chem.*, **30**, 947 (1958). With permission of the American Chemical Society.]

Effect of Other Complexing Agents on EDTA Titrations. Many EDTA titrations are complicated by the tendency on the part of the ion being titrated to precipitate as a basic oxide or hydroxide at the pH required for a satisfactory titration. To keep the metal ion in solution, particularly in the early stages of the titration, it is necessary to include an auxiliary complexing agent. Thus, for example, the titration of zinc(II) is ordinarily performed in the presence of high concentrations of ammonia and ammonium chloride. These species buffer the solution to an acceptable pH. In addition, the ammonia forms soluble ammine complexes with zinc ion and

prevents precipitation of zinc hydroxide. The titration reaction with EDTA is thus best represented as

$$Zn(NH_3)_4^{2+} + HY^{3-} \rightleftharpoons ZnY^{2-} + 3NH_3 + NH_4^+$$

In a system such as this the completeness of reaction, and thus the quality of the end point, depends not only upon pH but also upon the concentration of the auxiliary complexing agent (in this case ammonia).

The influence of an auxiliary complexing reagent can be treated in much the same way as the effect of pH on the concentration of EDTA species. In the context of the zinc titration we have been considering, a quantity β[6] is defined such that

$$\beta = \frac{[Zn^{2+}]}{C_{Zn}}$$

where C_{Zn} is the sum of the concentrations of species containing zinc(II), *exclusive of* ZnY^{2-}. The numerical value of β is readily calculated from the concentration of the auxiliary agent (in this case NH_3) as well as the formation constants for the various ammine complexes of zinc.

Substitution of βC_{Zn} for $[Zn^{2+}]$ in the conditional constant expression and rearrangement yield the new constant K''

$$K'' = \alpha_4 \beta K_{ZnY} = \frac{[ZnY^{2-}]}{C_{Zn} C_T}$$

Each value of K'' is unique to a system in which both the concentration of the complexing ligand and the pH are fixed and constant.

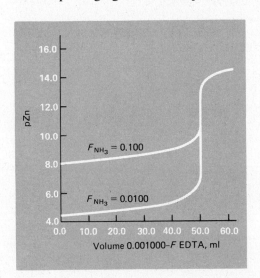

FIGURE 11-7 Influence of ammonia concentration upon the titration of 0.001000-F Zn^{2+} with 0.001000-F EDTA. Solutions are buffered to a pH of 9.0.

[6] An example illustrating the evaluation of β is to be found in D. A. Skoog and D. M. West, *Fundamentals of Analytical Chemistry*, 3d ed. New York: Holt, Rinehart and Winston, 1976, p. 280.

Figure 11-7 shows two theoretical curves for the titration of zinc(II) with EDTA at a pH of 9.00. The equilibrium concentration of ammonia was 0.100 F for the one titration and 0.0100 F for the other. The presence of ammonia clearly has the effect of decreasing the change in pZn in the equivalence point region. Thus it is desirable to keep the concentration of any auxiliary reagent to the minimum required to prevent hydroxide formation. Note that the presence of ammonia does not affect pZn beyond the equivalence point. On the other hand, it will be recalled (Figure 11-4) that α_4, and thus pH, plays an important role in defining this region of the titration curve.

END POINTS FOR EDTA TITRATIONS

Metal-Ion Indicators. A number of metal-ion indicators have been developed for use in complexometric titrations with EDTA and other chelating reagents. In general these materials are organic dyes that form colored chelates with metal ions in a pM range that is characteristic of the particular cation and dye. The complexes are often intensely colored, being discernible to the eye in the range of 10^{-6} to 10^{-7} M.

Most metal-ion indicators will also bond with protons to give species that impart to solutions colors that resemble those of the metal complexes. Thus these dyes function as acid-base indicators as well and are useful as indicators for metal ions only in pH ranges in which competition with the proton is negligible.

These properties are exhibited by *Eriochrome black T*, a widely used metal-ion indicator. In acidic and moderately basic solutions the predominant acid-base equilibrium exhibited by the indicator is as follows:

$$\underset{\text{red}}{H_2In^-} \rightleftharpoons \underset{\text{blue}}{HIn^{2-}} + H^+$$

At very high pH levels HIn^{2-} further dissociates to In^{3-}, which is orange.

The metal complexes of Eriochrome black T are generally red. To observe a color change with this indicator, then, it is necessary to adjust the pH to 7 or above so that the blue form of the species HIn^{2-} predominates. The end point reaction is then

$$\underset{\text{red}}{MIn^-} + HY^{3-} \rightleftharpoons \underset{\text{blue}}{HIn^{2-}} + MY^{2-}$$

Eriochrome black T forms red complexes with more than two dozen different metal ions, but with only certain of these cations are the stabilities of the products appropriate for end point detection. To be suitable for a titration with EDTA the formation constant of the metal-indicator complex should be less than one-tenth that of the metal-EDTA complex; otherwise a late end point will be observed. On the other hand, if this ratio becomes too small, as is the situation with calcium ion, early end points are observed. The applicability of a given indicator to an EDTA titration can be determined from the change in pM in the equivalence point region provided the formation constant for the metal-indicator complex is known.[7]

A limitation of Eriochrome black T is that its solutions decompose slowly with standing. It is claimed that solutions of Calmagite, an indicator that for all practical purposes is identical in behavior, do not suffer this disadvantage. The structure of Calmagite is similar to Eriochrome black T.

Many other metal indicators have been developed for EDTA titrations.[8] In contrast to Eriochrome black T, some can be employed in strongly acidic media.

TITRATION METHODS EMPLOYING EDTA

EDTA is used in several ways as a titrant for metal ions. The most common of these procedures are considered in the following paragraphs.

Direct Titration. Welcher[9] lists 25 metal ions that can be determined by direct titration with EDTA using metal-ion indicators for end point detection. Direct titration procedures are limited to those reactions for which a method for end point detection exists and to those metal ions that react rapidly with EDTA. When direct methods fail, it is often possible to achieve an analysis by a back titration or a displacement titration.

Back Titration. Back titration procedures are useful for the analysis of cations that form very stable EDTA complexes and for which a satisfactory indicator is not available. In such an analysis the excess EDTA is determined by back titration with a standard magnesium solution, using Eriochrome black T or Calmagite as the indicator. The metal-EDTA complex must be more stable than the magnesium-EDTA complex; otherwise the back titrant would displace the metal ion.

This technique is also useful for the titration of metals in the presence of anions that would otherwise form slightly soluble precipitates with the cation under the conditions of the analysis; the presence of EDTA prevents their precipitation.

---

[7] For a discussion of the principles of indicator choice in complex formation titrations, see C. N. Reilley and R. W. Schmid, *Anal. Chem.*, **31**, 887 (1959).

[8] See, for example, L. Meites, *Handbook of Analytical Chemistry.* New York: McGraw-Hill, 1963, p. 3-101.

[9] F. J. Welcher, *The Analytical Use of Ethylenediaminetetraacetic Acid.* New York: Van Nostrand, 1958, Chapter 3.

Displacement Titration. In displacement titrations an excess of a solution containing the magnesium (or zinc) complex of EDTA is introduced. If the metal ion forms a more stable complex than that of magnesium (or zinc), the following reaction occurs:

$$MgY^{2-} + M^{2+} \rightleftharpoons MY^{2-} + Mg^{2+}$$

The liberated magnesium is then titrated with a standard EDTA solution. This technique is useful where no satisfactory indicator is available for the metal ion being determined.

Alkalimetric Titration. In alkalimetric titrations an excess of Na_2H_2Y is added to a neutral solution of the metallic ion.

$$M^{2+} + H_2Y^{2-} \rightarrow MY^{2-} + 2H^+$$

The liberated hydrogen ions are then titrated with a standard solution of a base.

SCOPE OF COMPLEXOMETRIC TITRATIONS

Complexometric titrations with EDTA have been reported for the analysis of virtually every cation. Because the reagent tends to form chelates with most cations, EDTA would appear at first glance to be totally lacking in specificity. In fact, however, considerable control over the behavior of EDTA, as well as other chelating agents, can be achieved through pH regulation. Thus, for example, it is generally possible to determine trivalent cations without interference from divalent species by performing the titration in a medium with a pH of about 1; under these circumstances the less stable divalent chelates do not form to any significant extent, whereas the trivalent ions are quantitatively complexed. Similarly, cadmium, which forms a more stable EDTA chelate than magnesium, can be determined in the presence of the latter ion by buffering the mixture to pH 7 before titration. Eriochrome black T serves as an indicator for the cadmium end point without interference from magnesium because the magnesium-indicator chelate is not formed at this pH. Finally, interference from a particular cation can sometimes be eliminated by adding a suitable *masking agent*, an auxiliary ligand that preferentially forms highly stable complexes with the potential interference.[10] For example, cyanide ion is often employed as a masking agent to permit the titration of magnesium and calcium ions in the presence of ions such as cadmium, cobalt, copper, nickel, zinc, and palladium. All of the latter form sufficiently stable cyanide complexes to prevent reaction with EDTA.

[10] For further information see D. D. Perrin, *Masking and Demasking of Chemical Reactions.* New York: Wiley-Interscience, 1970.

Specific directions for the preparation and use of EDTA solutions are given in Section 4, Chapter 20.

DETERMINATION OF WATER HARDNESS

Historically "hardness" was defined in terms of the capacity of a water sample to precipitate soap, an undesirable quality. Soap is precipitated by most ions with multiple charges. In natural waters, however, the concentration of calcium and magnesium ions generally far exceeds that of any other cation; thus hardness has come to mean the total concentration of calcium and magnesium expressed in terms of the calcium carbonate equivalent.

The determination of hardness is a useful analytical process for measuring the quality of water for household and industrial uses. The importance for the latter is due to the fact that hard water, upon heating, precipitates calcium carbonate, which then clogs boilers and pipes.

Water hardness is ordinarily determined by an EDTA titration after the sample has been buffered to a pH of 10; Calmagite or Eriochrome black T serves as the indicator. Often a small concentration of the EDTA complex of magnesium is incorporated in the buffer to assure the presence of sufficient magnesium ions for satisfactory indicator action.

Test kits for determining the hardness of household water are available commercially. These consist of a vessel calibrated to contain a known volume of water, a measuring scoop to deliver an appropriate amount of a solid buffer mixture, an indicator solution, and a bottle of standard EDTA, which is equipped with a medicine dropper; a measure of the volume of standard reagent is obtained by counting drops to the end point. Usually the concentration of the EDTA solution is adjusted so that one drop corresponds to one grain of calcium carbonate per gallon of water.

PROBLEMS

*1. Mercury(II) reacts with thiocyanate ion to form the soluble complex $Hg(SCN)_2$. A solution containing 8.06 g of $Hg(NO_3)_2$ in 500 ml of water was prepared for the determination of SCN^-. Calculate the concentration of this solution in terms of its
   (a) formality.
   (b) normality.
   (c) titer for KSCN.
   (d) titer for $Ba(SCN)_2$.

2. Thiourea, $(H_2N)_2CS$ (gfw $= 76.1$), can be used for the titration of Hg^{2+}. Reaction:

$$Hg^{2+} + 4(H_2N)_2CS \rightarrow Hg[(H_2N)_2CS]_4^{2+}$$

A solution contains 0.843 g of thiourea in 250 ml. Calculate the concentration of this solution in terms of its
(a) formality.
(b) normality.
(c) titer for HgO.
(d) titer for Hg.

*3. An EDTA solution with a titer of 5.00 mg $CaCO_3$/ml is desired.
(a) Describe the preparation of 1.50 liters of such a solution from pure H_4Y (gfw $= 292$).
(b) What is the formality of this solution?
(c) What is its $Mg_2P_2O_7$ titer?

4. An EDTA solution has a ZnO titer of 4.50 mg/ml. Calculate
(a) its formal concentration.
(b) how many grams of $Na_2H_2Y \cdot 2H_2O$ are contained in 500 ml of this solution.
(c) its Fe_2O_3 titer.

*5. (a) Describe the preparation of 750.0 ml of a 0.01000-F solution of H_2Y^{2-} from $Na_2H_2Y \cdot 2H_2O$. Recall that this compound is commonly contaminated with 0.3% excess H_2O (p. 251).
(b) What is the $Ca_3(PO_4)_2$ titer of this solution?

6. An EDTA solution was prepared by dissolving 2.750 g of $Na_2H_2Y \cdot 2H_2O$ in water and diluting to 1.000 liter. The solid reagent contained 0.3% excess water (p. 251). Calculate its
(a) formality.
(b) $CaCl_2$ titer.
(c) $AlCl_3$ titer.

*7. A solution containing 3.76 g of KCN per liter was employed for the determination of Ni in an ore. A 1.64-g sample of the ore consumed 31.8 ml of the reagent.

$$Ni^{2+} + 4CN^- \rightarrow Ni(CN)_4^{2-}(aq)$$

(a) Calculate the normality of the KCN solution. 0.0289 N
(b) Calculate the percent Ni_2O_3 in the ore. 2.32%

8. The sodium salt of diethyldithiocarbamate, $(C_2H_5)_2NCS_2Na$ (gfw $= 171$), can be employed for the titration of Pd^{2+}. Reaction:

$$Pd^{2+} + 2(C_2H_5)_2NCS_2^- \rightarrow Pd[(C_2H_5)_2NCS_2]_2$$

A solution containing 4.69 g of $(C_2H_5)_2NCS_2Na$ per liter was employed to titrate a 50.0-ml sample containing Pd^{2+}. Exactly 26.4 ml were required.

(a) Calculate the formality of the reagent.

(b) Calculate the normality of the reagent.

(c) Calculate the milligrams of $PdCl_2$ per milliliter of the sample.

*9. A 24-hr urine specimen was diluted to exactly 2.00 liters. After being buffered to pH 10, a 25.0-ml aliquot was titrated with 18.1 ml of 0.0877-F EDTA. The calcium in a second 25.0-ml aliquot was isolated as $CaC_2O_4(s)$, redissolved in acid, and titrated with 14.1 ml of the EDTA solution. Calculate the milligrams of Ca^{2+} and Mg^{2+} in the 2.00-liter sample.

10. An EDTA solution was prepared by dissolving approximately 4 g of disodium salt in approximately 1 liter of water. An average of 37.6 ml of this solution was required to titrate 50.0-ml aliquots of a standard that contained 0.697 g of $MgCO_3$ per liter.

Titration of the Ca^{2+} and Mg^{2+} in a 50.0-ml mineral water sample at pH 10 required 30.2 ml of the EDTA solution.

A 100-ml aliquot of the mineral water was rendered strongly alkaline to precipitate the magnesium as $Mg(OH)_2$. Titrations with a calcium-specific indicator required 27.7 ml of the EDTA solution. Calculate

(a) the titer of the EDTA solution in terms of milligrams of $MgCO_3$ per milliliter as well as milligrams of $CaCO_3$ per milliliter.

(b) the parts per million of $CaCO_3$ in the mineral water sample.

(c) the parts per million of $MgCO_3$ in the water.

*11. The sulfate in a 1.839-g sample was homogeneously precipitated by adding 50.0 ml of a 0.0177-F Ba-EDTA solution and slowly increasing the acid concentration to liberate Ba^{2+}. When precipitation was complete, the solution was buffered to pH 10 and diluted to exactly 250 ml. The HY^{3-} in a 25.0-ml aliquot of the supernatant required a 23.1-ml titration with a standard 0.0154-F Mg^{2+} solution. Express the results of this analysis in terms of percent $Na_2SO_4 \cdot 10H_2O$.

12. Calamine, which is used for relief of skin irritations, is a mixture of zinc and iron oxides. A 0.864-g specimen of dried calamine was dissolved in acid and diluted to exactly 250 ml. Potassium fluoride was added to a 10.0-ml aliquot of the diluted solution to mask the iron; after suitable adjustment of the pH, 31.3 ml of 0.0134-F EDTA were required to titrate the Zn^{2+}. A second 50.0-ml aliquot was suitably buffered and titrated with 8.74 ml of a 0.00272-F ZnY^{2-} solution. Reaction:

$$Fe^{3+} + ZnY^{2-} = FeY^- + Zn^{2+}$$

Calculate the respective percentages of ZnO and Fe_2O_3 in the sample.

*13. A 2.63-g sample containing bromate and bromide was dissolved in sufficient water to give exactly 250 ml. Silver nitrate was introduced to a 25.0-ml aliquot that had been acidified; the AgBr produced was filtered, washed, and then redissolved in an ammoniacal solution of potassium tetracyanonickelate(II).

Reaction:

$$Ni(CN)_4^{2-} + 2AgBr(s) = 2Ag(CN)_2^- + Ni^{2+} + 2Br^-$$

The liberated nickel ion required 18.1 ml of 0.0184-F EDTA.

     The bromate in a 10.00-ml aliquot was reduced to bromide with arsenic(III) prior to the addition of silver nitrate; 19.6 ml of the EDTA solution were needed to titrate the nickel ion that was subsequently released. Calculate the respective percentages of NaBr and NaBrO$_3$ in the sample.

14. The chromium ($d = 7.10$ g/cm^3) plated on a 8.43-cm^2 surface was dissolved with hydrochloric acid; the solution was then diluted to exactly 100 ml. A 25.0-ml aliquot was buffered to pH 5 and 50.00 ml of 0.0109-F EDTA were added. Titration of the excess chelating reagent required 8.41 ml of 0.00899-F Zn^{2+}. Calculate the average thickness of the chromium plating.

*15. The silver ion in a 25.0-ml sample was converted to dicyanoargentate(I) ion by the addition of 25.0 ml 0.0914-F Ni(CN)$_4^{2-}$. Reaction:

$$Ni(CN)_4^{2-} + 2Ag^+ = 2Ag(CN)_2^- + Ni^{2+}$$

The liberated nickel ion was subsequently titrated with 43.7 ml of 0.0240-F EDTA. Calculate the formal concentration of the silver solution.

16. The potassium ion in a 250-ml mineral water sample was precipitated with sodium tetraphenylboron:

$$K^+ + (C_6H_5)_4B^- \rightarrow KB(C_6H_5)_4(s)$$

The precipitate was filtered, washed, and then redissolved in an organic solvent. An excess of the mercury(II)-EDTA chelate was added:

$$4HgY^{2-} + (C_6H_5)_4B^- + 4H_2O = H_3BO_3 + 4C_6H_5Hg^+ + 4HY^{3-} + OH^-$$

The liberated EDTA was titrated with 37.3 ml of 0.0342-F Mg^{2+}. Calculate the potassium ion concentration in parts per million.

*17. A 0.512-g sample containing lead, magnesium, and zinc was dissolved and treated with cyanide to complex the zinc:

$$Zn^{2+} + 4CN^- = Zn(CN)_4^{2-}$$

Titration of the lead and magnesium required 48.7 ml of 0.0297-F EDTA. The lead was next masked with BAL (2,3-dimercaptopropanol) and the released EDTA was titrated with 16.4 ml of 0.00765-F magnesium solution. Finally formaldehyde was introduced to demask the zinc:

$$Zn(CN)_4^{2-} + 4HCHO + 4H_2O = Zn^{2+} + 4HOCH_2CN + 4OH^-$$

which was subsequently titrated with 23.1 ml of 0.0297-F EDTA. Calculate the percentages of the three metals in the sample.

18. Chromel is an alloy composed of nickel, iron, and chromium. A 0.545-g sample was dissolved and diluted to exactly 250 ml. When 50.00 ml of 0.06260-F

EDTA were mixed with an equal volume of the diluted sample, all three ions were chelated, and a 20.90-ml back titration with 0.05850-F copper(II) was required. The chromium in a second 50.00-ml aliquot was masked through the addition of hexamethylenetetramine; titration of the iron and nickel required 26.73 ml of 0.0626-F EDTA. Iron and chromium were masked with pyrophosphate in a third 50.00-ml aliquot; 18.65 ml of the EDTA solution were needed to react with the nickel. Calculate the respective percentages of nickel, chromium, and iron in the alloy.

*19. A 0.337-g sample of brass was dissolved in nitric acid. The sparingly soluble metastannic acid was removed by filtration, and the combined filtrate and washings were then diluted to exactly 500 ml.

A 10.00-ml aliquot was suitably buffered; titration of the lead, zinc, and copper in this aliquot required 48.53 ml of 0.00200-F EDTA.

The copper in a 25.00-ml aliquot was masked with thiosulfate; the lead and zinc were then titrated with 33.27 ml of the EDTA solution.

Cyanide ion was used to mask the copper and zinc in a 100-ml aliquot; 11.42 ml of the EDTA solution were needed to titrate the lead ion.

Determine the composition of the brass sample; evaluate the percentage of tin by difference.

*20. Calculate the conditional constants for the formation of the EDTA complex of Mn^{2+} at (a) pH 7.0, (b) pH 9.0, and (c) pH 11.0.

21. Calculate the conditional constants for the formation of the EDTA complex of Sr^{2+} at (a) pH 6.0, (b) pH 8.0, and (c) pH 10.0.

22. The equilibrium constant for the process

$$Co^{2+} + Y^{4-} \rightleftharpoons CoY^{2-}$$

is 2.0×10^{16}. Assuming that a conditional constant of at least 10^{10} is needed for a successful titration, what is the minimum pH (expressed to the nearest whole number) that can be used?

*23. Derive a curve for the titration of 50.00 ml of 0.01500-F Sr^{2+} with 0.03000-F EDTA in a solution buffered to pH 10.0. Calculate pSr values after addition of 0.00, 10.00, 24.00, 24.90, 25.00, 25.10, 26.00, and 30.00 ml of titrant.

24. Derive a curve for the titration of 50.00 ml of 0.0100-F Fe^{2+} with 0.0200-F EDTA in a solution buffered to pH 8.0. Calculate pFe values after addition of 0.00, 10.00, 24.00, 24.90, 25.00, 25.10, 26.00, and 30.00 ml of titrant.

12

theory of oxidation-reduction titrations

Oxidation-reduction, or *redox*, processes involve the transfer of electrons from one reactant to another. Volumetric methods based upon electron transfer are more numerous and more varied than those for any other reaction type.

Oxidation involves the loss of electrons by a substance and *reduction* the gain of electrons. In any oxidation-reduction reaction the molar ratio between the substance oxidized and the substance reduced is such that the number of electrons lost by one species is equal to the number gained by the other. This fact must always be taken into account when balancing equations for oxidation-reduction reactions.

Oxidizing agents or *oxidants* possess a strong affinity for electrons and cause other substances to be oxidized by abstracting electrons from them. In the process the oxidizing agent accepts electrons and is thereby reduced. *Reducing agents* or *reductants* have little affinity for electrons and in fact readily give up electrons, thereby causing some other species to be reduced. A consequence of this electron transfer is the oxidation of the reducing agent. In its simplest form, then, an oxidation-reduction reaction can be written as

$$Ox_1 + Red_2 \rightleftharpoons Red_1 + Ox_2$$

The condition of equilibrium in an oxidation-reduction reaction depends upon the relative tendencies of the reactants to gain or to lose electrons. Thus the mixing of a strong oxidizing agent with a strong reducing agent will result in an equilibrium in which the products are overwhelmingly favored; reactions that are less complete yield correspondingly less favorable equilibria.

The relative tendencies for substances to gain or lose electrons can be measured quantitatively; such data permit evaluation of numerical values for redox equilibrium constants.

Separation of an oxidation-reduction reaction into its component parts (that is, into *half-reactions*) is a convenient way of indicating clearly the species that gains electrons and the one that loses them. Thus the overall reaction

$$5Fe^{2+} + MnO_4^- + 8H^+ \rightleftharpoons 5Fe^{3+} + Mn^{2+} + 4H_2O$$

is obtained by combining the half-reaction for the oxidation of iron(II),

$$5Fe^{2+} \rightleftharpoons 5Fe^{3+} + 5e$$

with that for the reduction of permanganate,

$$MnO_4^- + 5e + 8H^+ \rightleftharpoons Mn^{2+} + 4H_2O$$

It was necessary to multiply the first half-reaction through by 5 to eliminate the electrons from the overall equation.

The rules for balancing half-reactions are the same as those for ordinary reactions; that is, the number of atoms of each element as well as the net charge on both sides of the equation must be equal.

Oxidation-reduction reactions may result from the direct transfer of electrons from the donor to the receiver. Thus, upon immersing a strip of zinc in a copper(II) sulfate solution, copper(II) ions migrate to, and are reduced at, the surface of the zinc

$$Cu^{2+} + 2e \rightleftharpoons Cu(s)$$

and a chemically equivalent quantity of zinc is oxidized

$$Zn(s) \rightleftharpoons Zn^{2+} + 2e$$

The overall process is given by the sum of these half-reactions.

A unique aspect of oxidation-reduction reactions is that the transfer of electrons—and hence the same net reaction—can be accomplished when the donor and acceptor are remote from each other. With the arrangement illustrated in Figure 12-1, direct contact between zinc and copper(II) is prevented by a conducting *salt bridge*, which consists of a saturated potassium chloride solution in a U-tube. Although the reactants are physically separated from one another, the external conductor provides the means by which electrons are transferred from metallic zinc to copper(II) ions. Electron transfer will continue until the copper(II)- and

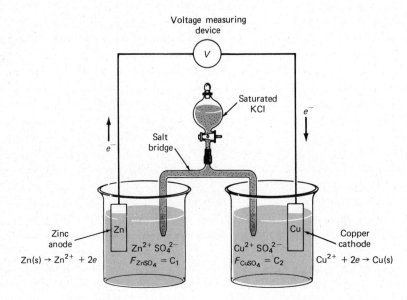

Voltage measuring
device

V

Saturated
KCl

Salt
bridge

Zinc
anode

e^-

Zn

$Zn^{2+}\ SO_4^{2-}$

$F_{ZnSO_4} = C_1$

$Zn(s) \rightarrow Zn^{2+} + 2e$

e^-

Cu

$Cu^{2+}\ SO_4^{2-}$

$F_{CuSO_4} = C_2$

Copper
cathode

$Cu^{2+} + 2e \rightarrow Cu(s)$

FIGURE 12-1   Schematic diagram of a galvanic cell.

zinc(II)-ion concentrations achieve levels corresponding to the equilibrium

$$Zn(s) + Cu^{2+} \rightleftharpoons Zn^{2+} + Cu(s)$$

When this condition is reached, no further net flow of electrons will occur. *It is essential to recognize that the overall process and the concentrations that exist at equilibrium are totally independent of the two routes that led to the condition of equilibrium.*

Fundamentals of Electrochemistry

The device illustrated in Figure 12-1 is, of course, an *electrochemical cell*; it develops an electrical potential from the tendency of the reacting species to transfer electrons and thereby approach the condition of equilibrium. The potential existing between the zinc and the copper electrodes is a measure of the driving force of the reaction and is easily evaluated with a suitable voltage-measuring device, V, placed in the circuit as shown. We shall see that the voltage produced by an electrochemical cell is directly related to the equilibrium constant for the particular oxidation-reduction process involved, as well as the extent to which the existing concentrations of the participants differ from their equilibrium values. Measurement of such potentials constitutes an important source of numerical values for equilibrium constants. It is desirable, therefore, to examine more closely the construction and behavior of electrochemical cells as well as the measurement of the potentials they develop.

CELLS

A cell consists of a pair of conductors or electrodes, usually metallic, each of which is immersed in an electrolyte. When the electrodes are arranged as shown in Figure 12-1 and electrons pass through the conductor, a chemical oxidation occurs at the surface of one electrode and a reduction at the surface of the other.

A *galvanic*, or *voltaic*, cell is one that provides electrical energy. In contrast, an *electrolytic* cell requires an external source of electrical energy for operation. The cell shown in Figure 12-1 provides a spontaneous flow of electrons from the zinc electrode to the copper electrode through the external circuit; it is thus a galvanic cell.

The same cell could also be operated as an electrolytic cell. Here it would be necessary to introduce a battery or some other electrical source in the external circuit to force electrons to flow in the opposite direction through the cell. Under these circumstances, zinc would deposit and copper would dissolve. These processes would consume energy from the battery.

Reversible and Irreversible Cells. Frequently, as with the cell in Figure 12-1, altering the direction of the current simply causes a reversal of the chemical reactions that occur at the electrodes. Such a cell is said to be electrochemically *reversible*. With other cells, reversing the current results in entirely different reactions at one or both electrodes. Such cells are called *irreversible*.

Currents in an Electrochemical Cell. Electricity is transported by three distinctly different processes in various parts of the cell shown in Figure 12-1. In the electrodes and the external conductor, electrons serve as carriers, moving from the zinc to the copper. Within the solutions the current involves migration of both positive and negative ions. Thus zinc ions, hydronium ions, and other positive species move away from the zinc electrode as it is oxidized; likewise, negative ions are attracted to this electrode by the excess positive ions produced in the electrochemical process. In the salt bridge the current consists largely of the movement of potassium ions in the direction of the copper electrode and chloride ions in the direction of the zinc electrode.

A third type of current occurs at the two electrode surfaces. Here an oxidation or a reduction provides the mechanism whereby the ionic conduction of the solution is coupled with the electron conduction of the electrodes to give a completed circuit wherein a current may exist.

ELECTRODE PROCESSES

We have thus far considered an electrochemical cell as being composed of two *half-cells*, each of which is associated with the process occurring at one of the electrodes. It must be stressed, however, that it is impossible to operate one half-cell independently of a second, nor can the potential of an individual half-cell be measured without reference to another.

Anode and Cathode. In any electrochemical cell the electrode at which oxidation occurs is called the anode, whereas the electrode at which reduction takes place is the cathode.

Common cathode reactions are illustrated by the following equations:

$$Ag^+ + e \rightarrow Ag(s)$$

$$2H^+ + 2e \rightarrow H_2(g)$$

$$Fe^{3+} + e \rightarrow Fe^{2+}$$

$$NO_3^- + 10H^+ + 8e \rightarrow NH_4^+ + 3H_2O$$

The last three of these reactions can occur at the surface of an inert cathode such as platinum. Hydrogen is frequently evolved from solutions that contain no other more easily reducible species.

Typical anodic reactions include

$$Cd(s) \rightarrow Cd^{2+} + 2e$$

$$2Cl^- \rightarrow Cl_2(g) + 2e$$

$$Fe^{2+} \rightarrow Fe^{3+} + e$$

$$2H_2O \rightarrow O_2(g) + 4H^+ + 4e$$

The last is often observed when a solution contains no easily oxidized species.

Electrode Signs. Electrodes are frequently provided with positive and negative signs to indicate the direction of electron flow during the operation of a cell. This practice leads to ambiguity, particularly when attempts are made to relate these signs to anodic or cathodic processes. In a galvanic cell, for example, the negative electrode is the one from which electrons flow to the external circuit (see Figure 12-1). In an electrolytic cell, on the other hand, the negative electrode is the one into which electrons are forced from an external source. Thus in a cell used for removal of zinc by electrolysis, the negative electrode is the one upon which metallic zinc is deposited, that is, the *cathode*. Clearly, the sign assumed by the anode, for example, depends upon whether we are considering an electrolytic or a galvanic cell. Irrespective of sign, however, the anode is the electrode at which oxidation occurs, and the cathode is the site for reduction. Use of these terms rather than positive and negative signs in describing electrode function is unambiguous and much preferred.

ELECTRODE POTENTIAL

Turning again to the galvanic cell shown in Figure 12-1, the driving force for the reaction

$$Zn(s) + Cu^{2+} \rightleftharpoons Cu(s) + Zn^{2+}$$

manifests itself as a measurable electrical force or voltage between the two electrodes. This force is the sum of two potentials called *half-cell potentials* or *single-electrode potentials*; one of these is associated with the half-reaction occurring at the anode and the other with the half-reaction taking place at the cathode.

It is not difficult to see how we might obtain information regarding the *relative* magnitudes of half-cell potentials. For example, if we replaced the left-hand side of the cell in Figure 12-1 with a cadmium electrode immersed in a solution of cadmium sulfate, the voltmeter reading would be approximately 0.4 V less than that of the original cell. Because the copper cathode remains unchanged, we logically ascribe this voltage decrease to the different anode reaction and thus conclude that the half-cell potential for the oxidation of cadmium is about 0.4 V less than that for the oxidation of zinc. Substitutions in the left-hand side of the cell would enable us to compare the chemical driving forces of other half-reactions to that of copper by means of similar electrical measurements.

Such a technique does not provide absolute values for half-cell potentials. Indeed, there is no way to determine such quantities because all voltage-measuring devices determine only *differences* in potential. To measure a potential difference, one lead of such a device is connected to the electrode in question; the second lead must then be brought in contact with the solution via another conductor. This latter contact, however, inevitably involves an interface between solid and solution and hence acts as a second half-cell at which a chemical reaction must also take place if the current required for the potential measurement is to exist. Thus we do not obtain an absolute measurement of the desired half-cell potential but rather a sum consisting of the potential of interest and the half-cell potential for the contact between the voltage-measuring device and the solution.

This inability to measure absolute potentials for half-cell processes is not a serious handicap because relative half-cell potentials are just as useful. Relative potentials can be combined to give cell potentials; in addition, they are useful for the calculation of equilibrium constants.

The Standard Hydrogen Electrode. To obtain consistent relative half-cell potential data it is necessary to compare all electrode potentials to a common reference. This reference electrode should be relatively easy to construct, be reversible, and give constant and reproducible potentials for a given set of experimental conditions.

The standard hydrogen electrode (SHE), illustrated in Figure 12-2,

FIGURE 12-2 The standard hydrogen electrode.

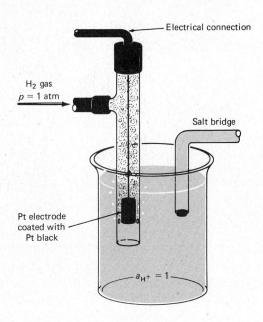

meets these requirements and is universally recognized as the ultimate reference. It is a typical *gas electrode*.

A hydrogen electrode consists of a piece of platinum foil (coated or *platinized* with finely divided platinum to increase its surface area) immersed in a solution of constant hydrogen ion activity and continuously swept by a stream of hydrogen at constant pressure. The platinum takes no part in the electrochemical reaction and serves only as the site for the transfer of electrons. The half-cell reaction responsible for the transmission of current across the interface is

$$H_2(g) \rightleftharpoons 2H^+ + 2e$$

This process undoubtedly involves the solution of molecular hydrogen; the overall equilibrium thus consists of two stepwise equilibria

$$H_2(g) \rightleftharpoons H_2(\text{sat'd})$$

$$H_2(\text{sat'd}) \rightleftharpoons 2H^+ + 2e$$

The continuous stream of gas at constant pressure provides the solution with a constant molecular hydrogen concentration.

The hydrogen electrode may act as an anode or a cathode, depending upon the half-cell with which it is coupled. Hydrogen is oxidized to hydrogen ions in the former; the reverse reaction takes place in the latter. Under proper conditions, then, the hydrogen electrode is electrochemically reversible.

The potential of a hydrogen electrode depends upon the temperature, the hydrogen ion concentration (more correctly, the activity) in the solution, and the pressure of the hydrogen at the surface of the electrode. Values for these parameters must be carefully defined in order for the half-cell process to serve as a reference. Specifications for the *standard* (or *normal*) *hydrogen electrode* call for a hydrogen ion activity of 1.00 and a partial pressure for hydrogen of 1.00 atm. *By convention the potential of the standard hydrogen electrode is assigned the value of exactly 0.0 V at all temperatures.*

Several secondary reference electrodes, which are more convenient to use on a routine basis, find extensive employment. Some of these are described in Chapter 14.

Measurement of Electrode Potentials. A cell such as that shown in Figure 12-3 can be used to measure relative half-cell potentials. Here one-half of the cell consists of the standard hydrogen electrode, and the other half is the electrode whose potential is to be determined. Connecting the two half-cells is a salt bridge to provide a conducting medium that also prevents direct interaction between the contents of the two half-cell compartments. Ordinarily the effect of the salt bridge upon the cell voltage is vanishingly small.

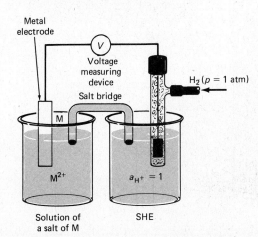

FIGURE 12-3 Schematic diagram of an arrangement for the measurement of electrode potentials against the standard hydrogen electrode.

The left-hand cell in Figure 12-3 consists of a pure metal in contact with a solution of its ions. The electrode reaction is given by

$$M(s) \rightleftharpoons M^{2+} + 2e$$

or the reverse. If the metal is cadmium and the solution is approximately 1 M in cadmium ions, the voltage indicated by the measuring device, V, will be about 0.4 V. Moreover, the cadmium will function as the anode; thus electrons pass from this electrode to the hydrogen electrode via the external circuit. The half-cell reactions for this galvanic cell can be written as

$$Cd(s) \rightleftharpoons Cd^{2+} + 2e \qquad \text{anode}$$

$$2H^+ + 2e \rightleftharpoons H_2(g) \qquad \text{cathode}$$

The overall cell reaction is the sum of these, or

$$Cd(s) + 2H^+ \rightleftharpoons Cd^{2+} + H_2(g)$$

If the cadmium electrode is replaced by a zinc electrode immersed in a solution that is about $1\,M$ in zinc ions, a potential of slightly less than 0.8 V will be observed. The metal electrode is again the anode in this cell. The larger voltage developed reflects the greater tendency of zinc to be oxidized. The difference between this potential and the one for cadmium is a quantitative measure of the relative strengths of these two metals as reducing agents.

Because a value of zero is assigned as the potential for the standard hydrogen electrode, the potentials for the two half-cells thus become 0.4 and 0.8 V with respect to this reference. It must be understood, however, that these are actually the potentials of electrochemical cells in which the standard hydrogen electrode serves as a common reference.

If the half-cell in Figure 12-3 consisted of a copper electrode in a 1-M solution of copper(II) ions, a potential of about 0.3 V would develop. However, in distinct contrast to the previous two examples, copper would tend to deposit and an external electron flow, if allowed, would be from the hydrogen electrode to the copper electrode. The spontaneous cell reaction, then, is the reverse of that in the two cells considered earlier:

$$Cu^{2+} + H_2(g) \rightleftharpoons Cu(s) + 2H^+$$

Thus metallic copper is a much less effective reducing agent than either zinc, cadmium, or *hydrogen*. As before the observed potential is a quantitative measure of this strength.

When the potential for the copper electrode is compared with those for zinc and cadmium, there emerges the further need to indicate the difference in direction of the spontaneous half-reaction with respect to the reference electrode. Positive and negative signs are used to make this distinction, the potential for half-cells such as copper being given a sign opposite to those of zinc and cadmium, which are spontaneous in the opposite sense. The choice as to which potential will be positive and which will be negative is *purely arbitrary*; however, the sign convention chosen must be used consistently.

Sign Conventions for Electrode Potentials. It is not surprising that the arbitrariness in the specification of signs has led to much controversy and confusion in the course of the development of electrochemistry. In 1953 the International Union of Pure and Applied Chemistry (IUPAC), meeting in Stockholm, attempted to resolve this controversy. The sign convention adopted at this meeting is sometimes called the IUPAC or Stockholm convention; there is hope for its general adoption in years to come. We shall always use the IUPAC sign convention.

Any sign convention must be based upon half-cell processes written in a single way—that is, as oxidations or as reductions. According to the IUPAC convention the term *electrode potential* (or more exactly, *relative electrode potential*) *is reserved exclusively for half-reactions written as reductions.* There is no objection to the use of the term *oxidation potential* to connote an electrode process written in the opposite sense, but an oxidation potential should never be called an electrode potential.

The sign of the electrode potential is determined by the actual sign of the electrode of interest when it is coupled with a standard hydrogen electrode in a galvanic cell. Thus a zinc or a cadmium electrode will behave as the anode from which electrons flow through the external circuit to the standard hydrogen electrode. These metal electrodes are thus the negative terminals of such galvanic cells, and their electrode potentials are *assigned* negative values. That is,

$$Zn^{2+} + 2e \rightleftharpoons Zn(s) \qquad E = -0.8 \text{ V}$$

$$Cd^{2+} + 2e \rightleftharpoons Cd(s) \qquad E = -0.4 \text{ V}$$

The potential for the copper electrode, on the other hand, is given a positive sign because the copper behaves as a cathode in a galvanic cell constructed from this electrode and the hydrogen electrode; electrons flow toward the copper electrode through the exterior circuit. It is thus the positive terminal of the galvanic cell and

$$Cu^{2+} + 2e \rightarrow Cu(s) \qquad E = +0.3 \text{ V}$$

It is important to note that each half-reaction describing the electrode potential *is written as a reduction.* For the first two, however, the spontaneous reactions are oxidations. It is evident, then, that the *sign of the electrode potential will indicate whether or not the reduction is spontaneous with respect to the standard hydrogen electrode.* That is, the positive sign for the copper electrode potential means that the reaction

$$Cu^{2+} + H_2 \rightleftharpoons 2H^+ + Cu(s)$$

proceeds toward the right under ordinary conditions. The negative electrode potential for zinc, on the other hand, means that the analogous reaction

$$Zn^{2+} + H_2 \rightleftharpoons 2H^+ + Zn(s)$$

does not ordinarily occur; indeed, the equilibrium favors the species on the left.

The IUPAC convention was adopted in 1953, but electrode potential data given in many texts and reference works are not always in accord with it. For example, in the prime source of oxidation potential data compiled by Latimer[1] one finds

$$Zn(s) \rightleftharpoons Zn^{2+} + 2e \qquad E = +0.76 \text{ V}$$

$$Cu(s) \rightleftharpoons Cu^{2+} + 2e \qquad E = -0.34 \text{ V}$$

To convert these oxidation potentials to electrode potentials as defined by the IUPAC convention, one must mentally (1) express the half-reactions as reductions and (2) change the signs of the potentials.

The sign convention employed in a table of standard potentials may not be explicitly stated. This information is readily determined, however, by referring to a half-reaction with which one is familiar and noting the direction of the reaction and the sign of the potential. Whatever changes, if any, are required to convert to the IUPAC convention are then applied to the remainder of the data in the table. For example, all one needs to remember is that strong oxidizing agents such as chlorine have large positive electrode potentials under the IUPAC convention. That is, the reaction

$$Cl_2(g) + 2e \rightleftharpoons 2Cl^- \qquad E = +1.36 \text{ V}$$

tends to occur spontaneously with respect to the standard hydrogen electrode. The sign and direction of this reaction in a given table can then serve as a key to any changes that may be needed to convert all data to the IUPAC convention.

Effect of Concentration on Electrode Potentials. Because it is a measure of the chemical driving force of a half-reaction, the electrode potential is affected by concentration. For example, the tendency of copper(II) ion to be reduced to the elemental state is substantially greater from a concentrated than from a dilute solution; therefore the electrode potential for this process must also be greater for the more concentrated solution. In general the concentrations of reactants and products in a half-reaction have a marked effect on the electrode potential; the quantitative aspects of this effect must now be examined.

Consider the generalized reversible half-cell reaction

$$a\text{A} + b\text{B} + \cdots + ne \rightleftharpoons c\text{C} + d\text{D} + \cdots$$

where the capital roman letters represent formulas of reacting species (whether charged or uncharged), e represents electrons, and the lowercase italic letters indicate the number of moles of each species participating in the reaction. It can be shown theoretically as well as experimentally that

[1]W. M. Latimer, *The Oxidation States of the Elements and Their Potentials in Aqueous Solutions*, 2d ed. Englewood Cliffs, N.J.: Prentice-Hall, 1952.

the potential E for this electrode process is governed by the relation

Nernst Rxn.

$$E = E^0 - \frac{RT}{nF} \ln \frac{[C]^c [D]^d \dots}{[A]^a [B]^b \dots}$$

(12-1)

where

E^0 = a constant, called the *standard electrode potential*
which is characteristic of the particular half-reaction
R = the gas constant = 8.314 volt coulomb $°K^{-1}$ $mole^{-1}$
T = the absolute temperature
n = number of electrons participating in the reaction
as defined by the equation describing the half-cell reaction
F = the faraday = 96,493 coulombs
$\ln$ = the natural logarithm = 2.303 $\log_{10}$

Upon substituting numerical values for the various constants and converting
to base 10 logarithms, Equation 12-1 at 25°C becomes

$$E = E^0 - \frac{0.0591}{n} \log \frac{[C]^c [D]^d \dots}{[A]^a [B]^b \dots}$$

(12-2)

The letters in brackets represent the activities of the various species.
Substitution of concentrations for activities does not introduce an excessive error in many calculations. Thus, where the species X is a solute,

$$[X] \cong \text{concentration in moles per liter}$$

If the reactant X is a gas,

$$[X] \cong \text{partial pressure of the gas, in atmospheres}$$

If the reactant X exists in a second phase as a *pure solid or liquid*, then by
definition

$$[X] = 1.00$$

The basis of this last statement is the same as that described
earlier, namely, that the concentration of a substance in the pure second
phase is constant; therefore so long as some solid (or pure liquid) is
present, its effect on an equilibrium is invariant and included in the
constant E^0. Similarly, although it may be involved in the half-cell process,
the concentration of water as a solvent is ordinarily so large with respect to
the other reactants that for all practical purposes it can be considered to
remain unchanged. Its contribution is customarily included in the constant
E^0. Thus, if water appears in Equation 12-2,

$$[H_2O] = 1.00$$

Equation 12-1 is called the *Nernst equation* in honor of a
nineteenth-century electrochemist. Application of the Nernst equation is
illustrated in the following examples.

1. $Zn^{2+} + 2e \rightleftharpoons Zn(s)$

$$E = E^0 - \frac{0.0591}{2} \log \frac{1}{[Zn^{2+}]}$$

The activity of the elemental zinc is unity by definition; the electrode potential thus varies with the logarithm of the reciprocal of the molar zinc ion concentration.

2. $Fe^{3+} + e \rightleftharpoons Fe^{2+}$

$$E = E^0 - \frac{0.0591}{1} \log \frac{[Fe^{2+}]}{[Fe^{3+}]}$$

This electrode potential can be measured with an inert metal electrode immersed in a solution containing iron(II) and iron(III). The potential is dependent upon the ratio between the molar concentrations of these ions.

3. $2H^+ + 2e \rightleftharpoons H_2(g)$

$$E = E^0 - \frac{0.0591}{2} \log \frac{p_{H_2}}{[H^+]^2}$$

In this example p_{H_2} represents the partial pressure of hydrogen, expressed in atmospheres, at the surface of the electrode. Ordinarily p_{H_2} will be very close to atmospheric pressure.

4. $Cr_2O_7^{2-} + 14H^+ + 6e \rightleftharpoons 2Cr^{3+} + 7H_2O$

$$E = E^0 - \frac{0.0591}{6} \log \frac{[Cr^{3+}]^2}{[Cr_2O_7^{2-}][H^+]^{14}}$$

Here the potential depends not only on the concentrations of chromium(III) and dichromate ions but also on the pH of the solution.

5. $AgCl(s) + e \rightleftharpoons Ag(s) + Cl^-$

$$E = E^0 - \frac{0.0591}{1} \log \frac{[Cl^-]}{1}$$

This half-reaction, which describes the behavior of a silver electrode immersed in a solution of chloride ions that is *saturated* in silver chloride, is the sum of two reactions, namely,

$$AgCl(s) \rightleftharpoons Ag^+ + Cl^-$$

$$Ag^+ + e \rightleftharpoons Ag(s)$$

Here the activities of both metallic silver and silver chloride are equal to unity by definition; the potential of the silver electrode will depend only upon the chloride ion concentration.

We need to make one additional observation with respect to the application of the Nernst equation. Although it is not obvious from this presentation, the quotient in the logarithm term is a unitless quantity. Each of the individual terms in the quotient is in fact a ratio of the present activity of the species compared to its activity in the standard state, which

then, are those species that appear in the lower right-hand part of a table of standard electrode potentials.

A compilation of standard potentials provides the chemist with qualitative information regarding the extent and direction of electron-transfer reactions between the tabulated species. On the basis of Table 12-1, for example, we see that zinc is more easily oxidized than cadmium, and we conclude that a piece of zinc immersed in a solution of cadmium ions will cause the deposition of metallic cadmium. On the other hand, cadmium has little tendency to reduce zinc ions. From Table 12-1 we also see that iron(III) is a better oxidizing agent than the triiodide ion, and we may therefore predict that in a solution containing an equilibrium mixture of iron(III), iodide, iron(II), and triiodide ions, the latter pair will pre-dominate.

Calculation of Half-Cell Potentials from E^0 Values. Some applications of the Nernst equation to the calculation of half-cell potentials are illus-trated in the following examples.

EXAMPLE

What is the potential for a half-cell consisting of a cadmium electrode immersed in a solution that is $0.0100 \, F$ in Cd^{2+}?

From Table 12-1 we find

$$Cd^{2+} + 2e \rightleftharpoons Cd(s) \qquad E^0 = -0.403 \, V$$

Thus

$$E = E^0 - \frac{0.0591}{2} \log \frac{1}{[Cd^{2+}]}$$

Substituting the Cd^{2+} concentration into this equation gives

$$E = -0.403 - \frac{0.0591}{2} \log \frac{1}{0.0100}$$

$$= -0.403 - \frac{0.0591}{2} (+2.0)$$

$$= -0.462 \, V$$

The sign for the potential indicates the direction of the reaction when this half-cell is coupled with the standard hydrogen electrode. The fact that it is negative shows that the reverse reaction

$$Cd(s) + 2H^+ \rightleftharpoons H_2(g) + Cd^{2+}$$

occurs spontaneously. Note that the calculated potential is a larger nega-tive number than the standard electrode potential itself. This follows from mass-law considerations because the half-reaction *as written* has less tendency to occur with the lower cadmium ion concentration.

EXAMPLE

Calculate the potential for a platinum electrode immersed in a solution prepared by saturating a 0.0100-F solution of KBr with Br_2.

Here the half-reaction is

$$Br_2(1) + 2e \rightleftharpoons 2Br^- \qquad E^0 = 1.065 \text{ V}$$

Note that the term (l) in the equation is employed to indicate that the aqueous solution is kept saturated by the presence of an excess of *liquid* Br_2. Thus the overall process is the sum of the two equilibria

$$Br_2(1) \rightleftharpoons Br_2(\text{sat'd aq})$$

$$Br_2(\text{sat'd aq}) + 2e \rightleftharpoons 2Br^-$$

The Nernst equation for the overall process is

$$E = 1.065 - \frac{0.0591}{2} \log \frac{[Br^-]^2}{1.00}$$

Here the activity of Br_2 in the pure liquid is constant and is 1.00 by definition. Thus

$$E = 1.065 - \frac{0.0591}{2} \log (0.0100)^2$$

$$= 1.065 - \frac{0.0591}{2} (-4.00)$$

$$= 1.183 \text{ V}$$

EXAMPLE

Calculate the potential for a platinum electrode immersed in a solution that is 0.0100 F in KBr and 1.00×10^{-3} F in Br_2.

Here the half-reaction used in the preceding example does not apply *because the solution is no longer saturated in Br_2.* In Table 12-1, however, we find the half-reaction

$$Br_2(\text{aq}) + 2e \rightleftharpoons 2Br^- \qquad E^0 = 1.087 \text{ V}$$

The term (aq) implies that all of the Br_2 present is in solution. That is, 1.087 is the electrode potential for the half-reaction when the Br^- and Br_2 *solution* activities are 1.00 mole/liter. It turns out, however, that the solubility of Br_2 in water at 25°C is only about 0.18 mole/liter. Therefore the recorded potential of 1.087 is based on a *hypothetical system that cannot be realized experimentally.* Nevertheless, this potential is useful because it provides a means by which potentials for undersaturated systems can be calculated. Thus

$$E = 1.087 - \frac{0.0591}{2} \log \frac{[Br^-]^2}{[Br_2]}$$

$$= 1.087 - \frac{0.0591}{2} \log \frac{(1.00 \times 10^{-2})^2}{1.00 \times 10^{-3}}$$

$$= 1.087 - \frac{0.0591}{2} \log 0.100$$

$$= 1.117 \text{ V}$$

Here the Br_2 activity is 1.00×10^{-3} rather than 1.00, as is the case when the solution is saturated.

Electrode Potentials in the Presence of Precipitation and Complex-forming Reagents. As shown by the following example, reagents that react with the participants of an electrode process have a marked effect on the potential for that process.

EXAMPLE

Calculate the potential of a silver electrode in a solution that is saturated with silver iodide and has an iodide ion activity of exactly 1.00.

$$Ag^+ + e \rightleftharpoons Ag(s) \qquad E^0 = +0.799 \text{ V}$$

$$E = +0.799 - 0.0591 \log \frac{1}{[Ag^+]}$$

We may calculate $[Ag^+]$ from the solubility product constant

$$[Ag^+] = \frac{K_{sp}}{[I^-]}$$

Substituting into the Nernst equation

$$E = +0.799 - \frac{0.0591}{1} \log \frac{[I^-]}{K_{sp}}$$

This equation may be rewritten as

$$E = +0.799 + 0.0591 \log K_{sp} - 0.0591 \log [I^-] \qquad (12\text{-}3)$$

If we substitute 1.00 for $[I^-]$ and use 8.3×10^{-17} for K_{sp}, the solubility product for AgI at 25.0°C, we obtain

$$E = -0.151 \text{ V}$$

The example just considered shows that the half-cell potential for the reduction of silver ion becomes smaller in the presence of iodide ions. Qualitatively this is the expected effect because decreases in the concentration of silver ions diminish the tendency for their reduction.

Equation 12-3 relates the potential of a silver electrode to the iodide ion concentration of a solution that is also saturated with silver iodide. *When the iodide ion activity is unity*, the potential is the sum of two constants; this sum is the standard electrode potential for the half-reaction

$$AgI(s) + e \rightleftharpoons Ag + I^- \qquad E^0 = -0.151 \text{ V}$$

where

$$E^0 = +0.799 + 0.0591 \log K_{sp} \qquad \textbf{(12-4)}$$

The Nernst relationship for the silver electrode in a solution saturated with silver iodide can then be written as

$$E = E^0 - 0.0591 \log [I^-] = -0.151 - 0.0591 \log [I^-]$$

Thus, when in contact with a solution *saturated with silver iodide*, the potential of a silver electrode can be described *either* in terms of the silver ion concentration (with the standard electrode potential for the simple silver half-reaction) *or* in terms of the iodide ion concentration (with the standard electrode potential for the silver–silver iodide half-reaction).

The potential of a silver electrode in a solution containing an ion that forms a soluble complex with silver ion can be treated in a fashion analogous to the foregoing. For example, in a solution containing thiosulfate ion the half-reaction can be written as

$$Ag(S_2O_3)_2^{3-} + e \rightleftharpoons Ag(s) + 2S_2O_3^{2-}$$

The standard electrode potential for this half-reaction will be the electrode potential when both the complex and the complexing anion are at unit activity. Using the same approach as in the previous example, we find

$$E^0 = +0.799 + 0.0591 \log \frac{1}{K_f}$$

where K_f is the formation constant for the complex ion.

Data for the potential of the silver electrode in the presence of selected ions are given in the tables of standard electrode potentials in Appendix 6 and Table 12-1. Similar information is also provided for other electrode systems. Such data often simplify the calculation of half-cell potentials.

CELLS AND CELL POTENTIALS

Schematic Representation of Cells. Chemists frequently use a short-hand notation to describe electrochemical cells. For example, the cell shown in Figure 12-1 can be represented as

$$Zn|ZnSO_4(C_1)\|CuSO_4(C_2)|Cu$$

where C_1 and C_2 are formal concentrations of the two salts. *Customarily the anode is listed on the left.* Single vertical lines represent phase boundaries in the cell. Potential differences arising across these are included in the measured cell potential. A double vertical line represents the two phase boundaries that exist at either end of the salt bridge. One of these boundaries is between the $ZnSO_4$ solution and saturated KCl; the other involves the KCl solution and the $CuSO_4$ solution. A potential, called a *liquid junction potential*, develops at each interface due to the different rates at which ions diffuse across the junction. This potential may be as large as several hundredths of a volt. The potentials at the two interfaces of a salt bridge, however, tend to cancel each other so that the net contribution of the total junction potential is a few millivolts or less. For our purposes the junction potential can be neglected without serious error.

Calculation of Cell Potentials. An important use of standard electrode potentials is the calculation of the potential obtainable from a galvanic cell or the potential required to operate an electrolytic cell. These calculated potentials (sometimes called *thermodynamic potentials*) are theoretical in the sense that they refer to cells in which there is essentially no current; additional factors must be taken into account where a current is involved.

The electromotive force of a cell is obtained by combining half-cell potentials as follows:

$$E_{cell} = E_{cathode} - E_{anode}$$

where E_{anode} and $E_{cathode}$ are the *electrode potentials* for the two half-reactions constituting the cell.

Consider the cell

$$Zn|ZnSO_4(a_{Zn^{2+}} = 1.00)\|CuSO_4(a_{Cu^{2+}} = 1.00)|Cu$$

The overall cell process involves the oxidation of elemental zinc to zinc ions and the reduction of copper(II) ions to the metallic state. Because the activities of the two ions are specified as unity, the standard potentials are also the electrode potentials. The cell diagram also specifies that the zinc electrode is the anode. Thus, using E^0 data from Table 12-1,

$$E_{cell} = +0.337 - (-0.763) = +1.100 \text{ V}$$

The positive sign for the cell potential indicates that the reaction

$$Zn(s) + Cu^{2+} \rightarrow Zn^{2+} + Cu(s)$$

occurs spontaneously and that this is a galvanic cell.

The foregoing cell, diagrammed as

$$Cu|Cu^{2+}(a_{Cu^{2+}} = 1.00)\|Zn^{2+}(a_{Zn^{2+}} = 1.00)|Zn$$

implies that the copper electrode is now the anode. Thus

$$E_{cell} = +0.763 + (-0.337) = -1.100 \text{ V}$$

$E_{cell} = \text{kethod} + \text{Anode.}$

The negative sign indicates the nonspontaneity of the reaction

$$Cu(s) + Zn^{2+} \rightarrow Cu^{2+} + Zn(s)$$

To cause this reaction to occur would require the application of an external potential greater than 1.100 V.

EXAMPLE

Calculate the theoretical potential of the cell   *thermodinamic potential*

$$Pt, H_2(0.800\ atm)|HI(0.200\ F), AgI(sat'd)|Ag$$

The two half-cell reactions and standard electrode potentials are (Table 12-1)

$$2H^+ + 2e \rightleftharpoons H_2(g) \qquad E^0 = 0.000\ V$$

$$AgI(s) + e \rightleftharpoons Ag(s) + I^- \qquad E^0 = -0.151\ V$$

If we make the approximation that the activities of the various species are identical to their molar concentrations, then for the hydrogen electrode

$$E = 0.000 - \frac{0.0591}{2} \log \frac{0.800}{(0.200)^2} = -0.038\ V$$

and for the silver–silver iodide electrode

$$E = -0.151 - 0.0591 \log 0.200 = -0.110\ V$$

The cell diagram specifies the hydrogen electrode as the anode and the silver electrode as the cathode. Therefore

$$E_{cell} = -0.110 - (-0.038) = -0.072\ V$$

The negative sign indicates that the cell reaction

$$H_2(g) + 2AgI(s) \rightarrow 2H^+ + 2Ag(s) + 2I^-$$

will require the application of an external potential greater than 0.072 V.

EXAMPLE

Calculate the potential required to initiate the deposition of metallic copper from a solution that is $0.010\ F$ in $CuSO_4$ and contains sufficient sulfuric acid to give a hydrogen ion concentration of $1.0 \times 10^{-4}\ M$.

The deposition of copper necessarily occurs at the cathode. Because no easily oxidizable species are present, the anode reaction will involve oxidation of H_2O to give O_2. From the table of standard potentials (Appendix 6) we find

$$Cu^{2+} + 2e \rightleftharpoons Cu(s) \qquad E^0 = +0.337\ V$$

$$\tfrac{1}{2}O_2(g) + 2H^+ + 2e \rightleftharpoons H_2O \qquad E^0 = +1.229\ V$$

Thus for the copper electrode

$$E = +0.337 - \frac{0.0591}{2} \log \frac{1}{0.010} = +0.278 \text{ V}$$

Assuming that O_2 is evolved at 1.00 atm, the potential for the oxygen electrode is

$$E = +1.229 - \frac{0.0591}{2} \log \frac{1}{(1.0 \times 10^{-4})^2 (1.00)^{1/2}} = +0.993 \text{ V}$$

The cell potential is then

$$E_{cell} = +0.278 - 0.993 = -0.715 \text{ V}$$

Thus to initiate the reaction

$$Cu^{2+} + H_2O \rightarrow \tfrac{1}{2}O_2(g) + 2H^+ + Cu(s)$$

would require the application of a potential greater than 0.715 V.

EQUILIBRIUM CONSTANTS FROM STANDARD ELECTRODE POTENTIALS

Changes in Potential during Discharge of a Cell. Consider again the galvanic cell based upon the reaction

$$Zn(s) + Cu^{2+} \rightleftharpoons Zn^{2+} + Cu(s)$$

At all times the cell potential is given by

$$E_{cell} = E_{cathode} - E_{anode}$$

As current is drawn from this cell, however, the zinc ion concentration increases and the copper(II) concentration decreases. These changes have the effect of rendering the electrode potential less negative for the zinc half-cell and less positive for the copper half-cell. The net effect, then, is a decrease in the potential of the cell. Ultimately the concentrations attain values such that there is no longer a tendency for electron transfer to occur. The potential of the cell thus becomes zero, and the system is in equilibrium. That is, at equilibrium

$$E_{cell} = E_{cathode} - E_{anode} = 0$$

or

$$E_{cathode} = E_{anode}$$

This equation expresses an important and general relationship: *for an oxidation-reduction system at chemical equilibrium, the electrode potentials (that is, the reduction potentials) of all half-reactions of the system will be equal.*

Calculation of Equilibrium Constants. Consider the oxidation-reduction equilibrium

$$a\,A_{red} + b\,B_{ox} \rightleftharpoons a\,A_{ox} + b\,B_{red}$$

for which the half-reactions may be written

$$a\,A_{ox} + ne \rightleftharpoons a\,A_{red}$$

$$b\,B_{ox} + ne \rightleftharpoons b\,B_{red}$$

When the components of this system are at chemical equilibrium,

$$E_A = E_B$$

where E_A and E_B are the electrode potentials for the two half-cells. With this equality expressed in terms of the Nernst equation, we find that *at equilibrium*

$$E^0_A - \frac{0.0591}{n} \log \frac{[A_{red}]^a}{[A_{ox}]^a} = E^0_B - \frac{0.0591}{n} \log \frac{[B_{red}]^b}{[B_{ox}]^b}$$

which yields upon rearrangement and combination of the log terms

$$E^0_B - E^0_A = \frac{0.0591}{n} \log \frac{[A_{ox}]^a [B_{red}]^b}{[A_{red}]^a [B_{ox}]^b} = \frac{0.0591}{n} \log K_{eq}$$

This relationship was derived for equilibrium conditions. The concentration terms are thus *equilibrium concentrations*; *hence the quotient is the equilibrium constant for the reaction.* That is,

$$\log K_{eq} = \frac{n(E^0_B - E^0_A)}{0.0591}$$

EXAMPLE

Calculate the equilibrium constant for the reaction

$$MnO_4^- + 5Fe^{2+} + 8H^+ \rightleftharpoons Mn^{2+} + 5Fe^{3+} + 4H_2O$$

From the table of standard electrode potentials (Appendix 6) we find

$$MnO_4^- + 8H^+ + 5e \rightleftharpoons Mn^{2+} + 4H_2O \qquad E^0_{MnO_4^-} = +1.51 \text{ V}$$

$$5Fe^{3+} + 5e \rightleftharpoons 5Fe^{2+} \qquad E^0_{Fe^{3+}} = +0.771 \text{ V}$$

Each half-reaction must contain the same number of moles of reactants and products as appear in the balanced net-ionic equation for the overall reaction. It is for this reason that the species in the iron half-reaction are multiplied by 5. *Note, however, that E^0 remains unchanged* (p. 281).

Because at equilibrium

$$E_{Fe^{3+}} = E_{MnO_4^-}$$

then

$$E^0_{Fe^{3+}} - \frac{0.0591}{5} \log \frac{[Fe^{2+}]^5}{[Fe^{3+}]^5} = E^0_{MnO_4^-} - \frac{0.0591}{5} \log \frac{[Mn^{2+}]}{[MnO_4^-][H^+]^8}$$

Rearranging this equation yields

$$\frac{0.0591}{5} \log \frac{[Mn^{2+}][Fe^{3+}]^5}{[MnO_4^-][Fe^{2+}]^5[H^+]^8} = E^0_{MnO_4^-} - E^0_{Fe^{3+}}$$

or

$$\log K_{eq} = \frac{5(1.51 - 0.77)}{0.0591}$$

$$= 62.6$$

$$K_{eq} = 10^{62.6} = 10^{0.6} \times 10^{62} = 4 \times 10^{62}$$

Here we are forced to round the result to one significant figure because the first two integers (6 and 2) of the logarithm of K equilibrium simply set the decimal point for the antilogarithm; the mantissa (0.6) then determines the number of significant figures in the result.

EXAMPLE

A piece of copper is placed in a 0.050-F solution of $AgNO_3$. What is the equilibrium composition of the solution?

The reaction is

$$Cu(s) + 2Ag^+ \rightleftharpoons Cu^{2+} + 2Ag(s)$$

We first calculate the equilibrium constant for the reaction and then use this to determine the solution composition. From the table of standard potentials we find

$$2Ag^+ + 2e \rightleftharpoons 2Ag(s) \qquad E^0_{Ag^+} = +0.799 \text{ V}$$

$$Cu^{2+} + 2e \rightleftharpoons Cu(s) \qquad E^0_{Cu^{2+}} = +0.337 \text{ V}$$

Because at equilibrium

$$E_{Cu^{2+}} = E_{Ag^+}$$

then

$$E^0_{Cu^{2+}} - \frac{0.0591}{2} \log \frac{1}{[Cu^{2+}]} = E^0_{Ag^+} - \frac{0.0591}{2} \log \frac{1}{[Ag^+]^2}$$

$$\log \frac{[Cu^{2+}]}{[Ag^+]^2} = \frac{2(E^0_{Ag^+} - E^0_{Cu^{2+}})}{0.0591} = 15.63$$

and

$$\frac{[Cu^{2+}]}{[Ag^+]^2} = K_{eq} = 4.3 \times 10^{15}$$

The magnitude of the equilibrium constant suggests that nearly all Ag^+ is used up. The molar concentration of Cu^{2+} is thus given by

$$[Cu^{2+}] = \tfrac{1}{2}(0.050 - [Ag^+])$$

Because the reaction is nearly complete, it is probably safe to assume that $[Ag^+]$ is small with respect to 0.050. Then

$$[Cu^{2+}] \cong \tfrac{1}{2}(0.050) = 0.025$$

Substituting,

$$\frac{0.025}{[Ag^+]^2} = 4.3 \times 10^{15}$$

$$[Ag^+] = 2.4 \times 10^{-9}$$

$$[Cu^{2+}] = \tfrac{1}{2}(0.050 - 2.4 \times 10^{-9}) \cong 0.025$$

DETERMINATION OF DISSOCIATION, SOLUBILITY PRODUCT, AND FORMATION CONSTANTS BY POTENTIAL MEASUREMENTS

We have seen that the potential of an electrode is determined by the concentrations of those species that participate in the electrode reaction; as a consequence, measurement of a half-cell potential often provides a convenient way to determine the concentration of solute species. One important virtue of this approach is that the measurement can be made without affecting appreciably any equilibria that may exist in the solution. For example, the potential of a silver electrode in a solution containing the cyanide complex of silver depends only upon the silver ion activity. With suitable equipment this potential can be measured with a negligible passage of current. Thus the concentration of silver ions in the solution is not sensibly altered during the measurement; the position of the equilibrium

$$Ag^+ + 2CN^- \rightleftharpoons Ag(CN)_2^-$$

is likewise undisturbed.

EXAMPLE

The solubility product constant for the sparingly soluble CuX_2 can be determined by saturating a 0.0100-F solution of NaX with solid CuX_2. After equilibrium is achieved, this solution is made part of the cell

$$Cu|CuX_2(sat'd), NaX(0.0100 \ F)\|SHE$$

where SHE is the standard hydrogen electrode.

Suppose that the potential of this cell is found to be 0.0103 V, with the copper electrode behaving as the anode, as indicated. We may then write

$$E_{cell} = E_{cathode} - E_{anode}$$

$$0.0103 = 0.000 - \left(E^0_{Cu^{2+}} - \frac{0.0591}{2} \log \frac{1}{[Cu^{2+}]}\right)$$

$$0.0103 = -0.337 - \frac{0.0591}{2} \log [Cu^{2+}]$$

Note the change of sign that accompanied inversion of the log term.

Thus

$$[Cu^{2+}] = 1.8 \times 10^{-12}$$

Because the X^- concentration is 0.0100 mole/liter,

$$K_{sp} = (1.8 \times 10^{-12})(0.0100)^2$$
$$= 1.8 \times 10^{-16}$$

Any electrode system that is sensitive to the hydrogen ion concentration of a solution can, in theory at least, be used to determine the dissociation constants of acids and bases. All half-cells in which hydrogen ion is a participant fall in this category. However, relatively few of these have been applied to the problem.

EXAMPLE

Calculate the dissociation constant of the acid HP if the cell

$$\text{Pt, } H_2(1.00 \text{ atm})|HP(0.010 \ F), \text{NaP}(0.030 \ F)\|\text{SHE}$$

develops a potential of 0.295 V.

As indicated by the diagram, the standard hydrogen electrode is the cathode. Thus $E_{cathode} = 0.000$ V. For the other half of the cell we may write

$$2H^+ + 2e \rightleftharpoons H_2(g)$$

$$E_{anode} = 0.000 - \frac{0.0591}{2} \log \frac{p_{H_2}}{[H^+]^2}$$

and

$$E_{cell} = E_{cathode} - E_{anode}$$

$$0.295 = 0.000 - \left(0.000 - \frac{0.0591}{2} \log \frac{p_{H_2}}{[H^+]^2}\right)$$

$$= \frac{0.0591}{2} \log \frac{1.00}{[H^+]^2} = -\frac{2 \times 0.0591}{2} \log [H^+]$$

After rearrangement

$$[H^+] = 1.0 \times 10^{-5}$$

$$K_a = \frac{[H^+][P^-]}{[HP]}$$

Because the HP present is largely undissociated,

$$K_a \cong \frac{(1.0 \times 10^{-5})(0.030)}{0.010} = 3.0 \times 10^{-5}$$

Formation constants for complex ions can be determined in an analogous fashion. Thus to evaluate the constant for the $Ag(CN)_2^-$ complex we might measure the potential of the cell

$$Ag|Ag(CN)_2^-(C_1), CN^-(C_2)\|SHE$$

Values for C_1 and C_2 would be known from the quantities used to prepare the cell. The silver ion concentration could be calculated from the potential of the cell. Combination of this concentration with C_1 and C_2 would then lead to a value for the desired equilibrium constant.

SOME LIMITATIONS TO THE USE OF STANDARD ELECTRODE POTENTIALS

The examples in the previous section illustrate typical applications of the Nernst equation to problems that are important to the analytical chemist. It frequently happens, however, that marked differences exist between calculated and experimentally measured potentials; the intelligent application of electrode potential calculations requires an appreciation of the sources of uncertainty and their effects upon a calculated result.

Use of Concentrations instead of Activities. It is usually expedient to substitute molar concentrations for activities in the Nernst equation even though differences between these quantities tend to increase with increasing ionic strength (Chapter 5). The typical electrochemical cell possesses a high ionic strength; as a consequence, potentials based on molar concentrations are likely to differ somewhat from those obtained by direct measurement.

Effect of Other Equilibria. The application of standard electrode potential data to many systems of interest is further complicated by solvolysis, dissociation, association, and complex formation equilibria involving species that appear in the Nernst equation. These phenomena can be accounted for only if their existence is known and equilibrium constant data are available. This information is frequently lacking; the chemist is then forced to hope that neglect of such effects will not seriously influence his calculations.

Formal Potential. To compensate partially for activity effects and errors resulting from side reactions, Swift[3] has proposed the use of a quantity called the *formal potential* in place of the standard electrode potential. The formal potential of a system is the potential of the half-cell with respect to the standard hydrogen electrode when the concentration of each reactant and product is exactly 1 F, and the concentrations of any other constituents of the solution are carefully specified. Formal potentials are available for many of the systems tabulated in Appendix 6. Thus, for example, the

[3]E. H. Swift, *A System of Chemical Analysis*. San Francisco: Freeman, 1939, p. 50.

formal potential for the reduction of iron(III) to iron(II) is +0.731 V in 1-F perchloric acid and +0.700 V in 1-F hydrochloric acid, as compared with the standard potential of 0.771 V. The lower formal potential in perchloric acid results from the fact that the activity coefficient of iron(III) is smaller than that of iron(II) in the highly ionic medium. The even larger effect of HCl results from the greater stability of chloride complexes of iron(III) than those of iron(II).

If formal potentials are employed in place of standard electrode potentials in the Nernst equation, better agreement is obtained between calculated and experimental potentials, provided, of course, that the electrolyte concentration of the solution approximates that for which the formal potential was measured. Needless to say, application of a formal potential to a system that differs greatly in kind or concentration of electrolyte often causes errors greater than those encountered with the use of standard potentials. Henceforth, we shall use whichever is the more appropriate.

Reaction Rate. Calculations with standard potentials will indicate whether or not a given oxidation-reduction reaction is complete enough for application to an analytical problem. Unfortunately, however, such calculations provide no information with respect to the rate at which the equilibrium state will be approached; frequently a reaction that appears extremely favorable from equilibrium considerations may be totally unacceptable from a kinetic standpoint. The oxidation of arsenic(III) with a solution of quadrivalent cerium in dilute sulfuric acid is an example:

$$H_3AsO_3 + 2Ce^{4+} + H_2O \rightleftharpoons H_3AsO_4 + 2Ce^{3+} + 2H^+$$

The respective formal potentials for the two systems are

$$2Ce^{4+} + 2e \rightleftharpoons 2Ce^{3+} \qquad E^f = +1.4 \text{ V}$$

$$H_3AsO_4 + 2H^+ + 2e \rightleftharpoons H_2O + H_3AsO_3 \qquad E^f = +0.56 \text{ V}$$

and an equilibrium constant of about 10^{28} can be deduced from these data. Despite this very favorable equilibrium constant, solutions of arsenic(III) cannot be titrated with cerium(IV) unless a catalyst is introduced because several hours are required for the attainment of equilibrium.

Oxidation-Reduction Titrations

The successful application of an oxidation-reduction reaction to volumetric analysis requires, among other things, the means for detecting the equivalence point. We must therefore examine the changes that occur in the course of such a titration and pay particular attention to those changes that are most pronounced in the region of the equivalence point.

TITRATION CURVES

In the titration curves considered thus far the negative logarithm of the concentration (the p-function) of some reacting species has been plotted against the volume of reagent added. The species chosen for each curve has been one to which the indicator for the reaction is sensitive. Most of the indicators used for oxidation-reduction titrations are themselves oxidizing or reducing agents that respond to changes in the potential of the system rather than to changes in concentration of any particular reactant or product. For this reason the usual practice is to plot the electrode potential for the system as the ordinate of the curve for an oxidation-reduction titration rather than a p-function for a reactant.

To develop a clear understanding of what is meant by the term "electrode potential for the system," consider the titration of iron(II) with cerium(IV):

$$Ce^{4+} + Fe^{2+} \rightleftharpoons Ce^{3+} + Fe^{3+}$$

Equilibrium is attained after each addition of titrant. After the first addition of cerium(IV), then, all four species will be present in amounts dictated by the equilibrium constant for the reaction. Recall now (p. 289) that the electrode potentials for the two half-reactions are identical at equilibrium; thus at any point in the titration

$$E_{Ce^{4+}} = E_{Fe^{3+}} = E_{system}$$

and it is this potential that we call the potential of the system. If a reversible oxidation-reduction indicator is present in the solution as well, its potential must also be the same as that for the system. That is, the ratio between the two forms of the indicator changes until

$$E_{In} = E_{Ce^{4+}} = E_{Fe^{3+}} = E_{system}$$

The potential of a system can be measured experimentally by determining the emf of a suitable cell. Thus for the titration of iron(II) with cerium(IV) the analyte solution could be made part of the cell:

$$SHE \| Ce^{4+}, Ce^{3+}, Fe^{3+}, Fe^{2+} | Pt$$

Here the potential of the platinum electrode (versus the SHE) is determined both by the affinity of Fe^{3+} for electrons

$$Fe^{3+} + e \rightleftharpoons Fe^{2+}$$

as well as that of Ce^{4+}

$$Ce^{4+} + e \rightleftharpoons Ce^{3+}$$

At equilibrium the concentration ratios of oxidized and reduced forms of each species are such that these two affinities (and thus their electrode potentials) are identical. Note that the concentration ratios for both species vary continuously as the titration proceeds; so also must E_{system}. It is the characteristic variation in this parameter that provides a means of end point detection.

In deriving E_{system} data for a titration curve, either $E_{Ce^{4+}}$ or $E_{Fe^{3+}}$ can be employed. We choose the more convenient one for any particular calculation. Short of the equivalence point, the concentrations of iron(II), iron(III), and cerium(III) are readily deduced from the amount of titrant that has been added; the concentration of cerium(IV), however, is vanishingly small. Thus application of the Nernst equation for the iron(III)/(II) couple provides a value for the potential of the system directly. The corresponding expression involving the cerium(IV)/(III) couple would give the same answer; however, it would first be necessary to calculate a value for the concentration of cerium(IV), which in turn would require evaluation of the equilibrium constant for the reaction. The situation is reversed after excess titrant has been introduced. Here the concentrations of cerium(IV), cerium(III), and iron(III) are immediately available, whereas the concentration for iron(II) would require a preliminary calculation involving the equilibrium constant. Thus the potential of the system is most directly evaluated from the cerium(IV)/(III) couple.

EQUIVALENCE POINT POTENTIAL

The potential of an oxidation-reduction system at the equivalence point is of particular importance from the standpoint of indicator selection. Calculation of the equivalence potential is also unique in that there is insufficient stoichiometric information to permit the direct use of the Nernst equation for either half-cell process. Using the iron(II)/cerium(IV) titration as an example, values for the formal concentrations of cerium(III) and iron(III) at the equivalence point are easily calculated. On the other hand, we know only that the concentrations of cerium(IV) and iron(II) are small and numerically equal. As before, the concentrations of the minor constituents can be calculated from the equilibrium constant expression. Alternatively, we may write that, in common with any other point in the titration, the potential of the system at equivalence, E_{eq}, is given by

$$E_{eq} = E^0_{Ce^{4+}} - 0.0591 \log \frac{[Ce^{3+}]}{[Ce^{4+}]}$$

and also by

$$E_{eq} = E^0_{Fe^{3+}} - 0.0591 \log \frac{[Fe^{2+}]}{[Fe^{3+}]}$$

Upon adding these expressions, we obtain

$$2E_{eq} = E^0_{Ce^{4+}} + E^0_{Fe^{3+}} - 0.0591 \log \frac{[Ce^{3+}][Fe^{2+}]}{[Ce^{4+}][Fe^{3+}]}$$

Note that the concentration quotient in this expression is *not* the equilibrium constant for the titration reaction.

We know from stoichiometric considerations that *at the equivalence point*

$$[Fe^{3+}] = [Ce^{3+}]$$

$$[Fe^{2+}] = [Ce^{4+}]$$

Thus

$$2E_{eq} = E^0_{Ce^{4+}} + E^0_{Fe^{3+}} - 0.0591 \log \frac{[Ce^{3+}][Fe^{2+}]}{[Ce^{4+}][Fe^{3+}]}$$

and

$$E_{eq} = \frac{E^0_{Ce^{4+}} + E^0_{Fe^{3+}}}{2}$$

The equilibrium concentrations of the reacting species can be readily calculated from the equivalence point potential.

EXAMPLE

Calculate the concentration of the various reactants and products at the equivalence point in the titration of $0.100\text{-}N$ Fe^{2+} with $0.100\text{-}N$ Ce^{4+} at 25°C if both solutions are $1.0\ F$ in H_2SO_4.

Here it is convenient to substitute formal potentials (see p. 294) into the derived expression for the equivalence point potential. That is,

$$E_{eq} = \frac{+1.44 + 0.68}{2} = 1.06\ V$$

The Nernst equation allows us to evaluate the molar ratio of Fe^{2+} to Fe^{3+} at the equivalence point

$$+1.06 = +0.68 - 0.0591 \log \frac{[Fe^{2+}]}{[Fe^{3+}]}$$

$$\log \frac{[Fe^{2+}]}{[Fe^{3+}]} = -\frac{0.38}{0.0591} = -6.4$$

$$\frac{[Fe^{2+}]}{[Fe^{3+}]} = 4 \times 10^{-7}$$

It is clear that most of the Fe^{2+} has been converted to Fe^{3+} at the equivalence point. As a consequence of dilution, then, the Fe^{3+} concentration will be essentially equal to one-half the original Fe^{2+} concentration. That is,

$$[Fe^{3+}] = \frac{0.100}{2} - [Fe^{2+}] \cong 0.050$$

Thus

$$[Fe^{2+}] = 4 \times 10^{-7} \times 0.050$$

$$= 2 \times 10^{-8}$$

Finally, the stoichiometry of the reaction requires that

$$[Ce^{4+}] = [Fe^{2+}] \cong 2 \times 10^{-8}$$

$$[Ce^{3+}] = [Fe^{3+}] \cong 0.050$$

Note that identical results would have been obtained if the Nernst equation had been applied to the cerium(IV)-cerium(III) system.

EXAMPLE

Derive the equivalence point potential for the somewhat more complicated reaction

$$5Fe^{2+} + MnO_4^- + 8H^+ \rightleftharpoons 5Fe^{3+} + Mn^{2+} + 4H_2O$$

The respective half-reactions may be written as

$$Fe^{3+} + e \rightleftharpoons Fe^{2+}$$

$$MnO_4^- + 8H^+ + 5e \rightleftharpoons Mn^{2+} + 4H_2O$$

The potential of this system is given by either

$$E = E^0_{Fe^{3+}} - \frac{0.0591}{1} \log \frac{[Fe^{2+}]}{[Fe^{3+}]}$$

or

$$E = E^0_{MnO_4^-} - \frac{0.0591}{5} \log \frac{[Mn^{2+}]}{[MnO_4^-][H^+]^8}$$

To combine the logarithmic terms so that the concentrations of various species cancel, it is necessary to multiply the permanganate half-reaction equation by 5.

$$5E_{eq} = 5E^0_{MnO_4^-} - 0.0591 \log \frac{[Mn^{2+}]}{[MnO_4^-][H^+]^8}$$

After addition we find that

$$6E_{eq} = E^0_{Fe^{3+}} + 5E^0_{MnO_4^-} - 0.0591 \log \frac{[Fe^{2+}][Mn^{2+}]}{[Fe^{3+}][MnO_4^-][H^+]^8}$$

The stoichiometry at the equivalence point requires that

$$[Fe^{3+}] = 5[Mn^{2+}]$$

$$[Fe^{2+}] = 5[MnO_4^-]$$

After substituting and rearranging,

$$E_{eq} = \frac{E^0_{Fe^{3+}} + 5E^0_{MnO_4^-}}{6} - \frac{0.0591}{6} \log \frac{5[MnO_4^-][Mn^{2+}]}{5[Mn^{2+}][MnO_4^-][H^+]^8}$$

$$= \frac{E^0_{Fe^{3+}} + 5E^0_{MnO_4^-}}{6} - \frac{0.0591}{6} \log \frac{1}{[H^+]^8}$$

Note that the equivalence point potential for this titration is dependent upon the pH.

EXAMPLE

Derive an expression for the equivalence point potential for the reaction

$$6Fe^{2+} + Cr_2O_7^{2-} + 14H^+ \rightleftharpoons 6Fe^{3+} + 2Cr^{3+} + 7H_2O$$

Proceeding as before, we obtain the expression

$$7E_{eq} = E^0_{Fe^{3+}} + 6E^0_{Cr_2O_7^{2-}} - 0.0591 \log \frac{[Fe^{2+}][Cr^{3+}]^2}{[Fe^{3+}][Cr_2O_7^{2-}][H^+]^{14}}$$

At the equivalence point

$$[Fe^{2+}] = 6[Cr_2O_7^{2-}]$$

$$[Fe^{3+}] = 3[Cr^{3+}]$$

Substitution of these quantities into the previous equation reveals that

$$E_{eq} = \frac{E^0_{Fe^{3+}} + 6E^0_{Cr_2O_7^{2-}}}{7} - \frac{0.0591}{7} \log \frac{2[Cr^{3+}]}{[H^+]^{14}}$$

Note that the equivalence point potential in this example is dependent not only upon the concentration of hydrogen ion but also upon that of a product ion. In general the equivalence point potential will depend upon the concentration of one of the participants in the reaction whenever there exists a molar ratio other than unity between the species containing that participant as a reactant and as a product.

VARIATION IN POTENTIAL AS A FUNCTION OF REAGENT VOLUME

The shape of the curve for an oxidation-reduction titration depends upon the nature of the system under consideration. The derivation of typical curves will illustrate the effects of several important variables.

EXAMPLE

Derive a curve for the titration of 50.00 ml of 0.0500-N iron(II) with 0.1000-N cerium(IV). Assume that both solutions are 1.0 F in H_2SO_4.

Formal potential data for both half-cell processes are available in Appendix 6 and will be used for these calculations.

1. Initial potential. The solution contains no cerium ions. It will have a small but unknown concentration of iron(III) owing to air oxidation of the iron(II). Thus we do not have sufficient information to calculate an initial potential.

2. Addition of 5.00 ml of cerium(IV). With the introduction of oxidant the solution acquires appreciable concentrations of three of the participating ions; that for the fourth, cerium(IV), will be small.

The concentration of cerium(III) is given by its formal concentration less that for the unreacted cerium(IV):

$$[Ce^{3+}] = \frac{5.00 \times 0.1000}{50.00 + 5.00} - [Ce^{4+}] \cong \frac{0.500}{55.00}$$

This approximation appears reasonable because the equilibrium constant for the reaction is large. Similarly,

$$[Fe^{3+}] = \frac{5.00 \times 0.1000}{55.00} - [Ce^{4+}] \cong \frac{0.500}{55.00}$$

$$[Fe^{2+}] = \frac{50.00 \times 0.0500 - 5.00 \times 0.100}{55.00} + [Ce^{4+}]$$

$$\cong \frac{2.00}{55.00}$$

As we have shown, the potential for the system can be calculated with the aid of *either of the two equations*

$$E = E^0_{Ce^{4+}} - 0.0591 \log \frac{[Ce^{3+}]}{[Ce^{4+}]}$$

$$= E^0_{Fe^{3+}} - 0.0591 \log \frac{[Fe^{2+}]}{[Fe^{3+}]}$$

The second equation is the more convenient for this calculation because the two concentrations that appear in it are known within acceptable limits. Therefore, substituting for the iron(III) and iron(II) concentrations,

$$E = +0.68 - 0.0591 \log \frac{2.00/\cancel{55.00}}{0.500/\cancel{55.00}}$$

$$= +0.64 \text{ V}$$

Had we used the formal potential for the cerium(IV)-cerium(III) system and the equilibrium concentrations of these ions, an identical potential would have been obtained.

Additional values for the potential needed to define the curve short of the equivalence point can be calculated in a fashion strictly analogous to that in step 2. Table 12-2 contains a number of these data; the student should confirm one or two to be sure he knows how they were obtained.

3. Equivalence point potential. We have seen that the potential at the equivalence point in this titration has a value of 1.06 V.

4. Addition of 25.10 ml of reagent. The solution now contains an excess of quadrivalent cerium in addition to equivalent quantities of iron(III) and cerium(III) ions. The concentration of iron(II) will be very small. Thus

$$[Fe^{3+}] = \frac{25.00 \times 0.1000}{75.10} - [Fe^{2+}] \cong \frac{2.500}{75.10}$$

$$[Ce^{3+}] = \frac{25.00 \times 0.1000}{75.10} - [Fe^{2+}] \cong \frac{2.500}{75.10}$$

$$[Ce^{4+}] = \frac{25.10 \times 0.1000 - 50.00 \times 0.0500}{75.10} + [Fe^{2+}] \cong \frac{0.010}{75.10}$$

These approximations should be reasonable in view of the favorable equilibrium constant. As before, the desired potential could be calculated from the iron(III)-iron(II) system. At this stage in the titration, however, it is more convenient to use the cerium(IV)-cerium(III) potential because the concentrations of these species are immediately available. Thus

$$E = +1.44 - 0.0591 \log \frac{[Ce^{3+}]}{[Ce^{4+}]}$$

$$= +1.44 - 0.0591 \log \frac{2.500/75.10}{0.010/75.10}$$

$$= +1.30 \text{ V}$$

TABLE 12-2
Electrode Potentials during Titrations of 50.0 ml of 0.0500-N Iron(II) Solutions[a]

VOLUME OF 0.1000-N REAGENT, ml	POTENTIAL VS. STANDARD HYDROGEN ELECTRODE, V	
	Titration with Ce^{4+}	Titration with MnO_4^-
5.00	0.64	0.64
15.00	0.69	0.69
20.00	0.72	0.72
24.00	0.76	0.76
24.90	0.82	0.82
25.00	1.06 ← Equivalence point → 1.37	
25.10	1.30	1.48
26.00	1.36	1.49
30.00	1.40	1.50

[a] H_2SO_4 concentration = 1.0 F throughout.

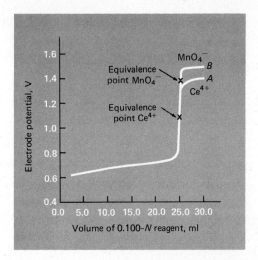

FIGURE 12-4 Titration curves for 50.00 ml of 0.0500-N Fe(III) with 0.1000-N Ce(IV) (curve A) and 0.1000-N KMnO$_4$ (curve B).

The additional postequivalence point potentials shown in Table 12-2 were calculated in a similar fashion.

The titration for iron(II) with cerium(IV) appears as curve A in Figure 12-4. Its shape is similar to the curves encountered in neutralization, precipitation, and complex formation titrations, the equivalence point being signaled by a large change in the ordinate function. A titration involving 0.01-F reactants will yield a curve that is, for all practical purposes, identical with the one that was derived because the electrode potentials are independent of dilution.

The titration curve just considered is symmetric about the equivalence point because the molar ratio of oxidant to reductant is equal to unity. As demonstrated by the following example, an asymmetric curve results if the ratio differs from this value.

EXAMPLE

Derive a curve for the titration of 50.00 ml of 0.0500-N iron(II) with 0.1000-N KMnO$_4$. For convenience assume that the solution contains sufficient H$_2$SO$_4$ so that [H$^+$] = 1.00 throughout.

No formal electrode potential is available for MnO$_4^-$; we must therefore employ its standard potential of 1.51 V.

The reaction is

$$5Fe^{2+} + MnO_4^- + 8H^+ \rightleftharpoons 5Fe^{3+} + Mn^{2+} + 4H_2O$$

It should be stressed that molar rather than normal concentrations are employed in the Nernst equations. For manganese species in the foregoing reaction the molarity is equal to one-fifth of the normality.

1. Preequivalence point potentials. The preequivalence point potentials are most easily calculated from the concentrations of iron(II) and iron(III) in the solution. Their values will be identical to those computed for the titration with cerium(IV).

2. Equivalence point potential. The equivalence point potential for this reaction is given by the equation

$$E = \frac{E^0_{Fe^{3+}} + 5E^0_{MnO_4^-}}{6} - \frac{0.0591}{6} \log \frac{1}{[H^+]^8}$$

It was specified that

$$[H^+] \cong 1.0$$

Therefore

$$E = \frac{0.68 + 5(+1.51)}{6} - \frac{0.0591}{6} \log \frac{1}{(1.0)^8}$$

$$= 1.37 \text{ V}$$

3. Postequivalence point potentials. When 25.10 ml of the 0.1000-N $KMnO_4$ have been added, stoichiometry requires that

$$[Fe^{3+}] = \frac{50.00 \times 0.0500}{75.10} - [Fe^{2+}] \cong \frac{2.500}{75.10}$$

$$[Mn^{2+}] = \frac{1}{5} \left(\frac{50.00 \times 0.0500}{75.10} - [Fe^{2+}] \right) \cong \frac{0.500}{75.10}$$

$$[MnO_4^-] = \frac{1}{5} \left(\frac{25.10 \times 0.1000 - 50.00 \times 0.0500}{75.10} + [Fe^{2+}] \right) \cong \frac{2.0 \times 10^{-3}}{75.10}$$

It is now advantageous to calculate the electrode potential from the standard potential of the manganese system. That is,

$$E = 1.51 - \frac{0.0591}{5} \log \frac{[Mn^{2+}]}{[H^+]^8[MnO_4^-]}$$

$$= 1.51 - \frac{0.0591}{5} \log \frac{0.500/\cancel{75.10}}{(1.0)^8(2.0 \times 10^{-3})/\cancel{75.10}}$$

$$= 1.48$$

Table 12-2 contains additional data obtained in this way.

Figure 12-4 depicts curves for the titration of iron(II) with permanganate and with cerium(IV). Note that the plots for both are alike to within 99.9% of the equivalence point; however, the equivalence point potentials are quite different. Furthermore, the permanganate curve is markedly asymmetric, the potential increasing only slightly beyond the equivalence point. Finally, the total change in potential associated with equivalence is somewhat greater with the permanganate titration, owing to the more favorable equilibrium constant for this reaction.

Effect of Concentration on Redox Titration Curves. It is of importance to note that the ordinate function, E_{system}, in the preceding calculations is ordinarily determined by the logarithm of a *ratio* of concentrations; it is therefore *independent* of dilution over a considerable range.[4] As a consequence, curves for oxidation-reduction titrations, in distinct contrast to those for other reaction types, tend to be independent of analyte and reagent concentrations.

Effect of the Completeness of Reaction on Redox Titration Curves. The change in the ordinate function in the equivalence point region of an oxidation-reduction titration becomes larger as the reaction becomes more complete; this effect is identical with that encountered for other reaction types. Thus in Figure 12-5 data are plotted for titrations involving a hypothetical analyte that has a standard potential of 0.2 V with several reagents that have standard potentials ranging from 0.4 to 1.2 V; the corresponding equilibrium constants for the reaction range from about 2×10^3 to 8×10^{16}. Clearly, an increase in completeness is accompanied by an increase in the electrode potential for the system.

The curves in Figure 12-5 were derived for reactions in which the oxidant and reductant each exhibit a one-electron change; where both reactants undergo a two-electron transfer, the change in potential in the region of 24.9 to 25.1 ml is larger by about 0.14 V.

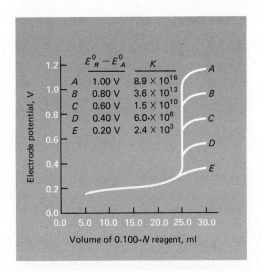

FIGURE 12-5 Titration of 50.0 ml of 0.0500-*N* A. $E^0{}_A$ is assumed to be 0.200 V. From the top, $E^0{}_R$ for the reagent is 1.20, 1.00, 0.80, 0.60, and 0.40 V, respectively. Both the reagent and the analyte are assumed to undergo a one-electron change.

[4] When the solution becomes sufficiently dilute so that the reaction is incomplete, E_{system} will in fact vary with further dilution. The magnitude of this effect can be determined by dispensing with the usual approximation that the formal and molar concentrations are identical.

Oxidation-Reduction Indicators

We have seen that the equivalence point in an oxidation-reduction titration is characterized by a marked change in the electrode potential of the system. Several methods exist for detecting such a change; these can serve to signal the end point in the titration.

CHEMICAL INDICATORS

Indicators for oxidation-reduction titrations are of two types. *Specific indicators* owe their behavior to a reaction with one of the participants in the titration. *True oxidation-reduction indicators*, on the other hand, respond to the potential of the system rather than to the appearance or disappearance of a particular species during the titration.

Specific Indicators. Perhaps the best known specific indicator is starch, which forms a dark-blue complex with triiodide ion. This complex serves to signal the end point in titrations in which iodine is either produced or consumed. Another specific indicator is potassium thiocyanate, which may be employed, for example, in the titration of iron(III) with solutions of titanium(III) sulfate; the end point involves the disappearance of the iron(III)-thiocyanate complex, owing to a marked decrease in the iron(III) concentration at the equivalence point.

True Oxidation-Reduction Indicators. True oxidation-reduction indicators enjoy wider application because their behavior is dependent only upon the change in the potential of the system.

The half-reaction responsible for color change in a typical true oxidation-reduction indicator can be written as

$$In_{ox} + ne \rightleftharpoons In_{red}$$

If the indicator reaction is reversible, we may write

$$E = E^0 - \frac{0.0591}{n} \log \frac{[In_{red}]}{[In_{ox}]} \tag{12-5}$$

Typically a change from the color of the oxidized form of the indicator to the reduced form involves a change in the ratio of reactant concentrations of about 100; that is, a color change can be seen when

$$\frac{[In_{red}]}{[In_{ox}]} \geqq \frac{1}{10}$$

changes to

$$\frac{[In_{red}]}{[In_{ox}]} \leqq 10$$

The conditions for the full color change for a typical indicator can be found by substituting these values into Equation 12-5[5]

$$E = E^0 \pm \frac{0.0591}{n} \qquad (12\text{-}6)$$

Equation 12-6 suggests that a typical indicator will exhibit a detectable color change when the titrant causes a shift from $(E^0 + 0.0591/n)$ V to $(E^0 - 0.0591/n)$, or about $(0.118/n)$ V in the potential of the system. With many indicators $n = 2$; a change of 0.059 V is thus sufficient.

The potential at which a color transition will occur depends upon the standard potential for the particular indicator system. Table 12-3 shows that indicators functioning in any desired potential range up to about +1.25 V are available.

Referring again to Figure 12-5, it is apparent that all of the indicators in Table 12-3 except for the first and the last could be employed

TABLE 12-3
A Selected List of Oxidation-Reduction Indicators[a]

INDICATORS	COLOR		TRANSITION POTENTIAL, V	CONDITIONS
	Oxidized	Reduced		
5-Nitro-1,10-phenanthroline iron(II) complex	Pale blue	Red-violet	+1.25	1-F H_2SO_4
2,3'-Diphenylamine dicarboxylic acid	Blue-violet	Colorless	+1.12	7–10-F H_2SO_4
1,10-Phenanthroline iron(II) complex	Pale blue	Red	+1.11	1-F H_2SO_4
Erioglaucin A	Bluish red	Yellow-green	+0.98	0.5-F H_2SO_4
Diphenylamine sulfonic acid	Red-violet	Colorless	+0.85	Dilute acid
Diphenylamine	Violet	Colorless	+0.76	Dilute acid
p-Ethoxychrysoidine	Yellow	Red	+0.76	Dilute acid
Methylene blue	Blue	Colorless	+0.53	1-F acid
Indigo tetrasulfonate	Blue	Colorless	+0.36	1-F acid
Phenosafranine	Red	Colorless	+0.28	1-F acid

[a] Data taken in part from I. M. Kolthoff and V. A. Stenger, *Volumetric Analysis*, 2d ed., vol. 1. New York: Interscience, 1942, p. 140.

[5] It should be noted that protons are involved in the reduction of many indicators; the transition potentials for these indicators will be pH-dependent.

with reagent A. On the other hand, only indigo tetrasulfonate could be employed with reagent D. The change in potential with reagent E is too small to be satisfactorily detected by an indicator. For a reaction involving a one-electron transfer the difference in standard potentials for the reagent and analyte must exceed about 0.4 V if the titration error is to be kept minimal; for a two-electron transfer a difference of about 0.25 V is necessary.

POTENTIOMETRIC END POINTS

End points in many oxidation-reduction titrations are readily observed by making the solution of the analyte a part of the cell:

$$\text{reference electrode} \| \text{analyte solution} | \text{Pt}$$

The potential of this cell, which is measured, will then vary in a way analogous to that shown in Figures 12-4 and 12-5. The end point can be determined from a plot of the measured potential as a function of titrant volume.

The reference electrode could be a standard hydrogen electrode. Usually, however, it is more convenient to use a secondary reference electrode. The experimental titration curves will take the same form as those in Figures 12-4 and 12-5 but will be displaced on the vertical axis by an amount corresponding to the difference between the potential of the reference electrode and the standard hydrogen electrode. The potentiometric end point is considered in detail in Chapter 14.

SUMMARY

Conclusions based upon the calculations in this chapter are helpful to the chemist as guides to the choice of reaction conditions and indicators for oxidation-reduction titrations. Thus, for example, the curves shown in Figures 12-4 and 12-5 clearly define the range of potentials within which an indicator must exhibit a color change for a successful titration. Nevertheless, it is important to emphasize that these calculations are theoretical and that they may not necessarily take into account all of the factors that actually determine the applicability and feasibility of a volumetric method. Also to be considered should be the rates at which both the principal and the indicator reactions occur, the effects of electrolyte concentration, pH and complexing agents, the presence of colored components other than the indicator in the solution, and the variation in color perception among individuals. The state of chemistry has not advanced to the point where the effects of these variables can be completely determined by computation. Theoretical calculations can and will eliminate useless experiments and act as guides to the ones most likely to be profitable. The final test must always come in the laboratory.

PROBLEMS

*1. Complete and balance the following equations, adding H^+, OH^-, and H_2O as required. Underline the oxidizing agent on the left side of each equation.

(a) $MnO_2(s) + HNO_2 \rightleftharpoons Mn^{2+} + NO_3^-$

(b) $VO^{2+} + I^- \rightleftharpoons V^{3+} + I_3^-$

(c) $Fe^{2+} + SeO_4^{2-} \rightleftharpoons Fe^{3+} + H_2SeO_3$

(d) $Mn^{2+} + MnO_4^- \rightleftharpoons MnO_2(s)$

(e) $O_2(g) + I^- \rightleftharpoons H_2O + I_3^-$

2. Complete and balance the following equations, adding H^+, OH^-, and H_2O as required. Underline the oxidizing agent on the left side of each equation.

(a) $Tl^{3+} + Mn^{2+} \rightleftharpoons Tl^+ + MnO_4^-$

(b) $Br_2(aq) + I^- \rightleftharpoons I_2(aq) + Br^-$

(c) $UO_2^{2+} + N_2H_5^+ \rightleftharpoons U^{4+} + N_2(g)$

(d) $PbO_2(s) + VO^{2+} \rightleftharpoons Pb^{2+} + V(OH)_4^+$

(e) $Pb(s) + SO_4^{2-} + V^{3+} \rightleftharpoons PbSO_4(s) + V^{2+}$

(f) $Cr_2O_7^{2-} + Cr^{2+} \rightleftharpoons Cr^{3+}$

*3. Calculate the potential of a copper electrode immersed in

(a) $0.250\text{-}F$ $Cu(NO_3)_2$.

(b) $0.0310\text{-}F$ KI that is saturated with CuI.

(c) $0.100\text{-}F$ KOH that is saturated with $Cu(OH)_2$ $(K_{sp} = 1.6 \times 10^{-19})$.

4. Calculate the potential of a mercury electrode immersed in

(a) $0.150\text{-}F$ $Hg_2(NO_3)_2$.

(b) $0.150\text{-}F$ $Hg(NO_3)_2$.

(c) $0.150\text{-}F$ Na_2SO_4 that is saturated with Hg_2SO_4.

(d) $0.150\text{-}F$ KI that is saturated with Hg_2I_2 $(K_{sp} = 4.5 \times 10^{-29})$.

*5. Calculate the potential of a platinum electrode immersed in a solution that is

(a) 0.120 F in $Tl_2(SO_4)_3$ and 0.0600 F in Tl_2SO_4.

(b) 0.160 F in H_2SeO_3, 0.0912 F in Na_2SeO_4, and 0.100 F in HCl.

(c) 0.200 F in NaBr and 0.00650 F in Br_2.

(d) 0.175 F in $NaBrO_3$, saturated with $Br_2(l)$ and having a pH of 3.50.

(e) equimolar in quinone and hydroquinone and having a pH of 8.62.

6. Calculate the potential of a platinum electrode immersed in a solution that is

(a) 0.0650 F in $Ti_2(SO_4)_3$ and 0.130 F in $TiSO_4$.

(b) 0.145 F in $NaClO_3$, saturated with Cl_2 at 1.00 atm and having a pH of 8.75.

(c) 0.154 F in HNO_2 and 0.0120 F in HNO_3.

(d) saturated with hydrogen at 1.00 atm and has a pH of 9.44.

(e) equimolar in UO_2^{2+} and U^{4+} and has a pH of 9.44.

*7. Calculate the theoretical electrode potential for the following half-cell when the pH of the electrolyte is (a) 1.00 and (b) 13.00.

$$Pt | VO^{2+}(0.342\ M),\ V^{3+}(0.101\ M),\ H^+$$

8. Calculate the theoretical electrode potential for the following half-cell when the pH of the electrolyte is (a) 1.00 and (b) 13.00.

$$Pb|PbO_2(s), Pb^{2+}(0.100\ M), H^+$$

*9. Indicate whether each of the following half-cells would behave as an anode or a cathode when coupled to a hydrogen electrode in a galvanic cell. Calculate the cell potential.
   (a) $Ag|AgI(sat'd), KI(2.00 \times 10^{-3}\ F)$
   (b) $Ag|AgI(sat'd), KI(0.200\ F)$
   (c) $Pt|S_4O_6^{2-}(0.800\ M), S_2O_3^{2-}(1.00 \times 10^{-4}\ M)$
   (d) $Pt|S_4O_6^{2-}(1.00 \times 10^{-4}\ M), S_2O_3^{2-}(0.800\ M)$
   (e) $Pt|TiO^{2+}(0.150\ M), Ti^{3+}(0.200\ M), HClO_4(0.500\ F)$
   (f) $Pt|TiO^{2+}(0.150\ M), Ti^{3+}(0.200\ M), NaOH(0.500\ F)$

10. Indicate whether each of the following half-cells would behave as an anode or a cathode when coupled to a hydrogen electrode in a galvanic cell. Calculate the cell potential.
   (a) $Ag|Ag(S_2O_3)_2^{3-}(0.100\ M), S_2O_3^{2-}(2.00 \times 10^{-4}\ M)$
   (b) $Ag|Ag(S_2O_3)_2^{3-}(2.00 \times 10^{-4}\ M), S_2O_3^{2-}(0.100\ M)$
   (c) $Pt|Sn^{4+}(0.120\ M), Sn^{2+}(1.20 \times 10^{-4}\ M)$
   (d) $Pt|Sn^{4+}(1.20 \times 10^{-4}\ M), Sn^{2+}(0.120\ M)$
   (e) $Pt|UO_2^{2+}(0.200\ M), U^{4+}(0.100\ M), HCl(1.00 \times 10^{-3}\ F)$
   (f) $Pt|UO_2^{2+}(0.200\ M), U^{4+}(0.100\ M), NaOH(1.00 \times 10^{-3}\ F)$

*11. Calculate the theoretical cell potential for each of the following. Is the cell galvanic or electrolytic?
   (a) $Ni|Ni^{2+}(0.0200\ M)\|KCl(0.0300\ F), AgCl(sat'd)|Ag$
   (b) $Pb|PbSO_4(sat'd), Na_2SO_4(0.0200\ F)\|Pb^{2+}(0.0100\ M), HClO_4(0.100\ F),$
       $PbO_2(sat'd)|Pb$
   (c) $Pt|VO^{2+}(0.0200\ M), V^{3+}(0.0500\ M), HClO_4(0.200\ F)\|Na_2CrO_4(0.100\ F),$
       $Ag_2CrO_4(sat'd)|Ag$
   (d) $Pt|H_2(1.00\ atm), HCl(0.200\ F)\|NaOH(0.200\ F), H_2(1.00\ atm)|Pt$
   (e) $Ag|AgI(sat'd), KI(0.100\ F), HClO_4(0.200\ F), H_2(1.00\ atm)|Pt$
   (f) $Ag|Ag(CN)_2^-(0.0150\ M), CN^-(0.300\ M)\|Tl^{3+}(0.250\ M), Tl^+(0.150\ M)|Pt$

12. Calculate the theoretical cell potential for each of the following. Is the cell galvanic or electrolytic?
   (a) $Ag|Ag(S_2O_3)_2^{3-}(0.175\ M), S_2O_3^{2-}(0.225\ M)\|Br_2(sat'd), KBr(0.0500\ F)|Pt$
   (b) $Hg|Hg_2SO_4(sat'd), Na_2SO_4(0.0360\ F)\|Na_2SO_4(0.0520\ F), PbSO_4(sat'd)|Pb$
   (c) $Cu|CuI(sat'd), I^-(0.0680\ M)\|IO_3^-(0.100\ M), HClO_4(0.0200\ F), I_2(sat'd)|Pt$
   (d) $Bi|BiO^+(0.100\ M), HClO_4(0.0100\ F)\|FeSO_4(0.100\ F), Fe_2(SO_4)_3(0.200\ F)|Pt$
   (e) $Pt|H_2(1.00\ atm), HClO_4(2.00\ F)\|NaOH(2.00\ F), H_2(1.00\ atm)|Pt$
   (f) $Pt|MnO_2(sat'd), Mn^{2+}(0.100\ M), HClO_4(0.500\ F)\|Cd^{2+}(0.200\ M)|Cd$

*13. Calculate equilibrium constants for each of the reactions in Problem 1.

14. Calculate equilibrium constants for each of the reactions in Problem 2.

*15. Calculate E^0 for the process

$$Pb(OH)_2(s) + 2e \rightleftharpoons Pb(s) + 2OH^-$$

16. Calculate E^0 for the process

$$Ag_2CO_3(s) + 2e \rightleftharpoons 2Ag(s) + CO_3^{2-}$$

*17. The formation constant for the complex $Ag(NH_3)_2^+$ is 1.4×10^7. Calculate E^0 for

$$Ag(NH_3)_2^+ + e \rightleftharpoons Ag(s) + 2NH_3$$

18. The formation constant for the complex $HgCl_2$ is 1.6×10^{13}. Calculate E^0 for the reaction

$$HgCl_2(aq) + 2e \rightarrow Hg(l) + 2Cl^-$$

*19. Calculate K_{sp} for $Ag_2C_2O_4$ given

$$Ag_2C_2O_4(s) + 2e \rightleftharpoons 2Ag(s) + C_2O_4^{2-} \qquad E^0 = 0.490 \text{ V}$$

20. Calculate K_{sp} for $Ni_2P_2O_7$ given

$$Ni_2P_2O_7(s) + 4e \rightarrow 2Ni + P_2O_7^{4-} \qquad E^0 = -0.439 \text{ V}$$

*21. Calculate the formation constant for $Ag(CN)_2^-$ given

$$Ag(CN)_2^- + e \rightleftharpoons Ag + 2CN^- \qquad E^0 = -0.31 \text{ V}$$

22. Calculate the formation constant for AlF_6^{3-} given

$$AlF_6^{3-} + 3e \rightleftharpoons Al + 6F^- \qquad E^0 = -2.069 \text{ V}$$

*23. The following cell was found to have a potential of 0.596 V

$$SHE \| OAc^- (0.0100 \ M), Hg_2(OAc)_2(sat'd) | Hg$$

where OAc^- is the acetate ion. Calculate K_{sp} for $Hg_2(OAc)_2$.

24. The following cell was found to have a potential of 0.416 V.

$$Pb | Pb_3(AsO_4)_2(sat'd), AsO_4^{3-} (1.00 \times 10^{-3} \ M) \| SHE$$

Calculate K_{sp} for $Pb_3(AsO_4)_2$.

*25. The potential of the following cell was found to be 0.936 V.

$$Zn | X^- (0.120 \ M), ZnX_4^{2-} (0.0320 \ M) \| SHE$$

Calculate the formation constant for ZnX_4^{2-}.

26. The potential of the following cell was 0.333 V.

$$SHE \| A^- (0.0646 \ M), AgA_2^- (0.100 \ M) | Ag$$

Calculate the formation constant for AgA_2^-.

*27. The following cell was employed to determine the dissociation constant of the weak acid HA:

$$Pt, H_2(1.00 \text{ atm})|NaA(0.130 \ F), HA(0.265 \ F)\|SHE$$

The potential was 0.421 V. Calculate K_a.

28. The following cell was employed to determine the dissociation constant of the amine RNH_2:

$$Pt, H_2(1.00 \text{ atm})|RNH_2(0.0414 \ F), RNH_3Cl(0.0633 \ F)\|SHE$$

where RNH_3Cl is the chloride salt of the amine. The potential of the cell was 0.475 V. Calculate K_b.

*29. Calculate the electrode potential of the system at the equivalence point for each of the following reactions. Where necessary assume that at equivalence, $[H^+] = 0.0100$. *Note that the equations are not balanced.*
(a) $Sn^{4+} + Cr^{2+} \rightleftharpoons Sn^{2+} + Cr^{3+}$
(b) $Tl^{3+} + HNO_2 \rightleftharpoons Tl^+ + NO_3^-$
(c) $Fe^{3+} + U^{4+} + H_2O \rightleftharpoons Fe^{2+} + UO_2^{2+} + H^+$
(d) $MnO_4^- + H_3AsO_3 + H^+ \rightleftharpoons Mn^{2+} + H_3AsO_4 + H_2O$

30. Calculate the electrode potential of the system at the equivalence point for each of the following reactions. Where necessary assume that at equivalence, $[H^+] = 0.0100$. *Note that the equations are not balanced.*
(a) $V(OH)_4^+ + H_2SO_3 \rightleftharpoons VO^{2+} + SO_4^{2-} + H_2O$
(b) $N_2H_5^+ + UO_2^{2+} + H_2O \rightleftharpoons N_2 + U^{4+} + H^+$
(c) $SeO_4^{2-} + Fe^{2+} + H^+ \rightleftharpoons H_2SeO_3 + Fe^{3+} + H_2O$
(d) $H_5IO_6 + H_2O_2 \rightleftharpoons IO_3^- + O_2 + H_2O$

31. Construct curves for the following titrations. Calculate potentials after addition of 10.00, 24.00, 24.90, 25.00, 25.10, 26.00, and 30.00 ml of the reagent. Where necessary assume that $[H^+] = 0.100$ throughout.
*(a) 50.00 ml of 0.05000-N V^{2+} with 0.1000-N Sn^{4+}
(b) 50.00 ml of 0.05000-N $Fe(CN)_6^{3-}$ with 0.1000-N Cr^{2+}
*(c) 50.00 ml of 0.05000-N $Fe(CN)_6^{4-}$ with 0.1000-N Tl^{3+}
(d) 50.00 ml of 0.05000-N Fe^{3+} with 0.1000-N Sn^{2+}
*(e) 50.00 ml of 0.05000-N U^{4+} with 0.1000-N MnO_4^-
(f) 50.00 ml of 0.05000-N SeO_4^{2-} with 0.1000-N Fe^{2+}

13
application of oxidation~reduction titrations

Several excellent oxidizing agents are available for the preparation of standard solutions. The number of standard reducing reagents is much more limited because their solutions are susceptible to air oxidation; storage is therefore inconvenient.

This chapter is concerned with the preparation and application of standard solutions of oxidants and reductants. In addition, consideration is given to auxiliary reagents which serve to convert an analyte to a single oxidation state prior to its titration.

Auxiliary Reagents

As often as not, the steps that precede an oxidation-reduction titration leave the analyte either in a mixture of oxidation states or else in its highest oxidation state. For example, when an iron alloy is dissolved in acid, a mixture of iron(II) and iron(III) results. Before the iron can be titrated, therefore, a reagent must be added that will convert the element quantitatively either to the divalent state for titration with a standard oxidizing agent or to the trivalent state for titration with a reducing agent. The electrode potential for an auxiliary reagent must be such as to assure quantitative conversion of the analyte to the desired oxidation state. Nevertheless, it should not be sufficient to convert other components of the solution into states that will also react with the titrant. Another requirement is that the unused portion of the auxiliary reagent be easily removed from the solution because it will almost inevitably interfere with the titration by reacting with the standard solution. Thus a reagent that would convert iron quantitatively to the divalent state for titration with a standard solution of permanganate would of necessity be a good reducing agent. Any excess remaining after the reduction would surely consume permanganate if it were not removed from the solution.

AUXILIARY REDUCING REAGENTS

Reagents that find general application to the prereduction of samples are described in the following paragraphs.

Metals. An examination of standard electrode potential data reveals a number of good reducing agents among the pure metals.[1] Such elements as zinc, cadmium, aluminum, lead, nickel, copper, mercury, and silver have proved useful for prereduction purposes. Where sticks or coils of the metal are used, the excess reductant is simply lifted from the solution and washed thoroughly. If granular or powdered forms of the metal are employed, filtration may be required. An alternative is to employ a *reductor* such as that shown in Figure 13-1. Here the granular or powdered metal is held in a vertical glass tube; the solution to be reduced is then drawn through the column with a moderate vacuum. Normally a reductor of this kind can be employed for several hundred reductions.

The *Jones reductor*, which finds wide application, employs amalgamated zinc as the reducing agent. Amalgamation is accomplished by allowing zinc granules to stand briefly in a solution of mercury(II) chloride:

$$Zn + Hg^{2+} \rightarrow Zn^{2+} + Zn(Hg)$$

The resulting zinc amalgam is nearly as good a reducing agent as the metal itself. It has the advantage, however, of inhibiting the reduction of hydrogen ions by zinc. Thus even quite acidic solutions can be reduced without significant formation of gaseous hydrogen. Reduction of hydrogen ions is, of course, undesirable not only because it consumes the reducing agent but also because it contaminates the solutions being analyzed with large amounts of zinc ions.

[1] For a discussion of metal reductants the reader should see I. M. Kolthoff and R. Belcher, *Volumetric Analysis*, vol. 3. New York: Interscience, 1957, pp. 11–23; W. I. Stephen, *Ind. Chemist*, **28**, 13, 55, 107 (1952).

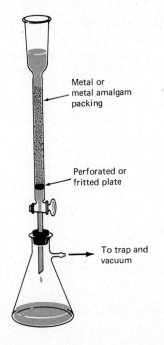

FIGURE 13-1 A metal or
metal amalgam reductor.

Metal or
metal amalgam
packing

Perforated or
fritted plate

To trap and
vacuum

For the typical Jones reductor, a 2-cm glass column is packed to a depth of 40 to 50 cm. The packing must be kept covered with liquid at all times to protect against air oxidation and the resultant formation of basic salts which tend to cause clogging.

Table 13-1 lists the principal applications of the Jones reductor; directions for preparation and use of the device are given in Chapter 20, Preparation 5-1.

As will be seen from Table 13-1, the *Walden reductor* has the advantage of being somewhat more selective than the Jones reductor. The Walden reductor employs metallic silver as the reductant. The metal is prepared by reducing a solution containing about 30 g of silver nitrate with metallic copper. The resulting suspension of finely divided silver is then poured into a narrow glass column to give about 10 cm of packing. When not in use, the silver is covered with 1-F HCl.

Reductions with a Walden reductor are always carried out in the presence of hydrochloric acid where the reducing strength of the metal is enhanced by the formation of silver chloride. As a consequence, the surface of the packing becomes coated with silver chloride with use. Regeneration is readily accomplished, however, by briefly dipping a zinc rod into the solution contained in the column.

Table 13-1 lists various uses of the Walden reductor.

Gaseous Reductants. Both hydrogen sulfide and sulfur dioxide are reasonably effective reducing reagents and have been used for prereduc-

TABLE 13-1
Uses of the Walden Reductor and the Jones Reductor[a]

WALDEN REDUCTOR $Ag(s) + Cl^- \rightleftarrows AgCl(s) + e$	JONES REDUCTOR $Zn(s) \rightleftarrows Zn^2 + 2e$
$e + Fe^{3+} \rightarrow Fe^{2+}$	$e + Fe^{3+} \rightarrow Fe^{2+}$
$e + Cu^{2+} \rightarrow Cu^+$	Cu^{2+} reduced to metallic Cu
$e + H_2MoO_4 + 2H^+ \rightarrow MoO_2^+ + 2H_2O$	$3e + H_2MoO_4 + 6H^+ \rightarrow Mo^{3+} + 4H_2O$
$2e + UO_2^{2+} + 4H^+ \rightarrow U^{4+} + 2H_2O$	$2e + UO_2^{2+} + 4H^+ \rightarrow U^{4+} + 2H_2O$
	$3e + UO_2^{2+} + 4H^+ \rightarrow U^{3+} + 2H_2O$[b]
$e + V(OH)_4^+ + 2H^+ \rightarrow VO^{2+} + 3H_2O$	$3e + V(OH)_4^+ + 4H^+ \rightarrow V^{2+} + 4H_2O$
TiO^{2+} not reduced	$e + TiO^{2+} + 2H^+ \rightarrow Ti^{3+} + H_2O$
Cr^{3+} not reduced	$e + Cr^{3+} \rightarrow Cr^{2+}$

[a] Taken from I. M. Kolthoff and R. Belcher, *Volumetric Analysis*, vol. 3. New York: Interscience, 1957, p. 12. With permission.
[b] A mixture of oxidation states is obtained. This does not, however, preclude the use of the Jones reductor for the analysis of uranium because any U^{3+} formed can be converted to U^{4+} by shaking the solution with air for a few minutes.

tion. Excesses of these reagents can be readily eliminated by boiling the acidified solution.

The reactions of both reagents are often slow, half an hour or more being required to complete the reduction and rid the solution of excess reagent. In addition to this time disadvantage, both gases are noxious and toxic. The employment of other reductants is much preferred.

Tin(II) Chloride. The reduction of iron(III) to the divalent state is conveniently accomplished with tin(II) chloride. The chemistry of this application is discussed on page 323.

AUXILIARY OXIDIZING REAGENTS

Sodium Bismuthate. Sodium bismuthate is an extremely powerful oxidizing agent capable, for example, of converting manganese(II) quantitatively to permanganate. It exists as a sparingly soluble solid of somewhat uncertain composition; its formula is usually written as $NaBiO_3$. Upon reaction, bismuth(V) is converted to the more common trivalent state. Ordinarily the solution to be oxidized is boiled in contact with an excess of the solid. The unused reagent is then removed by filtration.

Ammonium Peroxodisulfate. In acid solutions ammonium peroxodisulfate, $(NH_4)_2S_2O_8$, is a potent oxidizing agent that will convert chromium to dichromate, cerium(III) to the quadrivalent state, and manganese(II) ion to permanganate. The half-reaction is

$$S_2O_8^{2-} + 2e \rightleftharpoons 2SO_4^{2-} \qquad E^0 = 2.01 \text{ V}$$

The oxidations are catalyzed by traces of silver ion. The excess reagent is readily decomposed by boiling the solution for a few minutes.

$$2S_2O_8^{2-} + 2H_2O \rightleftharpoons 4SO_4^{2-} + O_2(g) + 4H^+$$

Sodium and Hydrogen Peroxide. Peroxide is a convenient oxidizing agent. Both the solid sodium salt and dilute solutions of the acid are used. The half-reaction for peroxide in acidic solution is

$$H_2O_2 + 2H^+ + 2e \rightleftharpoons 2H_2O \qquad E^0 = 1.77 \text{ V}$$

Any excess reagent is readily decomposed by brief boiling.

$$2H_2O_2 \rightleftharpoons 2H_2O + O_2(g)$$

Applications of Standard Oxidants

Table 13-2 summarizes the characteristics of oxidizing reagents that are commonly employed for titrimetric analysis. These reagents differ considerably in oxidizing strength, with standard potentials that range from 1.5 to 0.5 V. The choice among them depends upon the strength of the analyte as a reducing agent, the rate of reaction between the oxidant and the analyte, the stability of the standard oxidant solutions, and the availability of a satisfactory indicator for end point detection.

POTASSIUM PERMANGANATE

Potassium permanganate, a powerful oxidant, is perhaps the most widely used of all standard oxidizing agents. The color of a permanganate solution is so intense that an indicator is not ordinarily required. The reagent is readily available at modest cost. On the other hand, the tendency of permanganate to oxidize chloride ion is a disadvantage because hydrochloric acid is such a useful solvent. The multiplicity of possible reaction products (p. 150) can at times cause uncertainty regarding the stoichiometry of a permanganate oxidation. Finally, permanganate solutions have limited stability.

Reactions of Permanganate Ion in Acid Solution. Potassium permanganate is most commonly employed with solutions that are 0.1 N or

TABLE 13-2
Some Common Oxidants Employed for Standard Solutions

REAGENT AND FORMULA	REDUCTION PRODUCT	STANDARD POTENTIAL, V	PRIMARY STANDARD FOR	INDICATOR[a]	STABILITY[b]
Potassium permanganate $KMnO_4$	Mn^{2+}	1.51	$Na_2C_2O_4$, Fe, As_2O_3	MnO_4^-	(b)
Potassium bromate $KBrO_3$	Br_2	1.52	$KBrO_3$	(1)	(a)
Cerium(IV) Ce^{4+}	Ce^{3+}	1.44	$Na_2C_2O_4$, Fe, As_2O_3	(2)	(a)
Potassium dichromate $K_2Cr_2O_7$	Cr^{3+}	1.33	$K_2Cr_2O_7$, Fe	(3)	(a)
Iodine I_2	I^-	0.536	As_2O_3, $BaS_2O_3 \cdot H_2O$	starch	(c)

[a] (1) α-Naphthoflavone, (2) 1,10-orthophenanthroline iron(II) complex (ferroin), (3) diphenylamine sulfonic acid.
[b] (a) Indefinitely stable; (b) moderately stable, requires periodic standardization; (c) somewhat unstable, requires frequent standardization.

greater in mineral acid. Under these conditions the product is manganese(II):

$$MnO_4^- + 8H^+ + 5e \rightleftharpoons Mn^{2+} + 4H_2O \qquad E^0 = 1.51 \text{ V}$$

Although the mechanisms involved are frequently complicated, permanganate oxidizes most substances rapidly in acidic solution. Notable exceptions include the reaction with oxalic acid that requires elevated temperatures and with arsenic(III) oxide for which a catalyst such as osmium tetroxide or iodine monochloride is needed.

End Point. One of the obvious properties of potassium permanganate is its intense purple color, which commonly serves as the indicator for titrations. As little as 0.01 to 0.02 ml of a 0.02-F (0.1-N) solution is sufficient to impart a perceptible color to 100 ml of water. For very dilute permanganate solutions, diphenylamine sulfonic acid or the ortho-phenanthroline-iron(II) complex (Table 12-3) will give a sharper end point.

The permanganate end point is not permanent. Decolorization results from the reaction of the excess permanganate with the relatively large concentration of manganese(II) ion that is present at the end point:

$$2MnO_4^- + 3Mn^{2+} + 2H_2O \rightleftharpoons 5MnO_2(s) + 4H^+$$

The equilibrium constant for this reaction, which is readily calculated from the standard potentials for the two half-reactions, has a numerical value of about 10^{47}. Thus, even in a highly acidic solution, the concentration of permanganate in equilibrium with manganese(II) ion is small. Fortunately the rate at which this equilibrium is attained is slow, with the result that the end point fades only gradually.

The intense color of a permanganate solution complicates the measurement of reagent volumes in a buret. It is frequently more practical to use the surface of the liquid rather than the bottom of the meniscus as the point of reference.

Stability of Permanganate Solutions. Aqueous solutions of permanganate are not completely stable because the ion tends to oxidize water. The process may be depicted by the equation

$$4MnO_4^- + 2H_2O \rightleftharpoons 4MnO_2(s) + 3O_2(g) + 4OH^-$$

Although the constant for this equilibrium indicates that the products are favored in neutral solution, the reaction is so slow that a properly prepared solution is moderately stable. The decomposition of permanganate has been shown to be catalyzed by light, heat, acids, bases, manganese(II) ion, and manganese dioxide. To obtain a stable reagent for analysis it is necessary to minimize the influence of these effects.

The decomposition of permanganate solutions is greatly accelerated by the presence of manganese dioxide. Because it is also a product of the decomposition, this solid has an *autocatalytic* effect upon the process.

Photochemical catalysis of the decomposition is often observed when a permanganate solution is allowed to stand in a buret for an extended period. Manganese dioxide forms as a brown stain and serves to show that the concentration of the reagent has undergone a change.

In general the heating of acidic solutions containing an excess of permanganate should be avoided because of a decomposition error that cannot adequately be compensated for with a blank. At the same time, it is perfectly acceptable to titrate hot, acidic solutions of reductants directly with the reagent because at no time during the titration is the oxidant concentration large enough to cause a measurable uncertainty.

Preparation and Storage of Permanganate Solutions. A permanganate solution possessing reasonable stability can be obtained provided a number of precautions are observed. Perhaps the most important variable affecting stability is the catalytic influence of manganese dioxide. This compound is an inevitable contaminant in solid potassium permanganate; it is also produced when permanganate oxidizes organic matter in the water used to prepare the solution. Removal of manganese dioxide by filtration markedly enhances the stability of standard permanganate solutions. Sufficient time must be allowed for complete oxidation of contaminants in the water before filtration; the solution may be boiled to hasten the process. Paper cannot be used for the filtration because it reacts with permanganate to form the undesirable dioxide.

Standardized solutions should be stored in the dark. If any solid is detected in the solution, filtration and restandardization are necessary. In any event, restandardization every one to two weeks is a good precautionary measure.

Standardization against Sodium Oxalate. In acidic solution, permanganate oxidizes oxalic acid to carbon dioxide and water.

$$2MnO_4^- + 5H_2C_2O_4 + 6H^+ \rightleftharpoons 2Mn^{2+} + 10CO_2(g) + 8H_2O$$

This reaction is complex and proceeds slowly at room temperature; even at elevated temperatures it is not rapid unless catalyzed by manganese(II) ion. Thus several seconds are required to decolorize a hot oxalic acid solution at the outset of a permanganate titration. Later, when the concentration of manganese(II) ion has become appreciable, decolorization becomes rapid as a consequence of the autocatalysis.

The stoichiometry of the reaction between permanganate ion and oxalic acid has been investigated in great detail by McBride[2] and more recently by Fowler and Bright.[3] The former devised a procedure in which the oxalic acid is titrated slowly at a temperature between 60 and 90°C until the faint pink color of the permanganate persists. Fowler and Bright have demonstrated, however, that this titration consumes 0.1 to 0.4% too little permanganate, due perhaps to air oxidation of a small part of the oxalic acid.

$$H_2C_2O_4 + O_2(g) \rightleftharpoons H_2O_2 + 2CO_2$$

In the hot solution the peroxide is postulated to decompose spontaneously to oxygen and water.

Fowler and Bright devised a scheme for standardization in which 90 to 95% of the required permanganate is added rapidly to the cool oxalic acid solution. After all of this reagent has reacted, the solution is heated to 55 to 60°C and titrated as before. Although it minimizes the air oxidation of oxalic acid and gives data that appear to be in exact accord with the theoretical stoichiometry, this method suffers from the disadvantage of requiring a knowledge of the approximate normality of the permanganate solution to permit the proper initial addition of the reagent. The Fowler–

[2] R. S. McBride, *J. Amer. Chem. Soc.*, **34**, 393 (1912).

[3] R. M. Fowler and H. A. Bright, *J. Res. Nat. Bur. Stand.*, **15**, 493 (1935).

Bright procedure is not as convenient as the McBride method in this respect.

The method of McBride will give perfectly adequate data for many purposes (usually 0.2 to 0.3% too high). If a more accurate standardization is required, it is convenient to perform one titration by this procedure to obtain the approximate normality of the solution. Then a pair of titrations employing the Fowler and Bright method can be performed. Directions for both procedures are found in Chapter 20, Standardization 5-1.

Other Primary Standards. Several other primary standards can be employed to establish the normality of permanganate solutions. These include arsenic(III) oxide, potassium iodide, and metallic iron. Detailed procedures for the use of these standards are described by Kolthoff and Belcher.[4]

Applications of Permanganate Titrations. Table 13-3 indicates the multiplicity of analyses that make use of standard permanganate solutions in acidic media. Most of these reactions are rapid enough for direct titrations. Two typical applications, the determination of iron and the determination of calcium, are discussed in the paragraphs that follow.

Determination of Iron in an Ore. The common iron ores are hematite (Fe_2O_3), magnetite (Fe_3O_4), and limonite ($3Fe_2O_3 \cdot 3H_2O$). Volumetric methods for analysis of iron samples containing these substances consist of three steps: (1) solution of the sample, (2) reduction of the iron to the divalent state, and (3) titration with a standard oxidant.

Iron ores are often completely decomposed in concentrated hydrochloric acid. The rate of attack by this reagent is increased by the presence of a small amount of tin(II) chloride, which probably acts by reducing sparingly soluble iron(III) oxides on the surface of the particles to more soluble iron(II) species. Because iron(III) tends to form stable chloride complexes, hydrochloric acid is a much more efficient solvent than either sulfuric or nitric acid.

Many iron ores contain silicates that may not be entirely decomposed by treatment with hydrochloric acid. Where decomposition is complete, the white residue of hydrated silica that remains behind in no way interferes with the analysis. Incomplete decomposition is indicated by a dark residue remaining after prolonged treatment with the acid.

Part or all of the iron will exist in the trivalent state in the dissolved sample. Prereduction of the sample must therefore precede titration with the oxidant. Any of the methods described earlier may be used—the Jones or the Walden reductors, for example.

[4] I. M. Kolthoff and R. Belcher, *Volumetric Analysis*, vol. 3. New York: Interscience, 1957, pp. 41–59.

TABLE 13-3

Some Applications of Potassium Permanganate in Acidic Solution

SUBSTANCE SOUGHT	HALF-REACTION	CONDITION
Sn	$Sn^{2+} \rightleftarrows Sn^{4+} + 2e$	Prereduction with Zn
H_2O_2	$H_2O_2 \rightleftarrows O_2(g) + 2H^+ + 2e$	
Fe	$Fe^{2+} \rightleftarrows Fe^{3+} + e$	Prereduction with $SnCl_2$ or with Jones or Walden reductor
$Fe(CN)_6^{4-}$	$Fe(CN)_6^{4-} \rightleftarrows Fe(CN)_6^{3-} + e$	
V	$VO^{2+} + 3H_2O \rightleftarrows V(OH)_4^+ + 2H^+ + e$	Prereduction with Bi amalgam or SO_2
Mo	$Mo^{3+} + 4H_2O \rightleftarrows MoO_4^{2-} + 8H^+ + 3e$	Prereduction with Jones reductor
W	$W^{3+} + 4H_2O \rightleftarrows WO_4^{2-} + 8H^+ + 3e$	Prereduction with Zn or Cd
U	$U^{4+} + 2H_2O \rightleftarrows UO_2^{2+} + 4H^+ + 2e$	Prereduction with Jones reductor
Ti	$Ti^{3+} + H_2O \rightleftarrows TiO^{2+} + 2H^+ + e$	Prereduction with Jones reductor
$H_2C_2O_4$	$H_2C_2O_4 \rightleftarrows 2CO_2 + 2H^+ + 2e$	
Mg, Ca, Zn, Co, Pb, Ag	$H_2C_2O_4 \rightleftarrows 2CO_2 + 2H^+ + 2e$	Sparingly soluble metal oxalates filtered, washed, and dissolved in acid; liberated oxalic acid titrated
HNO_2	$HNO_2 + H_2O \rightleftarrows NO_3^- + 3H^+ + 2e$	15-min reaction time; excess $KMnO_4$ back-titrated
K	$K_2NaCo(NO_2)_6 + 6H_2O \rightleftarrows$ $Co^{2+} + 6NO_3^- + 12H^+ + 2K^+$ $+ Na^+ + 11e$	Precipitated as $K_2NaCo(NO_2)_6$; filtered and dissolved in $KMnO_4$; excess $KMnO_4$ back-titrated
Na	$U^{4+} + 2H_2O \rightleftarrows UO_2^{2+} + 4H^+ + 2e$	Precipitated as $NaZn(UO_2)_3 \cdot (OAc)_9$; filtered, washed, dissolved; U determined as above

Perhaps the most satisfactory of all prereductants for iron is tin(II) chloride. The only other common elements reduced by this reagent are vanadium, copper, molybdenum, tungsten, and arsenic. The excess reducing agent is removed from the solution by the addition of mercury(II) chloride:

$$Sn^{2+} + 2HgCl_2 \rightleftharpoons Hg_2Cl_2(s) + Sn^{4+} + 2Cl^-$$

The slightly soluble mercury(I) chloride produced will not consume permanganate, nor will the excess mercury(II) chloride reoxidize divalent iron. Care must be exerted, however, to prevent occurrence of the alternative reaction,

$$Sn^{2+} + HgCl_2 \rightleftharpoons Hg(1) + Sn^{4+} + 2Cl^-$$

Metallic mercury reacts with permanganate and causes a high result in an iron analysis. Formation of mercury is favored by an appreciable excess of tin(II); it is prevented by careful control of this excess and by the rapid addition of excess mercury(II) chloride. A proper reduction is indicated by the appearance of a slight, white precipitate after addition of the latter reagent. A gray precipitate indicates the presence of mercury; the total absence of a precipitate indicates that an insufficient amount of tin(II) chloride was added. In either of these events the sample must be discarded.

The reaction of iron(II) with permanganate proceeds smoothly and rapidly to completion. In the presence of hydrochloric acid, however, high results are obtained owing to the oxidation of chloride ion by permanganate. This reaction, which normally does not proceed rapidly enough to cause serious errors, is induced by the presence of divalent iron. Its effects are avoided by preliminary removal of chloride by evaporation with sulfuric acid or by use of the *Zimmermann-Reinhardt reagent*. The latter is a solution of manganese(II) in fairly concentrated sulfuric and phosphoric acids. Manganese(II) inhibits the oxidation of chloride ion, whereas phosphoric acid forms complexes with the iron(III) produced in the titration and prevents the intense yellow color of iron(III) chloride complexes from interfering with the end point.[5]

Directions for the analysis of iron in an ore are given in Chapter 20, Analyses 5-1 and 5-2.

Determination of Calcium. In common with a number of other cations, calcium is conveniently precipitated with oxalate. The solid is filtered, washed free of excess precipitating reagent, and then dissolved in dilute acid; the liberated oxalic acid is then titrated with a standard solution of permanganate or some other oxidizing agent. This method is applicable to samples that contain magnesium and the alkali metals. Most other cations must be absent, however, as they either precipitate or coprecipitate as oxalates and cause positive errors in the analysis.

[5] The mechanism of the action of Zimmermann-Reinhardt reagent has been the subject of much study. For a discussion of this work see H. A. Laitinen, *Chemical Analysis*. New York: McGraw-Hill, 1960, pp. 369–372.

To obtain satisfactory results from this procedure the mole ratio of calcium to oxalate must be exactly 1 in the precipitate and thus in the solution at the time of the titration. Several precautions are necessary to assure this condition. For example, calcium oxalate formed in a neutral or ammoniacal solution is likely to be contaminated with calcium hydroxide or a basic calcium oxalate; the presence of either causes low results. This problem can be eliminated by adding the oxalate to an acidic solution of the sample and slowly forming the precipitate by the dropwise addition of ammonia. The coarsely crystalline solid that is produced is readily filtered. Losses due to the solubility of calcium oxalate are negligible at a pH of about 4, provided washing is restricted to freeing the precipitate of excess oxalic acid.

A potential source of positive error in the analysis arises from the coprecipitation of sodium oxalate, which occurs when the sodium ion concentration exceeds that of calcium. The error from this source can be eliminated by double precipitation.

Magnesium, if present in high concentrations, may also contaminate the calcium oxalate precipitate. This interference is minimized if the excess of oxalate is sufficient to allow the formation of a soluble magnesium complex and if filtration is performed promptly after the completion of the precipitation. When the magnesium content exceeds that of calcium in the sample, a double precipitation may be required.

Directions for the determination of calcium with permanganate are given in Chapter 20, Analyses 5-3 and 5-4.

QUADRIVALENT CERIUM

A sulfuric acid solution of cerium(IV) is very nearly as potent an oxidizing reagent as permanganate and can be substituted for the latter in most of the applications just described. The reagent is indefinitely stable and does not oxidize chloride ion at a detectable rate. Furthermore, only a single reduction product, trivalent cerium, is possible; thus the stoichiometry of the reaction is less subject to uncertainty. In these respects cerium(IV) possesses considerable advantages over permanganate. On the other hand, the color of quadrivalent cerium solutions is not sufficiently intense to serve as an indicator. In addition, the reagent cannot be used in neutral or basic solutions. A final disadvantage is the relatively high cost of cerium compounds.

Properties of Quadrivalent Cerium Solutions. The electrode potential for a cerium(IV) solution depends upon the acid that is used for preparation. Solutions of the reagent prepared with sulfuric acid are roughly comparable in oxidizing power to those of permanganate; solutions containing nitric or perchloric acid are appreciably more potent. To a lesser degree the electrode potential is also influenced by the concentration of the acid.

Stability of Cerium(IV) Solutions. Sulfuric acid solutions of quadrivalent cerium are remarkably stable, remaining constant in titer for years. Solutions do not change appreciably when heated to 100°C for considerable periods. Perchloric and nitric acid solutions of the reagent are by no means as stable; these decompose water and decrease in normality by 0.3 to 1% during storage for one month. The decomposition reaction is catalyzed by light.

The oxidation of chloride is so slow that other reducing agents can be titrated without error in the presence of high concentrations of this ion. Hydrochloric acid solutions of cerium(IV), however, are not stable enough for use as standard solutions.

Indicators for Cerium(IV) Titrations. Several oxidation-reduction indicators are available for use with cerium(IV) solutions. The most widely used of these is the iron(II) complex of 1,10-orthophenanthroline; indicator solutions of the complex are sometimes given the trivial name "ferroin."

Preparation of Solutions. Several cerium(IV) salts are commercially available;[6] the most common are listed in Table 13-4. Cerium(IV) ammonium nitrate of primary standard quality can be purchased; standard solutions can thus be prepared directly by weight. More frequently, solutions of approximately the desired normality are prepared from one of the

TABLE 13-4
Analytically Useful Cerium(IV) Compounds

NAME	FORMULA	EQUIVALENT WEIGHT
Cerium(IV) ammonium nitrate	$Ce(NO_3)_4 \cdot 2NH_4NO_3$	548.2
Cerium(IV) ammonium sulfate	$Ce(SO_4)_2 \cdot 2(NH_4)_2SO_4 \cdot 2H_2O$	632.6
Cerium(IV) hydroxide	$Ce(OH)_4$	208.1
Cerium(IV) hydrogen sulfate	$Ce(HSO_4)_4$	528.4

[6] For further information regarding the preparation, standardization, and use of cerium(IV) solutions, see G. Frederick Smith, *Cerate Oxidimetry*. Columbus, Ohio: The G. Frederick Smith Chemical Company, 1942; I. M. Kolthoff and R. Belcher, *Volumetric Analysis*, vol. 3. New York: Interscience, 1957, pp. 121–167.

less expensive reagent-grade salts and then standardized. A stable and entirely satisfactory sulfuric acid solution of quadrivalent cerium is obtained from cerium(IV) ammonium nitrate; removal of the ammonium or nitrate ions is unnecessary.

Solutions of cerium(IV) tend to react with water, even in acid solutions, to produce slightly soluble basic salts. The acidity of solutions containing cerium(IV) must be $0.1 N$ or greater to prevent this precipitation. Neutral or basic solutions cannot be titrated with the reagent.

Standardization against Arsenic(III) Oxide. Arsenic(III) oxide is perhaps the most satisfactory primary standard for solutions of quadrivalent cerium. In the absence of a catalyst the reaction is so slow that iron(II) can be titrated without interference in the presence of trivalent arsenic. Fortunately good catalysts are available that permit standardization with this useful reagent. The best of these is osmium tetroxide, which is effective even at very low concentrations ($10^{-5} F$). Iodine monochloride also catalyzes the reaction. Directions for standardization with arsenic(III) oxide are found in Chapter 20, Standardization 6-1.

Standardization against Sodium Oxalate. Several methods exist for the standardization of sulfuric acid solutions of cerium(IV) against sodium oxalate.[7] The directions in Chapter 20, Standardization 6-2, call for titration at 50°C in hydrochloric acid solution. Iodine monochloride is used as a catalyst; orthophenanthroline is the indicator.

Applications of Quadrivalent Cerium Solutions. Many applications of cerium(IV) solutions are found in the literature. In general these parallel the uses of permanganate given in Table 13-3.[8] Directions for the determination of iron are given in Chapter 20, Analysis 6-1.

POTASSIUM DICHROMATE

In its analytical applications dichromate ion is reduced to the trivalent state:

$$Cr_2O_7^{2-} + 14H^+ + 6e \rightleftharpoons 2Cr^{3+} + 7H_2O \qquad E^0 = 1.33 \text{ V}$$

Potassium dichromate is more limited in application than either potassium permanganate or quadrivalent cerium owing to its lesser oxidizing strength and the slowness of some of its reactions. Despite these handicaps, the reagent is nonetheless useful because its solutions are

[7] I. M. Kolthoff and R. Belcher, *Volumetric Analysis*, vol. 3. New York: Interscience, 1957, pp. 132–134.

[8] Information about cerium(IV) methods can be found in I. M. Kolthoff and R. Belcher, *Volumetric Analysis*, vol. 3. New York: Interscience, 1957, pp. 136–158.

indefinitely stable and are also inert toward hydrochloric acid. Furthermore, the solid reagent can be obtained in high purity and at modest cost; standard solutions may be prepared directly by weight.

Standard solutions of potassium dichromate may be boiled for long periods without decomposition.

Preparation and Properties of Dichromate Solutions.

For most purposes commercial reagent-grade or primary standard-grade potassium dichromate can be used to prepare standard solutions with no prior treatment other than drying at 150 to 200°C. If desired, two or three recrystallizations of the solid from water will assure a high-quality, primary standard product.

Although solutions of dichromate are orange, the color is not sufficiently intense for end point determination. Diphenylamine sulfonic acid (Table 12-3) is an excellent indicator for titrations with the reagent; the color change is from the green of the chromium(III) ion to the violet color of the oxidized indicator. An indicator blank is not readily obtained because dichromate oxidizes the indicator only slowly in the absence of other oxidation-reduction systems. Ordinarily, however, the error resulting from neglect of the blank is vanishingly small. The reaction of diphenylamine sulfonic acid is reversible, and back-titration of small excesses of dichromate with iron(II) is possible. In the presence of large oxidant concentrations and at low acidities (above pH 2) the indicator is irreversibly oxidized to yellow or red compounds.

Directions for the preparation of a standard dichromate solution are given in Chapter 20, Preparation 7-2.

Determination of Iron.

The principal use of dichromate involves titration of iron(II):

$$6Fe^{2+} + Cr_2O_7^{2-} + 14H^+ \rightleftharpoons 6Fe^{3+} + 2Cr^{3+} + 7H_2O$$

Moderate amounts of hydrochloric acid do not affect the accuracy of the titration. Detailed procedures for the analysis of iron are found in Chapter 20, Analysis 7-1.

Other Applications.

A common method for the determination of oxidizing agents calls for treatment of the sample with a known excess of iron(II), followed by titration of the excess with standard dichromate. This technique has been successfully applied to the determination of nitrate, chlorate, permanganate, dichromate, and organic peroxides, among others.

POTASSIUM BROMATE SOLUTIONS AS A SOURCE OF BROMINE

Primary standard-grade potassium bromate is available commercially. Its solutions are stable indefinitely.

Although potassium bromate can be employed for direct titration of certain species, it is principally used as a convenient and stable source of bromine. In this application an excess of potassium bromide is added to an acidic solution of the analyte. The addition of standard bromate then releases an equivalent quantity of bromine for reaction with the analyte. That is,

$$BrO_3^- + 5Br^- + 6H^+ \rightarrow 3Br_2 + 3H_2O$$

standard   excess
solution

This indirect approach eliminates the principal disadvantage associated with the use of standard bromine solutions, namely, lack of stability. Note that each bromate is responsible for the formation of three molecules of bromine which in turn require six electrons for reduction to bromide. That is,

$$BrO_3^- \equiv 3Br_2 \equiv 6e$$

The equivalent weight of potassium bromate is thus one-sixth of its formula weight.

Indicators for Titrations Involving Bromine. Several organic indicators such as methyl orange and methyl red (Table 9-1) are readily brominated to yield products that differ in color from the original compounds. Unfortunately these reactions are totally nonreversible. This behavior precludes any sort of back-titration and, more important, makes direct titration more difficult because of the great need to avoid local excesses of the reagent.

Three indicators, α-naphthoflavone, p-ethoxychrysoidine, and quinoline yellow, are reversible with respect to bromine and make employment of the reagent more attractive; they are commercially available.

Applications of Standard Potassium Bromate Solutions. Potassium bromate is a convenient source of bromine in organic analysis. Few organic compounds react sufficiently rapidly for a direct titration. Instead, a measured excess of the bromate solution is added to the sample; after the addition of sufficient potassium bromide to convert all of the bromate to bromine, the mixture is acidified and allowed to stand until bromination is judged complete. The excess bromine is then back-titrated with a standard arsenic(III) solution. Alternatively, the analysis can be completed by adding an excess of potassium iodide and titrating the iodine liberated with a standard solution of sodium thiosulfate (pp. 336–338). Although the iodometric procedure appears complicated, it is actually quite simple.

In general bromine is incorporated into an organic molecule either by substitution or by addition.

Substitution Reactions. Substitution involves the replacement of hydrogen in an aromatic ring by atoms of the halogen. For example, three hydrogen atoms are replaced when phenol is brominated.

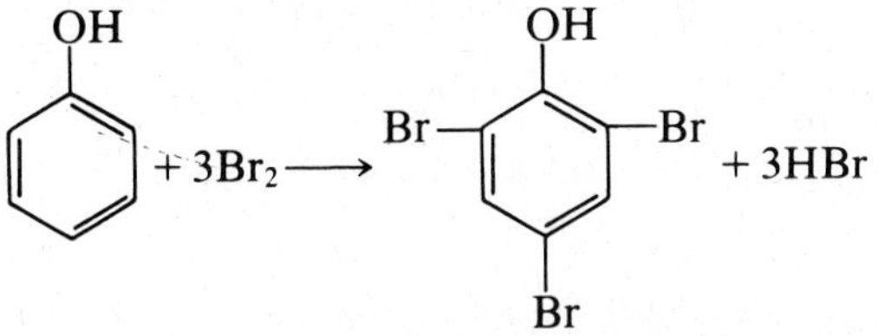

Here each molecule of phenol requires six atoms of bromine, each of which can be considered to undergo a one-electron change to the minus one state; thus the equivalence weight of phenol is one-sixth of its formula weight.

Substitution methods have been applied successfully to the analysis of aromatic compounds that contain strong ortho-para directing groups in the ring, particularly amines and phenols. An important application of the method is the titration of 8-hydroxyquinoline:

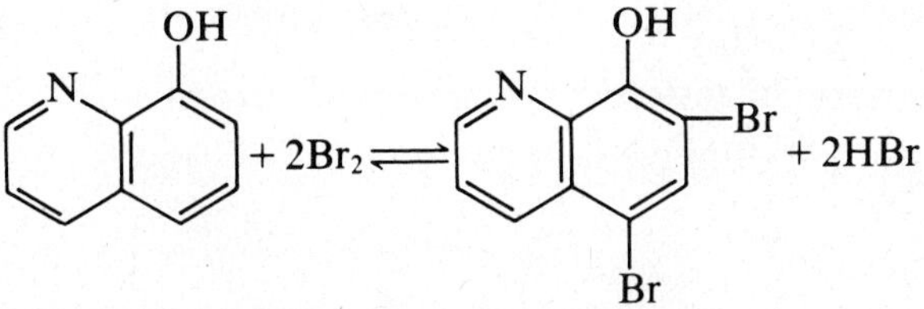

The reaction, which is sufficiently rapid in hydrochloric acid solution to permit direct titration, is of particular interest because 8-hydroxyquinoline is an excellent precipitating reagent for cations (Chapter 6). Thus, for example, aluminum can be determined by means of the following reaction sequence:

$$Al^{3+} + 3HOC_9H_6N \xrightarrow{pH\,4-9} Al(OC_9H_6N)_3(s) + 3H^+$$

8-hydroxyquinoline   (filter and wash)

$$Al(OC_9H_6N)_3(s) + 3H^+ \xrightarrow[HCl]{hot\,4-F} 3HOC_9H_6N + Al^{3+}$$

$$3HOC_9H_6N + 6Br_2 \rightarrow 3HOC_9H_4NBr_2 + 6HBr$$

Each mole of aluminum is indirectly responsible for the reaction of 6 moles (or 12 equivalents) of molecular bromine; the equivalent weight of the metal is thus one-twelfth of its formula weight.

Addition Reactions. Addition reactions involve the opening of an olefinic double bond. For example, one mole of molecular bromine (or two equivalents) reacts with each mole of ethylene:

$$H-\overset{\overset{\displaystyle H}{|}}{C}=\overset{\overset{\displaystyle H}{|}}{C}-H + Br_2 \longrightarrow H-\overset{\overset{\displaystyle H}{|}}{\underset{\underset{\displaystyle Br}{|}}{C}}-\overset{\overset{\displaystyle H}{|}}{\underset{\underset{\displaystyle Br}{|}}{C}}-H$$

The equivalent weight of ethylene will therefore be one-half of its formula weight.

A variety of methods, many involving the use of bromate-bromide mixtures, are found in the literature;[9] most of these were developed to estimate olefinic unsaturation in fats, oils, and petroleum products.

POTASSIUM IODATE

Potassium iodate is commercially available in a high state of purity and can be used without further treatment, other than drying, for the direct preparation of standard solutions. Iodate solutions, which are stable indefinitely, have a number of interesting and important uses in analytical chemistry.

Reactions of Iodate. The reaction of iodate with iodide is analogous to the bromate-bromide reaction. Thus the position of equilibrium for the reaction

$$IO_3^- + 5I^- + 6H^+ \rightleftharpoons 3I_2 + 3H_2O$$

lies far to the right in acidic solutions and far to the left in basic media.

This reaction is a convenient source for known amounts of iodine (see, for example, p. 339). A measured quantity of iodate is mixed with an excess of iodide in a solution that is 0.1 to $1 N$ in acid. Exactly six equivalents of iodine are liberated for each mole of iodate; the resulting solution can then be used for the standardization of thiosulfate solutions or for other analytical purposes.

In strongly acidic solutions iodate will oxidize iodide or iodine to the +1 state, provided some anion such as chloride, bromide, or cyanide is present to stabilize this oxidation state. For example, in solutions that are greater than $3 F$ in hydrochloric acid the following reaction proceeds essentially to completion:

$$IO_3^- + 2I_2 + 10Cl^- + 6H^+ \rightleftharpoons 5ICl_2^- + 3H_2O$$

A number of important iodate titrations are performed in concentrated hydrochloric acid where the iodate is initially reduced to iodine. As

---

[9] For example, see A. Polgar and J. L. Jungnickel, *Organic Analysis*, vol. 3. New York: Interscience, 1956.

the reducing agent is consumed, however, the reaction described in the previous paragraph takes place. Generally the end point for the process is signaled by the complete disappearance of the iodine. The equivalent weight for the iodate is one-fourth of its formula weight because the final reaction product is iodine in the +1 state. When used in this manner, iodate is a less powerful oxidant than either permanganate or cerium(IV).

End Points in Iodate Titrations.   The disappearance of iodine from the solution is often sufficient to indicate the end point in an iodate titration. Iodine is nearly always formed in the initial stages of the reaction; only at the equivalence point is it completely oxidized to the +1 state. Starch fails to function as an indicator in the highly acidic medium required for the production of ICl_2^-. Instead, a few milliliters of carbon tetrachloride, chloroform, or benzene are added at the start of the titration. After each addition of iodate the mixture is shaken thoroughly; the organic layer is examined after the phases have separated. The bulk of any unreacted iodine remains in the organic layer and imparts a violet-red color to it. The titration is judged complete when the minimum amount of iodate needed to discharge this color has been added. This excellent method for detecting iodine is quite comparable in sensitivity to the starch-iodine color; it suffers, however, from the disadvantage of being more time-consuming.

Directions for the iodate titration of iodide and iodine in solutions are given in Chapter 20, Analysis 11-1; further applications can be found in reference sources.[10]

IODIMETRIC METHODS

Many volumetric analyses are based on the half-reaction

$$I_3^- + 2e \rightleftharpoons 3I^- \qquad E^0 = 0.536 \text{ V}$$

These analyses fall into two categories. The first comprises procedures that use a standard solution of iodine to titrate easily oxidized substances. These *direct* or *iodimetric methods* have limited applicability because iodine is a relatively weak oxidizing agent. *Indirect* or *iodometric methods* employ a standard solution of sodium thiosulfate or arsenic(III) to titrate the iodine liberated when an oxidizing substance is allowed to react with an unmeasured excess of potassium iodide. The quantity of iodine formed is chemically equivalent to the amount of the oxidizing agent and thus serves as the basis for the analysis.

Iodine, which is a relatively weak oxidant, is used for the selective

[10] For details on the application of iodate titrations see I. M. Kolthoff and R. Belcher, *Volumetric Analysis*, vol. 3. New York: Interscience, 1957, pp. 449–473; R. Lang in W. Böttger, ed., *Newer Methods of Volumetric Analysis*. New York: Van Nostrand, 1938, pp. 69–98.

determination of strong reducing agents. The availability of a sensitive and reversible indicator for iodine is a great advantage. Disadvantages include the low stability of iodine solutions and the incompleteness of reactions between iodine and many reductants.

Preparation of Iodine Solutions.   A saturated aqueous iodine solution is only about $0.001\ F$ at room temperature. Much higher concentrations can be achieved, however, in the presence of iodide ion owing to the formation of the soluble triiodide complex.

$$I_2(s) + I^- \rightleftharpoons I_3^- \qquad K = 7.1 \times 10^2$$

Because the concentration of the species I_2 is low in such solutions, it would be more proper to refer to them as *triiodide solutions*. As a practical matter, however, they are usually called *iodine solutions* because of the convenience this affords in writing equations and in describing stoichiometric behavior.

The rate at which iodine dissolves in potassium iodide solution is slow, particularly where the iodide concentration is low. As a consequence, it is necessary to dissolve the solid completely in a small amount of a concentrated iodide solution before diluting to the desired volume. All of the element must be dissolved before dilution; otherwise the normality of the resulting reagent will increase continuously as the remaining iodine slowly passes into solution.

Stability.   Iodine solutions require restandardization every few days. This lack of stability has several sources, one being the volatility of the solute. Even though the excess of iodide is large, a measurable amount of iodine is lost in a relatively short period of time from an open container.

Iodine will slowly attack rubber or cork stoppers as well as other organic substances; reasonable precautions must, therefore, be taken to protect standard solutions of the reagent from contact with these materials. Exposure to organic dust and fumes must also be avoided.

Finally, changes in iodine normality result from air oxidation of iodide ions in the solution:

$$4I^- + O_2 + 4H^+ \rightleftharpoons 2I_2 + 2H_2O$$

This reaction is catalyzed by light, heat, and acids; consequently it is good practice to store the reagent in a dark, cool place. In contrast to the other effects, air oxidation of iodide causes an increase in normality.

Completeness of Iodine Oxidations.   Because iodine is such a weak oxidizing agent, the chemist frequently must take full advantage of those experimental variables that enhance its reduction to iodide by the analyte. Two effects, pH and the presence of complexing agents, are of particular importance.

The pH has little influence upon the electrode potential of the iodine-iodide couple in acidic solutions because hydrogen ions do not participate in the half-reaction. Many of the substances that react with iodine, however, produce hydrogen ions as they are oxidized. The position of equilibrium may therefore be markedly influenced by pH. The arsenic(III)-arsenic(V) system provides an important example. Its electrode potential differs by only 0.02 V from that for the iodide-iodine half-reaction:

$$H_3AsO_4 + 2H^+ + 2e \rightleftharpoons H_3AsO_3 + H_2O \qquad E^0 = 0.559 \text{ V}$$

In a strongly acidic medium arsenic(V) will quantitatively oxidize iodide to iodine. On the other hand, in a nearly neutral solution trivalent arsenic can be titrated successfully with iodine. Although iodine oxidations often become more nearly complete with lowered acidity, care must be taken to prevent the formation of hypoiodite, which tends to occur in alkaline solutions:

$$I_2 + OH^- \rightleftharpoons HOI + I^-$$

The hypoiodite may subsequently disproportionate to iodate and iodide:

$$3HOI + 3OH^- \rightleftharpoons IO_3^- + 2I^- + 3H_2O$$

The occurrence of these reactions will cause serious errors in an iodimetric analysis. In some titrations the reaction of iodate and hypoiodite with the reducing reagent is so slow that overconsumption of iodine is observed. In others, the presence of these two species can alter the reaction products; the oxidation of thiosulfate is an important example of such behavior. Thus, solutions to be titrated with iodine cannot have pH values much higher than 9. Occasionally a pH greater than 7 is detrimental.

Complexing reagents are also used to force certain iodine oxidations toward completion. For example, it is readily seen that the reduction potential of iodine is too low to permit the quantitative oxidation of iron(II) to the trivalent state. In the presence of reagents that strongly complex iron(III), however, complete conversion is achieved (see p. 285 for the effect of complexing agents on electrode potentials); pyrophosphate ion and ethylenediaminetetraacetate are useful for this purpose.

Starch Indicator. The most widely used indicator for iodimetry is an aqueous suspension of starch, which imparts an intense blue color to a solution containing a trace of triiodide ion. The nature of the colored species has been the subject of much speculation and controversy.[11]

Indicator solutions are most conveniently prepared from commercially available *soluble starch*; corn starch does not yield a satisfactory indicator solution. Aqueous starch suspensions decompose within a few

[11] See R. E. Rundle, J. F. Foster, and R. R. Baldwin, *J. Amer. Chem. Soc.*, **66**, 2116 (1944).

days, primarily because of bacterial action. The decomposition products may consume iodine as well as interfere with the indicator properties of the preparation. The rate of decomposition can be greatly inhibited by preparing and storing the indicator under sterile conditions and by the introduction of mercury(II) iodide or chloroform to act as a bacteriostat. Alternatively, a fresh indicator suspension can be prepared each day that an iodine titration is to be performed.

Starch added to a solution containing a high concentration of iodine is decomposed to products whose indicator properties are not entirely reversible. Thus the addition of the indicator to a solution containing an excess of iodine should be postponed until most of the iodine has been titrated, as indicated by a light yellow color of the solution.

Standardization of Iodine Solutions. Iodine solutions are most commonly standardized against arsenious oxide, barium thiosulfate monohydrate, or potassium antimony(III) tartrate.

Arsenious oxide, As_2O_3, is available commercially in primary standard quality. The oxide dissolves only slowly in water or in the common acids; solution occurs rapidly in 1-N NaOH, however.

$$As_2O_3(s) + 4OH^- \rightarrow 2HAsO_3^{2-} + H_2O$$

In strongly alkaline solution arsenic(III) is readily air-oxidized to arsenic(V), whereas neutral or slightly acidic solutions are indefinitely stable toward oxygen. Thus after the arsenious oxide has been dissolved in base, hydrochloric acid should be added immediately until the solution is slightly acidic. Standard solutions of arsenious acid are useful for periodic standardization of iodine solutions.

The iodimetric titration of arsenious acid must be carried out in a buffered system to use up the hydrogen ions formed in the reaction; otherwise the pH may decrease to a level where the reaction is not complete. Buffering is conveniently accomplished by acidifying the sample slightly and then saturating with sodium hydrogen carbonate. The carbonic acid-hydrogen carbonate buffer so established will hold the pH in a range between 7 and 8.

Directions for the standardization of iodine solutions against arsenic(III) are given in Chapter 20, Standardization 8-1.

Applications of Standard Iodine. Common analyses that make use of iodine as an oxidizing reagent are summarized in Table 13-5.

Determination of Antimony in Stibnite. The analysis of stibnite, a common antimony ore, illustrates the application of a direct iodimetric method. Stibnite is primarily antimony sulfide that contains silica and other contaminants. Provided the material is free of iron and arsenic, the determination of its antimony content is a straightforward process. The

TABLE 13-5
Analysis with Standard Iodine Solutions

SUBSTANCE ANALYZED	HALF-REACTION
As	$H_3AsO_3 + H_2O \rightleftarrows H_3AsO_4 + 2H^+ + 2e$
Sb	$H_3SbO_3 + H_2O \rightleftarrows H_3SbO_4 + 2H^+ + 2e$
Sn	$Sn^{2+} \rightleftarrows Sn^{4+} + 2e$
H_2S	$H_2S \rightleftarrows S(s) + 2H^+ + 2e$
SO_2	$SO_3^{2-} + H_2O \rightleftarrows SO_4^{2-} + 2H^+ + 2e$
$S_2O_3^{2-}$	$2S_2O_3^{2-} \rightleftarrows S_4O_6^{2-} + 2e$
N_2H_4	$N_2H_4 \rightleftarrows N_2 + 4H^+ + 4e$

sample is decomposed in hot, concentrated hydrochloric acid to eliminate the sulfide as H_2S. Some care is required in this step to prevent losses of the volatile antimony trichloride. The addition of potassium chloride increases the tendency for the formation of nonvolatile chloro complexes; these probably have the formulas $SbCl_4^-$ and $SbCl_6^{3-}$.

The reaction of trivalent antimony with iodine is quite analogous to that of trivalent arsenic. Here, however, an additional step is required to prevent precipitation of such basic salts as antimony oxychloride, SbOCl, as the solution is neutralized. These species react incompletely with iodine and cause erroneously low results. The problem is readily overcome by the addition of tartaric acid prior to dilution. The tartrate complex $(SbOC_4H_4O_6^-)$ that forms is rapidly and completely oxidized by iodine.

A procedure for determining antimony in a stibnite is found in Chapter 20, Analysis 8-2.

Application of Reductants

Solutions of reducing agents are often troublesome because of the readiness with which they react with atmospheric oxygen. As a consequence, titrations ordinarily must be carried out in and the reagents stored under an inert atmosphere. Alternatively, a stable standard oxidizing agent such as potassium dichromate is used as the reference substance; typically, then, an aliquot containing an excess of the reductant is added to the sample and the excess is quickly back-titrated with the standard oxidant. The current concentration of the reductant is determined by a similar titration of a blank. Very strong reducing agents such as titanium(III) and chromium(II), however, react too rapidly with oxygen to make this procedure feasible; for these, a blanket of an inert gas such as N_2 or CO_2 must be employed.

TABLE 13-6

Some Common Reductants Employed for Standard Solutions

REAGENT AND COMMON HALF-REACTION	OXIDATION POTENTIAL,[a] V	PRIMARY STANDARDS FOR	STABILITY OF SOLUTION
Iron(II) $Fe^{2+} \rightarrow Fe^{3+} + e$	-0.77	Fe, $K_2Cr_2O_7$	Unstable unless protected from oxygen
Arsenic(III) $H_3AsO_3 + H_2O \rightarrow$ $H_3AsO_4 + 2H^+ + 2e$	-0.56	As_2O_3, I_2	Indefinitely stable if acidic
Sodium thiosulfate $2S_2O_3^{2-} \rightarrow S_4O_6^{2-} + 2e$	-0.08	KIO_3, I_2	Frequent standardization required
Titanium(III) $Ti^{3+} + H_2O \rightarrow$ $TiO^{2+} + 2H^+ + e$	-0.1	$K_2Cr_2O_7$	Highly unstable unless protected from oxygen

[a] Note that these are potentials for the reactions written as oxidations; the sign, therefore, is the opposite of the corresponding electrode potential. The system with the smallest negative potential has the greatest tendency to occur. Thus Ti(III) is a considerably better reducing agent than is iron(II).

Table 13-6 is a list of common reductants employed for the determination of oxidizing substances.

SOLUTIONS OF IRON(II)

Solutions of iron(II) are readily prepared from *Mohr's salt*, $Fe(NH_4)_2(SO_4)_2 \cdot 6H_2O$, or *Oesper's salt*, $FeC_2H_4(NH_3)_2(SO_4)_2 \cdot 4H_2O$. Air oxidation of the iron(II) proceeds rapidly in neutral solution but is inhibited by acid; the most stable solutions are about 0.5 F in H_2SO_4. Such solutions should be standardized daily.

A variety of oxidizing agents such as Cr(VI), Ce(IV), Mo(VI), NO_3^-, NH_2OH, and organic peroxides are conveniently determined by reaction with a measured excess of a standard iron(II) solution; as mentioned earlier, standard potassium dichromate is frequently employed for the back-titration.

SODIUM THIOSULFATE; IODOMETRIC METHODS

Iodide ion is a moderately effective reducing agent that has been widely employed for the analysis of oxidants. In these applications a standard solution of sodium thiosulfate (or occasionally arsenious acid) is used to titrate the iodine liberated by reaction of the analyte with an unmeasured excess of potassium iodide.

The Reaction of Iodine with Thiosulfate Ion. The reaction between iodine and thiosulfate is described by the equation

$$2S_2O_3^{2-} + I_2 \rightleftharpoons S_4O_6^{2-} + 2I^-$$

The production of the tetrathionate ion requires the loss of two electrons from two thiosulfate ions; the equivalent weight of thiosulfate in this reaction must therefore be equal to its gram formula weight.

The quantitative conversion of thiosulfate to tetrathionate ion is unique with iodine; other oxidizing reagents tend to carry the oxidation, wholly or in part, to sulfate ion. The reaction with hypoiodous acid provides an important example of this stoichiometry:

$$4HOI + S_2O_3^{2-} + H_2O \rightleftharpoons 2SO_4^{2-} + 4I^- + 6H^+$$

The presence of hypoiodite in slightly alkaline solutions of iodine (p. 333), then, will seriously upset the stoichiometry of the iodine-thiosulfate reaction, causing too little thiosulfate or too much iodine to be used in the titration.

The equilibrium constant for the reaction

$$I_2 + H_2O \rightleftharpoons HOI + I^- + H^+$$

is small (about 3×10^{-13}).[12] It can be shown, however, that hypoiodite formation should become significant in media where the pH is greater than 7. Kolthoff[13] has shown that an error of about 4% will occur when 25 ml of 0.1-N iodine is titrated with 0.1-N thiosulfate in the presence of 0.5 g of sodium carbonate; this error becomes 10% or more in solutions containing about 2 g of this salt. Kolthoff recommends that the pH always be less than 7.6 for the titration of 0.1-N solutions, 6.5 or less for 0.01-N solutions, and less than 5 for 0.001-N solutions.

The titration of highly acidic iodine solutions with thiosulfate yields quantitative results, provided care is taken to prevent air oxidation of iodide ion.

The end point in the titration is readily established by means of a starch solution (p. 333). It should be emphasized that starch is partially decomposed in the presence of a large excess of iodine. For this reason the indicator is never added to an iodine solution until the bulk of that substance has been reduced. The change in color of the iodine from red to faint yellow signals the proper time for the addition of the indicator.

Stability of Thiosulfate Solutions. Principal among the variables affecting the stability of thiosulfate solutions are the pH, the presence of microorganisms and impurities, the concentration of the solution, the

[12] W. C. Bray and E. L. Connolly, *J. Amer. Chem. Soc.*, **33**, 1485 (1911).

[13] I. M. Kolthoff and R. Belcher, *Volumetric Analysis*, vol. 3. New York: Interscience, 1957, pp. 214–215.

presence of atmospheric oxygen, and exposure to sunlight. The decrease in iodine titer ordinarily amounts to as much as several percent in a few weeks. Occasionally, however, increases in normality are observed. Proper attention to detail will yield standard thiosulfate solutions that need only occasional restandardization.

The following reaction occurs at an appreciable rate when the pH of a thiosulfate solution is 5 or less:

$$S_2O_3^{2-} + H^+ \rightleftharpoons HS_2O_3^- \rightarrow HSO_3^- + S(s)$$

The velocity of this disproportionation increases with the hydrogen ion concentration; in a strongly acidic solution, elemental sulfur forms within a few seconds. The hydrogen sulfite ion produced is also oxidized by iodine, reacting with twice the quantity of that reagent as the thiosulfate from which it was derived. Clearly thiosulfate solutions cannot be allowed to stand in contact with acid. On the other hand, iodine solutions that are 3 to 4 F in acid may be titrated without error as long as care is taken to introduce the thiosulfate slowly and with good mixing. Under these conditions the thiosulfate is oxidized so rapidly by the iodine that the slower acid decomposition cannot occur to any measurable extent.

Experiments indicate that the stability of thiosulfate solutions is at a maximum in the pH range between 9 and 10, although for most purposes a pH of 7 is adequate. The addition of a small amount of a base such as sodium carbonate, borax, or disodium hydrogen phosphate is frequently recommended to preserve standard solutions of the reagent. If this procedure is followed, the iodine solutions to be titrated must be sufficiently acidic to neutralize the added base. Otherwise hypoiodite formation may occur before the equivalence point is attained and cause the partial oxidation of the thiosulfate to sulfate.

The most important single cause of instability can be traced to certain bacteria that metabolize the thiosulfate ion, converting it to sulfite, sulfate, and elemental sulfur.[14] Solutions that are free of bacteria are remarkably stable; it is common practice, therefore, to impose reasonably sterile conditions in the preparation of standard solutions. Substances such as chloroform, sodium benzoate, or mercury(II) iodide can be added to inhibit bacterial growth. Bacterial activity appears to be at a minimum at a pH between 9 and 10, which accounts, at least in part, for the maximum stability of thiosulfate solutions in this range.

Many other variables affect the stability of thiosulfate solutions. Decomposition is reported to be catalyzed by copper(II) ions as well as by the decomposition products themselves. Solutions that have become turbid from the formation of sulfur should be discarded. Exposure to sunlight increases the rate of decomposition as does atmospheric oxygen. Finally, the decomposition rate is greater in more dilute solutions.

[14] M. Kilpatrick, Jr., and M. L. Kilpatrick, *J. Amer. Chem. Soc.*, **45**, 2132 (1923); F. O. Rice, M. Kilpatrick, Jr., and W. Lemkin, *J. Amer. Chem. Soc.*, **45**, 1361 (1923).

Potassium Iodate.   As a primary standard for thiosulfate solutions iodate ion reacts rapidly with iodide in slightly acidic solution to give iodine.

$$IO_3^- + 5I^- + 6H^+ \rightleftharpoons 3I_2 + 3H_2O$$

$$3I_2 + 6e \rightarrow 6I^-$$

Three moles of iodine are furnished by each formula weight of potassium iodate for the standardization; thus its equivalent weight is one-sixth its formula weight because a six-electron change is associated with the reduction of three iodine molecules, the species actually titrated. That is,

$$IO_3^- \equiv 3I_2 \equiv 6I^-$$

The sole disadvantage of potassium iodate as a primary standard is its low equivalent weight (35.67). Only slightly more than 0.1 g can be taken for standardization of a 0.1-N thiosulfate solution; the relative error to be expected in weighing this quantity may, under some circumstances, be somewhat greater than that desirable for a standardization. This problem can be circumvented by dissolving a larger quantity of the solid in a known volume and taking aliquots of the resulting solution. This procedure, however, suffers from the disadvantage that it provides no duplicate check on the precision of the solution preparation process.

Other Primary Standards.   Other primary standards for sodium thiosulfate include potassium dichromate, potassium bromate, potassium hydrogen iodate [$KH(IO_3)_2$], potassium ferricyanide, and metallic copper. Details for the use of these may be found in various reference works.[15]

Directions for the preparation and standardization of a thiosulfate solution are given in Chapter 20, Section 9.

Errors in Iodometric Methods.   Three sources of errors in iodometric methods have already been mentioned, namely, the decomposition of thiosulfate solutions, the alteration of the stoichiometric relationship between iodine and thiosulfate ion in the presence of base, and the premature addition of starch. In addition, care must be taken to avoid loss of iodine by volatilization; significant volatility losses can generally be controlled by using stoppered containers when solutions must stand, by maintaining a goodly excess of iodide ion, and by avoiding elevated temperatures.

Air oxidation of iodide ion can also be a serious source of error in iodometric analyses (p. 332). Clearly the reaction is favored by hydrogen ions; from the standpoint of minimizing this source of error, reactions should be carried out at low acidities. Fortunately the reaction is slow under many circumstances. It is, however, catalyzed by acid, light, traces of Cu(II), and nitrogen oxides. The latter two are thus potential interferences in any iodometric procedure.

[15] For example, see I. M. Kolthoff and R. Belcher, *Volumetric Analysis*, vol. 3. New York: Interscience, 1957, pp. 234–243.

Applications of the Indirect Iodometric Method. Numerous substances can be determined iodometrically; some of the more common applications are summarized in Table 13-7. Detailed instructions for the iodometric determination of copper, a typical example of the indirect procedure, are found in Chapter 20, Analysis 9-1.

Determination of Copper. Copper(II) is quantitatively reduced to copper(I) by iodide ion. The reaction can be expressed as

$$2Cu^{2+} + 4I^- \rightleftharpoons 2CuI(s) + I_2$$

Consideration of the following standard potentials makes it evident that the formation of sparingly soluble CuI plays an important part in driving the reaction to completion.

$$Cu^{2+} + e \rightleftharpoons Cu^+ \qquad E^0 = 0.15 \text{ V}$$

$$I_2 + 2e \rightleftharpoons 2I^- \qquad E^0 = 0.54 \text{ V}$$

$$Cu^{2+} + I^- + e \rightleftharpoons CuI(s) \qquad E^0 = 0.86 \text{ V}$$

The first two potentials suggest that iodide would have little tendency to reduce Cu^{2+}; the formation of CuI, however, forces the reduction reaction.

Much systematic experimentation has been devoted to establishing

TABLE 13-7
Some Applications of the Iodometric Method

SUBSTANCE	HALF-REACTION	SPECIAL CONDITION
IO_4^-	$IO_4^- + 8H^+ + 7e \rightarrow \frac{1}{2}I_2 + 4H_2O$	Acidic solution
	$IO_4^- + 2H^+ + 2e \rightarrow IO_3^- + H_2O$	Neutral solution
IO_3^-	$IO_3^- + 6H^+ + 5e \rightarrow \frac{1}{2}I_2 + 3H^+$	Strong acid
BrO_3^-, ClO_3^-	$XO_3^- + 6H^+ + 6e \rightarrow X^- + 3H_2O$	Strong acid
Br_2, Cl_2	$X_2 + 2I^- \rightarrow I_2 + 2X^-$	
NO_2^-	$HNO_2 + H^+ + e \rightarrow NO + H_2O$	
Cu^{2+}	$Cu^{2+} + I^- + e \rightarrow CuI(s)$	
O_2	$O_2 + 4Mn(OH)_2(s) + 2H_2O \rightarrow$ $4Mn(OH)_3(s)$	Basic solution
	$Mn(OH)_3(s) + 3H^+ + e \rightarrow$ $Mn^{2+} + 3H_2O$	Acidic solution
O_3	$O_3 + 2H^+ + 2e \rightarrow O_2 + H_2O$	
Organic peroxide	$ROOH + 2H^+ + 2e \rightarrow ROH + H_2O$	

ideal conditions for a copper analysis.[16] These studies have revealed that the solution should be at least 4% with respect to potassium iodide. Furthermore, a pH less than 4 is expedient; at a higher pH formation of basic copper(II) species causes a slower and less complete oxidation of iodide ion. In the presence of copper(I) ion, hydrogen ion concentrations greater than about 0.3 M must be avoided to prevent air oxidation of iodide ion.

It has been found experimentally that the titration of iodine by thiosulfate in the presence of copper(I) iodide tends to yield slightly low results because small but appreciable quantities of iodine are physically adsorbed upon the solid. The adsorbed iodine is released only slowly, even in the presence of thiosulfate ion; transient and premature end points result. This difficulty is largely overcome by the addition of thiocyanate ion, which also forms a sparingly soluble copper(I) salt. Part of the copper(I) iodide is converted to the corresponding thiocyanate at the surface of the solid:

$$CuI(s) + SCN^- \rightleftarrows CuSCN(s) + I^-$$

Accompanying this reaction is the release of the adsorbed iodine, thus making it available for titration. Early addition of thiocyanate must be avoided, however, because of the tendency for that ion to reduce iodine slowly.

The iodometric method is convenient for the assay of copper in an ore. Ordinarily samples dissolve readily in hot, concentrated nitric acid. Care must be taken to volatilize any nitrogen oxides formed in the process because these catalyze the air oxidation of iodide. Some samples require the addition of hydrochloric acid to complete the solution step. The chloride ion must, however, be removed by evaporation with sulfuric acid because iodide ion will not reduce copper(II) quantitatively from its chloro complexes.

Of the elements ordinarily associated with copper in nature only iron, arsenic, and antimony interfere with the iodometric procedure. Fortunately difficulties caused by these elements are readily eliminated. Iron is rendered nonreactive by the addition of such complexing agents as fluoride or pyrophosphate; because these form more stable complexes with iron(III) than with iron(II), the potential for this system is altered to the point where appreciable oxidation of iodide cannot occur. Interference by arsenic and antimony is prevented by converting these elements to the +5 state during the solution step. Ordinarily the hot nitric acid used to dissolve the sample will cause conversion to the desired oxidation state, although a small amount of bromine water can be added in case of doubt; the excess bromine is then expelled by boiling. As has been pointed out, arsenic in the +5 state does not oxidize iodide ion, provided the solution is

[16] See E. W. Hammock and E. H. Swift, *Anal. Chem.*, **21**, 975 (1949).

not too acidic. Antimony exhibits a similar behavior. Thus, by maintaining the pH of the solution at 3 or greater, interference by these elements is avoided. We have seen, however, that oxidation of iodide by copper is incomplete at pH values greater than 4. Thus, when copper is to be determined in the presence of arsenic or antimony, control of the pH between 3 and 4 is essential. Ammonium hydrogen fluoride, NH_4HF_2, is a convenient buffer for this purpose. The anion of the salt dissociates as follows:

$$HF_2^- \rightleftharpoons HF + F^- \qquad K = 0.26$$

$$HF \rightleftharpoons H^+ + F^- \qquad K = 7.2 \times 10^{-4}$$

The first dissociation provides equal quantities of hydrogen fluoride and fluoride ions which then buffer the solution to a pH somewhat greater than 3. In addition to acting as a buffer, the salt also serves as a source of fluoride ions to complex any iron(III) that may be present.

A procedure for the determination of copper in an ore is given in Chapter 20, Analysis 9-1.

The iodometric method is also applicable to the determination of copper in brass, an alloy consisting principally of copper, zinc, lead, and tin. Several other elements, iron and nickel, for example, may be tolerated in minor amounts.

The method, which is described in detail in Chapter 20, Analysis 9-2, is relatively simple and applicable to the analysis of brasses containing less than 2% of iron. It involves solution in nitric acid, removal of nitrate by fuming with sulfuric acid, adjustment of the pH by neutralization with ammonia, acidification with a measured quantity of phosphoric acid, and finally the iodometric determination of the copper.

It is instructive to consider the fate of each major constituent during the course of this treatment. Tin is oxidized to the $+4$ state by the nitric acid and precipitates slowly as the slightly soluble hydrous tin(IV) oxide, $SnO_2 \cdot 4H_2O$. This precipitate, which tends to form as a colloid, is sometimes called *metastannic acid*. It has a tendency to adsorb copper(II) and other cations from the solution. Lead, zinc, and copper are oxidized to soluble divalent salts by the nitric acid, and iron is converted to the trivalent state. Evaporation and fuming with sulfuric acid redissolves the metastannic acid but may cause part of the lead to precipitate as the sulfate; the copper, zinc, and iron are unaffected. Upon dilution with water, lead is nearly completely precipitated as lead sulfate while the other elements remain in solution. None, except copper and iron, is reduced by iodide. Interference from the iron is eliminated by complexing with phosphate ion.

A procedure for the determination of copper in a brass is found in Chapter 20, Analysis 9-2.

Determination of Oxygen. The Winkler method for the determination of dissolved oxygen in natural water is an interesting example of iodometry.

In this procedure the sample is first treated with an excess of manganese(II), potassium iodide, and sodium hydroxide. The white manganese(II) hydroxide that forms reacts rapidly with oxygen to form brown manganese(III) hydroxide. That is,

$$4Mn(OH)_2(s) + O_2 + 2H_2O \rightarrow 4Mn(OH)_3(s)$$

When acidified, the manganese(III) oxidizes iodide to iodine. Thus

$$2Mn(OH)_3(s) + 2I^- + 6H^+ \rightarrow I_2 + 3H_2O + 2Mn^{2+}$$

The liberated iodine is titrated in the usual way. Instructions for the Winkler method are given in Chapter 20, Analysis 9-3.

PROBLEMS

*1. Write balanced equations to describe the following processes:
  (a) the oxidation of cerium(III) to cerium(IV) by ammonium peroxodisulfate.
  (b) the oxidation of manganese(II) to permanganate ion by sodium bismuthate.
  (c) the reaction of molybdic acid in a silver reductor.
  (d) the oxidation of vanadium(III) to vanadium(V) by hydrogen peroxide.
  (e) the oxidation of thiocyanate ion to sulfate and cyanide ion with bromine.
  (f) the titration of oxalic acid with potassium bromate to give carbon dioxide and bromide ion.
  (g) the titration of hydrogen peroxide with cerium(IV).
  (h) the reaction of hypochlorous acid with potassium iodide to give iodine and chloride ions.

2. Write balanced equations to describe the following processes:
  (a) the oxidation of potassium iodide to periodic acid (H_5IO_6) by ammonium peroxodisulfate.
  (b) the reaction of manganous ion with permanganate to give manganese dioxide in a slightly alkaline solution.
  (c) the oxidation of hydroquinone, $C_6H_4(OH)_2$, to quinone, $C_6H_4O_2$, by lead dioxide.
  (d) the reaction of vanadium(V) in a Jones reductor.
  (e) the reaction of uranium(VI) in a Walden reductor.
  (f) the air oxidation of arsenic(III) to arsenic(V) in a basic solution.
  (g) the oxidation of sulfurous acid by iodine.
  (h) the reaction of potassium dichromate with ferrocyanide ion.

*3. Describe the preparation of the following standard solutions from pure compounds:
  (a) 2.00 liters of 0.250-N $K_2Cr_2O_7$.
  (b) 500 ml of 0.150-N $KBrO_3$ (to be used with excess KBr as a source of Br_2).
  (c) 2.50 liters of 0.120-N I_3^-.

4. Describe the preparation of the following solutions:
   (a) 3.00 liters of 0.0105-N $S_2O_3^{2-}$ from primary standard $BaS_2O_3 \cdot H_2O$.

$$(2S_2O_3^{2-} \rightarrow S_4O_6^{2-} + 2e)$$

   (b) 1500 ml of 0.0500-N Ce^{4+} from primary standard $Ce(NO_3)_4 \cdot 2NH_4NO_3$.
   (c) 6.00 liters of 0.160-N H_3AsO_3 (as a reductant) from primary standard As_2O_3.

*5. What is the phenol titer of the solution described in Problem 3b?

6. What is the Fe_3O_4 titer of the solution described in Problem 3a (assuming that all the Fe is converted to Fe^{2+} by pretreatment)?

*7. What is the $K_2Cr_2O_7$ titer of the solution described in Problem 4a?

8. What is the HNO_2 titer of the solution described in Problem 4b?

$$(NO_2^- \rightarrow NO_3^-)$$

*9. To standardize a solution of $Na_2S_2O_3$, 0.0675 g of $K_2Cr_2O_7$ was dissolved in dilute HCl. An excess of KI was added, following which the liberated I_2 was titrated with 33.8 ml of the reagent. Calculate the normality of the $Na_2S_2O_3$.

10. To standardize a solution of I_3^-, 0.201 g of As_2O_3 was dissolved and titrated with 46.3 ml of the reagent. Calculate the normality of the I_3^-.

*11. Arsenic(III) oxide occurs in nature as the mineral claudetite. A 0.436-g sample of impure claudetite was found to require 31.3 ml of 0.112-N iodine solution. Calculate the percentage of As_2O_3 in the sample.

12. The potassium chlorate in a 0.175-g sample of high explosive was determined by reaction with 50.0 ml of 0.108-N Fe^{2+}:

$$ClO_3^- + 6Fe^{2+} + 6H^+ \rightarrow Cl^- + 3H_2O + 6Fe^{3+}$$

When reaction was complete, the excess iron(II) was back-titrated with 15.2 ml of 0.0755-N Ce^{4+}. Calculate the percentage of $KClO_3$ in the sample.

*13. The antimony(III) in a 1.964-g stibnite sample required a 37.3-ml titration with 0.0907-N I_2. Express the results of this analysis in terms of (a) percent antimony and (b) percent Sb_2S_3.

14. A 0.409-g sample of limestone was dissolved in dilute HCl. Ammonium oxalate was then introduced, and the pH of the resulting solution was adjusted to permit the quantitative precipitation of calcium oxalate. The solid was isolated by filtration, washed free of excess oxalate, and then redissolved in dilute sulfuric acid. Titration of the liberated oxalic acid required 29.7 ml of 0.196-N $KMnO_4$. Calculate the percentage of calcium oxide in the sample.

*15. A 7.50-g sample of an ant-control preparation was decomposed by wet-ashing with H_2SO_4 and HNO_3. The arsenic in the residue was reduced to the trivalent state with hydrazine. After removal of the excess reducing agent, the As(III) required 22.6-ml titration with 0.0399-N I_2 in a faintly alkaline medium. Express the results of this analysis in terms of the percent As_2O_3 in the original sample.

16. A 5.16-g sample containing the mineral tellurite was brought into solution and then treated with a 50.0-ml aliquot of 0.120-N dichromate. The reaction was

$$3TeO_2 + Cr_2O_7^{2-} + 8H^+ \rightarrow 3H_2TeO_4 + 2Cr^{3+} + H_2O$$

Upon completion of the reaction, the excess dichromate required a 15.1-ml back-titration with 0.0967-N Fe^{2+}. Calculate the percentage of TeO_2 in the sample.

*17. A 50.0-ml aliquot of a solution containing thallium(I) ion was treated with potassium chromate. The Tl_2CrO_4 was filtered, washed free of excess precipitating agent, and then redissolved in dilute sulfuric acid. The dichromate ion produced was then titrated with 31.7 ml of 0.0832-N iron(II) ammonium sulfate solution. What was the weight of thallium in the sample?

$$2Tl^+ + CrO_4^{2-} \rightarrow Tl_2CrO_4(s)$$

$$2Tl_2CrO_4(s) + 2H^+ \rightarrow 4Tl^+ + Cr_2O_7^{2-} + H_2O$$

$$Cr_2O_7^{2-} + 6Fe^{2+} + 14H^+ \rightarrow 6Fe^{3+} + 2Cr^{3+} + 7H_2O$$

18. A 0.449-g sample of an impure aluminum salt was dissolved in dilute acid, treated with an excess of ammonium oxalate, and slowly made alkaline through the addition of aqueous ammonia. The precipitated aluminum oxalate was filtered, washed, redissolved in dilute acid, and titrated with 37.8 ml of 0.164-N $KMnO_4$. Calculate the percentage of aluminum in the sample.

*19. The chromium in a 2.17-g sample of chromite ($FeO \cdot Cr_2O_3$) was oxidized to the +6 state by fusion with sodium peroxide. The fused mass was treated with water and boiled to destroy the excess peroxide. After acidification the sample was treated with 50.0 ml of 0.154-N Fe^{2+}. A back-titration of 7.76 ml of 0.0500-N $K_2Cr_2O_7$ was required to oxidize the excess iron(II).
   (a) What was the percentage of chromite in the sample?
   (b) What was the percentage of chromium in the sample?

20. A 0.360-g sample of pyrolusite (MnO_2) was reduced to Mn^{2+} with 50.0 ml of 0.0800-N sodium arsenite. The excess arsenite was then titrated with 5.17 ml of 0.107-N permanganate solution in the presence of a catalytic quantity of iodate. What was the percentage of pyrolusite in the original sample?

*21. In the presence of fluoride ion, Mn^{2+} can be titrated with MnO_4^-, both reactants being converted to a complex of Mn(III). A 0.816-g sample containing Mn_3O_4 was dissolved, and all the manganese was converted to Mn^{2+}. Titration in the presence of fluoride ion consumed 46.7 ml of $KMnO_4$ that was 0.150 N against oxalate.

(a) Write a balanced equation for the reaction, assuming that the complex is MnF_4^-.

(b) What was the normality of the $KMnO_4$ solution when MnF_4^- was the product?

(c) What was the percent Mn_3O_4 in the sample?

22. A 50.0-ml aliquot containing uranium(VI) was passed through a Jones reductor to reduce the U to the $+3$ state. Aeration converted the U(III) to U(IV), following which the latter was titrated with 37.7 ml of 0.0621-N $K_2Cr_2O_7$. What weight of uranium was contained in each liter of the sample solution?

*23. A 2.43-g sample of stainless steel was dissolved in HCl (this treatment converts the Cr present to Cr^{3+}) and diluted to 250 ml in a volumetric flask. One 25.00-ml aliquot was passed through a silver reductor (Table 13-1) and then titrated with 30.10 ml of 0.0970-N potassium permanganate solution. A second 25.00-ml aliquot was passed through a Jones reductor (Table 13-1) into 50.00 ml of 0.1260-F Fe^{3+}. Titration of the resulting solution required 26.91 ml of the standard permanganate solution. Calculate the percentages of iron and chromium in the alloy.

24. A 1.45-g sample containing both iron and vanadium was dissolved under conditions that converted the elements to Fe(III) and V(V). The solution was diluted to 250 ml; a 50.00-ml aliquot was passed through a Walden reductor and titrated with 16.82 ml of 0.1000-N Ce^{4+}. A second 50.00-ml aliquot was passed through a Jones reductor and required 39.66 ml of the Ce^{4+} to reach an end point. Calculate the percent Fe_2O_3 and V_2O_5 in the sample. See Table 13-1 for the reductor reactions.

*25. A 1.32-g sample of alkali-metal sulfates was dissolved and diluted to exactly 250 ml. The sodium in a 25.0-ml aliquot was precipitated as $NaZn(UO_2)_3(OAc)_9 \cdot 6H_2O$. After filtration and washing, the precipitate was dissolved in acid and the uranium reduced to U^{3+} in a Jones reductor. After aeration to convert the U^{3+} to U^{4+}, the solution was titrated with 36.3 ml of 0.113-N Ce^{4+}. Calculate the percent Na_2SO_4 in the sample.

26. A 5.07-g sample of an insecticide preparation was wet-ashed with a sulfuric-nitric acid mixture to destroy its organic components. The copper in the sample was then precipitated as the basic chromate, $CuCrO_4 \cdot 2CuO \cdot 2H_2O$, with an excess of potassium chromate. The solid was filtered, washed free of excess reagent, and then redissolved in acid:

$$2(CuCrO_4 \cdot 2CuO \cdot 2H_2O)(s) + 10H^+ \rightarrow 6Cu^{2+} + Cr_2O_7^{2-} + 9H_2O$$

Titration of the liberated dichromate required 29.6 ml of 0.0573-N Fe^{2+}. Express the results of this analysis in terms of the percentage of copper(II) oleate, $Cu(C_{18}H_{33}O_2)_2$ (gfw = 626), present in the sample.

*27. The barium ion in a 2.31-g sample was precipitated with 50.0 ml of 0.0800-N potassium iodate solution. The $Ba(IO_3)_2$ was filtered, after which the filtrate and washings were treated with an excess of potassium iodide. The iodine liberated required 12.1 ml of 0.0396-N $Na_2S_2O_3$ solution. Express the results of this analysis in terms of the percentage of barium oxide present in the sample.

28. A potassium permanganate solution was found to be 0.0616 N in terms of standardization against sodium oxalate in acidic solution. The oxidizing agent was then employed in a Volhard determination for manganese:

$$2MnO_4^- + 3Mn^{2+} + 4OH^- \rightarrow 5MnO_2(s) + 2H_2O$$

Calculate the percentage of manganese in a mineral specimen if a 0.266-g sample required a 41.3-ml titration with the permanganate solution.

*29. The H_2S and SO_2 concentrations of a gas were determined by passing a sample through three absorber solutions connected in series. The first contained a solution of Cd^{2+} to trap the H_2S as CdS. The second contained 25.0 ml of 0.0202-N I_2 to oxidize the SO_2 to SO_4^{2-}. The third contained 1.00 ml of 0.0345-N thiosulfate solution to retain any I_2 carried over from the second absorber. A 25.0-liter gas sample was passed through the apparatus, followed by an additional amount of pure N_2 to sweep the last traces of SO_2 from the first to the second absorber.

   The solution from the first absorber was made acidic, and 20.0 ml of 0.0202-N I_2 were added. The excess I_2 was back-titrated with 4.62 ml of the thiosulfate solution.

   The solutions in the second and third absorbers were combined, and the residual iodine was titrated with 3.13 ml of the thiosulfate solution.

   Calculate the concentrations (in milligrams per liter) of SO_2 and H_2S in the sample.

30. A 10.0-ml sample of a household bleach was diluted to 100 ml in a volumetric flask. An excess of KI was added to a 25.0-ml aliquot of the diluted sample. Titration of the liberated iodine required 37.1 ml of 0.101-N $Na_2S_2O_3$. Calculate the weight-volume percentage of NaClO in the bleach.

*31. The CO concentration of air was obtained by passing a 5.00-liter sample over iodine pentoxide heated at 150°C:

$$I_2O_5 + 5CO \rightleftharpoons 5CO_2 + I_2$$

The I_2 distilled at this temperature and was collected in a solution of iodide ion. The resulting triiodide was titrated with 12.9 ml of 0.00196-N thiosulfate. Calculate the milligrams of CO per liter of the gas.

32. Titration of a 1.19-g sample of a hair-setting preparation required 22.1 ml of 0.0545-N I_2. Calculate the percent thioglycolic acid (gfw = 92.1) in the sample. The reaction was

$$2HSCH_2COOH + I_2 \rightarrow HOOC-CH_2-S-S-CH_2-COOH + 2HI$$

*33. The ethyl mercaptan concentration of a mixture was determined by shaking a 2.12-g sample with 50.0 ml of 0.143-N iodine in a tightly stoppered flask:

$$2C_2H_5SH + I_2 \rightarrow C_2H_5SSC_2H_5 + 2I^- + 2H^+$$

The excess iodine was back-titrated with 9.42 ml of 0.101-N thiosulfate. Calculate the percent ethyl mercaptan in the sample.

34. A 0.404-g sample containing $BaCl_2 \cdot 2H_2O$ was dissolved and an excess of a K_2CrO_4 solution added. After a suitable period the $BaCrO_4$ was filtered, washed, and redissolved in HCl to convert the CrO_4^{2-} to $Cr_2O_7^{2-}$. An excess of KI was added, and the liberated iodine was titrated with 31.5 ml of 0.109-N thiosulfate. Calculate the percent $BaCl_2 \cdot 2H_2O$ in the sample.

*35. A 50.0-ml aliquot of a solution containing I_2 and KI required a 36.6-ml titration with 0.0875-N $Na_2S_2O_3$. Titration of 29.50 ml of the sample in strong HCl solution required 41.20 ml of 0.0200-F KIO_3. Calculate the milligrams of I_2 and KI in each milliliter of the sample (see p. 330 for the reaction).

36. Thallium(I) is quantitatively oxidized to the trivalent state by permanganate in the presence of excess fluoride ion:

$$2Tl^+ + MnO_4^- + 4F^- + 8H^+ \rightarrow 2Tl^{3+} + MnF_4^- + 4H_2O$$

Calculate the percentage of Tl_2SO_4 in a rodenticide preparation if a 10.7-g sample, after suitable pretreatment, required a 27.8-ml titration with a permanganate solution that was 0.0556 N when standardized against sodium oxalate.

*37. A 0.255-g sample of rocket fuel was diluted to 500 ml in a volumetric flask. Titration of the hydrazine in a 50.0-ml aliquot of the resulting solution required 39.4 ml of standard 0.0790-N potassium bromate. The reaction was

$$6H_2NNH_2 + 4BrO_3^- \rightarrow 6N_2 + 4Br^- + 12H_2O$$

Calculate the percentage of hydrazine in the sample.

38. A 20.0-ml sample of dinner wine was diluted to 1.00 liter in a volumetric flask. The ethanol in a 25.0-ml aliquot of the diluted solution was distilled into a receiver containing 50.0 ml of 0.1150-N $K_2Cr_2O_7$ in sulfuric acid. The reaction was

$$3CH_3CH_2OH + 2Cr_2O_7^{2-} + 16H^+ \rightarrow 3CH_3COOH + 4Cr^{3+} + 11H_2O$$

After oxidation of the alcohol was complete, the excess dichromate was titrated with 7.73 ml of standard 0.0643-N Fe(II). Calculate the weight-volume percentage of ethanol in the sample.

*39. The MnO_2 in a 0.305-g specimen of the mineral pyrolusite was permitted to react with an unmeasured excess of potassium iodide. The reaction was

$$2I^- + MnO_2(s) + 4H^+ \rightarrow I_2 + Mn^{2+} + 2H_2O$$

The liberated iodine required a 42.6-ml titration with 0.0560-N $Na_2S_2O_3$. Calculate the percentage of MnO_2 in the specimen.

40. A random sample of iron-supplement tablets was ground and thoroughly homogenized. A 9.47-g sample was treated to destroy the organic matter and free the iron. After conversion to the divalent state with tin(II) chloride (p. 323), the Fe(II) was titrated with 24.2 ml of standard 0.0497-N dichromate. Calculate the percentage of iron in the sample.

14 potentiometric methods

In Chapter 12 it was shown that the concentration (or, more correctly, the activity) of one or more species in a solution will determine the potential of an electrode that is responsive to such species. Thus potential measurements can be used to provide quantitative information about the composition of many solutions.

The potentiometric method consists of measuring the potential between a pair of suitable electrodes immersed in the solution to be analyzed. The apparatus required consists of an *indicator electrode*, a *reference electrode*, and a device for measuring potential.

Potential Measurement

The electromotive force developed by a galvanic cell cannot be measured accurately by placing a simple dc voltmeter across the electrodes because a significant current is required for operation of the meter. Accompanying this current is a voltage decrease due in part to changes in the concentrations of the reacting species as the cell discharges. In addition, the cell has an internal resistance that results in the development of an ohmic potential (equal to the current times the resistance) that opposes the potential of the two electrodes. Thus the measured potential is less than the actual cell potential. A truly significant value for the output of a cell can be attained only if the measurement is made with a negligible current. A *potentiometer* is one type of instrument that meets this specification.

POTENTIAL MEASUREMENTS WITH A POTENTIOMETER

Figure 14-1 is a schematic diagram of a simple potentiometer. The working battery, E_b, is connected to the terminals of a linear voltage divider, AB, whose electrical resistance, R_{AC}, from one end A to any point C is directly proportional to the length, AC, of the resistor; that is, $R_{AC} = kAC$, where k is a proportionality constant. In its simplest form the divider consists of a uniform resistance wire mounted on a meter stick. A sliding contact permits variation of the distance between A and C and thus the output voltage. More conveniently, the divider is a precision, wire-wound resistor formed in a helical coil. A continuously variable tap moves from one end

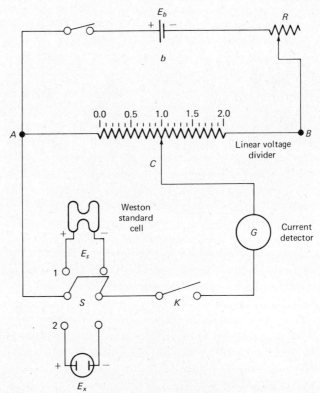

FIGURE 14-1 Circuit diagram for a laboratory potentiometer.

of the helix to the other to provide a variable voltage. Ordinarily the divider is powered by dry cells or mercury batteries that provide a potential, E_b, somewhat larger than that which is to be measured.

The potentiometer is also equipped with a sensitive current-detecting device, G, which may be either a galvanometer or a dc amplifier; a tapping key, K, by which the circuit can be momentarily closed; and a double-pole, double-throw switch, S, to permit the insertion of the unknown cell, E_x, or a standard cell of known potential, E_s, into the circuit.

Principles of the Potentiometer. A current, I, flows continuously from the battery through AB, causing a potential drop between A and B that, according to Ohm's law, is given by $E_{AB} = IR_{AB}$. The current passing between A and C is also I. Thus the potential drop between A and C is given by $E_{AC} = IR_{AC}$. We have noted, however, that the resistance along the resistor is linear. Thus we may write

$$E_{AB} = IR_{AB} = kIAB$$

$$E_{AC} = IR_{AC} = kIAC$$

Dividing the first equation by the second and rearranging yield

$$E_{AC} = E_{AB} \frac{AC}{AB} \tag{14-1}$$

If E_{AC} is greater than E_x (or E_s), electrons will be forced from right to left through the unknown cell when K is closed; the direction of flow will be reversed if E_{AC} is smaller. When E_{AC} is exactly equal to E_x, no current will exist in the circuit containing G and K. Note, however, that electrons continue to flow through AB under these circumstances.

When the standard cell and the unknown cell are placed in the circuit, Equation 14-1 becomes, respectively,

$$E_s = E_{AC_s} = E_{AB} \frac{AC_s}{AB}$$

and

$$E_x = E_{AC_x} = E_{AB} \frac{AC_x}{AB}$$

where AC_s and AC_x represent the linear distances corresponding to balance for the two cases. Dividing the two equations and rearranging give

$$E_x = E_s \frac{AC_x}{AC_s} \tag{14-2}$$

Thus E_x may be obtained from the two measured distances and the known emf of the standard cell.

The Weston Standard Cell. A common standard cell for potentiometers is the *Weston cell*, which can be represented by

$$Cd(Hg)|CdSO_4 \cdot \tfrac{8}{3}H_2O(\text{sat'd}), Hg_2SO_4(\text{sat'd})|Hg$$

where Cd(Hg) represents a solution of cadmium in mercury. The half-reactions, as they occur when current is drawn, are

$$Cd(Hg) \rightarrow Cd^{2+} + Hg(l) + 2e$$

$$Hg_2^{2+} + 2e \rightarrow 2Hg(l)$$

The potential of the Weston cell is 1.0183 V at 25°C.

For convenience the scale reading of AC_x is ordinarily calibrated to give the potential directly in volts by first switching the Weston cell into the circuit and positioning the contact so that AC_s corresponds numerically to the potential of this standard cell. The potential across AB is then adjusted by means of R until no current is indicated by the galvanometer. With the potentiometer adjusted thus, AC_s and E_s will be numerically equal; from Equation 14-2, then, AC_x and E_x will also be identical when balance is achieved with the unknown cell in the circuit.

The need for a working battery E_b may be questioned. In principle there is nothing to prevent a direct measurement of E_x by replacing E_b with the standard cell E_s. It must be remembered, however, that electrons are being continuously drawn from E_b; a standard cell would not maintain a constant potential for long under such usage.

Accuracy of a Potentiometer. The accuracy of a voltage measurement with a potentiometer equipped with a linear voltage divider depends upon several factors. For example, it is necessary to assume that the voltage of the working battery E_b remains constant during the time required to balance the instrument against the standard cell and to measure the potential of the unknown cell. This assumption does not ordinarily lead to appreciable error, provided E_b consists of one or two heavy-duty dry cells, mercury cells, or a lead storage battery. Nevertheless, the instrument should be calibrated against the standard cell before each voltage measurement to compensate for possible changes in the potential of E_b.

The linearity of the resistance AB and the precision with which distances along its length can be estimated both contribute to the accuracy of the potentiometer. Ordinarily, however, the ultimate precision of a good-quality instrument is determined by the sensitivity of the galvanometer relative to the resistance of the circuit. Suppose, for example, that the electrical resistance of the galvanometer plus the unknown cell is 1000 Ω (ohm), a typical figure. Furthermore, if we assume that the galvanometer is capable of detecting a minimum current of 1 μA (10^{-6} A), we can calculate from Ohm's law that the smallest distinguishable voltage difference will be $10^{-6} \times 1000 = 10^{-3}$ V, or 1 mV. By use of a galvanometer sensitive to 10^{-7} A, a difference of 0.1 mV will be detectable. A sensitivity of this order is found in an ordinary pointer-type galvanometer. More refined instruments with sensitivities up to 10^{-10} A are not uncommon.

From this discussion it is seen that the sensitivity of a potentiometric measurement decreases as the electrical resistance of the cell increases. As a matter of fact, potentials of cells with resistance much greater than a megohm (1 MΩ = $10^6\,\Omega$) cannot be measured accurately with an instrument employing a galvanometer as the current-sensing device.

Potentiometer with Electronic Amplification. The potentials of cells with resistances of several hundred megohms can be accurately measured by a potentiometer circuit like that in Figure 14-1 if the galvanometer G is replaced by an electronic circuit that amplifies the out-of-balance current by several orders of magnitude. The amplified current can then be detected by a rugged milliammeter. Commercial instruments that incorporate this feature are usually called *pH meters* because they have been designed to measure the potentials of cells that contain a high-resistance, pH-sensitive glass electrode. The slide wires of such instruments are calibrated both in millivolts and in the linearly related pH units. Meters of this type can be conveniently employed for measuring the potential of both high- and low-resistance cells.

DIRECT-READING INSTRUMENTS FOR POTENTIAL MEASUREMENTS

Electronic voltmeters can be designed to operate with such small currents (10^{-12} to 10^{-14} A) that the potential of a cell is unaffected when measured with such a device. A number of instrument manufacturers sell direct-reading voltmeters that employ electronic amplification to produce a signal that can be displayed in digital form or on a meter calibrated in both pH and millivolt units. These instruments are applicable to both high- and low-resistance cells and are also called pH meters.

Potentiometric-type pH meters offer somewhat greater accuracy and easier maintenance than most direct-reading instruments. On the other hand, the latter are entirely satisfactory for many applications.

Reference Electrodes

In many electroanalytical applications it is desirable that the half-cell potential of one electrode be known, constant, and completely insensitive to the composition of the solution under study. An electrode that fits this description is called a *reference electrode*. Employed in conjunction with the reference electrode will be an *indicator electrode*, whose response is dependent upon the analyte concentration.

A reference electrode should be easy to assemble and should maintain an essentially constant and reproducible potential in the presence of small currents. Several electrode systems meet these requirements.

CALOMEL ELECTRODES

Calomel half-cells may be represented as follows:

$$\|Hg_2Cl_2(\text{sat'd}), KCl(xF)|Hg$$

where x represents the formal concentration of potassium chloride in the solution. The electrode reaction is given by the equation

$$Hg_2Cl_2(s) + 2e \rightleftharpoons 2Hg(l) + 2Cl^-$$

The potential of this cell will vary with the chloride concentration x, and this quantity must be specified in describing the electrode.

Table 14-1 lists the composition and the electrode potentials for the three most commonly encountered calomel electrodes. Note that each solution is saturated with mercury(I) chloride and that the cells differ only with respect to the potassium chloride concentration. Note also that the potential of the normal calomel electrode is greater than the standard potential for the half-reaction because the chloride ion *activity* in a 1-F solution of potassium chloride is significantly smaller than 1. The last column in Table 14-1 gives expressions that permit the calculation of electrode potentials for calomel half-cells at temperatures t other than 25°C.

The saturated calomel electrode (SCE) is most commonly used by the analytical chemist because of the ease with which it can be prepared. Compared with the other two, its temperature coefficient is somewhat large.

A simple, easily constructed saturated calomel electrode is shown in Figure 14-2. The salt bridge, a tube filled with saturated potassium chloride, provides electric contact with the solution surrounding the in-

TABLE 14-1
Specifications of Calomel Electrodes

NAME	CONCENTRATION OF Hg_2Cl_2	KCl	ELECTRODE POTENTIAL (V) VS. STANDARD HYDROGEN ELECTRODE $[Hg_2Cl_2(s) + 2e \rightleftharpoons 2Hg(l) + 2Cl^-]$
Saturated	Saturated	Saturated	$+0.241 - 6.6 \times 10^{-4}(t-25)$
Normal	Saturated	$1.0\ F$	$+0.280 - 2.8 \times 10^{-4}(t-25)$
Decinormal	Saturated	$0.1\ F$	$+0.334 - 8.8 \times 10^{-5}(t-25)$

FIGURE 14-2 Diagram of a saturated calomel electrode.

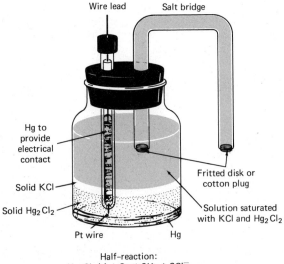

Wire lead

Salt bridge

Hg to provide electrical contact

Solid KCl

Solid Hg_2Cl_2

Pt wire

Hg

Fritted disk or cotton plug

Solution saturated with KCl and Hg_2Cl_2

Half-reaction:
$Hg_2Cl_2(s) + 2e \rightleftharpoons 2Hg + 2Cl^-$

dicator electrode. A fritted disk or a wad of cotton at one end of the salt bridge is often employed to prevent siphoning of the cell liquid and contamination of the solutions by foreign ions; alternatively, the tube can be filled with a 5% agar gel that is saturated with potassium chloride.

Several convenient calomel electrodes are available commercially. Typical of these is the one illustrated in Figure 14-3 that consists of a tube 5 to 15 cm in length and 0.5 to 1.0 cm in diameter. Mercury-mercury(I) chloride paste is contained in an inner tube that is connected to the saturated potassium chloride solution in the outer tube through a small opening. Contact with the second half-cell is made by means of a fritted disk or a porous fiber sealed in the end of the outer tubing. An electrode such as this has a relatively high resistance (2000 to 3000 Ω) and a limited current-carrying capacity.

SILVER-SILVER CHLORIDE ELECTRODES

A system analogous to the calomel electrode consists of a silver electrode immersed in a solution of potassium chloride that has been saturated with silver chloride:

$$\|AgCl(sat'd), KCl(xF)|Ag$$

The half-reaction is

$$AgCl(s) + e \rightleftharpoons Ag(s) + Cl^-$$

Normally this electrode is prepared with a saturated potassium chloride solution; its potential at 25°C is +0.197 V with respect to the standard hydrogen electrode.

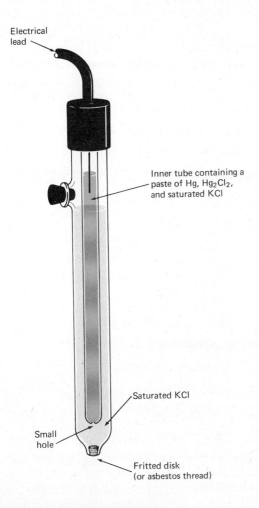

Electrical
lead

FIGURE 14-3 Diagram of a typical saturated calomel electrode.

Inner tube containing a
paste of Hg, Hg_2Cl_2,
and saturated KCl

Saturated KCl

Small
hole

Fritted disk
(or asbestos thread)

Metal Indicator Electrodes

Metallic indicator electrodes are constructed from coils of wire, flat metal plates, or heavy cylindrical billets. Generally the exposure of a large surface area to the solution will assure rapid attainment of equilibrium. Thorough cleaning of the metal surface before use is often important; a brief dip in concentrated nitric acid followed by several rinsings with distilled water is satisfactory for many metals.

CATION ELECTRODES

Several metals (such as silver, copper, mercury, lead, and cadmium) that show reversible half-reactions are satisfactory for the construction of indicator electrodes. The potentials of these metals reproducibly and predictably reflect the activities of their ions in solution. In contrast, some metals are not very satisfactory indicator electrodes because they tend to develop nonreproducible potentials that are influenced by strains or crystal deformations in their structures and by oxide coatings on their surfaces. Metals in this category include iron, nickel, cobalt, tungsten, and chromium.

SECOND-ORDER ELECTRODES FOR ANIONS

In addition to serving as an indicator electrode for its own cations, a metal electrode is also indirectly responsive to anions that form slightly soluble precipitates with cations of the metal. For this application it is only necessary to saturate the solution under study with the sparingly soluble salt. For example, the potential of a silver electrode will accurately reflect the concentration of iodide ion in a solution that is saturated with silver iodide. Here two equilibria govern the electrode behavior:

$$AgI(s) \rightleftharpoons Ag^+ + I^-$$

$$Ag^+ + e \rightleftharpoons Ag(s) \qquad E^0_{Ag} = 0.799 \text{ V}$$

Combining these equations (p. 285) yields

$$AgI(s) + e \rightleftharpoons Ag(s) + I^- \qquad E^0_{AgI} = -0.151 \text{ V}$$

Application of the Nernst equation to this half-reaction gives the relationship between the electrode potential and the anion concentration. Thus

$$E = -0.151 - 0.0591 \log [I^-]$$

A silver electrode serving as an indicator for iodide is an example of an *electrode of the second order* because it measures the concentration of an ion that is not directly involved in the electron-transfer process. The same electrode used as an indicator for silver ion functions as an *electrode of the first order* because its potential is directly dependent upon a participant in the electrode process.

An important second-order electrode for measuring the amount of the EDTA anion Y^{4-} is based upon the response of a mercury electrode in the presence of a small concentration of the stable EDTA complex of Hg(II) ion. The half-reaction for the electrode process can be written as

$$HgY^{2-} + 2e \rightleftharpoons Hg(l) + Y^{4-} \qquad E^0 = 0.21 \text{ V}$$

for which

$$E = 0.21 - \frac{0.0591}{2} \log \frac{[Y^{4-}]}{[HgY^{2-}]}$$

In employing this electrode system a small concentration of HgY^{2-} is introduced into the analyte solution at the outset. Because the complex is so very stable (for HgY^{2-}, $K_f = 6.3 \times 10^{21}$), dissociation to form Hg^{2+} is minimal; consequently the concentration of the complex remains essentially constant over a wide concentration range of Y^{4-}. Thus the foregoing equation can be written in the form

$$E = K - \frac{0.0591}{2} \log [Y^{4-}]$$

where the constant K is equal to

$$K = 0.21 - \frac{0.0591}{2} \log \frac{1}{[HgY^{2-}]}$$

This second-order electrode is useful for establishing end points of EDTA titrations.

INDICATORS FOR REDOX SYSTEMS

Electrodes fashioned from platinum or gold serve as indicator electrodes for oxidation-reduction systems. Of itself such an electrode is inert; the potential it develops depends solely upon the potential of the oxidation-reduction systems of the solution in which it is immersed.

Membrane Indicator Electrodes[1]

For many years the most convenient method for determining pH has involved measurement of the potential that develops across a thin glass membrane that separates two solutions with different hydrogen ion concentrations. The phenomenon, first reported by Cremer,[2] has been extensively studied by many investigators; as a result, the sensitivity and selectivity of glass membranes to pH is reasonably well understood. Furthermore, membrane electrodes have now been developed for the direct potentiometric determination of other ions such as K^+, Na^+, Li^+, F^-, and Ca^{2+}.[3]

It is convenient to divide membrane electrodes into four categories based upon the membrane composition. These include (1) glass electrodes, (2) liquid-membrane electrodes, (3) solid-state or precipitate electrodes, and (4) gas-sensing membrane electrodes. We shall consider the properties and behavior of the glass electrode in particular detail because of both its present and its historical importance.

[1]For further information on this topic see R. A. Durst, ed., *Ion-Selective Electrodes*, National Bureau of Standards Special Publication 314. Washington, D.C.: U.S. Government Printing Office, 1969.

[2]M. Cremer, *Z. Biol.*, **47**, 562 (1906).

[3]See G. A. Rechnitz, *Chem. Eng. News*, **45** (25), 146 (1967).

THE GLASS ELECTRODE FOR pH MEASUREMENT

The experimental observation fundamental to the development of the glass electrode is that a potential difference develops across a thin conducting glass membrane interposed between solutions of different pH. The potential can be detected by placing reference electrodes of known and *constant potential* in the solutions on each side of the membrane, as shown schematically in Figure 14-4.

The glass electrode itself consists of a very thin glass bulb formed by blowing out a sealed, heavy-walled soft-glass tube; see Figure 14-4. The solution held in this bulb is $1\,F$ in NaCl and saturated with AgCl. In addition, the solution contains sufficient acid to give a hydronium ion activity of a_2. Note that the solution on the outside of the membrane is identical in composition except that it has a hydronium ion activity of a_1. A silver wire is immersed in each of these solutions, thus forming two silver-silver chloride reference electrodes (p. 356) whose potentials are given by

$$E_{\text{ref}_1} = E_{\text{ref}_2} = E^0 - 0.0591 \log [\text{Cl}^-]$$

Because the chloride ion concentrations are identical, the two reference-electrode potentials are the same.

Over a half-century ago it was discovered empirically that a

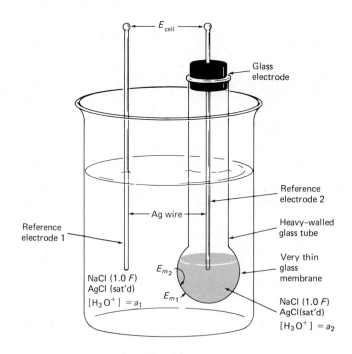

FIGURE 14-4 A cell containing a glass electrode.

potential, E_{cell}, develops between these two electrodes whenever a_1 and a_2 differ. The magnitude of this potential increases as the difference between a_1 and a_2 becomes larger. *Because the potentials of the two reference electrodes are identical and independent of* $[H_3O^+]$, *the observed potential must have its origins across the glass membrane.*

It is useful to picture this potential as the difference of two potentials, each being associated with one of the membrane surfaces. Thus

$$E_{cell} = (E_{ref_1} + E_{m_1}) - (E_{ref_2} + E_{m_2})$$

where E_{m_1} and E_{m_2} are surface potentials on the respective sides of the membrane. Because the two reference potentials are identical,

$$E_{cell} = E_{m_1} - E_{m_2} \qquad (14\text{-}3)$$

It is possible to demonstrate experimentally (and also derive theoretically) that

$$E_{m_1} = k_1 + \frac{0.0591}{1} \log a_1 \qquad (14\text{-}4)$$

and

$$E_{m_2} = k_2 + \frac{0.0591}{1} \log a_2 \qquad (14\text{-}5)$$

where k_1 and k_2 are constants. Substituting these relationships into Equation 14-3 gives

$$E_{cell} = k_1 + \frac{0.0591}{1} \log a_1 - k_2 - 0.0591 \log a_2 \qquad (14\text{-}6)$$

If we now *specify* that a_2, the hydronium ion activity inside the bulb, will remain constant, then Equation 14-6 converts to

$$E_{cell} = k + 0.0591 \log a_1 = k - 0.0591 \, pH \qquad (14\text{-}7)$$

where k is a new constant, whose numerical value can be ascertained experimentally by measuring E_{cell} for a solution in which a_1 is known from the way it was prepared.

The half-cell in Figure 14-4 consisting of a silver–silver chloride electrode and a solution of known hydronium activity contained in a thin glass bulb is a glass electrode. When it is combined with a second reference electrode, a cell is formed whose potential depends on a_1. Thus this cell can be employed for determining a_1.

The cell shown in Figure 14-4 is not practical for measuring pH in that it would require the solution in the outer compartment (whose pH is sought) to be saturated with AgCl and to have a chloride ion concentration of unity. To avoid this problem an arrangement such as the commercially available one shown in Figure 14-5 is commonly employed. The pH-sensitive membrane of the glass electrode is sealed on the end of the heavy-walled tubing and the resulting bulb contains a solution of HCl (often $0.1 \, F$) saturated with AgCl. The external reference is usually a

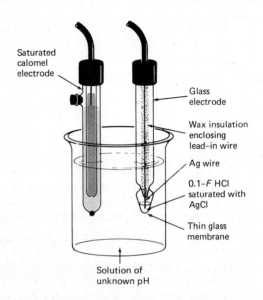

FIGURE 14-5 Typical electrode system for measuring pH.

Saturated calomel electrode

Glass electrode

Wax insulation enclosing lead-in wire

Ag wire

0.1-*F* HCl saturated with AgCl

Thin glass membrane

Solution of unknown pH

saturated calomel electrode (p. 355) as shown. In contrast to the cell shown in Figure 14-4 the two reference-electrode potentials are no longer identical. Thus k in Equation 14-7, while being constant, would now contain two additional terms and be equal to

$$k = k_1 - k_2 + E_{SCE} - E_{Ag,AgCl} - 0.0591 \log a_2$$

Composition of Glass Membranes. Much systematic investigation has been devoted to the effects of glass composition on the sensitivity of membranes to protons and other cations, and a variety of compositions are now used commercially.[4] For many years Corning 015 glass (consisting of approximately 22% Na_2O, 6% CaO, and 72% SiO_2) has been widely used. This glass shows an excellent specificity toward hydrogen ions up to a pH of about 9. At higher pH values, however, the membrane becomes somewhat sensitive to sodium and other alkali ions.

It has been shown that the surfaces of a glass must be hydrated in order for the membrane to function as a pH electrode. Nonhygroscopic glasses such as Pyrex and quartz show no pH function. Even Corning 015 glass shows little pH response after dehydration by storage over a desiccant, but its sensitivity is restored after standing for a few hours in water. This involves absorption of approximately 50 mg of water per cubic centimeter of hydrated glass.

It has been demonstrated experimentally that hydration of a pH-

[4]For a summary of this work see J. O. Isard, "The Dependence of Glass-Electrode Properties on Composition" in G. Eisenman, ed., *Glass Electrodes for Hydrogen and Other Cations.* New York: Marcel Dekker, 1967, Chapter 3.

sensitive glass membrane is accompanied by a chemical reaction in which cations of the glass are exchanged for protons of the solution. This exchange involves singly charged cations almost exclusively, inasmuch as the di- and trivalent cations in the silicate structure are much more strongly bonded. The ion-exchange can be written as

$$H^+ + Na^+Gl^- \rightleftharpoons Na^+ + H^+Gl^- \qquad \text{(14-8)}$$
$$\text{soln} \quad\ \text{solid} \qquad\ \text{soln} \quad\ \text{solid}$$

where Gl^- represents a bonding site in the glass. The equilibrium constant for this process favors incorporation of hydrogen ions into the silicate lattice; as a result the surface of a well-soaked membrane will ordinarily consist of a silicic acid gel. In highly alkaline solutions, however, sodium ions will occupy an appreciable number of the available bonding sites.

Electrical Resistance of Glass Membranes. The membrane in a typical commercial glass electrode will be between 0.03 and 0.1 mm in thickness and will have an electrical resistance of 50 to 500 MΩ. Current conduction across the membrane involves migration of singly charged cations. Across each solution interface passage of charge involves a transfer of protons; the direction of migration is from glass to solution at one interface and from solution to glass at the other. That is,

$$H^+Gl^- \rightleftharpoons Gl^- + H^+$$
$$\text{solid} \qquad \text{solid} \quad\ \text{soln}$$

or

$$H^+ + Gl^- \rightleftharpoons H^+Gl^-$$
$$\text{soln} \quad\ \text{solid} \qquad \text{solid}$$

The position of these two equilibria is determined by the hydrogen ion concentrations in the two solutions. When these positions differ, the surface at which the greater dissociation has occurred will be negative with respect to the other surface. Thus a potential develops, and it is this potential that is being measured. The potential, then, can be thought of as a measure of the relative driving force of the dissociation processes occurring at the two surfaces.

Conduction in the two silicic acid gel layers (which may have a thickness of 10^{-5} mm or more) is due exclusively to migration of hydrogen ions. In the dry center region of the membrane sodium ions take over this function.

Asymmetry Potential. If identical solutions and identical reference electrodes are placed on each side of the membrane shown in Figure 14-4, the measured potential of the cell should be zero. It is observed, however, that a small potential, called the *asymmetry potential*, often does develop when this experiment is performed. Moreover, the asymmetry potential associated with a given glass electrode changes slowly with time.

The causes of the asymmetry potential are obscure, but they undoubtedly include such factors as differences in strains established within the two surfaces during manufacture, mechanical and chemical attack of the external surface during use, and contamination of the outer surface by grease films and other adsorbed substances. The effect of the asymmetry potential on a pH measurement is eliminated by frequent calibration of the electrode against a standard buffer of known pH.

The Alkaline and Acid Errors. In solutions containing very low hydrogen ion concentrations (pH > 9) some glass membranes respond not only to changes in hydrogen ion concentration but also to the concentration of alkali-metal ions. The magnitude of the resulting error is indicated on the right-hand side of Figure 14-6. Note that the pH error is negative, indicating that the electrode is responding to sodium ions. Note also that this response becomes greater with decreasing hydrogen ion concentration; the error also increases with increasing sodium ion concentrations.

All singly charged cations cause alkaline errors; the magnitude of the error, however, depends upon the cation as well as the composition of the glass. Also shown in Figure 14-6 is the error associated with a commercial electrode designed especially for use in alkaline solutions.

Alkaline error is attributable to reversal of the reaction shown in Equation 14-8. At low hydronium and high alkali ion concentrations the equilibrium is shifted so that a significant fraction of the glass surface becomes populated by sodium ions. Under these circumstances the potential responds to both sodium and hydronium ions. Membranes that have a lower response to sodium ions (see dashed line in Figure 14-6) are fabricated from a glass in which the equilibrium constant for the reaction in Equation 14-8 is particularly large; thus replacement of hydronium ions on the surface by sodium ions is less favored.

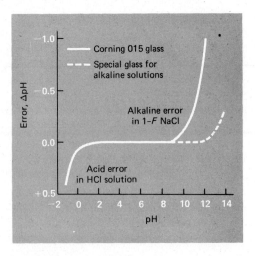

FIGURE 14-6 Alkaline and acid error with typical glass electrodes.

As shown in Figure 14-6, the typical glass electrode exhibits an error, opposite in sign to the alkaline error, in solutions of pH less than 1. As a consequence, pH readings tend to be too high in this region. The magnitude of the error is dependent upon many variables and is generally not very reproducible. The causes of the acid error are not well understood.

GLASS ELECTRODES FOR THE DETERMINATION OF OTHER CATIONS

The existence of the alkaline error led to studies concerned with the effect of membrane composition upon the magnitude of this error. One outcome of this work has been the development of glasses that have a negligible alkaline error below a pH of about 12. Another has been the discovery of glasses with responses that are dependent upon the concentration of cations other than the hydronium ion and are essentially insensitive to changes in pH. Glass electrodes for potassium ion and sodium ion are now available commercially.

LIQUID MEMBRANE ELECTRODES

Liquid membranes owe their response to the potential that is established across the interface between the solution to be analyzed and an immiscible liquid that selectively bonds with the ion being determined. Liquid membrane electrodes permit the direct potentiometric determination of the activities of several polyvalent cations and certain anions as well.

A liquid membrane electrode differs from a glass electrode only in that the solution of known and fixed activity is separated from the analyte by a thin layer of an immiscible organic liquid instead of a thin glass membrane. As shown schematically in Figure 14-7 a porous, hydrophobic (that is, water-repelling), plastic disk serves to hold the organic layer between the two aqueous solutions. By wick action the pores of the disk or membrane are kept filled with the organic liquid from the reservoir in the outer of the two concentric tubes. The inner tube contains an aqueous standard solution of MCl_2, where M^{2+} is the cation whose activity is to be determined. This solution is also saturated with AgCl to form a Ag-AgCl reference electrode with the silver lead wire.

The organic liquid is a nonvolatile, water-immiscible, organic ion exchanger that contains acidic, basic, or chelating functional groups. Between the liquid and an aqueous solution containing a divalent cation there is established an equilibrium that can be represented as

$$\underset{\substack{\text{organic} \\ \text{phase}}}{RH_{2x}} + \underset{\substack{\text{aqueous} \\ \text{phase}}}{xM^{2+}} \rightleftharpoons \underset{\substack{\text{organic} \\ \text{phase}}}{RM_x} + \underset{\substack{\text{aqueous} \\ \text{phase}}}{2xH^+}$$

FIGURE 14-7 Liquid membrane electrode sensitive to M^{2+}.

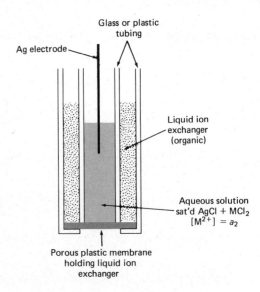

FIGURE 14-7 Liquid membrane electrode sensitive to M^{2+}.

By repeated treatment the exchange liquid can be converted essentially completely to the cationic form, RM_x; it is this form that is employed in electrodes for the determination of M^{2+}.

In order to determine the pM of a solution, the electrode shown in Figure 14-7 is immersed in a solution of the analyte which also contains a reference electrode—usually a saturated calomel electrode. The potential between the external and the internal reference electrodes is proportional to the pM of the analyte.

Figure 14-8 shows construction details of a commercial liquid membrane electrode that is selective for calcium ion. The ion exchanger is an aliphatic diester of phosphoric acid dissolved in a polar solvent. The chain lengths of the aliphatic groups in the former range from 8 to 16 carbon atoms. The diester contains a single acidic proton; thus two

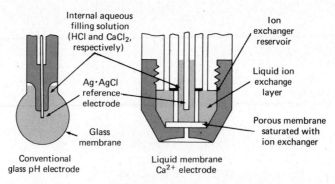

FIGURE 14-8 Comparison of a liquid membrane calcium ion electrode with a glass electrode. (Courtesy of Orion Research, Inc.)

molecules are required to bond a divalent cation, here, calcium. It is this affinity for calcium ions that imparts the selective properties to the electrode. The internal aqueous solution in contact with the exchanger contains a fixed concentration of calcium chloride; a silver–silver chloride reference electrode is immersed in this solution. When used for a calcium ion determination the porous disk containing the ion-exchange liquid separates the solution to be analyzed from the reference calcium chloride solution. The equilibrium established at each interface can be represented as

$$[(RO)_2POO]_2Ca \rightleftharpoons 2(RO)_2POO^- + Ca^{2+}$$

$$\text{organic} \qquad\qquad \text{organic} \qquad \text{aqueous}$$

The potential of this electrode is given by an equation analogous to Equation 14-7 for the glass electrode; that is,

$$E = k + \frac{0.0591}{2} \log a_1 \tag{14-9}$$

where a_1 here is the activity of Ca^{2+}.

The performance of the calcium electrode just described is reported to be independent of pH in the range between 5.5 and 11. At lower pH levels hydrogen ions undoubtedly exchange with calcium ions on the exchanger to a significant extent. The electrode then becomes pH- as well as pCa-dependent. Its sensitivity to calcium ion exceeds that for magnesium ion by a factor of 50 and that for sodium or potassium by a factor of 1000. It can be employed to measure calcium ion activities as low as $10^{-5} M$.

The calcium ion membrane electrode has proved to be a valuable tool for physiological studies because this ion plays important roles in nerve conduction, bone formation, muscle contraction, cardiac conduction and contraction, and renal tubular function. At least some of these processes are influenced more by calcium ion activity than calcium concentration; activity, of course, is the parameter measured by the electrode.

Another specific ion electrode of great consequence in physiological studies is that for potassium because the transport of nerve signals appears to involve movement of this ion across nerve membranes. Study of the process requires an electrode which can detect small concentrations of potassium ion in the presence of much larger concentrations of sodium. A number of liquid membrane electrodes show promise of meeting these needs. One employs a solution of valinomycin in diphenyl ether. Valinomycin is an antibiotic with a cyclical ether structure that has a much stronger affinity for potassium ion than sodium ion. A liquid membrane consisting of a valinomycin solution is found to be about 10^4 times as responsive to potassium ion as sodium ion.

Liquid membrane electrodes are commercially available for Ca^{2+}, Cl^-, BF_4^-, NO_3^-, ClO_3^-, K^+, and Mg^{2+} ions.

SOLID-STATE OR PRECIPITATE ELECTRODES

Considerable work has been devoted recently to the development of solid membranes that are selective toward anions in the way that some glasses behave toward specific cations. As we have seen, the selectivity of a glass membrane owes its origin to the presence of anionic sites on its surface that show particular affinity toward certain positively charged ions. By analogy, a membrane having similar cationic sites might be expected to respond selectively toward anions. To exploit this possibility, attempts have been made to prepare membranes of salts containing the anion of interest and a cation that selectively precipitates that anion from aqueous solutions; for example, barium sulfate has been proposed for sulfate ion and silver halides for the various halide ions. The problem encountered in this approach has been in finding methods for fabricating membranes from the desired salt with adequate physical strength, conductivity, and resistance to abrasion and corrosion.

Membranes prepared from cast pellets of silver halides have been successfully used in electrodes for the selective determination of chloride, bromide, and iodide ions. Other commercially available solid-state electrodes have been offered for the determination of Cu^{2+}, Pb^{2+}, Ag^+, Cd^{2+}, S^{2-}, and SCN^- ions.

A solid-state electrode selective for fluoride ion has been described. The membrane consists of a single crystal of lanthanum fluoride that has been doped with a rare earth to increase its electrical conductivity. The membrane, supported between a reference solution and the solution to be measured, shows the theoretical response to changes in fluoride ion activity (that is, $E = K + 0.0591 \log a_{F^-}$) to as low as $10^{-6} M$. The electrode is purported to be selective for fluoride over other common anions by several orders of magnitude. Only hydroxide ion appears to offer serious interference.

GAS-SENSING ELECTRODES

Figure 14-9 is a schematic diagram of a so-called gas-sensing electrode, which consists of a reference electrode, a membrane electrode, and an electrolyte solution housed in a cylindrical, plastic tube. A thin, replaceable, gas-permeable membrane is attached to one end of the tube and serves to separate the internal electrolyte solution from the analyte solution. The membrane is described by its manufacturers as being a thin, microporous film fabricated from a hydrophobic plastic; water and electrolytes are prevented from entering and passing through the pores of the film owing to its water-repellent properties. Thus the pores contain only air or other gases to which the membrane is exposed. When a solution containing a gaseous analyte such as sulfur dioxide is in contact with the membrane,

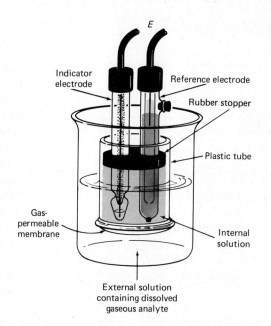

Indicator electrode

Reference electrode

Rubber stopper

Plastic tube

Gas-permeable membrane

Internal solution

External solution containing dissolved gaseous analyte

FIGURE 14-9 Schematic diagram of a gas-sensing electrode. A glass-calomel electrode system is depicted (see Figure 14-5); other ion-specific and/or reference electrodes can be used as well.

the SO_2 distills into the pores as shown by the reaction

$$SO_2(aq) \rightleftharpoons SO_2(g)$$
external   membrane
solution   pores

Because the pores are numerous, a state of equilibrium is rapidly approached. The SO_2 in the pores, however, is also in contact with the internal solution and a second reaction can readily take place; that is,

$$SO_2(g) \rightleftharpoons SO_2(aq)$$
membrane   internal
pores   solution

As a consequence of the two reactions, the film of internal solution adjacent to the membrane rapidly (in a few seconds to a few minutes) equilibrates with the external solution. Further, another equilibrium is established that causes the pH of the internal surface film to change, namely,

$$SO_2(aq) + 2H_2O \rightleftharpoons HSO_3^- + H_3O^+$$

A glass and calomel electrode pair immersed in the film of internal solution (see Figure 14-9) is then employed to detect the pH change.

The overall reaction for the process just described is obtained by adding the three chemical equations to give

$$SO_2(aq) + 2H_2O \rightleftharpoons H_3O^+ + HSO_3^-$$
external   internal
solution   solution

The equilibrium constant for the reaction is given by

$$\frac{[H_3O^+][HSO_3^-]}{[SO_2(aq)]_{ext}} = K$$

If the concentration of HSO_3^- in the internal solution is made relatively high so that its concentration is not altered significantly by the SO_2 which distills, then

$$\frac{[H_3O^+]}{[SO_2(aq)]_{ext}} = \frac{K}{[HSO_3^-]} = K_g \qquad \text{(14-10)}$$

which may be rewritten as

$$a_1 = [H_3O^+] = K_g[SO_2(aq)]_{ext} \qquad \text{(14-11)}$$

where a_1 is the internal hydrogen ion activity.

The potential of the electrode system in the internal solution is dependent upon a_1 as described by Equation 14-7. Substitution of Equation 14-10 and 14-11 into 14-7 yields

$$E = k + 0.0591 \log K_g[SO_2(aq)]_{ext}$$

or

$$E = k' + 0.0591 \log[SO_2(aq)]_{ext}$$

where $k' = k + 0.0591 \log K_g$. Thus the potential of the cell consisting of the internal reference and indicator electrode is determined by the SO_2 concentration of the external solution. Note that *no electrode comes directly in contact* with the analyte solution; for this reason it would perhaps be better to call the device a gas-sensing cell rather than a gas-sensing electrode. Note also that the only species that will interfere with the measurement are dissolved gases that can pass through the membrane and can additionally affect the pH of the internal solution.

It is possible to increase the selectivity of the gas-sensing electrode by employing an internal electrode that is sensitive to a species other than hydrogen ion; for example, a nitrate-sensing electrode could be used to provide a cell that would be sensitive to nitrogen dioxide. Here the equilibrium would be

$$2NO_2(aq) \; + \; H_2O \; \rightleftharpoons \; NO_2^- \; + \; NO_3^- \; + \; 2H^+$$

| external solution | internal solution | | internal solution | |

This electrode should permit the determination of NO_2 in the presence of gases such as SO_2, CO_2, and NH_3 which would also alter the pH of the internal solution.

Gas-sensing electrodes are commercially available for SO_2, NO_2, and NH_3. Others will undoubtedly be developed in the future.

Direct Potentiometric Measurements

Direct potentiometric measurements can be used to complete chemical analyses of those species for which an indicator electrode is available. The technique is simple, requiring only a comparison of the potential developed by the indicator electrode in the analyte solution with its potential when immersed in a standard solution of the analyte; insofar as the response of the electrode is specific for the analyte, no preliminary separation steps are required. In addition, direct potentiometric measurements are readily adapted to the continuous and automatic monitoring of analytical parameters.

Notwithstanding these attractive advantages the user of direct potentiometric measurements must be alert to limitations that are inherent to the method. One of these limitations arises from the existence in most potentiometric measurements of a *liquid junction* potential (see p. 287). For most electroanalytical methods this junction potential is inconsequential and can be neglected. Unfortunately, however, its existence places a limitation on the accuracy that can be attained with direct potentiometric measurements.

EQUATION FOR DIRECT POTENTIOMETRY

The observed potential of a cell employed for a direct potentiometric measurement can be expressed in terms of the potentials developed by the indicator electrode, the reference electrode, and a junction potential. That is,

$$E_{obs} = E_{ref} - E_{ind} + E_j \qquad \textbf{(14-12)}^5$$

Typically the junction potential E_j is made up of two potentials, one arising at the junction of the analyte solution and one end of a salt bridge; a second junction potential is found where the other end of this salt bridge interfaces with the reference electrode solution. The two potentials tend to cancel one another but seldom completely. Thus E_j in Equation 14-12 may be as large as 1 mV or more.

Ideally the potential of the indicator electrode is related to the activity, a_1, of M^{n+}, the ion of interest, by a Nernst-type equation

$$E_{ind} = k + \frac{0.0591}{n} \log a_1 \qquad \textbf{(14-13)}$$

where k is a constant; for many electrodes k is the standard potential for the indicator electrode. With membrane electrodes, however, k may also include an unknown, time-dependent asymmetry potential (p. 363).

---

[5]As written, Equation 14-12 has the reference electrode acting as cathode and the indicator electrode acting as anode. If, in a particular cell, the roles are reversed, the signs of the two electrodes are likewise reversed.

Combination of Equation 14-13 with Equation 14-12 and rearrangement yield

$$pM = -\log a_1 = \frac{E_{obs} - (E_{ref} + E_j - k)}{0.0591/n} \qquad \textbf{(14-14)}$$

$$= \frac{E_{obs} - K}{0.0591/n} \qquad \textbf{(14-15)}$$

The new constant K consists of three constants, of which at least one (E_j) has a magnitude that *cannot be evaluated by calculation*. Thus K must be determined experimentally with the aid of a standard solution of M before Equation 14-15 can be used for the measurement of pM.

Several methods for performing analyses by direct potentiometry have been developed; all are based, directly or indirectly, upon Equation 14-15.[6]

ELECTRODE CALIBRATION METHOD

In the electrode calibration procedure K in Equation 14-15 is determined by measuring E_{obs} for one or more standard solutions of known pM. The assumption is then made that K does not change during measurement of the analyte solution. Generally the calibration operation is performed at the time that pM for the unknown is determined; recalibration may be required if measurements extend over several hours.

The electrode calibration method offers the advantages of simplicity, speed, and applicability to the continuous monitoring of pM. Two important disadvantages attend its use, however. One of these is that the results of an analysis are in terms of activities rather than concentrations; the other is that the accuracy of a measurement obtained by this procedure is limited by the inherent uncertainty caused by the junction potential; unfortunately, this uncertainty can never be totally eliminated.

Activity versus Concentration. Electrode response is related to activity rather than analyte concentration. Ordinarily, however, the scientist is interested in concentration, and the determination of this quantity from a potentiometric measurement requires activity coefficient data. More often than not, activity coefficients will not be available because the ionic strength of the solution is either unknown or so high that the Debye–Hückel equation (Chapter 5) is not applicable. Unfortunately, the assumption that activity and concentration are identical may lead to errors of several percent, particularly when the analyte is polyvalent.

The difference between activity and concentration is shown in

---

[6]For additional information see R. A. Durst, *Ion-Selective Electrodes*. National Bureau of Standards Special Publication 314. Washington, D.C.: U.S. Government Printing Office, 1969, Chapter 11.

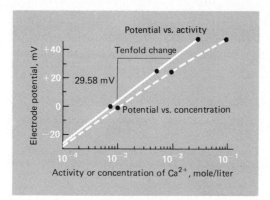

FIGURE 14-10 Response of a calcium ion electrode to variations of concentration and activity in solutions prepared from pure calcium chloride. (Courtesy of Orion Research, Inc.)

Figure 14-10 where the lower curve gives the change in potential of a calcium electrode as a function of chloride concentration (note that the concentration scale is logarithmic). The nonlinearity of the curve is due to the increase in ionic strength and the consequent decrease in the activity coefficient of the calcium ion as the electrolyte concentration becomes larger. When these concentrations are converted to activities, the upper curve is obtained; note that this straight line has the theoretical slope of 0.0296 (0.0591/2).

We have seen (p. 107) that activity coefficients for singly charged ions are less affected by changes in ionic strength than those for species with multiple charges. Thus the effect shown in Figure 14-10 will be less pronounced for electrodes that respond to H^+, Na^+, and other univalent ions.

In potentiometric pH measurements the pH of the standard buffer employed for calibration is generally based on the activity of hydronium ions. Thus the resulting hydrogen ion analysis is also on an activity scale. If the unknown sample has a high ionic strength, the hydronium ion *concentration* will differ appreciably from the activity measured.

Inherent Error in the Electrode Calibration Procedure. A serious disadvantage of the electrode calibration method is the existence of an inherent uncertainty that results from the assumption that K in Equation 14-15 remains constant after calibration. This assumption can seldom, if ever, be exactly true, because the electrolyte composition of the unknown will almost inevitably differ from that of the solutions employed for calibration. The junction potential contained in K will vary slightly as a consequence, even when a salt bridge is used. Often this uncertainty will be of the order of 1 mV or more; it is readily shown (see Problems 19 and 20 at the end of this chapter) that the relative error in activity or concentration associated with a 1 mV uncertainty in K is about 4% when n in Equation 14-15 is 1 and 8% when it is 2. *It is important to appreciate that* *this uncertainty is characteristic of all measurements of cells that contain a salt bridge, and that this error cannot be eliminated by even the most careful measurements of cell potentials;* nor does it appear possible to devise a method for completely eliminating the uncertainty in K that causes this error.

CALIBRATION CURVES FOR DIRECT POTENTIOMETRY

An obvious way of correcting potentiometric measurements to give results in terms of concentration is to make use of an empirical calibration such as the lower curve in Figure 14-10. For this approach to be successful, however, it is essential that the ionic composition of the standard closely approximate that of the analyte—a condition that is difficult to realize experimentally for complex samples.

Where electrolyte concentrations are not too great, it is often helpful to swamp both the samples and the calibration standards with a measured excess of an inert electrolyte. Under these circumstances the added effect of the electrolyte in the sample becomes negligible, and the empirical calibration curve yields results in terms of concentration. This approach has been employed for the potentiometric determination of fluoride in public water supplies. Here both samples and standards are diluted on a 1:1 basis with a solution containing sodium chloride, a citrate buffer, and an acetate buffer; the diluent is sufficiently concentrated so that the samples and standards do not differ significantly in ionic strength. The procedure permits a rapid measurement of fluoride ion in the 1 ppm range with a precision of about 5% relative.

STANDARD ADDITION METHOD

In the standard addition method the potential of the electrode system is measured before and after addition of a small volume of a standard to a known volume of the sample. The assumption is made that this addition does not alter the ionic strength and thus the activity coefficient f of the analyte. It is further assumed that the added standard does not alter the junction potential significantly.

If the potentials before and after the addition are E_1 and E_2, respectively, we may write with the aid of Equation 14-15

$$-\log C_x f = \frac{E_1 - K}{0.0591/n}$$

where $C_x f$ is equal to a_1, the activity of the analyte in the sample, f is its activity coefficient, and C_x is its molar concentration. If now V_s ml of a standard with a molar concentration of C_s is added to V_x ml of the sample, the potential will become E_2 and Equation 14-15 becomes

$$-\log\left[\frac{C_xV_x + C_sV_s}{V_x + V_s}f\right] = \frac{E_2 - K}{0.0591/n}$$

Subtracting the first equation from the second gives, upon rearrangement,

$$\log\left[\frac{C_x(V_x + V_s)}{C_xV_x + C_sV_s}\right] = \frac{E_2 - E_1}{0.0591/n} = \frac{\Delta E}{0.0591/n} \qquad \textbf{(14-16)}$$

Thus C_x is readily calculated from the concentration of the standard, the two volumes, and the potential difference ΔE.

The standard addition method has been applied to the determination of chloride and fluoride in samples of commercial phosphors.[7] Here two indicator electrodes and a reference electrode were used; the added standard contained known quantities of the two anions. The relative standard deviation for the measurement of replicate samples was found to be 0.7% for fluoride and 0.4% for chloride. When the standard addition method was not used, relative errors for the analyses appeared to range between 1 and 2%.

POTENTIOMETRIC pH MEASUREMENTS WITH A GLASS ELECTRODE[8]

The glass electrode is unquestionably the most important indicator electrode for hydronium ion. It is convenient to use and subject to few of the interferences that affect other pH-sensing electrodes.

Glass electrodes are available at relatively low cost and in many shapes and sizes. A common variety is illustrated in Figure 14-5; as shown here, the reference electrode is usually a commercial saturated calomel electrode.

The glass-calomel electrode system is a remarkably versatile tool for the measurement of pH under many conditions. The electrode can be used without interference in solutions containing strong oxidants, reductants, proteins, and gases; the pH of viscous or even semisolid fluids can be determined. Electrodes for special applications are available. Included among these are small electrodes for pH measurements in a drop (or less) of solution or in a cavity of a tooth, microelectrodes which permit the measurement of pH inside a living cell, systems for insertion in a flowing liquid stream to provide a continuous monitoring of pH, and a small glass electrode that can be swallowed to indicate the acidity of the stomach contents (the calomel electrode is kept in the mouth).

[7]L. G. Bruton, *Anal. Chem.*, **43**, 579 (1971).

[8]For a detailed discussion of potentiometric pH measurements see R. G. Bates, *Determination of pH, Theory and Practice*. New York: Wiley, 1964.

Summary of Errors Affecting pH Measurements with the Glass Electrode. The ubiquity of the pH meter and the general applicability of the glass electrode tend to lull the chemist into the attitude that any measurement obtained with such an instrument is surely correct. It is well to guard against this false sense of security since there are distinct limitations to the electrode system. These have been discussed in earlier sections and include the following:

1. *The alkaline error.* The ordinary glass electrode becomes somewhat sensitive to alkali-metal ions at pH values greater than 9.
2. *The acid error.* At a pH less than zero, values obtained with a glass electrode tend to be somewhat high.
3. *Dehydration.* Dehydration of the electrode may cause unstable performance and errors.
4. *Errors in unbuffered neutral solutions.* Equilibrium between the electrode surface layer and the solution is achieved only slowly in poorly buffered, approximately neutral solutions. Because the measured potential is determined by the surface layer of liquid, errors will arise unless time is allowed for equilibrium to be established; this process may take several minutes. In determining the pH of poorly buffered solutions, the glass electrode should first be thoroughly rinsed with water. Then, if the unknown is plentiful, the electrodes should be placed in successive portions until a constant pH reading is obtained. Good stirring is also helpful; several minutes should be allowed for the attainment of steady readings.
5. *Variation in junction potential.* It should be reemphasized that this is a fundamental uncertainty in the measurement of pH for which a correction cannot be applied. Absolute values more reliable than 0.01 pH unit are generally unobtainable. Even reliability to 0.03 pH unit requires considerable care. On the other hand, it is often possible to detect pH *differences* between similar solutions or pH *changes* in a single solution that are as small as 0.001 unit. For this reason many pH meters are designed to permit readings to less than 0.01 pH unit.
6. *Error in the pH of the buffer solution.* Because the glass electrode must be frequently calibrated, any inaccuracies in the preparation or changes in composition of the buffer during storage will be propagated as errors in pH measurements.

Potentiometric Titrations

The potential of a suitable indicator electrode is conveniently employed to establish the equivalence point for a titration (a *potentiometric titration*). A potentiometric titration provides different information from a direct potentiometric measurement. For example, the direct measurement of

0.100-F acetic and hydrochloric acid solutions with a pH-sensitive electrode would yield different values because the former is only partially dissociated. On the other hand, potentiometric titrations of equal volumes of the two acids would require the same amount of standard base.

The potentiometric end point is widely applicable and provides inherently more accurate data than the corresponding method employing indicators. It is particularly useful for titration of colored or opaque solutions and for detecting the presence of unsuspected species in a solution. Unfortunately it is more time-consuming than a titration performed with an indicator.

Figure 14-11 shows a typical apparatus for performing a potentiometric titration. Ordinarily the titration involves measuring and recording a cell potential (or a pH reading) after each addition of reagent. The titrant is added in large increments at the outset; as the end point is approached (as indicated by larger potential changes per addition), the increments are made smaller.

Sufficient time must be allowed for the attainment of equilibrium after each addition of reagent. Precipitation reactions may require several minutes for equilibration, particularly in the vicinity of the equivalence point. A close approach to equilibrium is indicated when the measured potential ceases to drift by more than a few millivolts. Good stirring is frequently effective in hastening the achievement of equilibrium.

The first two columns of Table 14-2 consist of typical poten-

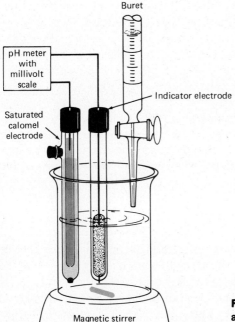

FIGURE 14-11 Apparatus for a potentiometric titration.

TABLE 14-2
Potentiometric Titration Data for 2.433 meq of Chloride with 0.1000-*N* Silver Nitrate

VOL AgNO₃, ml	E VS. SCE, V	ΔE/ΔV, V/ml
5.0	0.062	
		0.002
15.0	0.085	
		0.004
20.0	0.107	
		0.008
22.0	0.123	
		0.015
23.0	0.138	
		0.016
23.50	0.146	
		0.050
23.80	0.161	
		0.065
24.00	0.174	
		0.09
24.10	0.183	
		0.11
24.20	0.194	
		0.39
24.30	0.233	
		0.83
24.40	0.316	
		0.24
24.50	0.340	
		0.11
24.60	0.351	
		0.07
24.70	0.358	
		0.050
25.00	0.373	
		0.024
25.5	0.385	
		0.022
26.0	0.396	
		0.015
28.0	0.426	

tiometric titration data obtained with the apparatus illustrated in Figure 14-11. The data are plotted in Figure 14-12a. Note that this experimental plot closely resembles titration curves derived from theoretical considerations (p. 171).

END POINT DETERMINATION

Several methods can be used to determine the end point for a potentiometric titration. The most straightforward involves a direct plot of potential versus reagent volume as in the curve in Figure 14-12a. The midpoint in the steeply rising portion of the curve is then estimated visually and taken as the end point. Various mechanical methods to aid in the

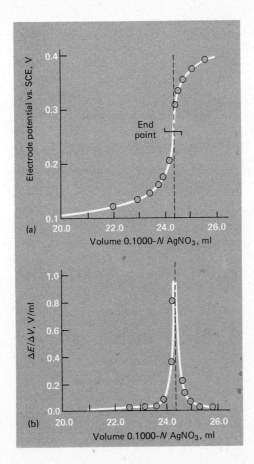

FIGURE 14-12 (a) Potentiometric titration curve of 2.433 meq of Cl^- with 0.1000-N $AgNO_3$. (b) First derivative curve.

establishment of the midpoint have been proposed; it is doubtful, however, that these significantly improve the accuracy.

A second approach is to calculate the change in potential per unit change in volume of reagent (that is, $\Delta E/\Delta V$) as has been done in column 3 of Table 14-2. A plot of this parameter as a function of average volume leads to a sharp maximum at the end point (see Figure 14-12b). Alternatively the ratio can be evaluated during the titration and recorded directly in lieu of the potential itself. Thus in column 3 of Table 14-2 it is seen that the maximum is located between 24.3 and 24.4 ml; selection of 24.35 ml would be adequate for most purposes.

PRECIPITATION TITRATIONS

Electrode Systems. The indicator electrode for a precipitation titration is frequently the metal from which the reacting cation was derived.

Membrane electrodes that are sensitive to one of the ions involved in the titration may also be employed.

Titration Curves. Theoretical curves for a potentiometric titration are readily derived. For example, we may describe the potential of a silver electrode during the titration of iodide with silver ion by the expression

$$E_{Ag} = -0.151 - 0.0591 \log [I^-]$$

where -0.151 is the standard potential for the half-reaction

$$AgI(s) + e \rightleftharpoons Ag(s) + I^-$$

Alternatively, the potential is given by

$$E_{Ag} = 0.799 - 0.0591 \log \frac{1}{[Ag^+]}$$

where 0.799 is the standard potential for the reduction of silver ion.

The first relationship is the more useful for deriving points prior to the equivalence point because the formal concentration of iodide ion will be known, and we may write

$$[I^-] \cong F_{I^-}$$

Beyond the equivalence point the second equation is more easily used; here we may assume

$$[Ag^+] \cong F_{Ag^+}$$

If the calomel electrode serves as the anode, combination of its potential with that of the silver electrode gives

$$E_{cell} = E_{Ag} - 0.242$$

A theoretically derived curve will depart slightly from the experimental because it is not possible to calculate a value for the junction potential.

Titration Curves for Mixtures. Multiple end points are often revealed by a potentiometric titration when the sample contains more than one reacting species. The titration of halide mixtures with silver nitrate is an important example. Figure 8-3 (p. 173) is a theoretically derived titration curve for a mixture of chloride and iodide with silver nitrate. Because the measured potential is directly proportional to the negative logarithm of the silver ion concentration, the experimental potentiometric curve should have the same shape although the ordinate units would be different.

Experimental curves for the titration of halide mixtures do not show the sharp discontinuity at the first equivalence point that appears in Figure 8-3. More important, the volume of silver nitrate required to reach the first end point is generally somewhat greater than theoretical; the total volume, however, approaches the correct amount. These observations apply to the titration of chloride-bromide, chloride-iodide, and bromide-iodide mixtures and can be explained by assuming that the more soluble silver halide is coprecipitated during formation of the less soluble compound. An overconsumption of reagent in the first part of the titration thus occurs.

Despite the coprecipitation error, the potentiometric method is useful for the analysis of halide mixtures. When approximately equal quantities are present, relative errors can be kept to about 1 to 2%.[9]

COMPLEX FORMATION TITRATIONS

Both metal electrodes and membrane electrodes can be used to detect end points in reactions that involve formation of a soluble complex. The mercury electrode, illustrated in Figure 14-13, is particularly useful for EDTA titrations.[10] It will function as an indicator electrode for the titration of cations forming complexes that are less stable than HgY^{2-} (see p. 358). Reilley and Schmid have undertaken a systematic study of this end point and describe conditions that are necessary for various EDTA titrations.[11]

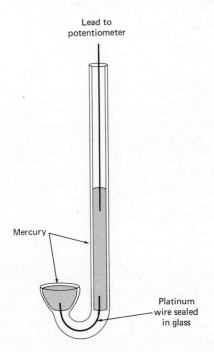

Lead to
potentiometer

Mercury

Platinum
wire sealed
in glass

FIGURE 14-13 A typical mercury electrode.

[9]For further details see I. M. Kolthoff and N. H. Furman, *Potentiometric Titrations*, 2d ed. New York: Wiley, 1931, pp. 154–158.

[10]These electrodes can be obtained from Kontes Manufacturing Corporation, Vineland, N.J.

[11]C. N. Reilley and R. W. Schmid, *Anal. Chem.*, **30**, 947 (1958).

NEUTRALIZATION TITRATIONS

In Chapter 9 we considered theoretical curves for various neutralization titrations in some detail. The shapes of these curves can be closely approximated experimentally. Often, however, the experimental curves will be displaced from the theoretical curves because the latter are usually derived employing concentrations rather than activities. A study of the theoretical curves will show that the small error inherent in the potentiometric measurement of pH is of no consequence insofar as locating the end point is concerned.

Potentiometric acid-base titrations are particularly useful for the analysis of mixtures of acids or polyprotic acids (or bases) because discrimination between the end points can often be made. An approximate numerical value for the dissociation constant of the reacting species can also be estimated from potentiometric titration curves. In theory this quantity can be obtained from any point along the curve; as a practical matter, it is most easily calculated from the pH at the point of half-neutralization. For example, in the titration of the weak acid HA we may ordinarily assume that at the midpoint

$$[HA] \cong [A^-]$$

and therefore

$$K_a = \frac{[H^+][A^-]}{[HA]} = [H^+]$$

or

$$pK_a = pH$$

It is important to note that a dissociation constant determined from a potentiometric titration curve may differ from that shown in a table of dissociation constants by a factor of 2 or more because the latter is based upon activities while the former is not.

OXIDATION-REDUCTION TITRATIONS

The derivation of theoretical titration curves for oxidation-reduction processes was considered in Chapter 12. In each example an electrode potential related to the concentration ratio of the oxidized and reduced forms of either of the reactants was determined as a function of the titrant volume. These curves can be duplicated experimentally (here again the curves may be displaced on the ordinate scale because of the effect of ionic strength), provided an indicator electrode responsive to at least one of the couples involved in the reaction is available. Such electrodes exist for most, but not all, of the reagents described in Chapter 13.

Indicator electrodes for oxidation-reduction are generally fabricated from platinum, gold, mercury, or silver. The metal chosen must be unreactive with respect to the components of the reaction—it is merely a

site for electron transfer. Without question the platinum electrode is most
widely used for oxidation-reduction titrations. Curves similar to those
shown in Figure 12-4 (p. 303) can be obtained experimentally with a
platinum-calomel electrode system. The end point can be established by
the methods already discussed. It should be recalled that curves for many
oxidation reduction titrations are not symmetric (p. 303); thus the
equivalence point potential may be significantly different from the potential
corresponding to the point of inflection in the curve.

DIFFERENTIAL TITRATIONS

We have seen that a derivative curve generated from the data of a
conventional potentiometric titration curve (Figure 14-12b) reaches a distinct maximum in the vicinity of the equivalence point. It is also possible to
acquire titration data directly in derivative form by means of suitable
apparatus.

   A differential titration requires the use of two identical indicator
electrodes, one of which is well shielded from the bulk of the solution.
Figure 14-14 illustrates a typical arrangement. Here one of the electrodes is
contained in a small sidearm test tube. Contact with the bulk of the
solution is made through a small (~ 1 mm) hole in the bottom of the tube.
Because of this restricted access, the composition of the solution surrounding the shielded electrode will not be immediately affected by an
addition of titrant to the bulk of the solution. The resulting difference in
solution composition gives rise to a difference in potential ΔE between

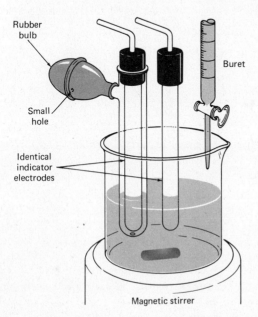

Rubber
bulb

Buret

Small
hole

Identical
indicator
electrodes

Magnetic stirrer

FIGURE 14-14 Apparatus for
differential potentiometric titrations.

the electrodes. After each potential measurement the solution is homogenized by squeezing the rubber bulb several times, whereupon ΔE again becomes zero. If the volume of solution in the tube that shields the electrode is kept small (say, 1 to 5 ml), the error arising from failure of the final addition of reagent to react with this portion of the solution can be shown to be negligibly small.

For oxidation-reduction titrations two platinum wires can be employed, with one enclosed in an ordinary medicine dropper.

The main advantage of a differential method is the elimination of the need for the reference electrode and salt bridge.

AUTOMATIC TITRATIONS

In recent years automatic titrators based on the potentiometric principle have come on the market. These are useful where many routine analyses are required. Such instruments cannot yield results that are more accurate than those obtained by manual potentiometric techniques; however, they do decrease the time needed to perform the titrations and thus may offer some economic advantages.

Basically two types of automatic titrators are available. The first yields a titration curve of potential versus reagent volume, or, in some instances, $\Delta E/\Delta V$ against volume. The end point is then obtained from the curve by inspection. In the second type the titration is stopped automatically when the potential of the electrode system reaches a predetermined value; the volume of reagent delivered is then read at the operator's convenience or printed on a tape.[12]

Automatic titrators normally employ a buret system with a solenoid-operated valve to control the flow; alternatively, use is made of a syringe, the plunger of which is activated by a motor-driven micrometer screw.

PROBLEMS

*1. (a) Calculate the standard potential for the reaction

$$CuBr(s) + e \rightleftharpoons Cu(s) + Br^-$$

(b) Give a schematic representation of a cell with a copper indicator electrode as an anode and a SCE as a cathode that could be used for the determination of Br^-.

(c) Derive an equation that relates the measured potential of the cell in (b) to pBr (assume that the junction potential is zero).

[12]For a description of such instruments see J. J. Lingane, *Electroanalytical Chemistry*, 2d ed. New York: Interscience, 1958, Chapter 8.

(d) Calculate the pBr of a bromide-containing solution that is saturated with CuBr and employed in conjunction with a copper electrode in the type of cell in (b) if the resulting potential is 0.0176 V.

2. (a) Calculate the standard potential for the reaction

$$AgSCN(s) + e \rightleftharpoons Ag(s) + SCN^-$$

(b) Give a schematic representation of a cell with a silver indicator electrode as the cathode and a SCE as an anode that could be used for determining SCN^-.

(c) Derive an equation that relates the measured potential of the cell in (b) to pSCN (assume that the junction potential is zero).

(d) Calculate the pSCN of a solution that is saturated with AgSCN and then employed with the type of cell in (b) if the resulting potential is 0.308 V.

*3. For each of the following cells give a schematic representation that might be suitable and an equation for the relationship between the cell potential and the desired quantity. Assume that the junction potential is negligible, and specify any necessary concentrations as 1.00×10^{-4} M. Treat the indicator electrode as the cathode in each cell.

(a) A cell with a Hg indicator electrode for the determination of pBr.

(b) A cell with a Ag indicator electrode for the determination of pC_2O_4.

(c) A cell with a platinum electrode for the determination of pSn(II).

4. For each of the following cells give a schematic representation that might be suitable and an equation for the relationship between the cell potential and the desired quantity. Assume that the junction potential is negligible, and specify any necessary concentration as 1.00×10^{-4} M. Treat the indicator electrode as the anode in each cell.

(a) A cell with a Pb electrode for the determination of pSO_4.

(b) A cell with a Ag indicator electrode for the determination of pCO_3.

(c) A cell with a platinum electrode for the determination of pTl(I).

*5. The following cell was employed for the determination of $pCrO_4$:

$$SCE \| Ag_2CrO_4(sat'd), CrO_4^{2-}(xM) | Ag$$

Calculate $pCrO_4$ when the cell potential was 0.356 V.

6. Calculate the potential of the cell (neglecting the junction potential)

indicator electrode $\|$ SCE

where the indicator electrode is mercury in a solution that is
(a) 4.00×10^{-6} M Hg^{2+}.
(b) 4.00×10^{-6} M Hg_2^{2+}.
(c) saturated with Hg_2Br_2 and is 3.00×10^{-3} M in Br^-.
(d) 2.50×10^{-5} F in $Hg(NO_3)_2$ and 0.040 F in KCl.

$$Hg^{2+} + 2Cl^- \rightleftharpoons HgCl_2 \qquad K_f = 6.1 \times 10^{12}$$

(e) $2.50 \times 10^{-5} F$ in $Hg(NO_3)_2$ and $0.0040 F$ in KCl.

*7. The formation constant for the mercury(II) cyanide complex is

$$Hg^{2+} + 2CN^- \rightleftharpoons Hg(CN)_2(aq) \qquad K_f = 5.0 \times 10^{34}$$

Calculate the standard potential for the half-reaction

$$Hg(CN)_2 + 2e \rightleftharpoons Hg(l) + 2CN^-$$

8. Calculate the standard potential for the half-reaction

$$HgY^{2-} + 2e \rightleftharpoons Hg(l) + Y^{4-}$$

where Y^{4-} is the anion of EDTA. The formation constant for the reaction $Hg^{2+} + Y^{4-} \rightleftharpoons HgY^{2-}$ is $K_f = 6.3 \times 10^{21}$.

*9. Calculate the potential of the cell (neglecting the junction potential)

$$Hg|HgY^{2-}(4.50 \times 10^{-5}\,M),\ Y^{4-}(xM)\|SCE$$

where Y^{4-} is the EDTA anion [for $HgY^{2-} + 2e \rightleftharpoons Hg(l) + Y^{4-}$, $E^0 = 0.210$]. The concentration of Y^{4-} is

(a) $3.33 \times 10^{-1}\,M$.

(b) $3.33 \times 10^{-3}\,M$.

(c) $3.33 \times 10^{-5}\,M$.

10. Calculate the potential of the cell

$$Hg|HgY^{2-}(5.00 \times 10^{-4}\,M),\ EDTA(C_T = 4.65 \times 10^{-2}\,F)\|SCE$$

if the pH of the solution on the left is (a) 7.00, (b) 8.00, (c) 9.00. See Table 11-3 for values of α_4 and Problem 9 for E^0.

11. The following cell was found to have a potential of 0.209 V when the solution in the left compartment was a buffer of pH 5.21:

$$\text{glass electrode}|H^+(a = x)\|SCE$$

The following potentials were obtained when the buffered solution was replaced with unknowns. Calculate the pH and the hydrogen ion activity of each unknown.

*(a) 0.064 V   (c) 0.510 V

*(b) 0.329 V   (d) 0.677 V

12. The following cell was found to have a potential of 0.411 V:

$$\text{membrane electrode for } Mg^{2+}|Mg^{2+}(a = 1.77 \times 10^{-3}\,M)\|SCE$$

When the solution of known magnesium activity was replaced with an unknown solution, the potential was found to be 0.439 V. What was the pMg of this unknown solution? Neglect the junction potential.

*13. The following cell was found to have a potential of 0.893 V:

$$Cd|CdX_2(\text{sat'd}),\ X^-(0.0200\,M)\|SCE$$

Calculate the solubility product of CdX_2, neglecting the junction potential.

14. The following cell was found to have a potential of 0.693 V:

$$Pt, H_2(1.00\ atm)|HA(0.200\ F),\ NaA(0.300\ F)\|SCE$$

Calculate the dissociation constant of HA, neglecting the junction potential.

15. A 40.00-ml aliquot of 0.1000-N U^{4+} is diluted to 75.0 ml and titrated with 0.0800-N Ce^{4+}. The pH of the solution is maintained at 1.00 throughout the titration. (Use 1.44 V for the formal potential of the cerium system.)
  *(a) Calculate the potential of the indicator cathode with respect to a saturated calomel reference electrode after the addition of 5.00, 10.00, 15.00, 25.00, 40.00, 49.00, 50.00, 51.00, 55.00, and 60.00 ml of cerium(IV).
  (b) Draw a titration curve for these data.

*16. Calculate the potential of a mercury cathode (vs. SCE) after the addition of 5.00, 15.00, 25.00, 30.00, 35.00, 39.00, 40.00, 41.00, 45.00, and 50.00 ml of 0.0500-F $Hg_2(NO_3)_2$ to 50.00 ml of 0.0800-F NaCl. Construct a titration curve from these data (for $Hg_2Cl_2 \rightleftharpoons Hg_2^{2+} + 2Cl^-$, $K_{sp} = 1.3 \times 10^{-18}$).

17. Calculate the potential (vs. SCE) of a lead anode after the addition of 0.00, 10.00, 20.00, 24.00, 24.90, 25.00, 25.10, 26.00, and 30.00 ml of 0.2000-F $NaIO_3$ to 50.00 ml of 0.0500-F $Pb(NO_3)_2$ [for $Pb(IO_3)_2$, $K_{sp} = 3.2 \times 10^{-13}$].

18. Calculate the equivalence point cell potential for each of the following potentiometric titrations. Treat the indicator-electrode system as the cathode. The reference electrode is a saturated calomel electrode. Assume that the solutions of the reagent and analyte are 0.100 F at the outset.
  *(a) The titration of IO_3^- with a standard solution of $Hg_2(NO_3)_2$ [for $Hg_2(IO_3)_2$, $K_{sp} = 1.9 \times 10^{-14}$], employing a mercury indicator electrode.
  *(b) The titration of Sn^{2+} with I_3^-, employing a platinum indicator electrode (assume that $[I^-] = 0.300$ at the equivalence point).
  (c) The titration of I^- with $Pb(NO_3)_2$, using a lead electrode [product: $PbI_2(s)$].
  (d) The titration of HNO_2 with MnO_4^-, employing a platinum electrode (assume that $[H^+] = 0.200$ at the equivalence point).

*19. A glass calomel electrode system was found to develop a potential of 0.0620 V when used with a buffer of pH 7.00; with an unknown solution the potential was observed to be 0.2794 V.
  (a) Calculate the pH and $[H^+]$ of the unknown.
  (b) Assume that K is uncertain by ± 0.001 V as a consequence of a difference in the junction potential between standardization and measurement. What is the range of $[H^+]$ associated with this uncertainty?
  (c) What is the relative error in $[H^+]$ associated with the uncertainty in E_j?

20. The following cell was found to have a potential of 0.3674 V:

$$\text{membrane electrode for } Mg^{2+}|Mg^{2+}(a = 6.87 \times 10^{-3}\ M)\|SCE$$

  (a) When the solution of known magnesium activity was replaced with an unknown solution, the potential was found to be 0.4464 V. What was the pMg of this unknown solution?

(b) Assuming an uncertainty of ± 0.002 V in the junction potential, what is the range of Mg^{2+} activities within which the true value might be expected?

(c) What is the relative error in $[Mg^{2+}]$ associated with the uncertainty in E_j?

*21. The sodium ion concentration of a solution was determined by a differential measurement with a glass membrane electrode. The electrode system developed a potential of 0.2331 V when immersed in 10.0 ml of the unknown. After addition of 1.00 ml of a standard solution that was 2.00×10^{-2} M in Na^+, the potential decreased to 0.1846 V. Calculate the sodium ion concentration and the pNa of the original solution.

22. The calcium ion concentration of a solution was determined by a differential measurement with a liquid membrane electrode. The electrode system developed a potential of 0.4965 V when immersed in 25.0 ml of the sample. After addition of 2.00 ml of 5.45×10^{-2} F $CaCl_2$, the potential changed to 0.4117 V. Calculate the calcium concentration and the pCa of the sample.

15 additional electroanalytical methods

In contrast to potentiometry, the electrochemical methods considered in this chapter all require a current through a solution of the analyte. Therefore, before treating these methods in detail, it is desirable to examine the behavior of electrochemical cells during current passage.

Behavior of Cells during Current Passage

When a current is passed through a cell, the overall potential may be influenced by three phenomena which have not been discussed thus far, namely, ohmic potential, concentration polarization, and kinetic polarization.

OHMIC POTENTIAL; *IR* DROP

To develop a current in either a galvanic or an electrolytic cell, a driving force or a potential is required to overcome the resistance of the ions to movement toward the anode or the cathode. Just as in metallic conduction, this force follows Ohm's law and is equal to the product of the current in amperes and the resistance of the cell in ohms. The force is generally referred to as the *ohmic potential*, or the *IR drop*.

In general the net effect of *IR* drop is to increase the potential required to operate an electrolytic cell and to decrease the measured potential of a galvanic cell. Therefore the *IR* drop is always *subtracted* from the theoretical cell potential. That is,

$$E_{cell} = E_{cathode} - E_{anode} - IR \qquad (15\text{-}1)[1]$$

EXAMPLE

1. Calculate the potential when 0.100 A is drawn from the galvanic cell

$$Cd|Cd^{2+}(1.00\ M)||Cu^{2+}(1.00\ M)|Cu$$

Assume a cell resistance of 4.00 Ω.

Because both cation concentrations are 1.00 *M*, the respective half-cell potentials are equal to the standard potentials; thus

$$E = E^0_{Cu} - E^0_{Cd} = 0.337 - (-0.403) = 0.740 \text{ V}$$

$$E_{cell} = 0.740 - IR = 0.740 - 0.100 \times 4.00$$

$$= 0.340 \text{ V}$$

Thus the emf of this cell drops dramatically at the first instant a current is drawn.

2. Calculate the potential required to generate a current of 0.100 A in the reverse direction in the foregoing cell.

$$E = E^0_{Cd} - E^0_{Cu} = -0.403 - 0.337 = -0.740 \text{ V}$$

$$E_{cell} = -0.740 - 0.100 \times 4.00 = -1.140 \text{ V}$$

Here an external potential greater than 1.140 V would be needed to cause Cd^{2+} to deposit and Cu to dissolve at a rate required for a current of 0.100 A.

POLARIZATION EFFECTS

The linear relationship between potential and the instantaneous current in a cell (Equation 15-1) is frequently observed experimentally when *I* is small; at high currents, however, marked departures from linear behavior occur. Under these circumstances the cell is said to be *polarized* (see Figure 15-1).

---

[1] Here and in the subsequent discussion we neglect the junction potential because it is sufficiently small, relative to other potentials, to be of no consequence.

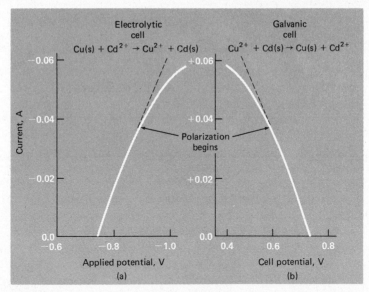

FIGURE 15-1 Current-voltage curves for a cell: (a) Cu|Cu^{2+}-(1.00 M)‖Cd^{2+}(1.00 M)|Cd. (b) Cd|Cd^{2+}(1.00 M)‖Cu^{2+}(1.00 M)|Cu.

Thus a polarized electrolytic cell requires application of potentials larger than theoretical for a given current; similarly, a polarized galvanic cell develops potentials that are smaller than predicted. Under some conditions the polarization of a cell may be so extreme that the current becomes essentially independent of the voltage; under these circumstances polarization is said to be complete.

Polarization is an electrode phenomenon; either or both electrodes in a cell can be affected. Included among the factors influencing the extent of polarization are the size, shape, and composition of the electrodes; the composition of the electrolyte solution; the temperature and the rate of stirring; the magnitude of the current; and the physical states of the species involved in the cell reaction. Some of these factors are sufficiently understood to permit quantitative statements concerning their effects upon cell processes. Others, however, can be accounted for on an empirical basis only.

For purposes of discussion polarization phenomena are conveniently classified into the two categories of *concentration polarization* and *kinetic polarization* (also called *overvoltage* or *overpotential*).

Concentration Polarization. When the reaction at an electrode is rapid and reversible, the concentration of the reacting species in the layer of solution immediately adjacent to the electrode is always that which would be predicted from the Nernst equation. Thus the cadmium ion concentration in the immediate vicinity of a cadmium electrode will always be given by

$$E = E_{Cd}^0 - \frac{0.0591}{2} \log \frac{1}{[Cd^{2+}]}$$

irrespective of the concentration of this cation in the bulk of the solution. Because the reduction of cadmium ions is rapid and reversible, *the concentration of this ion in the film of liquid surrounding the electrode is determined at any instant by the potential of the cadmium electrode* at that instant. If the potential is changed, deposition or dissolution of cadmium occurs such that the concentration of cadmium ion in the surface film is rapidly brought to that required by the Nernst equation.

In contrast to this substantially instantaneous surface process, the rate at which equilibrium between the electrode and the bulk of the solution is attained can be very slow, depending upon the magnitude of the current as well as the volume of the solution and its solute concentration.

When a sufficient potential is applied to an electrolytic cell such as the one just considered, cadmium ions are reduced, and an instantaneous current is generated. For current to continue at a level predicted by Equation 15-1, however, an additional supply of the cation must be transported into the surface film surrounding the electrode at a suitable rate. If the demand for reactant cannot be met by this mass transfer, concentration polarization will set in, and *lowered currents must result.* This type of polarization, then, occurs when the rate of transfer of reactive species between the bulk of the solution and the electrode surface is inadequate to maintain the current at the level required by Ohm's law. A departure from linearity such as that shown in Figure 15-1a is the consequence. Inadequate material transport will also cause concentration polarization in a galvanic cell such as that described by the curve in Figure 15-1b; here, however, the current would be limited by the transport of copper ions.

Ions or molecules can be transported through a solution by (1) diffusion, (2) electrostatic attraction or repulsion, and (3) mechanical or convection forces. We must therefore consider the variables that influence these forces as they relate to electrode processes.

Whenever a concentration gradient develops in a solution, molecules or ions diffuse from the more concentrated to the more dilute region. The rate at which transfer occurs is proportional to the concentration difference. In an electrolysis a gradient is established as a result of ions being removed from the film of solution next to the cathode. Diffusion then occurs, the rate being expressed by the relationship

$$\text{rate of diffusion to cathode surface} = k(C - C_0) \qquad \textbf{(15-2)}$$

where C is the reactant concentration in the bulk of the solution, C_0 is its equilibrium concentration at the cathode surface, and k is a proportionality constant. *The value of C_0 is fixed by the potential of the electrode and can* *be calculated from the Nernst equation.* As larger potentials are applied to the electrode, C_0 becomes smaller and smaller and the diffusion rate greater and greater.

Electrostatic forces also influence the rate at which an ionic reactant migrates to or from an electrode surface. The electrostatic attraction (or repulsion) between a particular ionic species and the electrode becomes smaller as the total electrolyte concentration of the solution is increased. It may approach zero when the reactive species is but a small fraction of the total concentration of ions with a given charge.

Clearly, reactants can be transported to an electrode by mechanical means. Thus stirring or agitation will aid in decreasing concentration polarization. Convection currents due to temperature or density differences are also effective.

To summarize, then, concentration polarization occurs when the forces of diffusion, electrostatic attraction, and mechanical mixing are insufficient to transport the reactant to or from an electrode surface at a rate demanded by the theoretical current. Concentration polarization causes the potential of a galvanic cell to be smaller than the value predicted on the basis of the theoretical potential and the *IR* drop (Figure 15-1b). In an electrolytic cell a potential more negative than theoretical is required in order to maintain a given current (Figure 15-1a).

Concentration polarization is important in several electroanalytical methods. In some applications steps are taken to eliminate it; in others, however, it is essential to the method, and every effort is made to promote its occurrence. The degree of concentration polarization is influenced experimentally by (1) the reactant concentration, (2) the total electrolyte concentration, (3) mechanical agitation, and (4) the size of the electrodes; as the area toward which a reactant is transported becomes greater, polarization effects become smaller.

Kinetic Polarization. Kinetic polarization results when the rate at which the electrochemical reaction occurring at one or both electrodes is slow; here an additional potential (the *overvoltage* or *overpotential*) is required to overcome the energy barrier to the half-reaction. In contrast to concentration polarization the current is controlled by the *rate of the electron-transfer process* rather than by the rate of mass transfer.

While exceptions can be cited, some empirical generalizations can be made regarding the magnitude of overvoltage.

1. Overvoltage increases with current density (current density is defined as the amperes per square centimeter of electrode surface).
2. It usually decreases with increases in temperature.
3. Overvoltage varies with the chemical composition of the electrode, often being most pronounced with softer metals such as tin, lead, zinc, and particularly mercury.

4. Overvoltage is most marked for electrode processes that yield gaseous products such as hydrogen or oxygen; it is frequently negligible where a metal is being deposited or where an ion is undergoing a change of oxidation state.

5. The magnitude of overvoltage in any given situation cannot be predicted exactly because it is determined by a number of uncontrollable variables.[2]

The high overvoltage associated with the formation of hydrogen permits the electrolytic deposition of several metals that require potentials at which hydrogen would otherwise be expected to interfere. For example, it is readily shown from their standard potentials that rapid formation of hydrogen should occur well before a potential sufficient for the deposition of zinc from a neutral solution is reached. In fact, however, a quantitative deposition of the metal can be achieved, provided a mercury or a copper electrode is used; because of the high overvoltage of hydrogen on these metals as well as on zinc itself, little or no gas is evolved during the process.

Electrodeposition

Electrodeposition of metals and metal oxides provides a convenient way of isolating a number of elements for analytical purposes. Methods based upon these processes have been employed since the middle of the nineteenth century. Often the weight of the deposit can be readily obtained, thus providing an *electrogravimetric* method for completion of an analysis.

APPARATUS

Figure 15-2 shows the components in a typical electrodeposition assembly.

Electrodes. Electrodes are usually constructed of platinum, although copper, brass, and other metals find occasional use. Platinum electrodes have the advantage of being relatively nonreactive; moreover, they can be ignited to remove any grease, organic matter, or gases that could have a deleterious effect upon the physical properties of the deposit. Certain metals (notably bismuth, zinc, and gallium) cannot be deposited directly onto platinum without causing permanent damage to the electrode. A protective coating of copper should always be deposited on a platinum

---

[2]Overvoltage data for various gaseous species at different electrode surfaces are to be found in L. Meites, ed., *Handbook of Analytical Chemistry.* New York:McGraw-Hill, 1963, p. 5-184.

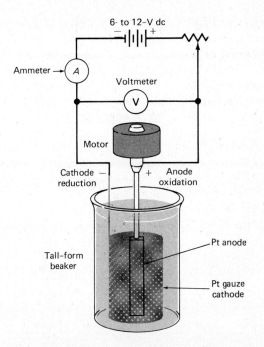

6- to 12-V dc

Ammeter

Voltmeter

Motor

Cathode — reduction   + Anode oxidation

Tall–form beaker

Pt anode

Pt gauze cathode

FIGURE 15-2 Apparatus for the electrodeposition of metals.

electrode before undertaking the electrolysis of these metals. The presence of chloride ion will result in attack on a platinum anode; the chloroplatinate complexes that form ultimately migrate to and are reduced at the cathode, causing high results.

Cathodes are usually formed as gauze cylinders 2 to 3 cm in diameter and perhaps 6 cm in length. This construction minimizes the likelihood of concentration polarization by providing a large surface area to which the solution can freely circulate. The anode may also be a gauze cylinder of somewhat smaller diameter so that it can be fitted inside the cathode; alternatively it may take the form of a heavy wire spiral or a solid paddle.

Electrolytic depositions are ordinarily carried out in tall-form beakers. Efficient stirring is useful in minimizing concentration polarization. Frequently the anode is rotated with an electric motor for this purpose.

Typical Equipment. The apparatus shown in Figure 15-2 is typical of that employed for most electrolytic analyses. The direct-current power supply may consist of a storage battery, a generator, or an alternating-current rectifier. A rheostat is used to control the applied potential. An ammeter and a voltmeter are provided to show the approximate current and applied voltage. An entirely adequate electrolysis unit can be assembled from components found in most laboratories; more elaborate equipment can be obtained from laboratory suppliers.

PHYSICAL PROPERTIES OF ELECTROLYTIC PRECIPITATES

Ideally an electrolytic deposit should be strongly adherent, dense, and smooth so that the processes of washing, drying, and weighing can be performed without mechanical loss or reaction with the atmosphere. Good metallic deposits are fine grained and have a metallic luster; spongy, powdery, or flaky precipitates are likely to be less pure and less adherent.

The principal factors that influence the physical characteristics of deposits include current density, temperature, and the presence of complexing agents. Ordinarily the best deposits are formed at current densities that are less than $0.1 \, A/cm^2$. Stirring generally improves the quality of a deposit. The effects of temperature are unpredictable and must be determined empirically.

It is also found that many metals form smoother and more adherent films when deposited from solutions in which their ions exist primarily as complexes. Cyanide and ammonia complexes often provide the best deposits. The reasons for this effect are not obvious.

Codeposition of hydrogen during electrolysis is likely to cause the formation of nonadherent deposits, which are unsatisfactory for analytical purposes. The evolution of hydrogen can be controlled by introduction of a *cathode depolarizer*, which is a substance that is reduced in preference to hydrogen ion. Nitrate functions in this manner, being reduced to ammonium ion:

$$NO_3^- + 10H^+ + 8e \rightleftharpoons NH_4^+ + 3H_2O$$

TABLE 15-1

Common Elements That Can Be Determined by Electrogravimetric Methods

ION	WEIGHED AS	CONDITIONS
Cd^{2+}	Cd	Alkaline cyanide solution
Co^{2+}	Co	Ammoniacal sulfate solution
Cu^{2+}	Cu	HNO_3—H_2SO_4 solution
Fe^{3+}	Fe	$(NH_4)_2C_2O_4$ solution
Pb^{2+}	PbO_2	HNO_3 solution
Ni^{2+}	Ni	Ammoniacal sulfate solution
Ag^+	Ag	Cyanide solution
Sn^{2+}	Sn	$(NH_4)_2C_2O_4$—$H_2C_2O_4$ solution
Zn^{2+}	Zn	Ammoniacal or strong NaOH solution

APPLICATIONS OF ELECTROLYTIC METHODS

Table 15-1 lists the common elements that can be determined electrogravimetrically. Directions for the electrogravimetric analysis of copper are given in Chapter 20, Analysis 13-1.

Coulometry

Coulometry encompasses a group of analytical methods that involve measuring the quantity of electricity (in coulombs) needed to convert the analyte to a different chemical state. Like gravimetric analysis, coulometry offers the advantage that the proportionality constant between the measured quantity in the analysis and the amount of analyte can be derived from known physical constants; thus a calibration or standardization step is not ordinarily required. Coulometric methods are often as accurate as gravimetric or volumetric procedures; they are usually faster and more convenient than the former. In addition, coulometric procedures are readily adapted to automation.[3]

QUANTITY OF ELECTRICITY

The quantity of electricity is measured in terms of the *coulomb* (C) and the *faraday* (F). The coulomb is that amount of electricity that flows during the passage of a constant current of 1 A for 1 sec. Thus for a constant current of i amperes flowing for t seconds, the number of coulombs, q, is given by the expression

$$q = i \times t \qquad (15\text{-}3)$$

For a variable current the number of coulombs is given by the integral

$$q = \int_0^t i\,dt \qquad (15\text{-}4)$$

The coulomb represents 6.24×10^{18} electrons.

The faraday is the quantity of electricity that will produce 1 equivalent of chemical change at an electrode. Because the equivalent in an oxidation-reduction reaction corresponds to the change brought about by 1 mole of electrons (p. 150), the faraday is equal to 6.02×10^{23} electrons. One faraday is also equal to 96,493 coulombs.

These definitions can be employed to calculate the quantity of chemical change that occurs at an electrode.

[3]For summaries of coulometric methods see H. L. Kies, *J. Electroanal. Chem.*, **4**, 257 (1962); J. J. Lingane, *Electroanalytical Chemistry*, 2d ed. New York: Interscience, 1958, Chapters 19–21; G. W. C. Milner and G. Phillips, *Coulometry in Analytical Chemistry*. New York: Pergamon Press, 1967.

EXAMPLE

The iodide ion in 30.0 ml of an aqueous solution was determined by coulometric titration with Ag^+ generated at a silver anode ($Ag \rightarrow Ag^+ + e$). A constant current of 40.2 mA, passed for 283 sec, produced enough Ag^+ to precipitate the I^- quantitatively. Calculate the molarity of I^-.

$$\text{no. coulombs} = \frac{40.2\,\text{mA}}{1000\,\text{mA/A}} \times 283\,\text{sec} \times \frac{1\,\text{C}}{\text{A-sec}} = 11.38$$

$$\text{no. faradays} = \text{no. eq}\,I^- = \frac{11.38\,\text{C}}{96,493\,\text{C/F}} = 1.179 \times 10^{-4}$$

$$M_{I^-} = \frac{1.179 \times 10^{-4}\,\text{eq}\,I^-}{30.0\,\text{ml}} \times \frac{1000\,\text{ml}}{\text{liter}} \times \frac{1\,\text{mole}}{\text{eq}} = 3.93 \times 10^{-3}\,\text{mole/liter}$$

TYPES OF COULOMETRIC METHODS

Two general techniques are used for coulometric analyses. The first involves maintaining the potential of the working electrode at a constant level such that quantitative oxidation or reduction of the analyte occurs without involvement of less reactive species in the sample or solvent. Here the current is initially high but decreases rapidly and approaches zero as the analyte is removed from the solution. The quantity of electricity required is measured with a chemical coulometer or by integration of the current-time curve. A second coulometric technique makes use of a constant current that is continued until an indicator signals completion of the reaction. The quantity of electricity required to attain the end point is then calculated from the magnitude of the current and the time of its passage. The latter method has a wider variety of applications than the former; it is frequently called a *coulometric titration.*

A fundamental requirement of all coulometric methods is that the species determined interact with 100% current efficiency. That is, each faraday of electricity must bring about a chemical change corresponding to 1 equivalent of the analyte. This requirement does not, however, imply that the species must necessarily participate directly in the electron-transfer process at the electrode. Indeed, more often than not, the substance being determined is involved wholly or in part in a reaction that is secondary to the electrode process. For example, at the outset of the oxidation of iron(II) at a platinum anode, all current transfer results from the reaction

$$Fe^{2+} \rightleftharpoons Fe^{3+} + e$$

As the concentration of iron(II) decreases, concentration polarization may cause the anode potential to rise until decomposition of water occurs as a competing process. That is,

$$2H_2O \rightleftharpoons O_2(g) + 4H^+ + 4e$$

The current required to complete the oxidation of iron(II) would then exceed that demanded by theory. To avoid the consequent error an excess of cerium(III) can be introduced at the start of the electrolysis. This ion is oxidized at a lower anode potential than is water:

$$Ce^{3+} \rightleftharpoons Ce^{4+} + e$$

The cerium(IV) produced diffuses rapidly from the electrode surface, where it then oxidizes an equivalent amount of iron(II):

$$Ce^{4+} + Fe^{2+} \rightarrow Ce^{3+} + Fe^{3+}$$

The net effect is an electrochemical oxidation of iron(II) with 100% current efficiency even though only a fraction of the iron(II) is directly oxidized at the electrode surface.

COULOMETRIC TITRATIONS

A coulometric titration involves the electrolytic generation of a reagent that in turn reacts with the analyte. The electrode reaction may involve only reagent preparation, as in the formation of silver ion for the precipitation of halides. In other titrations the analyte may also be directly involved at the generator electrode; the coulometric oxidation of iron(II)—in part by electrolytically generated cerium(IV) and in part by direct electrode reaction—is an example. Under any circumstance the net process must approach 100% current efficiency with respect to a single chemical change in the analyte.

The current in a coulometric titration is carefully maintained at a constant and accurately known level; the product of this current in amperes and the time in seconds required to reach the equivalence point yields the number of coulombs and thus the number of equivalents involved in the electrolysis. The constant current aspect of this operation precludes the quantitative oxidation or reduction of the unknown species entirely at the generator electrode; as the solution is depleted of analyte, concentration polarization is inevitable. The electrode potential must then rise if a constant current is to be maintained. Unless this potential rise produces a reagent that can react with the analyte, the current efficiency will be less than 100%. In a coulometric titration at least part (and frequently all) of the analytical reaction thus occurs away from the surface of the working electrode.

A coulometric titration, like the more conventional titration, requires some means of detecting the point of chemical equivalence. Most of the end points applicable to volumetric analysis are equally satisfactory here; the color change of indicators, potentiometric, amperometric (p. 421), and conductance measurements have all been successfully applied.

The analogy between a volumetric and a coulometric titration extends well beyond the common requirement of an observable end point. In both the amount of unknown is determined through evaluation of its combining capacity—in the one case for a standard solution and in the other for a quantity of electricity. Similar demands are made of the reactions; that is, they must be rapid, essentially complete, and free of side reactions.

Electrical Apparatus. Coulometric titrators are available from several laboratory supply houses. In addition they can be readily assembled from components available in most laboratories.

Figure 15-3 is a block diagram of a typical coulometric titrator. It consists of a source of constant current, a switch that simultaneously starts the flow of current and the operation of an electric stopclock, and a means for accurately measuring the current. In the apparatus in Figure 15-3 the current is determined by measuring the potential drop across the standard resistor, R_{std}.

Many electronic or electromechanical constant-current sources are described in the literature. The ready availability of inexpensive operational amplifiers makes their construction a relatively simple matter.

A nonelectronic source, capable of delivering currents of about 20 mA (mA = 10^{-3} A) that are constant to 0.5% relative, can be assembled by employing sufficient B-batteries to give a potential of 90 to 180 V; these

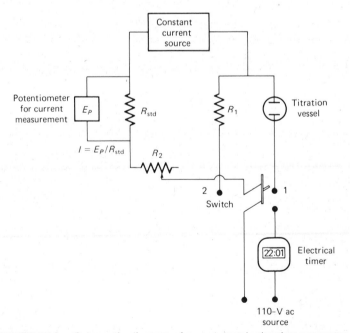

FIGURE 15-3   Schematic diagram of a coulometric titration apparatus.

batteries could serve as the source in Figure 15-3. Experience has shown that high-capacity B-batteries will provide relatively constant currents, provided that current is drawn from them more or less continuously. This requirement is met by the switching arrangement shown in Figure 15-3. The resistor R_1 is introduced into the circuit whenever the switch is moved from 1 to 2; the resistance of R_1 should be about the same as that of the cell. The variable resistor R_2 controls the magnitude of the current; its resistance should be large relative to the other components of the system.

It is useful to point out the close analogy between the various components of the apparatus shown in Figure 15-3 and the apparatus and solutions employed in a volumetric analysis. The constant current source of known magnitude serves the same function as the standard solution in a volumetric method. The electric clock and switch correspond closely to the buret, the switch performing the same function as a stopcock. During the early phases of a coulometric titration the switch is kept closed for extended periods; as the end point is approached, however, small additions of "reagent" are achieved by closing the switch for shorter and shorter intervals. The similarity to the operation of a buret is obvious.

APPLICATIONS OF COULOMETRIC TITRATIONS[4]

Coulometric titrations have been developed for all types of volumetric reactions. Selected applications are described in the following paragraphs.

Neutralization Titrations. Both weak and strong acids can be titrated with a high degree of accuracy using electrogenerated hydroxide ions. The most convenient method, where applicable, involves generation of hydroxide ion at a platinum cathode within the solution.

$$2H_2O + 2e \rightarrow 2OH^- + H_2(g)$$

In this application the platinum anode must be isolated by some sort of diaphragm (see Figure 15-4) to eliminate potential interference from the hydrogen ions produced at that electrode. A convenient alternative involves substitution of a silver wire as the anode and the addition of chloride or bromide ions to the solution of the analyte. The anode reaction then becomes

$$Ag + Br^- \rightleftharpoons AgBr + e$$

Clearly the silver bromide will not interfere with the neutralization reaction.

Both potentiometric and indicator end points can be employed for

[4]Applications of the coulometric procedure are summarized in J. J. Lingane, *Electroanalytical Chemistry*, 2d ed. New York: Interscience, 1958, pp. 536–613; and in H. L. Kies, *J. Electroanal. Chem.*, **4**, 257 (1962).

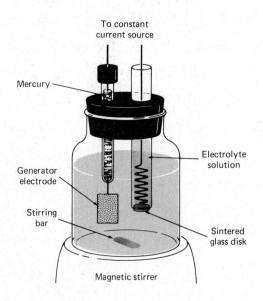

FIGURE 15-4 A typical coulometric titration cell.

To constant
current source

Mercury

Generator
electrode

Stirring
bar

Electrolyte
solution

Sintered
glass disk

Magnetic stirrer

these titrations. The problems associated with the estimation of the equivalence point are identical with those encountered in a conventional volumetric analysis. A real advantage to the coulometric method, however, is that interference by carbonate ion is far less troublesome; it is only necessary to eliminate carbon dioxide from the solution containing the analyte by aeration with a carbon dioxide-free gas before beginning the analysis.

Coulometric titration of strong and weak bases can be performed with hydrogen ions generated at a platinum anode.

$$H_2O \rightleftharpoons \tfrac{1}{2}O_2 + 2H^+ + 2e$$

Here the cathode must be isolated from the solution to prevent interference from the hydroxide ions produced at that electrode.

Precipitation and Complex Formation Titrations. A variety of coulometric titrations have been developed involving anodically generated silver ions (see Table 15-2). A cell, such as that shown in Figure 15-4, can be employed with a generator electrode constructed from a length of heavy silver wire. End points are detected potentiometrically or with chemical indicators. Similar analyses, based upon the generation of mercury(I) ion at a mercury anode, have been described.

An interesting coulometric titration makes use of a solution of the ammine mercury(II) complex of ethylenediaminetetraacetic acid (H_4Y).[5] The complexing agent is released to the solution as a result of the following

[5]C. N. Reilley and W. W. Porterfield, *Anal. Chem.*, **28**, 443 (1956).

TABLE 15-2

Summary of Applications of Coulometric Titrations Involving Neutralization, Precipitation, and Complex Formation Reactions

SPECIES DETERMINED	GENERATOR ELECTRODE REACTION	SECONDARY ANALYTICAL REACTION
Acids	$2H_2O + 2e \rightleftharpoons 2OH^- + H_2$	$OH^- + H^+ \rightleftharpoons H_2O$
Bases	$H_2O \rightleftharpoons 2H^+ + \frac{1}{2}O_2 + 2e$	$H^+ + OH^- \rightleftharpoons H_2O$
Cl^-, Br^-, I^-	$Ag \rightleftharpoons Ag^+ + e$	$Ag^+ + Cl^- \rightleftharpoons \underline{AgCl(s)}$, etc.
Mercaptans	$Ag \rightleftharpoons Ag^+ + e$	$Ag^+ + RSH \rightleftharpoons \underline{AgSR(s)} + H^+$
Cl^-, Br^-, I^-	$2Hg \rightleftharpoons Hg_2^{2+} + 2e$	$Hg_2^{2+} + 2Cl^- \rightleftharpoons \underline{Hg_2Cl_2(s)}$, etc.
Zn^{2+}	$Fe(CN)_6^{3-} + e \rightleftharpoons$ $Fe(CN)_6^{4-}$	$3Zn^{2+} + 2K^+ + 2Fe(CN)_6^{4-} \rightleftharpoons$ $\underline{K_2Zn_3[Fe(CN)_6]_2(s)}$
Ca^{2+}, Cu^{2+}, Zn^{2+} and Pb^{2+}	$HgNH_3Y^{2-} + NH_4^+ + 2e \rightleftharpoons$ $Hg + 2NH_3 + HY^{3-}$ (where Y^{4-} is ethylene-diaminetetraacetate ion)	$HY^{3-} + Ca^{2+} \rightleftharpoons CaY^{2-} + H^+$, etc.

reaction at a mercury cathode:

$$HgNH_3Y^{2-} + NH_4^+ + 2e \rightleftharpoons Hg + 2NH_3 + HY^{3-}$$

Because the mercury chelate is more stable than the corresponding complexes with calcium, zinc, lead, or copper, complexation of these ions will not occur until the electrode process frees the ligand.

Oxidation-Reduction Titrations. Table 15-3 indicates the variety of reagents that can be generated coulometrically and the analyses to which they have been applied. Electrogenerated bromine has proved to be particularly useful among the oxidizing agents, and it forms the basis for a host of methods. Of interest also are some of the unusual reagents not ordinarily encountered in volumetric analysis because of the instability of their solutions; these include dipositive silver ion, tripositive manganese, and the chloride complex of unipositive copper.

Comparison of Coulometric and Volumetric Titrations. Some real advantages can be claimed for a coulometric titration in comparison with the classical volumetric process. Principal among these is the elimination of problems associated with the preparation, standardization, and storage of standard solutions. This advantage is particularly important with labile reagents such as chlorine, bromine, or titanium(III) ion. Owing to their

TABLE 15-3

Summary of Applications of Coulometric Titrations Involving Oxidation-Reduction Reactions

REAGENT	GENERATOR ELECTRODE REACTION	SUBSTANCE DETERMINED
Br_2	$2Br^- \rightleftharpoons Br_2 + 2e$	As(III), Sb(III), U(IV), Tl(I), I^-, SCN^-, NH_3, N_2H_4, NH_2OH, phenol, aniline, mustard gas; 8-hydroxyquinoline
Cl_2	$2Cl^- \rightleftharpoons Cl_2 + 2e$	As(III), I^-
I_2	$2I^- \rightleftharpoons I_2 + 2e$	As(III), Sb(III), $S_2O_3^{2-}$, H_2S
Ce^{4+}	$Ce^{3+} \rightleftharpoons Ce^{4+} + e$	Fe(II), Ti(III), U(IV), As(III), I^-, $Fe(CN)_6^{4-}$
Mn^{3+}	$Mn^{2+} \rightleftharpoons Mn^{3+} + e$	$H_2C_2O_4$, Fe(II), As(III)
Ag^{2+}	$Ag^+ \rightleftharpoons Ag^{2+} + e$	Ce(III), V(IV), $H_2C_2O_4$, As(III)
Fe^{2+}	$Fe^{3+} + e \rightleftharpoons Fe^{2+}$	Cr(VI), Mn(VII), V(V), Ce(IV)
Ti^{3+}	$TiO^{2+} + 2H^+ + e \rightleftharpoons Ti^{3+} + H_2O$	Fe(III), V(V), Ce(IV), U(VI)
$CuCl_3^{2-}$	$Cu^{2+} + 3Cl^- + e \rightleftharpoons CuCl_3^{2-}$	V(V), Cr(VI), IO_3^-
U^{4+}	$UO_2^{2+} + 4H^+ + 2e \rightleftharpoons U^{4+} + 2H_2O$	Cr(VI), Ce(IV)

instability, these species are inconvenient as volumetric reagents; their utilization in coulometric analysis is straightforward, however, because they undergo reaction almost immediately after being generated.

Where small quantities of reagent are required, a coulometric titration offers a considerable advantage. By proper choice of current, micro quantities of a substance can be introduced with ease and accuracy, whereas the equivalent volumetric process requires small volumes of very dilute solutions, a recourse that is always difficult.

A single constant current source can be employed to generate precipitation, oxidation-reduction, or neutralization reagents. Furthermore, the coulometric method is readily adapted to automatic titrations, because current control is easily accomplished.

Coulometric titrations are subject to five potential sources of error: (1) variation in the current during electrolysis, (2) departure of the process from 100% current efficiency, (3) error in the measurement of current, (4) error in the measurement of time, and (5) titration error due to the difference between the equivalence point and the end point. The last of these difficulties is common to volumetric methods as well; where the indicator error is the limiting factor, the two methods are likely to have comparable reliability.

With simple instrumentation, currents constant to 0.2 to 0.5% relative are easily achieved; with somewhat more sophisticated apparatus control to 0.01% is obtainable. In general, then, errors due to current fluctuations need not be serious.

Although generalizations concerning the magnitude of uncertainty associated with the electrode process are difficult, current efficiencies of 99.5 to better than 99.9% are often reported in the literature. Currents are readily measured to $\pm 0.1\%$ relative or better. Similarly, a good quality electric stopclock or digital timer permits measurement of time to within $\pm 0.1\%$ relative.

To summarize, then, the current-time measurements required for a coulometric titration are inherently as accurate or more accurate than the comparable volume-normality measurements of classical volumetric analysis, particularly where small quantities of reagent are involved. Often, however, the accuracy of a titration is not limited by these measurements but by the sensitivity of the end point; in this respect the two procedures are equivalent.

VARIABLE CURRENT COULOMETRIC METHODS[6]

In variable current coulometry the potential of the working electrode is maintained at a level that will cause the analyte to react quantitatively without involvement of other components in the sample. A variable current analysis possesses all the advantages of an electrogravimetric method and is not subject to the limitation imposed by the need for a weighable product. The technique can therefore be applied to systems that yield deposits with poor physical properties as well as to reactions that yield no solid product at all. For example, arsenic may be determined coulometrically by electrolytic oxidation of arsenious acid (H_3AsO_3) to arsenic acid (H_3AsO_4) at a platinum anode. Similarly, the analytical conversion of iron(II) to iron(III) can be accomplished with suitable control of the anode potential.

EXAMPLE

The quantity of Fe^{3+} in a solution was determined coulometrically by quantitative reduction to Fe^{2+} at a platinum electrode. The quantity of electricity required for this reaction was determined by means of a hydrogen-oxygen *chemical coulometer* arranged in series with the cell containing the analyte. This device consists of a pair of platinum electrodes immersed in a solution of K_2SO_4; the oxygen evolved at the anode and the hydrogen at the cathode are collected in a gas buret, and their combined volume is measured.

[6]For a good discussion of this method see J. J. Lingane, *Electroanalytical Chemistry*, 2d ed. New York: Interscience, 1958, pp. 450–483.

Calculate the grams of $Fe_2(SO_4)_3$ (gfw = 400) in the analyte solution if the volume of hydrogen and oxygen was 39.3 ml at 23°C; the barometric pressure was 765 mm Hg (after correction for water vapor).

Converting the gas volume to standard conditions gives

$$V = 39.3 \text{ ml} \times \frac{765 \text{ mm Hg}}{760 \text{ mm Hg}} \times \frac{273°K}{296°K} = 36.48 \text{ ml}$$

The reactions in the coulometer are

$$4H^+ + 4e \rightarrow 2H_2(g)$$

$$2H_2O \rightleftharpoons O_2(g) + 4H^+ + 4e$$

Thus 3 mole of gas are formed by passage of 4 mole of electrons, or 0.750 mole of gas is produced per faraday. Therefore

$$\text{no. faraday} = \frac{36.48 \text{ ml gas}}{22,400 \text{ ml/mole}} \times \frac{1}{0.750 \text{ mole gas/F}}$$

$$= 2.171 \times 10^{-3}$$

$$\text{wt } Fe_2(SO_4)_3 = 2.171 \times 10^{-3} \text{ F} \times \frac{1.00 \text{ eq}}{\text{F}}$$

$$\times \frac{400 \text{ g } Fe_2(SO_4)_3}{\text{mole}} \times \frac{1}{2 \text{ eq/mole } Fe_2(SO_4)_3}$$

$$= 0.434 \text{ g}$$

The coulometric method has been advantageously applied to the deposition of metals at a mercury cathode for which a gravimetric completion of the analysis is inconvenient. Excellent methods have been described for the analysis of lead in the presence of cadmium, copper in the presence of bismuth, and nickel in the presence of cobalt.

The controlled-potential coulometric procedure also offers possibilities for the electrolytic determination of organic compounds. For example, Meites and Meites[7] have demonstrated that trichloroacetic acid and picric acid are quantitatively reduced at a mercury cathode whose potential is suitably controlled:

$$Cl_3CCOO^- + H^+ + 2e \rightleftharpoons Cl_2HCCOO^- + Cl^-$$

[7] T. Meites and L. Meites, *Anal. Chem.*, **27**, 1531 (1955); **28**, 103 (1956).

Coulometric measurements permit the analysis of these compounds with an accuracy of a few tenths of a percent.

Variable-current coulometric methods are frequently used to monitor continuously and automatically the concentration of constituents of gas or liquid streams. An important example is the determination of small concentrations of oxygen in gases or liquids.[8] A schematic diagram of the apparatus is shown in Figure 15-5. The porous silver cathode serves to break up the incoming gas into small bubbles; the reduction of oxygen

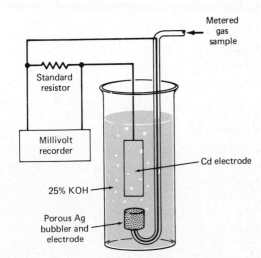

FIGURE 15-5 An instrument for continuously recording the O_2 content of a gas stream.

takes place quantitatively within the pores. That is,

$$O_2(g) + 2H_2O + 4e \rightleftharpoons 4OH^-$$

The anode is a heavy cadmium sheet; here the half-cell reaction is

$$Cd(s) + 2OH^- \rightleftharpoons Cd(OH)_2(s) + 2e$$

Note that a galvanic cell is formed so that no external supply is required. The current produced is passed through a standard resistor and the potential drop is recorded on a millivolt recorder. The oxygen concentration can then be calculated by Equation 15-4. The chart paper, of course, can be made to display the instantaneous oxygen concentration directly. The instrument is reported to provide oxygen concentration data in the range from 1 ppm to 1%.

[8]For further details see F. A. Keidel, *Ind. Eng. Chem.*, **52**, 491 (1960).

Polarography[9]

The polarographic method of analysis was discovered by the Czechoslovakian chemist Jaroslav Heyrovsky in 1922. This method proved to be of sufficient importance to merit award of the 1959 Nobel Prize in chemistry to its inventor.

Several variants of the original polarographic procedure have been developed, and the term *voltammetry* is often used to describe these as well as the original polarographic method. A discussion of one of these variants, *amperometric titrations*, is included near the end of this chapter.

POLAROGRAPHIC MEASUREMENTS

In a polarographic analysis a solution of the analyte is made part of an electrochemical cell that employs a tiny, easily polarized electrode (a *microelectrode*) and a large nonpolarizable reference electrode. Known and variable potentials are applied to the cell and the resulting currents are recorded. The plot of the current as a function of voltage, called a *polarogram* (see Figure 15-8), provides both qualitative and quantitative information regarding the analyte solution.

For successful polarography the electrode at which the analyte reacts must be small; typically its surface area will range from 1 to 10 mm^2. In addition, the electrode must be chemically inert. Thus microelectrodes are fabricated of conducting materials such as mercury, platinum, gold, silver, and graphite. With the exception of mercury, the electrodes usually are fine wires or disks that are sealed into glass tubing.

The most important microelectrode for polarography, and the one used in Heyrovsky's original work, is the *dropping mercury electrode*. Most of the discussion in this chapter will therefore be concerned with this electrode, which has unique properties that make it particularly well suited for polarographic analyses.

The Dropping Mercury Electrode. A typical dropping mercury electrode is shown as part of the polarographic cell in Figure 15-6. Here mercury is forced through a 5- to 10-cm capillary tubing having an internal diameter of approximately 0.05 mm (capillaries of this type are available

---

[9]The principles and applications of polarography are considered in detail in a number of monographs. See, for example, I. M. Kolthoff and J. J. Lingane, *Polarography*, 2d ed. New York: Interscience, 1952; L. Meites, *Polarographic Techniques*, 2d ed. New York: Interscience, 1965; J. Heyrovsky and J. Kůta, *Principles of Polarography*. New York: Academic Press, 1966; P. Zuman, *Organic Polarographic Analysis*. Oxford: Pergamon Press, 1964; P. Zuman, *The Elucidation of Organic Electrode Processes*. New York: Academic Press, 1969; H. W. Nurnberg, ed., *Electroanalytical Chemistry*. New York: Wiley, 1974, Chapters 1–5.

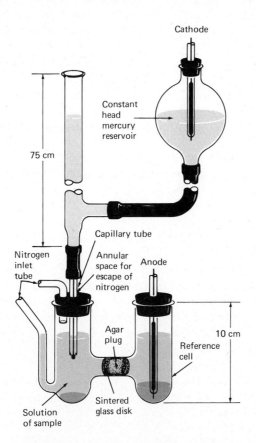

Cathode

Constant
head
mercury
reservoir

75 cm

Capillary tube

Nitrogen
inlet
tube

Annular
space for
escape of
nitrogen

Anode

Agar
plug

10 cm

Reference
cell

Solution
of sample

Sintered
glass disk

FIGURE **15-6** A dropping mercury electrode and cell. [From J. J. Lingane and H. A. Laitinen, *Ind. Eng. Chem., Anal. Ed.*, **11**, 504 (1939). With permission of the American Chemical Society.]

commercially). Under a head of about 50 cm of mercury a continuous flow of identical droplets with diameters of 0.5 to 1 mm results; typically the lifetime of an individual drop is 2 to 6 sec.

The success of a polarographic experiment is critically dependent upon the reproducible behavior of the dropping electrode. Care must therefore be taken to be sure that the tip is as nearly square as possible and that the electrode is mounted in a vertical position. Furthermore, the mounting must be freed as much as possible from vibrations.

With reasonable care a capillary will last for several months or even years. The precautions needed to ensure such performance involve the use of scrupulously clean mercury and the maintenance of a slight mercury flow under all circumstances. Malfunction of the electrode can be expected if the head of mercury is ever diminished to the point where solution comes in contact with the inner surface of the tip. Thus a sufficient head of mercury should always be provided before the tip of the electrode is immersed in a solution.

To prevent its becoming contaminated when not in use, the electrode should be rinsed thoroughly with water and dried. The mercury head should then be decreased until the flow in air just ceases. Care must be taken to avoid lowering the mercury too far. Before use, the head is increased. The tip is immersed in 1 : 1 nitric acid for a minute or so and then washed with distilled water.

Reference Electrodes. Any of the reference electrodes discussed in Chapter 14 (p. 354) can be employed for polarographic measurements. The cell shown in Figure 15-5 employs a saturated calomel electrode separated from the analyte solution by means of a sintered glass disk and an agar plug.

For some applications, a pool of mercury in the bottom of the cell serves as an adequate reference electrode.

Electrical Apparatus. For polarographic measurements it is necessary to have the means for varying a voltage continuously over the range from 0 to 3.0 V. The applied potential should usually be known to about 0.01 V. In addition, it must be possible to measure cell currents over the range between 0.01 and perhaps 100 μA with an accuracy of about 0.01 μA. A manual apparatus meeting these requirements is easily constructed from equipment available in most laboratories. More elaborate devices that record polarograms automatically are commercially available.

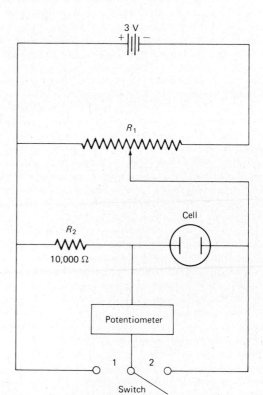

FIGURE 15-7 A simple polarographic circuit. [From J. J. Lingane, *Anal. Chem.*, **21**, 47 (1949). With permission of the American Chemical Society.]

Figure 15-7 shows a circuit diagram of a simple instrument for polarographic work. Two 1.5-V batteries provide a voltage across the 100-Ω potential divider R_1 to permit variation of the potential that is applied to the cell. The magnitude of this voltage can be determined by means of the potentiometer by shifting the switch of the instrument to position 2. The current is measured by determining the potential drop across a precision 10,000-Ω resistance R_2, with the same potentiometer and the switch in position 1.

Polarograms. A polarogram is a plot of current as a function of the potential applied to a polarographic cell. The microelectrode is ordinarily connected to the negative terminal of the power supply; *by convention* the applied potential is given a negative sign under these circumstances. By convention also, the currents are designated as positive when the flow of electrons is from the power supply into the microelectrode—that is, when that electrode behaves as a cathode.

Figure 15-8 shows two polarograms. The lower one is for a solution that is 0.1 F in potassium chloride; the upper one is for a solution that is additionally 1×10^{-3} F in cadmium chloride. The step-shaped current-

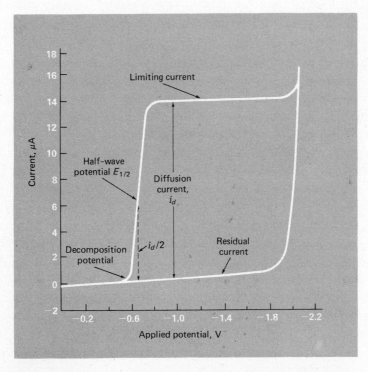

FIGURE 15-8 Polarogram for cadmium ion. The upper curve is for a solution that is 1×10^{-3} F with respect to Cd^{2+} and 0.1 F with respect to KCl. The lower curve is for a solution that is 0.1 F in KCl only.

voltage curve, called a *polarographic wave*, is produced as a result of the reaction

$$Cd^{2+} + 2e + Hg \rightleftharpoons Cd(Hg)$$

The sharp increase in current at about -2 V in both plots is associated with reduction of potassium ions to give a potassium amalgam.

For reasons to be considered presently, a polarographic wave suitable for analysis is obtained in the presence of a large excess of a *supporting electrolyte*; potassium chloride serves this function in the present example. Examination of the polarogram for the supporting electrolyte alone reveals that a small *residual current* exists in the cell even in the absence of cadmium ions. The voltage at which the polarogram for the electrode-reactive species departs from the residual current curve is called the *decomposition potential*.

A characteristic feature of a polarographic wave is the region in which the current, after increasing sharply, becomes essentially independent of the applied voltage; under these circumstances it is called a *limiting current*. We shall see that the limiting current is the result of a restriction in the rate at which the participant in the electrode process can be brought to the surface of the microelectrode; with proper control over experimental conditions this rate is determined exclusively for all points on the wave by the velocity at which the reactant diffuses. A diffusion-controlled limiting current is given a special name, the *diffusion current*, and is assigned the symbol i_d. Ordinarily the diffusion current is directly proportional to the concentration of the reactive constituent and is thus of prime importance from the standpoint of analysis. As shown in Figure 15-8, the diffusion current is the difference between the limiting and the residual currents.

One other important quantity, the *half-wave potential*, is the potential at which the current is equal to one-half the diffusion current. The half-wave potential is given the symbol $E_{1/2}$; it may permit qualitative identification of the reactant.

INTERPRETATION OF POLAROGRAPHIC WAVES

This section provides a qualitative description of the electrode phenomena that lead to the characteristic, step-shaped polarographic wave. The reduction of cadmium ion to yield cadmium amalgam at a dropping mercury electrode will be used as a specific example; the conclusions, however, are applicable to other types of electrodes, oxidation processes, and reactions that yield other types of products.

The half-reaction

$$Cd^{2+} + Hg + 2e \rightleftharpoons Cd(Hg)$$

is said to be reversible. In the context of polarography, reversibility implies that the electron transfer process is sufficiently rapid that the activities of reactants and products in the liquid film at the interface between the solution and the mercury electrode are determined by the electrode potential alone; thus for the reversible reduction of cadmium ion, it may be assumed that at any instant the activities of reactant and product at this interface are given by

$$E_{applied} = E^0_A - \frac{0.0591}{2} \log \frac{[Cd]_0}{[Cd^{2+}]_0} - E_{ref} \qquad \textbf{(15-5)}$$

Here $[Cd]_0$ is the activity of metallic cadmium dissolved in the surface film of the mercury, and $[Cd^{2+}]_0$ is the activity of the ion in the aqueous phase. Note that the subscript zero has been employed for the activity terms to emphasize that this relationship *applies to surface films of the two media only*; the activity of cadmium ion in the bulk of the solution and of elemental cadmium in the interior of the mercury drop *will ordinarily be quite different from the surface activities*. The films we are concerned with are no more than a few molecules thick.

The term $E_{applied}$ in Equation 15-5 is the potential applied to the dropping electrode, and E^0_A is the standard potential for the half-reaction in which a saturated cadmium amalgam is the product; the difference between E^0_A and the standard potential when the product is elemental cadmium is about $+0.05$ V.

Consider now what occurs when $E_{applied}$ is sufficiently negative to cause appreciable reduction of cadmium ion. Because the reaction is reversible, the activity of cadmium ion in the film surrounding the electrode decreases, and the activity of cadmium in the outer layer of the mercury drop increases instantaneously to the level demanded by Equation 15-5; a momentary current results. This current would rapidly decay to zero were it not for the fact that cadmium ions are mobile in the aqueous medium and can migrate to the surface of the mercury as a consequence. Because the reduction reaction is essentially instantaneous, *the magnitude of the current depends upon the rate at which the cadmium ions move from the bulk of the solution to the surface where reaction occurs*. That is,

$$i = k' \times v_{Cd^{2+}}$$

where i is the current at an applied potential $E_{applied}$, $v_{Cd^{2+}}$ is the rate of migration of cadmium ions, and k' is a proportionality constant.

We have noted (p. 392) that the ions or molecules in a cell migrate as a consequence of diffusion, thermal or mechanical convection, or electrostatic attraction. In polarography every effort is made to eliminate the latter two effects. Thus vibration or stirring of the solution is avoided, and an excess of a nonreactive supporting electrolyte is employed; if the concentration of supporting electrolyte is 50 times (or more) greater than that of the reactant, the attractive (or repulsive) force between the electrode and the reactant becomes negligible.

With mechanical mixing and electrostatic attraction eliminated, diffusion becomes the only force responsible for the transport of cadmium ions to the electrode surface. Because the rate of diffusion is directly proportional to the concentration (strictly activity) difference between two parts of a solution, we may write

$$v_{Cd^{2+}} = k''([Cd^{2+}] - [Cd^{2+}]_0)$$

where $[Cd^{2+}]$ is the concentration *in the bulk of the solution* from which ions are diffusing and $[Cd^{2+}]_0$ is the concentration in the aqueous film surrounding the electrode. As long as diffusion is the only process bringing cadmium ions to the surface, it follows that

$$i = k'v_{Cd^{2+}} = k'k''([Cd^{2+}] - [Cd^{2+}]_0)$$
$$= k([Cd^{2+}] - [Cd^{2+}]_0)$$

Note that $[Cd^{2+}]_0$ becomes smaller as $E_{applied}$ is made more negative. Thus the rate of diffusion as well as the current increases with increases in applied potential. When the applied potential becomes sufficiently negative, however, the concentration of cadmium ion in the surface film approaches zero with respect to the concentration in the bulk of the solution; under this circumstance the rate of diffusion and thus the current becomes constant, and the expression for current becomes

$$i_d = k[Cd^{2+}]$$

where i_d is the potential-independent diffusion current. Note that *the magnitude of the diffusion current is directly proportional to the concentration of the reactant in the bulk of the solution.* Quantitative polarography is based upon this fact.

A state of *complete concentration polarization* is said to exist when the current in a cell is limited by the rate at which a reactant can be brought to the surface of an electrode. With a microelectrode the current required to reach this condition is small—typically 3 to 10 μA (microampere) for a 10^{-3} M solution. It is important to note that such current levels do not significantly alter the reactant concentration.

Half-Wave Potential. An equation relating the applied potential, $E_{applied}$, and current, i, is readily derived.[10] For the reduction of cadmium ion to cadmium amalgam the equation takes the form

$$E_{applied} = E_{1/2} - \frac{0.0591}{n} \log \frac{i}{i_d - i} \tag{15-6}$$

where

$$E_{1/2} = E^0_A - \frac{0.0591}{n} \log \frac{f_{Cd}k_{Cd}}{f_{Cd^{2+}}k_{Cd^{2+}}} - E_{ref} \tag{15-7}$$

[10] I. M. Kolthoff and J. J. Lingane, *Polarography*, vol. 1. New York: Interscience, 1952, p. 190.

Here f_{Cd} and $f_{Cd^{2+}}$ are activity coefficients for the metal in the amalgam and the ion in the solution; k_{Cd} and $k_{Cd^{2+}}$ are proportionality constants related to the rates at which the two species diffuse in the respective media.

Examination of Equation 15-7 reveals that the half-wave potential is a reference point on a polarographic wave; it is independent of the reactant concentration but directly related to the standard potential for the half-reaction. In practice the half-wave potential can be a useful quantity for identification of the species responsible for a given polarographic wave.

It is important to note that the half-wave potential may vary considerably with concentration for electrode reactions that are not rapid and reversible; in addition, Equation 15-6 no longer describes such waves.

Effect of Complex Formation on Polarographic Waves. We have already seen (p. 285) that the potential for the oxidation or reduction of a metallic ion is greatly affected by the presence of species that form complexes with that ion. It is not surprising, therefore, that similar effects are observed with polarographic half-wave potentials. The data in Table 15-4 indicate that the half-wave potential for the reduction of a metal complex is generally more negative than that for reduction of the corresponding simple metal ion.

Lingane[11] has shown that the shift in half-wave potential as a function of the concentration of complexing agent can be employed to determine the formula and the formation constant for complexes, provided that the cation involved reacts reversibly at the dropping electrode.

TABLE 15-4
Effect of Complexing Agents on Polarographic Half-Wave Potentials at the Dropping Mercury Electrode

ION	NONCOMPLEXING MEDIA	1-F KCN	1-F KCl	1-F NH$_3$, 1-F NH$_4$Cl
Cd^{2+}	-0.59	-1.18	-0.64	-0.81
Zn^{2+}	-1.00	NR^a	-1.00	-1.35
Pb^{2+}	-0.40	-0.72	-0.44	-0.67
Ni^{2+}	—	-1.36	-1.20	-1.10
Co^{2+}	—	-1.45	-1.20	-1.29
Cu^{2+}	$+0.02$	NR^a	$+0.04$	-0.24
			and -0.22	and -0.51

^{a}No reduction occurs before involvement of the supporting electrolyte.

[11]J. J. Lingane, *Chem. Rev.*, **29**, 1 (1941).

Polarograms for Mixtures of Reactants. The reactants of a mixture ordinarily behave independently of one another at a microelectrode so that a polarogram for a mixture is simply a composite of the waves for the individual components. Figure 15-9 shows polarograms for a pair of two-component mixtures. The half-wave potentials of the two reactants differ by about 0.1 V in curve *A* and by about 0.2 V in curve *B*. Figure 15-9 suggests that a single polarogram may permit the analysis for several components in a mixture. Success depends upon the existence of a sufficient difference between succeeding half-wave potentials to permit evaluation of individual diffusion currents. Approximately 0.2 V is required if the more reducible species undergoes a two-electron reduction; a minimum between 0.2 and 0.3 V is needed if the first reduction is a one-electron process.

Polarograms for Irreversible Reactions. Many polarographic electrode processes, particularly those associated with organic systems, are irreversible; drawn-out and less well-defined waves result. Although half-wave potentials for irreversible reactions ordinarily show a dependence upon concentration, diffusion currents remain linearly related to this variable, and such processes are readily adapted to quantitative analysis.

PROPERTIES OF THE DROPPING MERCURY ELECTRODE

Most polarographic work has been performed with a dropping mercury electrode; therefore let us consider some of the unique features of this device.

Current Variations during the Lifetime of a Drop. The current passing through a cell containing a dropping electrode undergoes periodic fluctuations corresponding in frequency to the drop rate. As a drop breaks, the

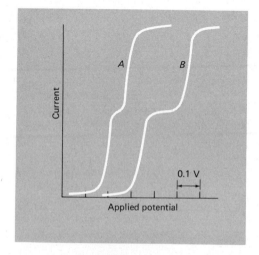

FIGURE 15-9 Polarograms for two-component mixtures. Half-wave potentials differ by 0.1 V in curve *A* and by 0.2 V in curve *B*.

current falls to zero; it then increases rapidly as the electrode area grows because of the greater surface to which diffusion can occur. In order to measure such rapidly fluctuating currents, a well-damped galvanometer must be employed. As shown in Figure 15-10, the oscillations under these circumstances are limited to a reasonable magnitude, and the average current is readily determined provided the drop rate is reproducible. Note the effect of irregular drops in the center of the limiting current region, probably caused by vibration of the apparatus.

Advantages and Limitations of the Dropping Mercury Electrode.

The dropping mercury electrode offers several advantages over other types of microelectrodes. The first is the large overvoltage for the formation of hydrogen from hydrogen ions. As a consequence, the reduction of many substances from acidic solutions can be studied without interference. Second, because a new metal surface is continuously generated, the behavior of the electrode is independent of its past history. Thus reproducible current-voltage curves are obtained regardless of how the electrode has been used previously. The third useful feature of a dropping electrode is that reproducible average currents are immediately achieved at any given applied potential.

The most serious limitation to the dropping electrode is the ease with which mercury is oxidized; this property severely restricts the use of mercury as an anode. At applied potentials much above +0.4 V (versus the

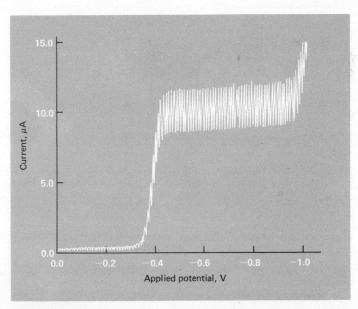

FIGURE 15-10 Typical polarogram produced by an instrument that continuously and automatically records current-voltage data.

saturated calomel electrode) formation of mercury(I) occurs; the resulting current masks the polarographic waves of other oxidizable species in the solution. Thus the dropping mercury electrode can be employed only for the analysis of reducible or very easily oxidizable substances. Other disadvantages are that it is cumbersome to use and tends to malfunction by clogging.

SCOPE OF POLAROGRAPHY

Virtually every element in one form or another is amenable to polarographic analysis. In addition, the method can be extended to the determination of numerous organic functional groups. Because the polarographic behavior of any species is unique for a given set of experimental conditions, the technique offers attractive possibilities for selective analysis.

Most polarographic analyses are performed in aqueous solution, but other solvent systems may be substituted if necessary. For quantitative analyses the optimum concentration range lies between 10^{-2} and 10^{-4} M; by suitably modifying the basic polarographic procedure, however, concentration determinations in the parts-per-billion range become possible. An analysis can be easily performed on 1 to 2 ml of solution; with a little effort a volume as small as one drop is sufficient. The polarographic method is thus particularly useful for the determination of quantities in the milligram to microgram range.

Relative errors ranging between 2 and 3% are to be expected in routine polarographic work. This order of uncertainty is comparable to or smaller than errors affecting other methods for the analysis of small quantities.

ANALYTICAL DETAILS

Several important variables that affect the applicability and the accuracy of polarography are considered in this section.

Temperature Control. The diffusion currents for most substances are found to increase by about 2.5% per degree Celsius. Thus solutions must be thermostated to a few tenths of a degree for accurate polarographic analysis.

Oxygen Removal. Dissolved oxygen undergoes a two-step irreversible reduction at the dropping electrode. The H_2O_2 produced in the first step is reduced to H_2O in the second. Two waves of equal size result, the first with a half-wave potential at about -0.14 V and the second at about -0.9 V. The two half-reactions are somewhat slow. As a consequence, the waves are drawn out over a considerable potential range.

While these polarographic waves are convenient for the determination of oxygen in solutions, the presence of this element often interferes with the accurate determination of other species. Thus oxygen removal is ordinarily the first step in a polarographic analysis. Aeration of the solution for several minutes with an inert gas accomplishes this end. A stream of the same gas, usually nitrogen, is passed over the surface during the analysis to prevent reabsorption.

Current Maxima. Polarographic waves are frequently distorted by so-called current maxima (see Figure 15-11). Maxima are troublesome because they interfere with the accurate evaluation of diffusion currents and half-wave potentials. The cause or causes of maxima are not fully understood; fortunately, however, considerable empirical knowledge exists for their elimination. The addition of traces of such high molecular weight substances as gelatin, Triton X-100 (a commercial surface-active agent), methyl red, other dyes, or carpenter's glue is effective. The first two of these additives are particularly useful.

Determination of Diffusion Currents. In measuring currents obtained with the dropping electrode, it is common practice to use the *average* value of the galvanometer or recorder oscillations rather than the maximum or minimum values; the measurement is thus less dependent upon the damping employed.

For analytical work, limiting currents must always be corrected for the residual current. A residual current curve is ordinarily obtained immediately before or after the curve for the analyte. The diffusion current is taken as the difference between the two at some potential in the limiting current region (see Figure 15-8).

Because the residual current usually increases nearly linearly with applied voltage, it is often possible to dispense with a separate residual-current curve and obtain the correction by extrapolation of the residual-current portion of the curve for the sample.

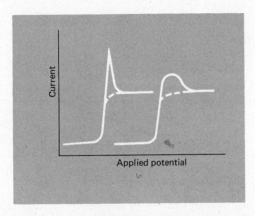

Applied potential

FIGURE 15-11 Typical current maxima.

Concentration Determination. The best and most straightforward method for quantitative polarographic analysis involves preliminary calibration with a series of standard solutions. As nearly as possible these standards should be identical with the samples being analyzed and cover a concentration range within which the unknown samples will likely fall. From such data the linearity of the current-concentration relationship can be assessed; if this relationship is nonlinear, the analysis can be based upon the calibration curve.

The standard addition method (p. 374) is also applicable to polarographic analysis. This procedure is particularly effective when the diffusion current is sensitive to other components of the solution that occur with the sample.

INORGANIC POLAROGRAPHIC ANALYSIS

The polarographic method is generally applicable to the analysis of inorganic substances. Most metallic cations, for example, are reduced at the dropping electrode to form a metal amalgam or a lower oxidation state. Even the alkali- and alkaline-earth metals are reducible, provided the supporting electrolyte used does not decompose at the high potentials required. The tetraalkyl ammonium halides serve this function well.

ORGANIC POLAROGRAPHIC ANALYSIS

Almost from its inception the polarographic method has been used for the study and analysis of organic compounds, and a large number of papers have been devoted to this subject. Several common functional groups are oxidized or reduced at the dropping electrode; compounds containing these groups can be analyzed polarographically.

Organic compounds containing any of the following functional groups can be expected to react at the dropping mercury electrode and produce one or more polarographic waves.

1. *The carbonyl group* includes aldehydes, ketones, and quinones. Aldehydes tend to be reduced at lower potentials than ketones. Conjugation of the carbonyl double bond also leads to lower half-wave potentials.
2. *Certain carboxylic acids* are reduced polarographically, although the simple aliphatic and aromatic monocarboxylic acids are not. Dicarboxylic acids such as fumaric, maleic, or phthalic acid, in which the carboxyl groups are conjugated with one another, give characteristic polarograms. The same is true of certain keto and aldehydo acids.
3. *Most peroxides and epoxides* yield useful polarograms.
4. *Nitro, nitroso, amine oxide,* and *azo groups* are generally reduced at the dropping electrode.

5. *Most organic halogen groups* produce a polarographic wave as the halogen is replaced with an atom of hydrogen.
6. *The carbon-carbon double bond* is reduced when it is conjugated with another double bond, an aromatic ring, or an unsaturated group.
7. *Hydroquinones and mercaptans* produce anodic waves.

In addition, numerous applications to biological systems have been reported.[12]

Amperometric Titrations with One Microelectrode

The polarographic method can be used to estimate the equivalence point of a reaction, provided at least one of the participants or products of the titration is oxidized or reduced at a microelectrode. Here the current passing through a polarographic cell at some fixed potential is measured as a function of titrant volume (or of time in a coulometric titration). Plots of the data on either side of the equivalence point are straight lines of differing slopes; the end point can be fixed by extrapolation to their intersection.

The amperometric method is inherently more accurate than the polarographic method and less dependent upon the characteristics of the electrode and the supporting electrolyte. Furthermore, the temperature need not be fixed accurately, although it must be kept constant during the titration. Finally, the substance being determined need not be reactive at the electrode; a reactive reagent or product will suffice as well.

TITRATION CURVES

Amperometric titration curves typically take one of the forms shown in Figure 15-12. Figure 15-12a represents a titration in which the analyte reacts at the electrode while the reagent does not. The titration of lead with sulfate or oxalate ions may be cited as an example. Here a sufficiently high potential is applied to give a diffusion current for lead; a linear decrease in current is observed as lead ions are removed from the solution by precipitation. The curvature near the equivalence point reflects the incompleteness of the analytical reaction in this region. The end point is obtained by extrapolation of the linear portions, as shown.

The curve of Figure 15-12b is typical of a titration in which the reagent reacts at the microelectrode and the analyte does not. An example of this would be the titration of magnesium with 8-hydroxyquinoline. A

[12]M. Brezina and P. Zuman, *Polarography in Medicine, Biochemistry and Pharmacy.* New York: Interscience, 1958; P. Zuman, *Organic Polarographic Analysis.* Oxford: Pergamon Press, 1964.

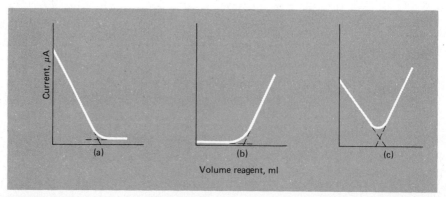

FIGURE 15-12 Typical amperometric titration curves. (a) The substance analyzed is reduced; the reagent is not. (b) The reagent is reduced; the substance analyzed is not. (c) Both the reagent and the substance analyzed are reduced.

diffusion current for the latter is obtained at -1.6 V (versus the saturated calomel electrode), whereas magnesium ion is inert at this potential.

The curve of Figure 15-12c corresponds to the titration of lead ion with a chromate solution at an applied potential greater than -1.0 V. Both lead and chromate ions give diffusion currents at this potential, and a minimum in the curve signals the end point. This titration would yield a curve like that of Figure 15-12b at zero applied potential because only chromate ions are reduced under these conditions.

In order to obtain plots with linear regions before and after the equivalence point (Figure 15-12), it is necessary to apply corrections for the volume change that results from the added titrants. This correction is made by multiplying the measured diffusion current by $(V + v)/V$, where V is the original volume and v is the volume of reagent; thus all measured currents are corrected back to the original volume. An alternative, which is often satisfactory, is to make v negligibly small through use of a titrant solution that is 20 or more times as concentrated as the solution being titrated. This approach, however, does require the use of a microburet so that a total reagent volume of 1 or 2 ml can be measured with a suitable accuracy.

EQUIPMENT

For amperometric titrations a simple manual polarograph is entirely adequate. Ordinarily the applied voltage does not have to be known any closer than about ± 0.05 V because it is only necessary to select a potential within the diffusion current region of at least one reactant or product in the titration.

Many amperometric titrations can be carried out conveniently with a dropping mercury electrode. For reactions involving oxidizing agents that attack mercury (bromine, silver ion, iron(III), and so on) a rotating platinum electrode is preferable. This microelectrode consists of a short length of platinum wire sealed into the side of a glass tube. Mercury inside the tube provides electrical contact between the wire and the lead to the polarograph. The tube is held in the hollow chuck of a synchronous motor and rotated at a constant speed in excess of 600 rpm. Commercial models of the rotating electrode are available. A typical apparatus is shown in Figure 15-13.

Polarographic waves, similar in appearance to those observed with the dropping mercury electrode, can be obtained with the rotating platinum electrode. Here, however, the reactive species is brought to the electrode surface not only by diffusion but also by mechanical mixing. As a consequence, the limiting currents are as much as 20 times larger than those obtained with a microelectrode that is supplied by diffusion only. With a rotating electrode steady currents are instantaneously obtained. This behavior is in distinct contrast to that of a solid microelectrode in an unstirred solution.

Several limitations restrict the widespread application of the rotating platinum electrode to polarography. The low hydrogen overvoltage prevents its use as a cathode in acidic solutions. In addition, the high currents obtained with the electrode make it particularly sensitive to traces of oxygen in the solution. These two factors have largely confined its employment to anodic reactions. Limiting currents from a rotating electrode are often influenced by the previous history of the electrode and are seldom as reproducible as the diffusion currents obtained with a dropping mercury electrode. These limitations, however, do not seriously restrict the use of the rotating electrode for amperometric titrations.

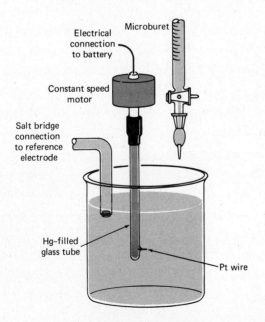

Electrical
connection
to battery

Microburet

Constant speed
motor

Salt bridge
connection
to reference
electrode

Hg–filled
glass tube

Pt wire

FIGURE 15-13 Typical cell arrangement for amperometric titrations with a rotating platinum electrode.

APPLICATION OF AMPEROMETRIC TITRATIONS WITH ONE MICROELECTRODE

The amperometric end point has been largely confined to titrations in which a slightly soluble precipitate or a stable complex is the reaction product. Selected applications are listed in Table 15-5. Some organic precipitants that are reduced at the dropping electrode appear in this table.

TABLE 15-5
Applications of Amperometric Titrations

REAGENT	REACTION PRODUCT	TYPE ELECTRODE[a]	SUBSTANCE DETERMINED
K_2CrO_4	Precipitate	DME	Pb^{2+}, Ba^{2+}
$Pb(NO_3)_2$	Precipitate	DME	SO_4^{2-}, MoO_4^{2-}, F^-, Cl^-
8-Hydroxy-quinoline	Precipitate	DME	Mg^{2+}, Zn^{2+}, Cu^{2+}, Cd^{2+}, Al^{3+}, Bi^{3+}, Fe^{3+}
Cupferron	Precipitate	DME	Cu^{2+}, Fe^{3+}
Dimethyl-glyoxime	Precipitate	DME	Ni^{2+}
α-Nitroso-β-naphthol	Precipitate	DME	Co^{2+}, Cu^{2+}, Pd^{2+}
$K_4Fe(CN)_6$	Precipitate	DME	Zn^{2+}
$AgNO_3$	Precipitate	RP	Cl^-, Br^-, I^-, CN^-, RSH
EDTA	Complex	DME	Bi^{3+}, Cd^{2+}, Cu^{2+}, Ca^{2+}, and so on
$KBrO_3$, KBr	Substitution, addition, or oxidation	RP	Certain phenols, aromatic amines, olefins; N_2H_4, As(III), Sb(III)

[a] DME = dropping mercury electrode; RP = rotating platinum electrode.

A few applications of the amperometric method to oxidation-reduction reactions exist. For example, the technique has been applied to various titrations involving iodine and bromine (in the form of bromate) as reagents.

Amperometric Titrations with Two Polarized Microelectrodes

A convenient modification of the amperometric method involves the use of two identical, stationary microelectrodes immersed in a well-stirred solution of the sample. A small potential (say, 0.1 to 0.2 V) is applied between these electrodes, and the current is followed as a function of the volume of added reagent. The end point is marked by a sudden current rise from zero, a decrease in the current to zero, or a minimum (at zero) in a V-shaped curve.

Although the use of two polarized electrodes for end-point detection was first proposed before 1900, almost 30 years passed before chemists came to appreciate the potentialities of the method.[13] The name *dead-stop end point* was used to describe the technique, and this term is still occasionally encountered. It was not until about 1950 that a clear interpretation of dead-stop titration curves was made.[14]

APPLICATIONS TO TITRATIONS WITH SILVER IONS

Twin silver microelectrodes can be employed to mark the end point for the various titrations employing silver nitrate shown in Table 8-2 (p. 181). Consider what happens, for example, when a small potential is applied between two such electrodes during the titration of bromide with silver ions. Short of the equivalence point essentially no current can exist because no easily reduced species are present in the solution. Consequently electron transfer at the cathode is precluded and that electrode is completely polarized. Note that the anode is not polarized because the reaction

$$Ag \rightleftharpoons Ag^+ + e$$

could occur in the presence of a suitable cathodic reactant.

After equivalence has been passed, the cathode becomes depolarized owing to the presence of significant amount of silver ions which can react to give silver. That is,

$$Ag^+ + e \rightleftharpoons Ag$$

Current is permitted as a result of this half-reaction and the corresponding oxidation of silver at the anode. The magnitude of the current is, as in other amperometric methods, directly proportional to the concentration of the excess reagent. Thus the titration curve is similar in form to that shown in Figure 15-12b.

[13] C. W. Foulk and A. T. Bawden, *J. Amer. Chem. Soc.*, **48**, 2045 (1926).

[14] For an excellent analysis of this type of end point see J. J. Lingane, *Electroanalytical Chemistry*, 2d ed. New York: Interscience, 1958, pp. 280–294.

APPLICATION TO TITRATIONS WITH BROMINE AND IODINE

An amperometric titration with twin platinum microelectrodes has also been applied to titrations in which iodine or bromine is the titrant. When, for example, this technique is employed for the titration of arsenious acid, a curve similar to Figure 15-12b is again obtained. In the early stages of the titration no current is observed because of cathodic polarization. In contrast to the previous example, however, polarization here is of the kinetic type. That is, even though a cathodic half-reaction involving H_3AsO_4, the titration product, can be written

$$H_3AsO_4 + 2e + 2H^+ \rightleftharpoons H_3AsO_3 + H_2O$$

this process occurs so slowly at the electrode surface that no current can be detected.

The principal advantage of the twin microelectrode procedure is its simplicity. One can dispense with a reference electrode; the only instrumentation needed is a simple voltage divider, powered by a dry cell, and a galvanometer or microammeter for current detection.

PROBLEMS

*1. Electrodeposition is to be used to separate the cations in a solution that is buffered to a pH of 8.00 and is $5.00 \times 10^{-3} F$ with respect to Zn(II) and $6.00 \times 10^{-3} F$ with respect to Cd(II). Oxygen is evolved at the anode at a partial pressure of 1.00 atm; the cell has a resistance of 2.40 Ω.
   (a) Which metal ion will deposit first?
   (b) Calculate the theoretical potential needed to commence deposition with a current of 0.300 A.

*2. The cell
   $$Pt|Fe^{2+}(0.180\ F),\ Fe^{3+}(0.0100\ F)\|Pd^{2+}(0.0900\ F)|Pd$$

   has a resistance of 6.50 Ω. Calculate the potential needed to pass a current of 0.0500 A through this cell.

3. The cell
   $$Cu|Cu^{2+}(0.0300\ F)\|Ag^+(0.120\ F)|Ag$$

   has a resistance of 8.00 Ω. Calculate the initial potential when 0.0100 A is drawn from this cell.

*4. Deposition of 0.0764 g of PbO_2 on the anode of an electrolysis cell required a constant current of 0.240 A for 19.6 min. Calculate the current efficiency of the anode with respect to the process

   $$Pb^{2+} + 2H_2O \rightleftharpoons PbO_2(s) + 4H^+ + 2e$$

5. Passage of a constant 0.300 A through an electrolysis cell for 26.8 min resulted

in the deposition of 0.0744 g of copper. Calculate the current efficiency of the cathode with respect to the process

$$Cu^{2+} + 2e \rightleftharpoons Cu(s)$$

*6. The Cl^- in 50.0 ml of an aqueous sample was precipitated by silver ion generated at a silver anode. Quantitative precipitation required the passage of a constant current of 0.142 A for 3 min and 45 sec. Calculate the milligrams per milliliter of $BaCl_2$ in the sample.

7. An apparatus similar to that shown in Figure 15-5 was employed to determine the oxygen content of a light hydrocarbon gas stream having a density of 0.00140 g/ml. Passage of a 25.0-liter sample of the gas through the cell consumed 5.50 C of electricity. Calculate the parts per million O_2 in the sample on a weight basis.

*8. The iron in a 0.1120-g sample of ore was converted to the +2 state by suitable treatment and then oxidized quantitatively at a Pt anode maintained at −1.0 V (vs. SCE). The quantity of electricity required to complete the oxidation was determined with a chemical coulometer equipped with a platinum anode immersed in an excess of iodide ion. The iodine liberated by the passage of current required 33.3 ml of 0.0101-N sodium thiosulfate to reach a starch end point. What was the percentage of Fe_3O_4 in the sample?

9. The calcium content of a water sample was determined by adding an excess of an $HgNH_3Y^{2-}$ solution to a 25.0-ml sample. The anion of EDTA was then generated at a mercury cathode (see Table 15-2). A constant current of 0.0303 A was needed to reach an end point after 3 min and 47 sec. Calculate the milligrams of $CaCO_3$ per liter of sample.

*10. The phenol content of water downstream from a coking furnace was determined by coulometric analysis. A 100-ml sample was rendered slightly acidic and an excess of KBr was introduced. To produce Br_2 for the reaction

$$C_6H_5OH + 3Br_2 \rightarrow Br_3C_6H_2OH(s) + 3HBr$$

a steady current of 28.6 mA for 6 min and 2.5 sec was required. Express the results of this analysis in terms of parts phenol per million parts of water (assume that the density of water = 1.00 g/ml).

11. The equivalent weight of an organic acid was obtained by dissolving 0.0145 g of the purified compound in an alcohol-water mixture and titrating with coulometrically generated hydroxide ions. With a current of 0.0712 A, 286 sec were required to reach a phenolphthalein end point. Calculate the equivalent weight of the compound.

*12. The cyanide ion concentration in a 10.0-ml sample of a plating solution was determined by titration with electrogenerated hydrogen ions to a methyl orange end point. A color change occurred after 6 min and 10 sec with a current of 35.4 mA. Calculate the grams of NaCN per liter of solution.

*13. A 2.16-g sample of an ant-control preparation was decomposed by wet-ashing with H_2SO_4 and HNO_3. The arsenic in the residue was reduced to the trivalent state with hydrazine. After the excess reducing agent was removed, the arsenic(III) was oxidized with electrolytically generated I_2 in a faintly alkaline medium:

$$HAsO_3^{2-} + I_2 + 2HCO_3^- \rightarrow HAsO_4^{2-} + 2I^- + 2CO_2 + H_2O$$

The titration was complete after a constant current of 87.6 mA had been passed for 4 min and 42 sec. Express the results of this analysis in terms of the percentage As_2O_3 in the original sample.

*14. The nitrobenzene in a 181-mg sample of an organic mixture was reduced to phenylhydroxylamine at a constant potential of -0.96 V (vs. SCE) applied to a Hg cathode:

$$C_6H_5NO_2 + 4H^+ + 4e \rightarrow C_6H_5NHOH + H_2O$$

The sample was dissolved in 100 ml of methanol; after electrolysis for 30 min the reaction was judged complete. An electronic coulometer in series with the cell indicated that the reduction required 22.29 C. Calculate the percent nitrobenzene in the sample.

15. The H_2S content of a water sample was assayed with electrolytically generated iodine. After 3 g of potassium iodide had been introduced to a 50.0-ml portion of the water, the titration required a constant current of 31.9 mA for a total of 6.36 min. The reaction was

$$H_2S + I_2 \rightarrow S(s) + 2H^+ + 2I^-$$

Express the concentration of H_2S in terms of milligrams per liter of sample.

*16. The chromium deposited on one surface of a 10.0-cm^2 test plate was dissolved by treatment with acid and oxidized to the $+6$ state with ammonium peroxodisulfate:

$$3S_2O_8^{2-} + 2Cr^{3+} + 7H_2O = Cr_2O_7^{2-} + 14H^+ + 6SO_4^{2-}$$

The solution was boiled to remove the excess peroxodisulfate, cooled, and the $Cr_2O_7^{2-}$ subjected to coulometric titration with Cu(I) generated from 50 ml of 0.10-F Cu^{2+}. Calculate the weight of chromium that was deposited on each square centimeter of the test plate if the titration required a steady current of 11.7 mA for a period of 4 min and 10 sec. The reaction was

$$Cr_2O_7^{2-} + 6CuCl_3^{2-} + 14H^+ \rightarrow 2Cr^{3+} + 6Cu^{2+} + 18Cl^- + 7H_2O$$

17. Ascorbic acid (gfw = 176) is oxidized to dehydroascorbic acid with Br_2:

$$
\begin{array}{c}
\text{O} \\
\parallel \\
\text{C} \\
|\quad \\
\text{C--OH} \\
\quad\quad\quad \text{O} + Br_2 \longrightarrow \\
\text{C--OH} \\
| \\
\text{H--C} \\
| \\
\text{HO--C--H} \\
| \\
\text{CH}_2\text{OH}
\end{array}
\qquad
\begin{array}{c}
\text{O} \\
\parallel \\
\text{C} \\
|\quad \\
\text{C}=\text{O} \\
\quad\quad\quad \text{O} + 2Br^- + 2H^+ \\
\text{C}=\text{O} \\
| \\
\text{H--C} \\
| \\
\text{HO--C--H} \\
| \\
\text{CH}_2\text{OH}
\end{array}
$$

A vitamin C tablet was dissolved in sufficient water to give exactly 250 ml of solution. A 5.00-ml aliquot was then mixed with an equal volume of 0.100-F KBr. Calculate the weight of ascorbic acid in the tablet if the bromine generated by a steady current of 40.2 mA for a total of 3 min and 26.2 sec was required for the titration.

*18. The odorant concentration of household gas can be monitored by passage of a fraction of the gas stream through a solution containing an excess of bromide ion. Electrogenerated bromine reacts rapidly with the mercaptan odorant:

$$2RSH + Br_2 \rightarrow RSSR + 2H^+ + 2Br^-$$

Continuous analysis is made possible by means of an electrode system that signals the need for additional Br_2 to oxidize the mercaptan. Ordinarily the current required to react with the odorant is automatically plotted as a function of time. Calculate the average odorant concentration (as parts per million C_2H_5SH on a weight basis) from the following information:

Average density of the gas	0.00185 g/ml
Rate of gas flow	13.2 liters/min
Average current during a 10.0-min sampling period	1.66 mA

*19. The cadmium and zinc in a 1.78-g sample were dissolved and subsequently deposited from an ammoniacal solution with a mercury cathode. When the cathode potential was maintained at -0.95 V (vs. SCE), only the cadmium deposited. When the current ceased at this potential, a hydrogen-oxygen coulometer in series with the cell was found to have evolved 67.3 ml of gas (corrected for water vapor) at a temperature of 19.0°C and a barometric pressure of 753 mm Hg. The potential was raised to about -1.3 V, whereupon zinc ion was reduced. Upon completion of this electrolysis an additional 40.6 ml of gas were produced under the same conditions.

   Calculate the percent Cd and Zn in the sample.

20. A 1.56-g sample of a solid containing BaBr$_2$, KI, and inert species was dissolved, made ammoniacal, and placed in a cell equipped with a silver anode. By maintaining the potential at -0.06 V (vs. SCE), I$^-$ was quantitatively

precipitated as AgI without interference from Br^-. The volume of H_2 and O_2 formed in a gas coulometer connected in series with the cell was 27.2 ml (corrected for water vapor) at 29.4°C and at a pressure of 768 mm Hg.

After precipitation of iodide was complete, the solution was acidified, and the Br^- was precipitated from solution as AgBr, at a potential of 0.016 V. The volume of gas formed under the same conditions was 31.3 ml. Calculate the percent $BaBr_2$ and KI in the sample.

*21. When salicylic acid is allowed to stand in the presence of excess Br_2, the following quantitative reaction takes place:

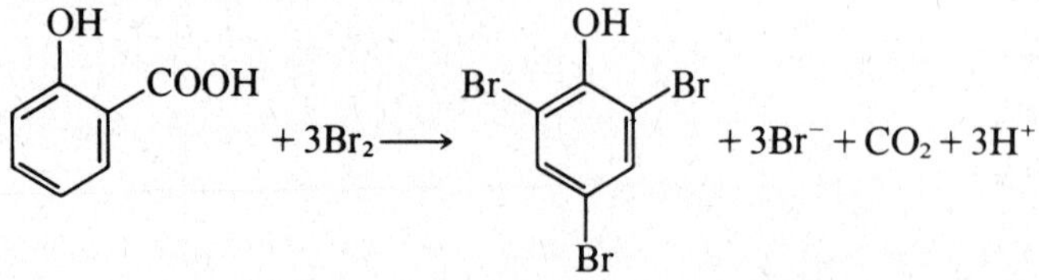

A 0.0668-g sample of a medicinal preparation was dissolved in an acidic solution containing NaBr. The salicylic acid from the sample was brominated by bromine produced at a platinum anode. After 100.0 C of electricity had passed through the solution, the current was turned off and the mixture was allowed to stand for 2 min. The polarity of the working electrode was then reversed and current was passed through the cell until an indicator showed the disappearance of Br_2. Here the reaction was

$$Br_2 + 2e \rightleftharpoons 2Br^-$$

The quantity of electricity consumed by the second reaction was 13.6 C. Calculate the percent salicylic acid in the preparation.

22. Quinone can be reduced to hydroquinone with an excess of electrolytically generated Sn(II)

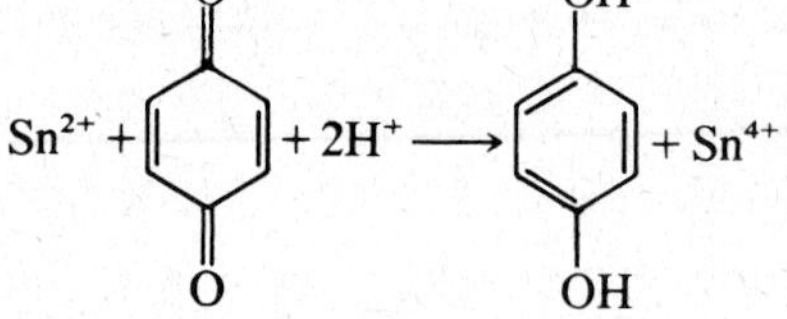

The polarity of the working electrode is then reversed, and the excess Sn(II) is oxidized by Br_2 in a coulometric titration

$$Sn^{2+} + Br_2 \rightarrow Sn^{4+} + 2Br^-$$

Appropriate quantities of $SnCl_4$ and KBr were added to a 50.0-ml aliquot of sample. Calculate the milligrams of quinone (gfw = 108) in each milliliter of sample from the accompanying data:

WORKING ELECTRODE FUNCTIONING AS	GENERATION TIME (min) WITH A CONSTANT CURRENT OF 1.30 mA
Cathode	7.05
Anode	0.75

*23. In a methanol solution, carbon tetrachloride is reduced to chloroform at a Hg cathode, -1.0 V (vs. SCE) being required for the reduction:

$$2CCl_4 + 2H^+ + 2e + 2Hg(l) \rightarrow 2CHCl_3 + Hg_2Cl_2(s)$$

At -1.80 V the chloroform produced further reacts to give methane:

$$2CHCl_3 + 6H^+ + 6e + 6Hg(l) \rightarrow 2CH_4 + 3Hg_2Cl_2(s)$$

A 0.821-g sample containing CCl_4, $CHCl_3$, and inert organic species was dissolved in methanol and electrolyzed at -1.0 V until the current approached zero. A coulometer indicated that 12.16 C were used. The reduction was then continued at -1.80 V; an additional 93.12 C were required to complete the reaction. Calculate the percent CCl_4 and $CHCl_3$ in the mixture.

24. A 0.1027-g sample containing only $CHCl_3$ and CH_2Cl_2 was dissolved in methanol and electrolyzed in a cell containing a Hg cathode; the potential of the cathode was held constant at -1.80 V (vs. SCE). Both compounds were reduced to CH_4 (see Problem 23 for the reaction type). Calculate the percent $CHCl_3$ and CH_2Cl_2 if 245.6 C were required to complete the reduction.

*25. Calculate the bismuth concentration (M) on the basis of the following polarographic data:

SOLUTION COMPOSITIONS, ml				CURRENT AT
0.20-F $HClO_4$	Sample	7.82×10^{-3} M BiO^+	H_2O	-1.0 V, μA
10.0	0.00	0.00	15.0	1.4
10.0	5.00	0.00	10.0	33.7
10.0	5.00	5.00	5.00	66.2

26. Calculate the cadmium concentration (M) on the basis of the following polarographic data:

SOLUTION COMPOSITIONS, ml				CURRENT AT
0.200-F NaCl	Sample	2.00×10^{-2} M Cd^{2+}	H_2O	-0.6 V, μA
25.0	0.00	0.00	25.0	2.7
25.0	10.0	0.00	15.0	19.9
25.0	10.0	5.00	10.0	83.5

27. Predict the general shape of the curve for each of the following amperometric titrations. Label coordinates with care.

	CURRENT DUE TO INVOLVEMENT OF		
Titration	Analyte	Reagent	Product
*(a)	0	0	+
(b)	0	+	0
*(c)	+	0	0
(d)	+	+	0
*(e)	0	+	+

16 absorptiometric methods of analysis

All chemical species interact with electromagnetic radiation and, in so doing, diminish the intensity or power of the radiant beam. Absorption spectroscopy, a major branch of analytical chemistry, is based upon the measurement of this decrease in power of the radiation brought about by the analyte.

It is convenient to characterize *absorptiometric methods* according to the type of electromagnetic radiation absorbed. The categories include X-ray, ultraviolet, visible, infrared, microwave, and radio-frequency radiation. We will be concerned mainly with the absorption of ultraviolet and visible radiation but will make occasional reference to the other types as well.

Properties of Electromagnetic Radiation

Electromagnetic radiation is a kind of energy that is transmitted through space at enormous velocities. It takes many forms, the most easily recognizable being visible light and radiant heat. Less obvious manifestations include X-ray, ultraviolet, microwave, and radio radiations.

The propagation of electromagnetic radiation through space is most conveniently described in terms of wave parameters such as velocity, frequency, wavelength, and amplitude. In contrast to other wave phenomena such as sound, electromagnetic radiation requires no supporting medium for transmission; thus it readily passes through a vacuum.

This wave model fails to account for phenomena associated with the absorption or the emission of radiant energy; for these processes it is necessary to view electromagnetic radiation as discrete particles of energy called *photons*. The energy of a photon is proportional to the frequency of the radiation. These dual views of radiation as particles and waves are not mutually exclusive. Indeed, the apparent duality is rationalized by wave mechanics and found to apply to other phenomena such as the behavior of streams of electrons or other elementary particles.

WAVE PROPERTIES

For many purposes electromagnetic radiation is conveniently treated as an electrical force field which oscillates at right angles with respect to the direction of propagation. The electrical force involved is a vector quantity; it can be represented at any instant by an arrow whose length is proportional to the magnitude of the force and whose direction is parallel to that of the force. It is seen in Figure 16-1 that a plot of the vector as a function of distance along the axis of propagation is sinusoidal.[1] This electrical force is responsible for such phenomena as the transmission, reflection, refraction, and absorption of radiation by matter.

The *wavelength*, λ, of a beam of radiation is the linear distance between successive maxima or minima of a wave (see Figure 16-1).[2] The *frequency*, ν, of radiation is the number of oscillations of the field that occur per second.[3] Multiplication of the frequency (in cycles per second or hertz) by the wavelength (in centimeters per cycle) gives the velocity at which the radiation is propagated. In a vacuum this velocity c is at its maximum and has a value of 3.00×10^{10} cm/sec; that is,

$$\nu\lambda = c = 3.00 \times 10^{10} \text{ cm/sec} \tag{16-1}$$

[1]Figure 16-1 is a two-dimensional representation of monochromatic (that is, a single wavelength) radiation. A better representation would be three-dimensional with a circular cross section that periodically fluctuates in radius from zero to the maximum amplitude a.

[2]The units commonly used for describing wavelength differ considerably in the various spectral regions. For example, the ångström unit Å (10^{-10} m) is convenient for X-ray and short ultraviolet radiation; the nanometer nm (10^{-9} m) is employed with visible and ultraviolet radiation; the micrometer μm (10^{-6} m) is useful for the infrared region. (In the older literature the nanometer is often termed the millimicron mμ, and the micrometer, the micron μ.)

[3]The common unit of frequency is the *hertz*, Hz, which is equal to one cycle per second.

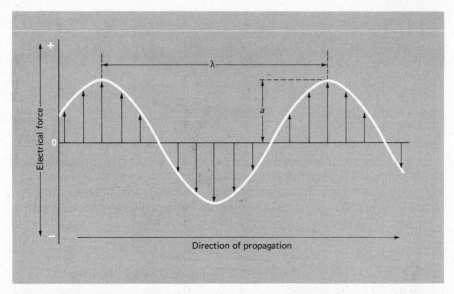

FIGURE 16-1 Representation of a beam of monochromatic radiation of wavelength λ
and amplitude a. The arrows represent the electrical vector of the radiation.

Radiation is propagated through matter at velocities smaller than c, owing
to interactions between the electrical vector and the bound electrons of the
medium. The frequency, on the other hand, is invariant and fixed by the
source. Thus a decrease in velocity must also be accompanied by a
corresponding decrease in wavelength (Equation 16-1).

The *wave number* σ is defined as the number of waves per
centimeter and is yet another way of describing electromagnetic radiation.
When the wavelength *in vacuo* is expressed in centimeters, the wave
number is equal to $1/\lambda$.

The *power P* of radiation is the energy of the beam that reaches a
given area per second; the *intensity I* is the power per unit solid angle.
Both quantities are related to the square of the amplitude a (see Figure
16-1). Although it is not strictly correct, power and intensity are often used
synonymously.

PARTICLE PROPERTIES OF RADIATION

Energy of Electromagnetic Radiation. In certain interactions with
matter it is necessary to consider radiation as packets of energy called
photons or *quanta*. The energy of a photon depends upon the frequency of
the radiation and is given by

$$E = h\nu \qquad \textbf{(16-2)}$$

where h is Planck's constant (6.63×10^{-27} erg sec). In terms of wavelength and wave number,

$$E = \frac{hc}{\lambda} = hc\sigma \qquad (16\text{-}3)$$

Note that the wave number, like the frequency, is directly proportional to energy.

THE ELECTROMAGNETIC SPECTRUM

The electromagnetic spectrum covers an immense range of wavelengths, or energies. For example, an X-ray photon ($\lambda \sim 10^{-8}$ cm) is approximately 10,000 times more energetic than a photon emitted by an incandescent tungsten wire ($\lambda \sim 10^{-4}$ cm).

Figure 16-2 depicts qualitatively the major divisions of the electromagnetic spectrum. Note that the energy (or wavelength) scale is logarithmic; note also that the region to which the human eye is perceptive (the *visible spectrum*) is very small. Such diverse radiations as gamma rays

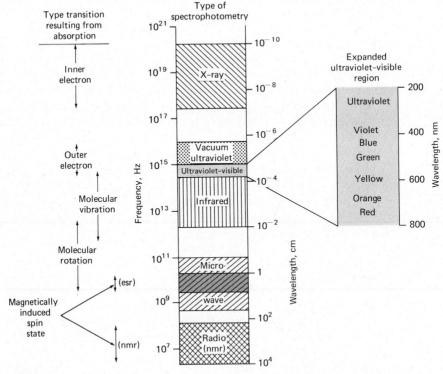

FIGURE 16-2 Parts of the electromagnetic spectrum which are useful for spectrophotometric measurements.

or radio waves differ from visible light only in the matter of frequency and hence energy (Equation 16-2).

Figure 16-2 also indicates the regions of the spectrum that are useful for analytical purposes and the molecular or atomic transitions responsible for absorption of radiation in each region.

Generation of Electromagnetic Radiation

Atoms and molecules have a limited number of discrete or quantized energy levels, the lowest of which is called the *ground state.* When sufficient energy is added to a system, the atoms or molecules are *excited* to higher energy levels. The lifetime of an excited species is generally transitory, and *relaxation* to a lower energy level or to the ground state takes place with a release of energy in the form of heat or of electromagnetic radiation, or perhaps both. When radiation is the product, the energy of each emitted photon, $h\nu$, is equal to the corresponding energy difference between the excited state and the lower energy level.

Radiating particles that are well separated from one another, as in the gaseous state, behave as independent bodies and often emit relatively few specific wavelengths. The resulting spectrum is thus discontinuous and termed a *line spectrum.* In contrast, a *continuous spectrum* contains all wavelengths over an appreciable range; here the individual wavelengths are so closely spaced that they cannot be resolved by ordinary means. Continuous spectra result from the excitation of (1) solids or liquids, in which the atoms are so closely packed as to be incapable of independent behavior, or (2) complicated molecules that possess many closely related energy states. Continuous spectra also occur when the energy changes involve particles with unquantized kinetic energies.

Both continuous spectra and line spectra are of importance in analytical chemistry. The former are valuable as sources in methods based on the interaction of radiation with matter, such as spectrophotometry. Line spectra, on the other hand, are useful for the identification and determination of the emitting species.

ABSORPTION OF RADIATION

Absorption refers to a process by which the species in a transparent medium selectively removes certain frequencies of electromagnetic radiation. In this process the energy of the photon converts the absorbing atom, molecule, or ion M to a more energetic or *excited* form M*; excitation can be depicted by the equation

$$M + h\nu \rightarrow M^*$$

After a brief period (10^{-8} to 10^{-9} sec) the excitation energy is lost, ordinarily as heat, and the species *relaxes* to its former state; that is,

$$M^* \rightarrow M + heat$$

Relaxation may also result from decomposition of M^* to form new species; such a process is called a *photochemical reaction.* Alternatively, relaxation may involve the fluorescent or phosphorescent reemission of radiation. It is important to note that the lifetime of M^* is so very short that its concentration at any instant is ordinarily negligible. Furthermore, the amount of thermal energy created is usually so small as to be undetectable. Thus, absorption measurements have the advantage of creating minimal disturbance of the system under study.

Quantitative Aspects of Absorption Measurements

The principles and laws that govern the absorption of radiation apply to all wavelengths from the X-ray region to the radio-frequency range. The absorption measurement involves determination of the decrease (the *attenuation*) in power suffered by a beam of radiation as it passes through an absorbing medium of known dimensions.

BEER'S LAW

When a beam of monochromatic radiation passes through a population of absorbing species, the radiant power of the beam is progressively decreased as part of its energy is absorbed by the particles of that species. The diminution in power is dependent upon the concentration of the substance responsible for the absorption as well as upon the length of the path traversed by the beam. These relationships are expressed by *Beer's law.*

Let P_0 be the radiant power of a beam incident upon a section of solution that contains c moles of an absorbing substance per liter. Further, let P be the power of the beam after it has passed through b cm of the solution (Figure 16-3). As a consequence of absorption, P will be smaller than P_0. Beer's law relates these quantities as follows:

$$\log \frac{P_0}{P} = \epsilon b c = A \qquad (16\text{-}4)$$

FIGURE 16-3 Attenuation of a beam with initial power P_0 by a solution containing c moles/liter of absorbing solute and a path length of b cm; $P < P_0$.

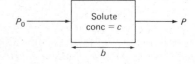

In this equation ϵ is a constant called the *molar absorptivity*. The logarithm (to the base 10) of the ratio of the incident power to the transmitted power is called the *absorbance* of the solution; this quantity is given the symbol A. Clearly the absorbance of the solution increases directly with the concentration of the absorbing substance and the path length traversed by the beam.[4]

Beer's law applies equally well to a solution containing more than one kind of absorbing substance, provided there is no interaction among the various species. Thus for a multicomponent system we may write

$$A_{total} = A_1 + A_2 + \cdots + A_n$$

$$= \epsilon_1 b c_1 + \epsilon_2 b c_2 + \cdots + \epsilon_n b c_n \tag{16-5}$$

where the subscripts refer to absorbing components $1, 2, \ldots, n$.

MEASUREMENT OF ABSORPTION

Beer's law, as given by Equation 16-4, is not directly applicable to chemical analysis. Neither P nor P_0, as defined, can be measured easily in the laboratory because the solution to be studied must be held in some sort of container. Interaction between the radiation and the walls is inevitable, producing a loss in power at each interface as a consequence of reflection (and possibly absorption). Reflection losses can be appreciable. For example, about 4% of a beam of visible radiation is reflected upon vertical passage across an air-to-glass or glass-to-air interface. In addition, the beam may suffer a diminution in power during its passage through the solution as a result of scattering by large molecules or inhomogeneities.

To compensate for these effects, the power of the beam transmitted through the solution of interest is generally compared with the power of the same beam passing through an identical cell containing the solvent for the sample. An experimental absorbance that closely approximates the true absorbance of the solution is thus measurable. That is,

$$A = \log \frac{P_0}{P} = \log \frac{P_{solvent}}{P_{solution}} \tag{16-6}$$

The term P_0, when used henceforth, refers to the power of radiation after it has passed through a cell containing the solvent for the component of interest.

---

[4]For a derivation of Beer's law see D. F. Swinehart, *J. Chem. Educ.*, **39**, 333 (1962); F. C. Strong, *Anal. Chem.*, **24**, 338 (1952).

TERMINOLOGY ASSOCIATED WITH ABSORPTION MEASUREMENTS

In recent years the attempt has been made to develop a standard nomenclature for the various quantities related to the absorption of radiation. The recommendations of the American Society for Testing Materials are given in Table 16-1 along with some of the alternative names and symbols that are frequently encountered, particularly in older literature. An important term in this table is the *transmittance T*, which is defined as

$$T = \frac{P}{P_0}$$

The transmittance is the fraction of incident radiation transmitted by the solution; it is often expressed as a percentage. The transmittance is related to the absorbance as follows:

$$-\log T = A$$

TABLE 16-1
Important Terms and Symbols Employed in Absorption Measurement

TERM AND SYMBOL[a]	DEFINITION	ALTERNATIVE NAME AND SYMBOL
Radiant power, P, P_0	Energy of radiation reaching a given area of a detector per second	Radiation intensity, I, I_0
Absorbance, A	$\log \frac{P_0}{P}$	Optical density, D; extinction, E
Transmittance, T	$\frac{P}{P_0}$	Transmission, T
Path length of radiation, in cm, b	—	l, d
Molar absorptivity,[b] ϵ	$\frac{A}{bc}$	Molar extinction coefficient
Absorptivity,[c] a	$\frac{A}{bc}$	Extinction coefficient, k

[a] Terminology recommended by the American Chemical Society in *Anal. Chem.*, **24**, 1349 (1952).
[b] c expressed in units of mole/liter.
[c] c may be expressed in g/liter or other specified concentration units; b may be expressed in cm or in other units of length.

LIMITATIONS TO THE APPLICABILITY OF BEER'S LAW

The linear relationship between absorbance and path length with a fixed concentration of absorbing substances is a generalization for which no exceptions are known. On the other hand, deviations from the direct proportionality between absorbance and concentration when b is constant are frequently encountered. Some of these deviations are fundamental and represent real limitations of the law; others occur as a consequence of the manner in which the absorbance measurements are made or as a result of chemical changes associated with concentration changes; the latter two are sometimes known, respectively, as *instrumental deviations* and *chemical deviations.*

Real Limitations to Beer's Law. Beer's law is successful in describing the absorption of dilute solutions only (usually $< 0.01\,M$) where concentration-dependent interactions among absorbing species are at a minimum (these interactions often alter the molar absorptivity of the species). Likewise, it is only in dilute solutions that variations in refractive index with concentration are negligibly small. With increasing concentrations these effects become sufficiently large to cause departures from the proportionality between absorbance and concentration.

Chemical Deviations. Apparent deviations from Beer's law are frequently encountered as a consequence of association, dissociation, or reaction of the absorbing species with the solvent. A classic example of a chemical deviation occurs in unbuffered potassium dichromate solutions, in which the following equilibria exist:

$$Cr_2O_7{}^{2-} + H_2O \rightleftarrows 2HCrO_4{}^- \rightleftarrows 2H^+ + 2CrO_4{}^{2-}$$

At most wavelengths the molar absorptivities of the dichromate ion and the two chromate species are quite different. Thus the total absorbance of any solution depends upon the concentration ratio between the dimeric and the monomeric forms. This ratio, however, changes markedly with dilution and causes a pronounced deviation from linearity between the absorbance and the total concentration of chromium(VI). Nevertheless, the absorbance due to the dichromate ion remains directly proportional to its molar concentration; the same is true for chromate ions. This fact is easily demonstrated by measuring the absorbance of chromium(VI) solutions in a strongly acidic medium, where dichromate is the principal species, and in strongly alkaline solution, where chromate predominates. Thus deviations in the absorbance of this system from Beer's law are more apparent than real, because they result from shifts in chemical equilibria. These deviations can in fact be readily predicted from the equilibrium constants for the reactions and the molar absorptivities of the dichromate and chromate ions.

Instrumental Deviations. Strict adherence of an absorbing system to Beer's law is observed only when the radiation employed is monochromatic. This observation is another manifestation of the limiting character of the relationship. Use of truly monochromatic radiation for absorbance measurements is seldom practical, and a polychromatic beam may cause departures from Beer's law.

Deviations from Beer's law resulting from the use of a polychromatic beam are not appreciable in practice, provided the radiation used does not encompass a spectral region in which the absorber exhibits large changes in absorbance as a function of wavelength; this effect is illustrated in Figure 16-4.

SPECTRAL CURVES

The relationship between absorbance and the frequency or wavelength of radiation is often employed to identify chemical species. The graphical representation of this relationship is called an *absorption spectrum*. In such plots the frequency, the wave number, or the wavelength is commonly employed for the abscissa. The ordinate is expressed in units of transmittance (or percent transmittance), absorbance, or the logarithm of absorbance.

Spectral data for three permanganate solutions are presented in three different ways in Figure 16-5. The concentrations stand in the ratio of 1:2:3. Note that employment of absorbance provides the greatest

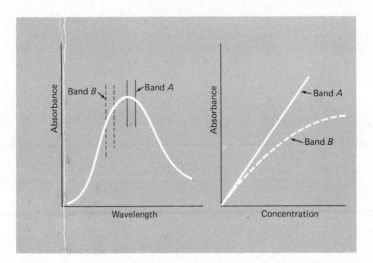

FIGURE 16-4 The effect of polychromatic radiation upon the Beer's law relationship. Band *A* shows little deviation because ε does not change greatly throughout the band. Band *B* shows marked deviations because ε undergoes significant changes in this region.

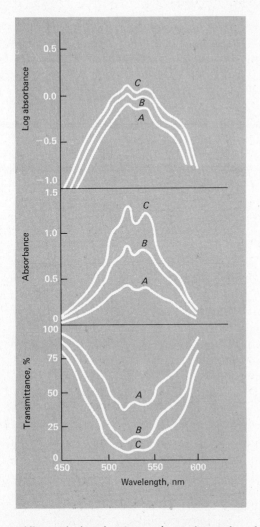

differentiation in the region where the absorbance is high (0.8 to 1.3) and the transmittance is low ($<20\%$). In contrast, greater differences occur in the transmittance curves when the transmittances lie in the range of 20 to 60%. With a plot of log A spectral detail tends to be lost; on the other hand, this type of plot is particularly convenient for comparing curves of solutions that have different concentrations because their curves are displaced an equal amount along the ordinate scale at all wavelengths.

Components of Instruments for Absorption Measurements

Regardless of the spectral region, instruments that measure the transmittance or absorbance of solutions contain five basic components: (1) a stable source of radiant energy, (2) a device that permits isolation and use of a restricted wavelength region, (3) transparent containers for sample and solvent, (4) a radiation detector or *transducer* that converts the radiant energy to a measurable signal (usually electrical), and (5) a signal indicator. The block diagram in Figure 16-6 shows the usual arrangement of these components.

The nature and the complexity of the several components in absorption instruments will depend upon the wavelength region involved and how the data are to be used; regardless of the degree of sophistication, however, the function of each component remains the same.

The signal indicator for many instruments that measure absorption consists of a linear scale that covers a range from 0 to 100 units. Direct percent transmittance readings can be obtained by first adjusting the indicator to read zero when radiation is blocked from the detector by a shutter (the 0% *T* adjustment). The indicator is then brought to 100 with a blank in the light path (the 100% *T* adjustment); this adjustment is accomplished either by varying the intensity of the source or by amplifying the detector signal. When the sample container is placed in the beam, the indicator gives percent transmittance directly, provided the detector responds linearly to changes in radiant power. Clearly a logarithmic scale can be scribed on the indicator to permit direct absorbance readings as well.

Three types of instruments are employed for absorptiometric measurements in the visible region; in increasing order of sophistication these are *colorimeters*, *photometers*, and *spectrophotometers*. Most ultraviolet and infrared absorptiometric instruments are spectrophotometers.

RADIATION SOURCES

To be suitable for absorption measurements, the source of radiation must generate a beam with sufficient power for ready detection and measurement. In addition, it should provide continuous radiation; that is, its spectrum should contain all wavelengths over the region in which it is to be used. Finally, the source should be stable; the power of the radiant beam

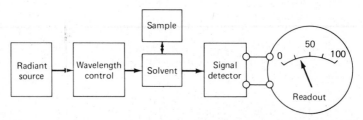

FIGURE 16-6 Components of instruments for measuring absorption of radiation.

must remain constant for the period needed to measure both P and P_0. Only under these conditions are absorbance measurements reproducible. Some instruments are designed to measure P and P_0 simultaneously; here fluctuations in the output of the source present no problem.

Sources of Visible Radiation. The most common source of visible radiation is the tungsten filament lamp, which is useful for the wavelength region between 320 and 2500 nm. The output of a tungsten lamp varies approximately as the fourth power of the operating voltage. As a consequence, voltage control is required to assure stability. Constant voltage transformers or electronic voltage regulators are often employed for this purpose. As an alternative, the lamp can be operated from a 6-V storage battery, which provides a remarkably stable voltage if it is maintained in good condition.

Ultraviolet Sources. The most common source of continuous radiation in the ultraviolet region is the hydrogen (or deuterium) lamp. Two types are encountered. The high-voltage variety employs potentials of 2000 to 6000 V to cause a discharge between aluminum electrodes; water cooling of the lamp is required if high radiation intensities are to be produced. In low-voltage lamps an arc is formed between a heated, oxide-coated filament and a metal electrode. About 40-V dc are required to maintain the arc.

Both high- and low-voltage lamps produce a continuous spectrum in the region between 180 and 375 nm. Quartz windows are required because glass absorbs strongly in this wavelength region.

Deuterium lamps produce higher intensities of ultraviolet radiation than hydrogen lamps under the same operating conditions.

Infrared Sources. Continuous infrared radiation is produced by electric heating of an inert solid. A silicon carbide rod, called a *Globar*, provides radiant energy in the region of 1 to 40 μm when heated to perhaps 1500°C between a pair of electrodes. A *Nernst glower* produces radiation in the region between 0.4 and 20 μm. This source is a rod of zirconium and yttrium oxides that is heated to about 1500°C by passage of current. A heated coil of nichrome wire is also a useful source of infrared radiation.

WAVELENGTH CONTROL

Both photometers and spectrophotometers employ devices that restrict the wavelength region used for an analysis. A narrow band of radiation offers three advantages. The probability that the absorbing system will adhere to Beer's law is greatly enhanced (see p. 442). In addition, a greater selectivity is assured because substances that absorb in other wavelength regions are less likely to interfere. Finally, a greater change in absorbance per increment of concentration will be observed if only wavelengths that are strongly absorbed are employed; thus a greater sensitivity is attained.

Devices for restricting radiation fall into two categories. *Filters*, which are employed in photometers, function by absorbing large portions of the spectrum while transmitting relatively limited wavelength regions. *Monochromators*, which are used in spectrophotometers, are more sophisticated devices that permit the continuous variation of wavelength.

When illuminated with a continuous source, filters do not produce radiation of a single wavelength; instead, a band encompassing a narrow region of the spectrum results. The distribution of wavelengths within the region typically takes the form of the curves shown in Figure 16-7a with the peak corresponding to the stated wavelength of the filter. The quality of the filter is measured by its *effective band width*, which is the width of the band (in wavelength units) at half intensity (see Figure 16-7a).

Absorption Filters. Absorption filters limit radiation by absorbing certain portions of the spectrum. The most common type consists of colored glass or of a dye suspended in gelatin and sandwiched between glass plates; the former has the advantage of greater thermal stability.

Absorption filters have effective band widths ranging from perhaps

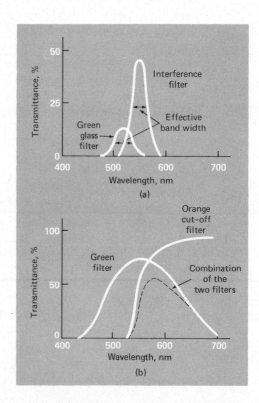

FIGURE 16-7 Comparison of transmittance characteristics of two filters and a filter combination.

30 to 250 nm (see Figure 16-7a). Filters that provide the narrowest band widths also absorb a significant fraction of the desired radiation and may have a transmittance that is 0.1 or less at their band peaks. Glass filters with transmittance maxima throughout the entire visible region are available commercially and are relatively inexpensive.

A *cut-off filter* has a transmittance of nearly 100% over a portion of the visible spectrum but then rapidly decreases to zero transmittance over the remainder. A narrow spectral band can be isolated by coupling a cut-off filter with a second filter (see Figure 16-7b).

Interference Filters. As the name implies, interference filters rely on optical interference to produce relatively narrow bands of radiation (see Figure 16-7a). An interference filter consists of a transparent dielectric layer (frequently calcium fluoride or magnesium fluoride) that occupies the space between two semitransparent metallic films coated on the inside surfaces of two glass plates. The thickness of the dielectric layer is carefully controlled and determines the wavelength of the transmitted radiation. When a perpendicular beam of collimated (that is, parallel) radiation strikes this array, a fraction passes through the first metallic layer while the remainder is reflected. The portion that is passed undergoes a similar partition upon striking the second metallic film. If the reflected portion from this second interaction is of the proper wavelength, it is partially reflected from the inner side of the first layer in phase with incoming light of the same wavelength. The result is that this particular wavelength is reinforced, while most others, being out of phase, suffer destructive interference.

Interference filters provide significantly narrower band widths (as low as 10 nm) and greater transmission of the desired wavelength than do absorption-type filters (see Figure 16-7a). Interference filters that provide radiation bands from the ultraviolet up to about 14 μm in the infrared can be purchased.

Monochromators. A monochromator is the device in a spectrophotometer that disperses the radiation from the source into its component wavelengths. Associated with the dispersing element is a system of lenses, mirrors, and slits that isolate and direct radiation of the desired wavelength through the sample container and toward the detector. The wavelength capability of a spectrophotometer is determined by the composition of the components through which radiation is transmitted.

Prism Monochromators. Figure 16-8 is a schematic representation of a *Bunsen* monochromator, which employs a 60-deg prism for dispersion. Radiation is admitted through an entrance slit, is collimated by a lens, and then strikes the surface of the prism at an angle. Refraction occurs at both faces of the prism; the dispersed radiation is then focused on a slightly

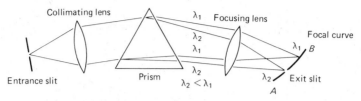

FIGURE 16-8 A prism monochromator.

curved surface containing the exit slit. The desired wavelength can be caused to pass through this slit by rotation of the prism.

The radiation emerging from a monochromator is a narrow rectangle or line which is a replica of the slit system. The distribution of wavelengths within this line is similar to that shown in Figure 16-7a, except that the band width is generally much less. The effective band width becomes smaller as the thickness of the base of the prism is increased. To decrease bulk and still maintain a minimal band width, a *Littrow* prism is frequently employed in monochromator construction. This prism is obtained by halving the prism shown in Figure 16-8 along its vertical axis. The back face is then silvered so that radiation both enters and emerges from the same face; thus the same dispersion is achieved as with a full prism. Figure 16-13 illustrates a monochromator that employs a Littrow prism.

Most monochromators are equipped with adjustable slits to permit some control of the band width. A narrow slit width decreases the effective band width but also diminishes the power of the emergent beam. Thus the minimum effective band width is often limited by the sensitivity of the detector.

For fixed slit widths the effective band width of a prism mono-chromator increases continuously as the wavelength of radiation increases because the refractive index of any prism material becomes less at longer wavelengths. For example, the quartz monochromator employed in one widely used instrument has an effective band width of 1.5 nm per milli-meter of slit at 250 nm compared with 50 at 700 nm. Thus to obtain a spectrum at a fixed band width (which may be desirable), it is necessary to decrease the slit width of a prism monochromator continuously as the wavelengths become longer.

In the visible region glass is a more effective dispersing agent than quartz; monochromators employing glass optics are restricted to the visible region, however, because this material strongly absorbs ultraviolet radiation.

The distribution or *dispersion* of radiation from a continuous source along the line *AB* in Figure 16-8 for comparable glass and quartz prisms is shown in Figure 16-9.

For the infrared region, where glass and quartz are not transparent, prisms are constructed from materials such as sodium chloride, lithium

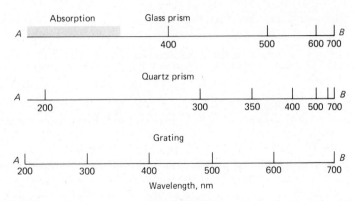

FIGURE 16-9 Dispersion characteristics of three types of monochromators.

fluoride, calcium fluoride, or potassium bromide. All of these substances are susceptible to mechanical abrasion and attack by water vapor; consequently their use requires special precautions.

Grating Monochromators. Dispersion of ultraviolet, visible, and infrared radiation can be brought about by passage of a beam through a *transmission grating* or by reflection from a *reflection grating*. A transmission grating consists of a series of parallel grooves ruled on a piece of glass or other transparent material. A grating suitable for use in the ultraviolet and visible region has about 15,000 lines/in. It is vital that these lines be equally spaced throughout the length of the grating. Such gratings require elaborate apparatus for their production and are consequently expensive. Replica gratings are less costly. They are manufactured by employing a master grating as a mold for the production of numerous plastic replicas; the products of this process, although inferior in performance to an original grating, suffice for many applications.

When a transmission grating is illuminated by radiation from a slit, each groove acts as an individual source; interference among the multitude of beams results in dispersion of the radiation into its component wavelengths. A spectrum consisting of images of the entrance slit is produced when the dispersed radiation is focused on a plane surface.

Reflection gratings are produced by ruling a polished metal surface or by evaporating a thin film of aluminum onto the surface of a replica grating. The radiation is reflected from each of the unruled portions, and interference among the reflected beams produces dispersion.

One advantage of a grating monochromator is that the dispersion along the focal plane of the exit slit is very nearly *independent of wavelength*; thus a given slit setting will provide the same band width regardless of spectral region. Figure 16-9 contrasts the spectral dispersion of a typical grating monochromator with two prism monochromators.

Double Monochromators. Many modern monochromators contain two dispersing elements, that is, two prisms, two gratings, or a prism and a grating. This arrangement markedly reduces the amount of stray radiation[5] and also provides greater dispersion and spectral resolution.

SAMPLE CONTAINERS

In common with the optical elements of monochromators, the *cells* or *cuvettes* that hold the samples must be made of material that passes radiation in the spectral region of interest. Thus quartz or fused silica is required for work below 350 nm in the ultraviolet region. These materials are also transparent throughout the visible region and to about 3 μm in the infrared. Silicate glasses can be employed between 350 nm and 2 μm. Plastic containers have also found application in the visible region. For wavelengths longer than 2 to 3 μm, cells with windows of polished sodium chloride or silver chloride are regularly employed.

In general the best cells have windows that are exactly perpendicular to the direction of the beam in order to minimize reflection losses. Most instruments are provided with a pair of cells that have been matched with respect to light path and transmission characteristics to permit an accurate comparison of the radiant power transmitted through the sample and the solvent. The most common cell length for ultraviolet and visible radiation is 1 cm; matched, calibrated cells of this size are available from several commercial sources. Other path lengths from 0.1 cm and shorter to 10 cm can also be purchased. Transparent spacers for shortening the path length of 1-cm cells to 0.1 cm are also available.

For reasons of economy cylindrical cells are sometimes employed in the ultraviolet and visible regions. Care must be taken to duplicate the position of such cells with respect to the beam; otherwise, variations in reflection losses and path length introduce appreciable error.

In contrast to the ultraviolet and visible regions, few solvents are transparent throughout large regions of the infrared. As a consequence, pure samples or concentrated solutions of the samples must often be employed. This restriction makes necessary the use of very thin layers (0.1 to 1 mm) of sample; otherwise, the radiation from the source is completely absorbed (that is, $T \rightarrow 0$).

The quality of absorbance data is critically dependent upon the way matched cells are used and maintained. Fingerprints, grease, or other deposits on the cell wall alter the transmission characteristics markedly; thus thorough cleaning before and after use is imperative. The windows

[5]Stray radiation refers to radiation reaching the exit slit of a monochromator as a result of reflections off the surfaces of the various components within the device. Any such radiation with a wavelength different from that which has been isolated by the mono-chromator is undesirable.

must not be touched during the handling of cells. Matched cells should never be dried by heating in an oven or over a flame, for such treatment may cause physical damage or a change in path length. They should be calibrated against each other regularly with an absorbing solution.

RADIATION DETECTORS

Photoelectric detectors include the photovoltaic cell and various types of phototubes. All provide an electrical signal in response to radiant energy.

To be useful a radiation detector must respond over a broad wavelength range. It should, in addition, be sensitive to low levels of radiant power, respond rapidly to the radiation, produce an electrical signal that can be readily amplified, and have a relatively low noise level.[6] Finally, it is essential that the signal produced be directly proportional to the power of the beam striking it; that is,

$$G = k'P + k'' \tag{16-7}$$

where G is the electrical response of the detector in units of current, resistance, or emf. The constant k' measures the sensitivity of the detector in terms of electrical response per unit of radiant power. Many detectors exhibit a small constant response, known as a *dark current* k'', when no radiation impinges on their surfaces. Instruments with detectors that have a dark-current response are ordinarily equipped with a compensating circuit that permits application of a countersignal to reduce k'' to zero. Thus under most circumstances

$$P = \frac{G}{k'} \tag{16-8}$$

and

$$P_0 = \frac{G_0}{k'} \tag{16-9}$$

where G and G_0 represent the electrical response of the detector to radiation passing through the solution and the blank, respectively. Substitution of Equation 16-8 and Equation 16-9 into the Beer's law expression gives

$$\log \frac{P_0}{P} = \log \frac{k'G_0}{k'G} = \log \frac{G_0}{G} = A \tag{16-10}$$

Photovoltaic or Barrier-Layer Cells. A photovoltaic cell consists of a flat copper or iron electrode upon which a layer of semiconducting material such as copper(I) oxide or selenium has been deposited. On the surface of

---

[6]Noise in an electronic instrument refers to small random fluctuations which occur in the signal source, the detector, the amplifier, or the readout device. Common causes for these fluctuations include vibration, pickup from 60-Hz electrical lines, temperature variations, and frequency or voltage fluctuations in the power supply.

the semiconductor is a transparent metallic film of gold, silver, or lead, which serves as the second or collector electrode; the entire array is protected by a transparent envelope. The interface between the semiconductor and the metal film serves as a barrier to the passage of electrons. Irradiation with light, however, provides some electrons in the oxide layer with sufficient energy to overcome this barrier, and electrons flow from the semiconductor to the metal film. If the film is connected via an external circuit to the plate on the other side of the semiconducting layer, and if the resistance is not too great, a flow of electrons occurs. Ordinarily this current is large enough to be measured with a galvanometer or microammeter; provided the resistance of the external circuit is small, the magnitude of the current is directly proportional to the power of the radiation striking the cell. Currents on the order of 10 to 100 μA are typical.

Photovoltaic cells are used primarily for the detection and measurement of radiation in the visible region. The typical cell has a maximum sensitivity at about 550 nm, and the response falls off continuously to perhaps 10% of the maximum at 250 and 750 nm. The wavelength response of a typical photovoltaic cell closely resembles that of the human eye.

The barrier-layer cell is a rugged, low-cost means for measuring radiant power. No external source of electrical energy is required. On the other hand, the cell has a low internal resistance; as a result, amplification of its output is difficult. Thus, although the barrier-layer cell provides a readily measured response at high levels of illumination, it suffers from lack of sensitivity at low levels. Finally, a photovoltaic cell exhibits fatigue, its response falling off upon prolonged illumination; proper circuit design and choice of experimental conditions largely eliminate this source of difficulty.

Phototubes. A second type of photoelectric device is the *phototube*, which consists of a semicylindrical cathode and a wire anode sealed inside an evacuated transparent envelope. The concave surface of the cathode supports a layer of material that emits electrons upon being irradiated. When a potential is applied across the electrodes, the emitted electrons flow to the wire anode and a photocurrent results. For a given radiant intensity the current produced is approximately one-fourth as great as that from a photovoltaic cell. In contrast, however, amplification is easily accomplished since the phototube has a very high electrical resistance. Figure 16-10 is a schematic diagram of a typical phototube arrangement.

A potential of about 90 V, impressed across the electrodes, is sufficient to cause all photoemitted electrons to be captured and thereby assure an electronic response that is proportional to the radiant power of the beam that strikes the phototube.

The photoemissive cathode surfaces of phototubes ordinarily consist of alkali metals or alkali-metal oxides, alone or combined with

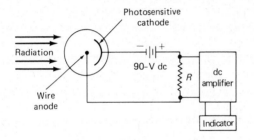

FIGURE 16-10 Schematic
diagram of a phototube and its
accessory circuit. The current
induced by the radiation
causes a potential drop across
the resistor R; this is amplified
and measured by the indicator.

other metal oxides. The coating on the cathode determines the spectral
response of a phototube.

Photomultiplier Tubes. For the measurement of low radiant power the
photomultiplier tube offers advantages over the phototube. Figure 16-11 is
a schematic diagram of such a device. The cathode surface has a similar
composition to that of a phototube, electrons being emitted upon exposure
to radiation. The tube also contains additional electrodes (labeled 1 to 9 in
Figure 16-11) called *dynodes*. Dynode 1 is maintained at a potential 90 V
more positive than the cathode, and electrons are accelerated toward it as a
consequence. Upon striking the dynode, each photoelectron causes emission of several additional electrons; these in turn are accelerated toward
dynode 2, which is 90 V more positive than dynode 1. Again, several
electrons are emitted for each electron striking this dynode. By the time
the process has been repeated nine times, 10^6 to 10^7 electrons per photon
will have been generated; this cascade is finally collected at the anode. The

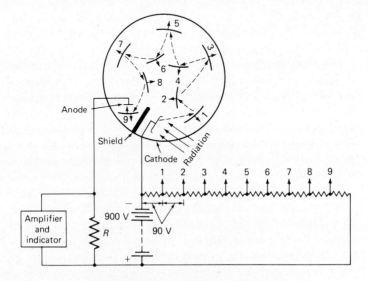

FIGURE 16-11 A photomultiplier tube.

resulting enhanced current is then passed through the resistor R and can be further amplified and measured.

Photomultiplier tubes are easily damaged by exposure to strong radiation and can be used only for measurement of low radiant power. To avoid irreversible changes in their performance, the tubes must be mounted in light-tight housings, and care must be taken to avoid even momentary exposure to strong light.

INSTRUMENTS AND INSTRUMENT DESIGN

The instrumental components discussed in the previous section have been combined in various ways to produce dozens of commercial instruments that differ in complexity, performance characteristics, and cost. Some of the simpler instruments can be purchased for $100 to $200, whereas the most sophisticated cost a hundred times this amount or more. No single instrument is best for all purposes. Selection is governed by the type of work for which the instrument is to be used.

DESIGN VARIABLES

Wavelength Selection and Control. A *spectrophotometer* has a prism or grating monochromator that permits the use of a narrow band of radiation that can be varied continuously in wavelength. A *photometer*, on the other hand, employs filters to provide bands of radiation that usually encompass broader wavelength spans than those obtained with a spectrophotometer. The other components are similar to those of a spectrophotometer. Photometers are largely, but not exclusively, confined to the visible region and, as a consequence, are sometimes called *colorimeters* or *photoelectric colorimeters*.

We shall limit the use of the term colorimeter to those instruments that employ the human eye as the detector for radiation. Colorimetric methods represent the simplest form of absorption analysis.

Single- and Double-Beam Designs. Both spectrophotometers and photometers are of single-beam and double-beam design. In double-beam instruments the beam is split within the monochromator or after exiting from it by any of several means; the one beam then passes through the sample and the other through the solvent. In some instruments the power of the two beams is compared by twin detector-amplifier systems so that transmittance or absorbance data are obtained directly. In other instruments radiation from the source is mechanically chopped so that pulses pass alternately through the sample and the blank. The resulting beams are then recombined and focused on a single detector. The pulsating electrical output from the detector is fed into an amplifier system that is programmed to compare the magnitude of the pulses and convert this information into transmittance or absorbance data.

Because comparison of the beam passing through the solvent with that passing through the sample is made simultaneously or nearly simultaneously, a split-beam instrument compensates for all but the most short-term electrical fluctuations, as well as for irregular performance in the source, the detector, and the amplifier. Therefore the electrical components of a double-beam photometer need not be of as high quality as those for a single-beam instrument. Offsetting this advantage, however, is the greater number and complexity of components associated with double-beam instruments. Moreover, in photometers equipped with twin detectors and amplifiers, a close match between the components of the two systems is essential.

Single-beam instruments are particularly suited for the quantitative analysis that involves an absorbance measurement at a single wavelength. Here simplicity of instrumentation and the concomitant ease of maintenance offer real advantages. The greater speed and convenience of measurement, on the other hand, make the double-beam instrument particularly useful for qualitative analyses, where absorbance measurements must be made at numerous wavelengths in order to obtain a spectrum. Furthermore, the double-beam device is readily adapted to continuous monitoring of absorbance; for this reason most modern recording spectrophotometers employ twin beams.

Both single-beam and double-beam instruments are available for ultraviolet and visible radiation. Commercial infrared spectrophotometers are always double-beam because they are ordinarily employed to scan and record a large spectral region. We shall limit our discussion of spectrophotometric design to two widely used single-beam instruments.

COLORIMETERS AND PHOTOMETERS

Colorimeters and photometers are the simplest instruments for absorption analysis. They provide entirely adequate quantitative data for many purposes.

Colorimeters. Colorimeters employ the human eye as a detector and the brain as a transducer and signal detector. The eye and brain, however, can only match colors; thus they are incapable of providing numerical information about the relative power of two beams of light and therefore about absorbance. As a consequence, colorimetric methods always require the use of one or more standards for color matching with the analyte solutions.

The simplest colorimetric methods involve the comparison of the sample with a set of standards until a match is found. Flat-bottomed *Nessler tubes* are frequently employed for this purpose. These tubes are calibrated so that a uniform light path is achieved. Daylight commonly serves as a radiation source; ordinarily no attempt is made to restrict the portion of the spectrum employed.

Colorimetric methods suffer from several disadvantages. A standard or a series of standards must always be available. The human eye responds to a limited spectral region (400 to 700 nm) and additionally is unable to match absorbances if the solution contains a second colored substance. Finally, the eye is not as sensitive to small differences in absorbance as a photoelectric device; as a consequence, concentration differences smaller than about 5% relative cannot be detected.

Despite their limitations visual comparison methods find extensive application for routine analyses in which the requirements for accuracy are modest. For example, simple but useful colorimetric test kits are sold for determining the pH and the chlorine content of swimming pool water; kits are also available for the analysis of soils. Filtration plants commonly employ color comparison tests for the estimation of iron, silicon, fluorine, and chlorine in city water supplies. For such analyses a colorimetric reagent is introduced to the sample, and the resulting color is compared with permanent standard solutions or with colored glass disks. Accuracies of perhaps 10 to 50% relative are to be expected and suffice for the purposes intended.

Photometers. The photometer provides a simple, relatively inexpensive tool for the performance of absorption analyses. Convenience, ease of maintenance, and ruggedness are properties of a filter photometer that may not be found in the more sophisticated spectrophotometer. Moreover, where high spectral purity is not important to a method (and often it is not), analyses can be performed as adequately with this instrument as with more complex instrumentation.

Figure 16-12 presents schematic diagrams for two photometers. The first is a single-beam, direct-reading instrument consisting of a tungsten filament lamp, a lens to provide a parallel beam of light, a filter, and a photovoltaic cell. The current produced is indicated with a microammeter, the face of which is ordinarily scribed with a linear scale from 0 to 100. In some instruments the 100% T adjustment involves changing the voltage applied to the lamp. In others the aperture size of a diaphragm located in the light path is altered. Because the signal from the photovoltaic cell is linear with respect to the radiation it receives, the scale reading with the sample in the light path will be the percent transmittance (that is, the percent of full scale). Clearly a logarithmic scale could be substituted to give the absorbance of the solution directly.

Also shown in Figure 16-12 is a schematic representation of a double-beam, null-type photometer. Here the light beam is split by a mirror, a part passing through the sample, and thence to a photovoltaic cell; the other half passes through the solvent to a similar detector. The

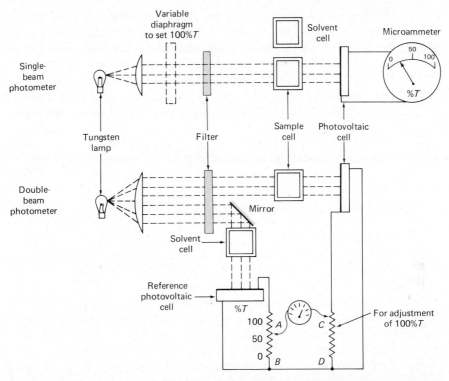

FIGURE 16-12   Schematic diagram for a single-beam and a double-beam photometer.

currents from the two photovoltaic cells are passed through variable
resistances; one of these is calibrated as a transmittance scale in linear
units from 0 to 100. A sensitive galvanometer, which serves as a null
indicator, is connected across the two resistances. When the potential drop
across AB is equal to that across CD, no current passes through the
galvanometer; under all other circumstances a current is indicated. At the
outset the solvent is introduced into both cells, and contact A is set at 100;
contact C is then adjusted until no current is indicated. Introduction of the
sample into the sample cell results in a decrease in radiant power reaching
the working phototube and a corresponding decrease in the potential drop
across CD; this lack of balance is compensated for by moving A to a lower
value. At balance the percent transmittance is read directly from the scale.

Commercial photometers usually cost a few hundred dollars. The
majority employ the double-beam principle because this design largely
compensates for fluctuations in the source intensity due to voltage variations.

Filter Selection for Photometric Analysis.   Photometers are generally
supplied with several filters, each of which transmits a different portion of

the spectrum. Selection of the proper filter for a given application is important inasmuch as the sensitivity of the measurement is directly dependent upon the choice. For example, a liquid appears red because it transmits unchanged the red portion of the spectrum but absorbs the green. It is the intensity of green radiation that varies with concentration; a green filter should thus be employed. In general, then, the most suitable filter for a colorimetric analysis will be the color complement of the solution being analyzed. If several filters possessing the same general hue are available, the one that causes the sample to exhibit the greatest absorbance (or least transmittance) should be used.

SPECTROPHOTOMETERS

Spectrophotometers, in contrast to photometers, contain prism- or grating-monochromators that permit a continuous choice of wavelengths. Instruments exist for measurements in the ultraviolet, visible, and infrared regions. Although the materials used for the optics and cells differ according to the spectral region for which the instrument is intended, the basic design of a spectrophotometer is largely independent of wavelength.

Figure 16-13 is a schematic diagram of a high-quality, single-beam spectrophotometer—the Beckman DU-2. The first versions of this instrument appeared on the market over 30 years ago; it is one of the most widely used spectrophotometers of its kind.

The DU-2 spectrophotometer is equipped with quartz optics and can be operated in both the ultraviolet and visible regions of the spectrum. The instrument is provided with interchangeable radiation sources, consisting of a deuterium or hydrogen discharge tube for the lower wavelengths and a tungsten filament lamp for the visible and near infrared regions. A pair of mirrors reflect radiation through an adjustable slit into the monochromator compartment. After traversing the length of the instrument, the radiation is reflected into a Littrow prism; by adjusting the position of the prism, light of the desired wavelength can be focused on the slit. The optics are so arranged that the entrance and the exit beams are displaced from one another on the vertical axis; thus the exit beam passes above the entrance mirror as it enters the cell compartment.

Ordinarily the cell compartment will accommodate as many as four rectangular 1-cm cells, any one of which can be positioned in the path of the beam by movement of a carriage arrangement. Compartments are also available that will hold both cylindrical cells and cells with 10-cm light paths.

The detectors are housed in a phototube compartment; control over the incoming radiation is achieved with a manually operated shutter in the path of the beam. Interchangeable detectors are provided—a red-sensitive phototube for the wavelength region beyond 625 nm and a pho-

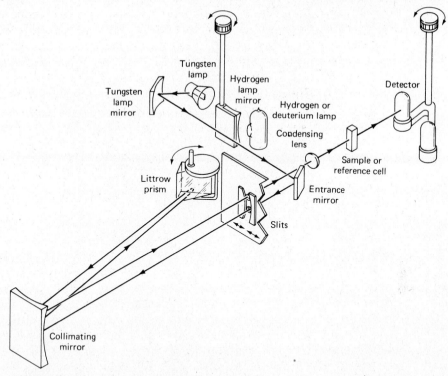

FIGURE 16-13   Schematic diagram of the Beckman DU-2 spectrophotometer. (By permission of Beckman Instruments, Inc.)

tomultiplier tube for the range between 190 and 625 nm. The current from the phototube in the light path is passed through a fixed resistance of large magnitude; the potential drop across this resistor then gives a measure of the radiant power reaching the detector.

The DU-2 design achieves photometric accuracies as good as ±0.2% transmittance by employing high-quality electronic components that are operated well below their rated capacities. Narrow effective band widths (less than 0.5 nm) can be obtained throughout the spectral region by suitable adjustment of the slit. The instrument is particularly well adapted for research and quantitative analytical measurements that require absorbance data at a limited number of wavelengths.

The Bausch and Lomb Spectronic 20, shown schematically in Figure 16-14, may be considered as representative of instruments in which a degree of photometric accuracy is sacrificed in return for simplicity of operation and low cost. Its normal range is 350 to 650 nm, although measurements can be extended to 900 nm by the use of a red-sensitive phototube. The monochromator system consists of a reflection grating, lenses, and a pair of fixed slits. Because the grating produces a dispersion

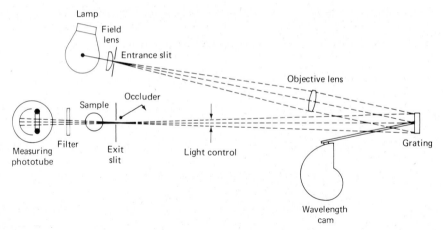

FIGURE 16-14 Schematic diagram of the Bausch and Lomb Spectronic 20 spectrophotometer. (By permission of Bausch and Lomb, Inc.)

that is independent of wavelength, a constant band width of 20 nm is obtained throughout the entire operating region.[7]

The Spectronic 20 is an example of a direct-reading, single-beam spectrophotometer. The photocurrent from the detector is amplified and indicated by the position of a needle on the scale of a current-indicating meter. Because the amplified current is directly proportional to radiant power, the meter scale can be scribed to read transmittance and absorbance.

The Absorption Process

Atoms, molecules, and ions have a limited number of discrete, quantized energy levels. For absorption to occur the energy of the exciting photon must match exactly the energy difference between the ground state and an excited energy level for the absorbing species. Because these energy differences are unique for each species, a study of the frequencies that are absorbed provides a means for characterizing the constituents in a sample of matter.

Typical absorption spectra are shown in Figure 16-15. Clearly great diversity exists in the general appearance of such spectra.

---

[7]Newer versions of this instrument have a second phototube whose function, in conjunction with a feedback circuit, is to monitor and maintain a constant output from the source.

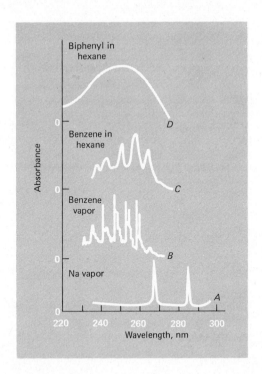

FIGURE 16-15 Some typical
ultraviolet absorption spectra.

ATOMIC ABSORPTION

When polychromatic ultraviolet and visible radiation is passed through a medium containing monatomic particles (for example, mercury or sodium in the vapor state), only a few well-defined frequencies are removed by absorption (see curve A in Figure 16-15) because the number of possible energy states is small. With atomic particles the only possible excitation process involves transition of electrons from lower to higher energy levels.[8]

Ultraviolet and visible radiation have sufficient energy to cause transitions of the outermost or bonding electrons only. X-Ray frequencies, on the other hand, are several orders of magnitude more energetic and capable of interacting with electrons closest to the nuclei of atoms. Absorption peaks corresponding to electronic transitions of the innermost electrons are thus observed in the X-ray region. Because the electrons involved do not participate in bonding, the X-ray absorption spectrum for an element tends to be independent of its chemical combination.

Regardless of the wavelength region involved, atomic absorption spectra typically consist of a limited number of discrete peaks (see curve A

---

[8]For example, sodium atoms show an absorption peak at 589 nm that results from the transition of the single $2s$ electron of unexcited or ground state sodium to the $3p$ excited state.

in Figure 16-15). These peaks are useful for both qualitative identification of elements and their quantitative determination. Atomic absorption spectroscopy is discussed in Chapter 17.

MOLECULAR ABSORPTION

Absorption by polyatomic molecules, particularly in the condensed state, is a considerably more complex process because the number of possible energy states is greatly increased. Here the total energy of a molecule is given by

$$E = E_{electronic} + E_{vibrational} + E_{rotational} \qquad \text{(16-11)}$$

where $E_{electronic}$ describes the energy associated with the various orbitals of the outer electrons of the molecule, $E_{vibrational}$ refers to the energy of the molecule as a whole due to interatomic vibrations, and the third term accounts for the energy associated with the rotation of the molecule around its center of gravity.

Absorption in the Infrared and Microwave Regions. The three terms on the right in Equation 16-11 are ordered according to decreasing energy, with the average value for each term differing from the next by roughly two orders of magnitude. Pure rotational absorption spectra, which are free from electronic and vibrational transitions, can be brought about by microwave radiation which is less energetic than infrared. Spectroscopic studies of gaseous species in this region are important in gaining fundamental information concerning molecular behavior; applications of microwave absorption to analytical problems, however, have been limited.

Vibrational absorption occurs in the infrared region, where the energy of the radiation is insufficient for electronic transitions. Infrared spectra typically exhibit narrow, closely spaced absorption peaks resulting from transitions among the various vibrational quantum levels (see Figure 16-16). Variations in rotational levels may also give rise to a series of peaks for each vibrational state; with liquid or solid samples, however, rotation is often hindered or prevented, and the effects of these small energy differences are not detected. Thus a typical infrared spectrum for a liquid such as that in Figure 16-16 consists of a series of vibrational peaks.

The number of individual ways a molecule can vibrate is largely related to the number of atoms, and thus the number of bonds, it contains. Even for a simple molecule the number is large. Thus n-butanal ($CH_3CH_2CH_2CHO$) has 33 vibrational modes, most differing from one another in energy. Not all of these vibrations give rise to infrared peaks; nevertheless, as shown in Figure 16-16, the spectrum for n-butanal is relatively complex.

Infrared absorption is not confined to organic molecules. Covalent metal-ligand bonds are also infrared active in the longer wavelength region.

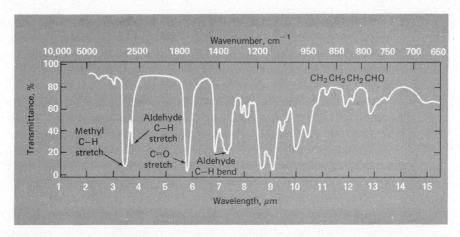

FIGURE 16-16 Infrared spectrum for *n*-butanal (*n*-butyraldehyde). Note that transmittance rather than absorbance is plotted. [Spectrum from B. J. Zwolinski et al., "Catalog of Selected Infrared Spectral Data," Serial No. 225, Thermodynamics Research Center Data Project, Thermodynamics Research Center, Texas A&M University, College Station, Texas (loose-leaf data sheets extant, 1964).]

Infrared spectrophotometric studies have thus provided much useful information about complex metal ions.

Absorption in the Ultraviolet and Visible Region. The first term in Equation 16-11 is ordinarily larger than the other two; electronic transitions of outer electrons generally require energies corresponding to ultraviolet or visible radiation. In contrast to atomic absorption spectra, molecular spectra are often characterized by absorption bands that encompass a substantial wavelength range (see curves *B*, *C*, and *D* in Figure 16-15). Here numerous vibrational and rotational energy states exist for each electronic state. Thus for a given value of $E_{electronic}$ in Equation 16-11, values for E will exist that differ only slightly due to variations in $E_{vibrational}$ and/or $E_{rotational}$. As a consequence, the spectrum for a molecule often consists of a series of closely spaced absorption bands such as that shown for benzene vapor in curve *B* of Figure 16-15. Unless a high resolution instrument is employed, individual bands may be undetected; as a result, the spectra appear as smooth curves. Furthermore, in the condensed state and in the presence of solvent molecules the individual bands tend to broaden to give the type of spectra shown in the upper two curves in Figure 16-15.

Absorption Induced by a Magnetic Field. When electrons or the nuclei of certain elements are subjected to a strong magnetic field, additional quantized energy levels are produced; these owe their origins to magnetic properties possessed by such elementary particles. The difference in energy between the induced states is small, and transitions between them are brought about only by absorption of radiation of long wavelengths or low frequency. For nuclei, radio waves ranging from 10 to 200 MHz are generally employed; for electrons, microwaves with frequencies of 1000 to 25,000 MHz are absorbed.

ABSORBING SPECIES

Absorption in the visible and ultraviolet regions can yield both quantitative and qualitative information about the species responsible for the absorption.

Absorption of Ultraviolet and Visible Radiation by Organic Compounds. The electrons responsible for absorption of ultraviolet and visible radiation by organic molecules are of two types: (1) those that participate directly in bond formation and are thus associated with more than one atom; (2) unshared outer electrons that are largely localized about such atoms as oxygen, the halogens, sulfur, and nitrogen.

The wavelength at which an organic molecule absorbs depends upon how tightly its various electrons are bound. Thus the shared electrons in single bonds such as carbon-carbon or carbon-hydrogen are so firmly held that their excitations require energies corresponding to wavelengths in the vacuum ultraviolet region ($\lambda < 180$ nm). This region is not readily accessible because components of the atmosphere also absorb; as a result, absorption by single bonds is not important for analytical purposes.

The unshared electrons in sulfur, bromine, and iodine are less strongly held than the shared electrons of a saturated bond. Organic molecules containing these elements frequently exhibit useful peaks in the ultraviolet region as a consequence.

Electrons of double and triple bonds in organic molecules are relatively easily excited by radiation; thus species containing unsaturated bonds generally exhibit useful absorption peaks. Unsaturated organic functional groups that absorb in the ultraviolet and visible regions are termed *chromophores*. Table 16-2 lists common chromophores and the approximate location of their absorption maxima. The data for position and peak intensity can serve only as a rough guide for identification purposes because the position of a maximum is also affected by solvent as well as structural details. Furthermore, when two chromophores are conjugated, shifts in peak maxima to longer wavelengths usually occur. Finally, peaks in the ultraviolet and visible regions are ordinarily broad because of vibrational effects; the precise determination of the position of the maximum is thus difficult.

Absorption of Ultraviolet and Visible Radiation by Inorganic Species. The absorption spectra for most inorganic ions or molecules

TABLE 16-2
Absorption Characteristics of Some Common Organic Chromophores

CHROMOPHORE	EXAMPLE	SOLVENT	λ_{max} nm	ϵ_{max}
Alkene	$C_6H_{13}CH{=}CH_2$	n-Heptane	177	13,000
Conjugated alkene	$CH_2{=}CHCH{=}CH_2$	n-Heptane	217	21,000
Alkyne	$C_5H_{11}C{\equiv}C{-}CH_3$	n-Heptane	178	10,000
			196	2,000
			225	160
Carbonyl	$CH_3\overset{\text{O}}{\overset{\|}{C}}CH_3$	n-Hexane	186	1,000
			280	16
	$CH_3\overset{\text{O}}{\overset{\|}{C}}H$	n-Hexane	180	Large
			293	12
Carboxyl	$CH_3\overset{\text{O}}{\overset{\|}{C}}OH$	Ethanol	204	41
Amido	$CH_3\overset{\text{O}}{\overset{\|}{C}}NH_2$	Water	214	60
Azo	$CH_3N{=}NCH_3$	Ethanol	339	5
Nitro	CH_3NO_2	Isooctane	280	22
Nitroso	C_4H_9NO	Ethyl ether	300	100
			665	20
Nitrate	$C_2H_5ONO_2$	Dioxane	270	12
Aromatic	Benzene	n-Hexane	204	7,900
			256	200

resemble those for organic compounds, with broad absorption maxima and little fine structure. The spectra for ions of the lanthanide and actinide series represent an important exception. The electrons responsible for absorption by these elements ($4f$ and $5f$ respectively) are shielded from external influences by electrons that occupy orbitals with larger principal quantum numbers. As a consequence, the bands are narrow and relatively unaffected by the nature of the species bonded by the outer electrons.

With few exceptions the ions and complexes of the 18 elements in the first two transition series are colored in one if not all of their oxidation states. Here absorption involves transitions between filled and unfilled d-orbitals which differ in energy as a consequence of ligands bonded to the metal ions. The energy difference between the d-orbitals (and thus the position of the corresponding absorption peak) depends upon the oxidation state of the element, its position in the periodic table, and the kind of ligand bonded to its ion.

Charge-Transfer Absorption.[9] For analytical purposes *charge-transfer absorption* by inorganic species is of particular importance because the molar absorptivities of the band peaks are very large ($\epsilon_{max} > 10,000$). Thus a highly sensitive means for detecting and determining the absorbing species is provided. Many inorganic and organic complexes exhibit charge-transfer absorption and are therefore called *charge-transfer complexes.* Common examples include the thiocyanate and phenolic complexes of iron(III), the *o*-phenanthroline complex of iron(II), the iodide complex of molecular iodine, and the ferro-ferricyanide complex responsible for the color of Prussian blue.

A requirement for a charge-transfer spectrum is the existence of an electron-donor group as well as an electron acceptor within the complex. Absorption of radiation involves transition of an electron from the donor group to an orbital that is largely associated with the acceptor. The excited state is thus the product of an internal oxidation-reduction process.

Applications of Absorption Spectroscopy to Qualitative Analysis

Absorption spectroscopy is a useful tool for qualitative analysis. Identification of a pure compound by this method involves an empirical comparison of the spectral details of the unknown (maxima, minima, and inflection points) with those of authentic compounds; a close match is considered good evidence of chemical identity, particularly if the spectrum of the unknown contains a number of sharp and well-defined features. Absorption in the infrared region is particularly useful for qualitative purposes because of the wealth of fine structure that exists in the spectra of many compounds (see Figure 16-16). The application of ultraviolet and visible spectrophotometry to qualitative analysis is more limited because the absorption bands tend to be broad and hence lacking in detail. Nevertheless, spectral investigations in this region frequently provide useful qualitative information concerning the presence or absence of certain functional groups (such as carbonyl, aromatic, nitro, or conjugated diene) in organic compounds. A further important application involves the detection of highly absorbing impurities in nonabsorbing media. If an absorption peak for the contaminant has a sufficiently high absorptivity, its presence in trace amounts can be readily established.

To obtain an absorption spectrum useful for qualitative comparison, absorbance data should be collected with the narrowest possible band width. Otherwise significant details of the spectrum may be lost. This effect is demonstrated in Figure 16-17.

---

[9]For a brief discussion of this type of absorption see C. N. R. Rao, *Ultra-Violet and Visible Spectroscopy, Chemical Applications.* New York: Plenum Press, 1967, Chapter 11.

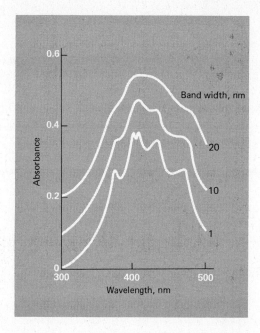

FIGURE 16-17 The effect of band width upon spectra of identical solutions. Note that spectra have been displaced from one another by 0.1 in absorbance.

Quantitative Analysis by Absorption Measurements

Absorption spectroscopy, particularly in the ultraviolet and visible regions, is one of the most useful tools available to the chemist for quantitative analysis. Important characteristics of spectrophotometric and photometric methods include the following:

1. *Wide applicability.* Many inorganic and organic species absorb in the ultraviolet and visible regions and are thus susceptible to quantitative determination. In addition, many nonabsorbing species can be caused to absorb after suitable chemical treatment.
2. *High sensitivity.* Molar absorptivities in the range of 10,000 to 40,000 are common, particularly for the charge-transfer complexes of inorganic species. Thus analyses for concentrations in the range of 10^{-4} to $10^{-5} M$ are commonplace. With suitable procedural modifications, the range can be extended to 10^{-6} or even $10^{-7} M$.
3. *Moderate to high selectivity.* Through the judicious choice of conditions, it may be possible to locate a wavelength region in which the analyte is the only absorbing component in a sample. Furthermore, where overlapping absorption bands do occur, corrections based on additional measurements at other wavelengths are sometimes possible. As a consequence, the separation step can be omitted.
4. *Good accuracy.* The relative error for the typical spectrophotometric

or photometric procedure lies in the range of 1 to 3%. Errors can be decreased to a few tenths of a percent with special techniques.

5. *Ease and convenience.* Spectrophotometric or photometric measurements are easily and rapidly performed with modern instruments. Furthermore, the methods are often readily automated. As a consequence, absorption methods are widely used for routine chemical analysis, continuous monitoring of atmospheric and water pollutants, and control of industrial processes.

SCOPE

The applications of quantitative absorption methods are not only numerous but also touch upon every area in which quantitative chemical information is required. The reader can obtain a notion of the scope of spectrophotometry by consulting the series of review articles published in *Analytical Chemistry*[10] as well as monographs on the subject.[11]

Applications to Absorbing Species. Table 16-2 lists many common organic chromophoric groups. If a molecule contains several conjugated chromophores, absorption may be shifted to the visible region of the spectrum. Spectrophotometric analysis for any organic compound containing one or more of these groups is potentially feasible; many examples of this type of analysis are found in the literature.

A number of inorganic species also absorb and are thus susceptible to direct determination; we have already mentioned the various transition metals. In addition, other species show characteristic absorption; examples include nitrate ion, the oxides of nitrogen, the elemental halogens, and ozone.

Applications to Nonabsorbing Species. Numerous reagents react with nonabsorbing inorganic species to yield products that absorb strongly in the ultraviolet or visible regions. The successful application of such reagents to quantitative analysis usually requires that the color-forming reaction be forced near completion. It should be noted that these reagents are also frequently employed for the determination of an absorbing species such as a transition metal ion; the molar absorptivity of the product can be orders of magnitude greater than that of the uncombined ion.

A host of complexing agents have been employed for the deter-

---

[10]See, for example, J. A. Howell and L. G. Hargis, *Anal. Chem.*, **50**, 243R (1978); D. F. Boltz and M. G. Mellon, *Anal. Chem.*, **48**, 216R (1976).

[11]See, for example, E. B. Sandell, *Colorimetric Determination of Traces of Metals*, 4th ed. New York: Interscience, 1978; D. F. Boltz, ed., *Colorimetric Determination of Nonmetals*. New York: Interscience, 1958; F. D. Snell and C. T. Snell, *Colorimetric Methods of Analysis*, 3d ed., 4 vols. Princeton, N. J.: Van Nostrand, 1959.

mination of inorganic species. Typical inorganic reagents include thiocyanate ion for iron, cobalt, and molybdenum, the anion of hydrogen peroxide for titanium, vanadium, and chromium, and iodide ion for bismuth, palladium, and tellurium. Of even more importance are organic chelating agents which form stable, colored complexes with cations. Examples include o-phenanthroline for the determination of iron, dimethylglyoxime for nickel, diethyldithiocarbamate for copper, and diphenylthiocarbazone for lead.

Certain nonabsorbing organic functional groups are also determined by absorption procedures. For example, low molecular weight aliphatic alcohols react with cerium(IV) to produce a red 1:1 complex that can be employed for quantitative purposes.

Quantitative Infrared Methods. Quantitative measurements in the infrared region are not different in principle from similar measurements in the ultraviolet and visible region. Several practical problems, however, prevent the attainment of comparable accuracies. These include the necessity of using very narrow cell widths that are difficult to reproduce, the high probability of overlap in absorption among the components in the sample (owing to the complexity of their spectra), and the narrowness of the peaks that often leads to deviations from Beer's law.

PROCEDURAL DETAILS

Before a photometric or spectrophotometric analysis can be undertaken, it is necessary to develop a set of working conditions and prepare a calibration curve that relates concentration to absorbance.

Selection of Wavelength. For quantitative spectrophotometric analyses, selection of a wavelength corresponding to an absorption peak has several advantages. The change in absorbance per unit change in concentration is greatest at this wavelength. In addition, the measurements are less sensitive to uncertainties resulting from inability to reproduce precisely the wavelength setting of the instrument. Finally, good adherence to Beer's law can be expected (p. 442). Occasionally a wavelength other than a peak may be appropriate for a particular analysis to avoid interference from other absorbing substances. In this event the region selected should, if possible, be one in which the change in absorptivity with wavelength is not too great.

Variables That Influence Absorbance. Common variables that influence the absorption spectrum of a substance include the nature of the solvent, the pH of the solution, the temperature, high electrolyte concentrations, and the presence of interfering species. The effects of these variables must be known, and a set of analytical conditions must be chosen such that the absorbance will not be materially influenced by small, uncontrolled variations in their magnitudes.

Determination of the Relationship between Absorbance and Concentration. The calibration standards for a photometric or spectrophotometric analysis should approximate the overall composition of the actual samples and should cover a reasonable analyte concentration range. Seldom, if ever, is it safe to assume adherence to Beer's law and use only a single standard to determine the molar absorptivity. It is folly to base the results of an analysis on a literature value for the molar absorptivity.

The difficulties that attend production of a set of standards whose overall composition closely resembles that of the sample can be formidable if not insurmountable. Under these circumstances a standard addition approach (p. 374) may prove useful. Here a known quantity of standard is added to an additional aliquot of the sample. Provided Beer's law is obeyed, the analyte concentration can be calculated.

EXAMPLE

A 2.00-ml urine specimen was treated with reagents to develop color with phosphate and diluted to 100 ml. Photometric analysis for the phosphate in a 25.0-ml aliquot yielded an absorbance of 0.428. Addition of 1.00 ml of a solution containing 0.050 mg of phosphate to a second 25.0-ml aliquot resulted in an absorbance of 0.517. Calculate the milligrams phosphate per milliliter of sample.

We must first correct the second measurement for dilution. Thus

$$\text{corrected absorbance} = 0.517 \times \frac{26.0}{25.0} = 0.538$$

$$\text{absorbance due to 0.050 mg phosphate} = 0.538 - 0.428 = 0.110$$

$$\text{mg phosphate in } \frac{25.0}{100} \text{ of specimen} = \frac{0.428}{0.110} \times 0.050 = 0.195$$

Thus

$$\text{mg phosphate/ml of specimen} = \frac{100}{25.0} \times 0.195 \times \frac{1}{2.00} = 0.39$$

Analysis of Mixtures of Absorbing Substances. The total absorbance of a solution at a given wavelength is equal to the sum of the absorbances of the individual components present. This relationship makes possible the analysis of the individual components in a mixture even if an overlap in their spectra exists. Consider, for example, the spectra of M and N, shown in Figure 16-18. There is obviously no wavelength at which the absorbance of this mixture is due simply to one of the components; thus an analysis for either M or N is impossible by a single measurement. However, the absorbances of the mixture at two wavelengths λ' and λ'' may be expressed as follows:

$$A' = \epsilon'_M bc_M + \epsilon'_N bc_N$$

$$A'' = \epsilon''_M bc_M + \epsilon''_N bc_N$$

The four molar absorptivities ϵ'_M, ϵ'_N, ϵ''_M, and ϵ''_N can be evaluated from individual standard solutions of M and of N, or better, from the slopes of their Beer's law plots. The absorbances, A' and A'', for the mixture are experimentally determinable as is b, the cell thickness. Thus from these two equations the concentration of the individual components in the mixture, c_M and c_N, can be readily calculated. These relationships are valid only if Beer's law is followed. The greatest accuracy in an analysis of this sort is attained by choosing wavelengths at which the differences in molar absorptivities are large.

Mixtures containing more than two absorbing species can be analyzed, in principle at least, if a further absorbance measurement is made for each added component. The uncertainties in the resulting data become greater, however, as the number of measurements increases.

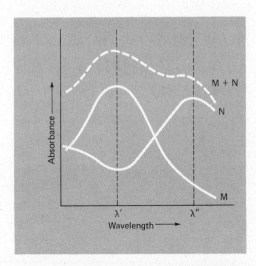

FIGURE 16-18 Absorption spectrum of a two-component mixture.

Analytical Errors in Absorption Measurements

In many (but certainly not all) absorptiometric methods the major source of indeterminate error lies in the absorbance measurement. The determination of this quantity involves three steps: the 0% T adjustment, the 100% T adjustment, and the measurement of T with the sample in the beam. The standard deviation for the transmittance, σ_T, is a combination of the individual standard deviations associated with these three measurement steps. Because of the logarithmic relationship between transmittance and concentration (or absorbance), the effect of the uncertainty in transmittance on the indeterminate error in these quantities is somewhat complex; it is given by the equation

$$\frac{\sigma_c}{c} = -\frac{0.434}{T \log T} \sigma_T \tag{16-12}$$

where σ_c/c is the relative standard deviation in concentration c that results from the absolute standard deviation σ_T in the measurement of T.[12]

TYPES OF INSTRUMENTAL UNCERTAINTIES

Three general categories of instrumental uncertainties have been recognized. They include (1) those for which σ_T is constant and independent of T ($\sigma_T = k_1$); (2) those for which σ_T is directly proportional to T ($\sigma_T = k_2 T$); and (3) those for which σ_T is approximately proportional to $\sqrt{T + T^2}$ ($\sigma_T = k_3\sqrt{T + T^2}$) where k_1, k_2, and k_3 are constants. Common sources of these three types of uncertainties are listed in Table 16-3.[13]

THE EFFECT OF INSTRUMENTAL UNCERTAINTIES ON THE PRECISION OF SPECTROPHOTOMETRIC ANALYSIS

The accuracy of measurements with most spectrophotometers is limited by one of the three types of uncertainties shown in Table 16-3. The magnitude

TABLE 16-3
Sources of Instrumental Uncertainties in Transmittance Measurements

CATEGORY	SOURCES
$\sigma_T = k_1$	Dark current and amplifier uncertainties (noise) Thermal detector uncertainties in infrared detectors Limited resolution of the readout device
$\sigma_T = k_2 T$	Variation in output of the source Uncertainties in cell positioning
$\sigma_T = k_3\sqrt{T^2 + T}$	Photon detector uncertainties (shot noise) in ultraviolet and visible detectors

[12]For a discussion of error propagation in a relationship such as $c = -\log T/\epsilon b$, see H. A. Strobel, *Chemical Instrumentation*, 2d ed. Reading, Mass.: Addison-Wesley, 1973, pp. 11–13.

[13]For a detailed discussion of the various errors associated with spectrophotometric measurement in the ultraviolet and visible regions, see L. D. Rathman, S. R. Crouch, and J. D. Ingle, Jr., *Anal. Chem.*, **44**, 1375 (1972).

of the resulting concentration error, however, differs from category to category. Thus it is worthwhile examining how each type of uncertainty affects the precision of analytical results as well as the relationship between types of instruments and uncertainty categories.

Concentration Errors When $\sigma_T = k_1$. For many photometers and spectrophotometers the standard deviation in the measurement of T is constant and independent of the magnitude of T. A particularly common source of this behavior is found in direct reading instruments that employ a meter equipped with a 5- to 7-in. scale. Usually the resolution of such a meter is a few tenths percent of full scale, and the absolute uncertainty in T is the same from one end of the scale to the other. A similar limitation is found in some digital panel meters.

The other common type of instrument which may exhibit an uncertainty that is independent of transmittance is the infrared spectrophotometer. Here the limiting indeterminate error lies in the thermal detector. Fluctuations in the output of this type of transducer are independent of the output; indeed, fluctuations are observed even in the absence of radiation and therefore current.

The precision of instruments for which $\sigma_T = k_1$ can be readily obtained by measuring the transmittance of 20 or 30 portions of a solution and calculating the standard deviation σ_T of the measurements. The precision of analyses made with the instrument can then be derived from Equation 16-12. Clearly the precision for a particular analysis depends upon the magnitude of T. The third column of Table 16-4 shows data obtained with this equation when an absolute standard deviation σ_T of ± 0.003 or $\pm 0.3\%$ T was assumed. A plot of the data is shown by curve A in Figure 16-19.

An indeterminate uncertainty of 0.3% T is typical of many moderate priced spectrophotometers or photometers. Clearly, concentration errors of 1 to 2% relative are to be expected with these instruments. It is also evident that indeterminate errors at this level can only be realized if the absorbance of the sample lies between about 0.1 and 1.

Concentration Errors When $\sigma_T = k_2 T$. Uncertainties of this type commonly arise from failure to position the sample cell reproducibly with respect to the beam during replicate measurements. All cells have minor imperfections. As a consequence, reflection and scattering losses vary as different sections of the cell window are exposed to the beam. Small variations in transmittance result. Rathman, Crouch, and Ingle[13] have shown that this uncertainty is probably the most common limitation to the accuracy of high-quality spectrophotometers.

Fluctuation in source intensity also yields standard deviations that are proportional to transmittance.

TABLE 16-4

Relative Concentration Error as a Function of Transmittance and Absorbance for Various Types of Uncertainties

TRANSMITTANCE, T	ABSORBANCE, A	RELATIVE ERROR, $\frac{\sigma_c}{c} \times 100$		
		FOR σ_T EQUAL TO		
		k_1	$k_2 T$	$k_3\sqrt{T^2 + T}$
0.95	0.022	$\pm 6.2^a$	$\pm 25.3^b$	$\pm 8.4^c$
0.90	0.046	± 3.2	± 12.3	± 4.1
0.80	0.097	± 1.7	± 5.8	± 2.0
0.60	0.222	± 0.98	± 2.5	± 0.96
0.40	0.398	± 0.82	± 1.4	± 0.61
0.20	0.699	± 0.93	± 0.81	± 0.46
0.10	1.00	± 1.3	± 0.56	± 0.43
0.032	1.50	± 2.7	± 0.38	± 0.50
0.010	2.00	± 6.5	± 0.28	± 0.65
0.0032	2.50	± 16.3	± 0.23	± 0.92
0.0010	3.00	± 43.4	± 0.19	± 1.4

[a] From Equation 16-12 employing $k_1 = \sigma_T = \pm 0.003$.

[b] From Equation 16-13 employing $k_2 = \pm 0.013$.

[c] From Equation 16-14 employing $k_3 = \pm 0.003$.

Substitution of $\sigma_T = k_2 T$ into Equation 16-12 yields

$$\frac{\sigma_c}{c} = -\frac{0.434\, k_2}{\log T} \tag{16-13}$$

Column 4 of Table 16-4 contains data obtained from Equation 16-13 with k_2 assumed to be 0.013. The data are plotted as curve B in Figure 16-19.

The value of k_2 used in the foregoing calculation was obtained experimentally by Rathman, Crouch, and Ingle[13] with a high-quality spectrophotometer. They were able to demonstrate that the source of this uncertainty was a variation in cell positioning. The uncertainty could be eliminated by leaving the cell in position at all times; samples and standards were then introduced by means of a syringe.

Concentration Error When $\sigma_T = k_3\sqrt{T^2 + T}$. This type of uncertainty often limits the accuracy of the highest quality instruments. It arises from the so-called *shot noise* that causes the output of photomultipliers and

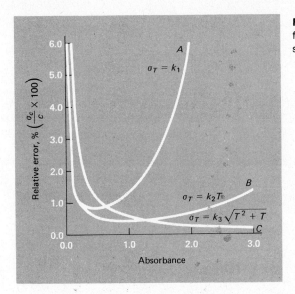

FIGURE 16-19 Error curves for various categories of instrumental uncertainties.

phototubes to fluctuate in a random way about a mean value. Substitution of this relationship between σ_T and T into Equation 16-12 gives

$$\frac{\sigma_c}{c} = -\frac{0.434\,k_3}{T \log T}\sqrt{T^2 + T} = -\frac{0.434\,k_3}{\log T}\sqrt{1 + \frac{1}{T}} \qquad \textbf{(16-14)}$$

The last column in Table 16-4 demonstrates the effect of shot noise on the indeterminate errors associated with an analysis. Here k_3 was assumed to have a value of 0.003. The data are plotted as curve C in Figure 16-19.

The value used for k_3 in the foregoing calculations is typical of high-quality spectrophotometers. Note that with such instruments the most accurate analyses are obtained in an absorbance range of about 0.6 to 2.5.

PROBLEMS

*1. Use the data provided to evaluate the missing quantities.

	ABSORB-ANCE, A	TRANS-MITTANCE, $T, \%$	MOLAR ABSORP-TIVITY, ϵ	PATH LENGTH, b, cm	CONCEN-TRATION, concn
(a)	0.474			0.100	6.51×10^{-4} M
(b)		88.4		2.50	3.40×10^{-5} M
(c)	0.126		2.67×10^4	1.50	M
(d)		31.2	2.67×10^4	1.50	ppm (gfw = 325)

ABSORB-ANCE, A	TRANS-MITTANCE, $T, \%$	MOLAR ABSORP-TIVITY, ϵ	PATH LENGTH, b, cm	CONCEN-TRATION, concn
(e) 0.946		9.96×10^3		$5.11 \times 10^{-5}\ M$
(f)	9.64	7.00×10^2		$1.42 \times 10^{-3}\ M$
(g)		1.60×10^4	1.50	10.0 ppm (gfw = 180)
(h)		7.63×10^3	2.00	$5.00 \times 10^{-5}\ M$

2. Use the data provided to evaluate the missing quantities.

ABSORB-ANCE, A	TRANS-MITTANCE, $T, \%$	MOLAR ABSORP-TIVITY, ϵ	PATH LENGTH, b, cm	CONCEN-TRATION, concn
(a) 0.084			2.50	$1.23 \times 10^{-4}\ M$
(b)	74.3		0.250	$4.46 \times 10^{-4}\ M$
(c) 0.555		2.22×10^4	1.50	M
(d)	18.2	1.54×10^3	1.50	ppm (gfw = 283)
(e) 0.897		8.74×10^3		$8.32 \times 10^{-5}\ M$
(f)	10.6	3.36×10^4		$6.76 \times 10^{-5}\ M$
(g)		4.16×10^3	0.500	10.0 ppm (gfw = 225)
(h)		1.45×10^4	10.00	$5.00 \times 10^{-6}\ M$

*3. A solution containing 6.76 ppm $KMnO_4$ has a transmittance of 0.145 in a 0.843-cm cell at a certain wavelength. Calculate the molar absorptivity for $KMnO_4$ at this wavelength.

4. A solution containing 4.64 mg/250 ml of M (gfw = 220) has a transmittance of 41.2% in a 1.50-cm cell at 480 nm. Calculate the molar absorptivity ϵ for M at this wavelength.

*5. A solution containing 8.43 ppm of B has a transmittance of 37.5% in a 1.50-cm cell. Calculate the absorptivity a for B.

6. A solution containing the complex formed between Bi(III) and thiourea has a molar absorptivity of 9.3×10^3 liter cm^{-1} mole^{-1} at 470 nm.
   (a) What will be the absorbance of a $3.95 \times 10^{-5}\ M$ solution of the complex when measured at 470 nm in a 2.50-cm cell?
   (b) What is the percent transmittance of the solution described in (a)?
   (c) What will be the concentration of the complex in a solution that has the same absorbance described in (a) when measured at 470 nm in a 1.00-cm cell?

*7. At 580 nm, the wavelength of its maximum absorption, the complex $FeSCN^{2+}$ has a molar absorptivity of 7.00×10^3 liter cm^{-1} mole^{-1}. Calculate
   (a) the absorbance of a $3.64 \times 10^{-5}\ M$ solution of the complex at 580 nm when measured in a 0.500-cm cell.

(b) the transmittance of the solution described in (a).

(c) the absorbance of a solution that has half the transmittance of that described in (a).

8. A 25.0-ml aliquot of a solution that contains 5.45 ppm iron(III) is treated with an appropriate excess of KSCN and diluted to a final volume of 50.0 ml. What will be the absorbance and percent transmittance of the resulting solution when measured at 580 nm in a 1.25-cm cell? See Problem 7 for absorptivity data.

*9. Zinc(II) and the ligand L form a product that absorbs strongly at 600 nm. As long as the formal concentration of L exceeds that of zinc(II) by a factor of 5 (or more), the absorbance is dependent only on the cation concentration. Neither zinc(II) nor L absorbs at 600 nm.

A solution that is $1.00 \times 10^{-4}\ F$ in zinc(II) and $1.00 \times 10^{-3}\ F$ in L has an absorbance of 0.771 when measured in a 1.25-cm cell at 600 nm. Calculate

(a) the percent transmittance of this solution.

(b) the percent transmittance of this solution when it is contained in a 2.50-cm cell.

(c) the molar absorptivity of the complex.

*10. A 3.65-g petroleum specimen was decomposed by wet-ashing and subsequently diluted to 500 ml in a volumetric flask. Analysis for cobalt was performed by treating aliquots as indicated:

VOLUME OF DILUTED SAMPLE, ml	REAGENT VOLUMES			ABSORBANCE (1.00-cm CELLS)
	Co(II), 3.00 ppm	Ligand	H₂O	
50.0	0.0	20.0	10.0	0.264
50.0	10.0	20.0	0.0	0.598

With the added information that the Co(II)-ligand chelate obeys Beer's law, calculate the percentage of Co in the original sample.

11. A four-tablet sample of vitamin supplement, weighing a total of 4.6 g, was wet-ashed to eliminate organic matter and then diluted to 2.00 liters in a volumetric flask. Calculate the average weight of iron in each tablet, based upon the accompanying information.

VOLUME OF DILUTED SAMPLE, ml	REAGENT VOLUMES			ABSORBANCE (1.00-cm CELLS)
	Fe(III), 1.00 ppm	Ligand	H₂O	
5.00	0.00	25.00	15.00	0.501
5.00	15.00	25.00	0.00	0.613

12. The equilibrium constant for the conjugate acid-base pair

$$HIn + H_2O \rightleftarrows H_3O^+ + In^-$$

is 5.00×10^{-3}. Additional information:

SPECIES	ABSORPTION MAXIMUM, nm	MOLAR ABSORPTIVITY AT	
		430 nm	600 nm
HIn	430	1720	0
In$^-$	600	0	1960

Calculate the absorbance (1.00-cm cells) at 430 nm and at 600 nm for the following formal indicator concentrations: *(a) 10.0×10^{-4}, (b) 5.00×10^{-4}, *(c) 2.50×10^{-4}, (d) 1.00×10^{-4}, and *(e) 0.500×10^{-4}. Plot the data.

13. The absolute error in transmittance for a particular instrument is estimated to be ± 0.005. Calculate the percent relative error in concentration due to this source when $\sigma_T = k_1$ and

*(a) $T = 0.016$.   (d) $A = 0.936$.

*(b) $A = 0.412$.   (e) $T = 0.799$.

*(c) $\%T = 63.7$.   (f) $\%T = 50.0$.

14. The response of a photometer with respect to a particular solution yielded the accompanying transmittance data: 0.377, 0.371, 0.370, 0.375, 0.378. Assume that the indeterminate error in the measurement is independent of the magnitude of T. For solutions having transmittances of 10, 50, and 90%, respectively, estimate the relative error in concentration (in parts per thousand) if

*(a) the absolute indeterminate error is assumed to be twice the average deviation from the mean of the five data.

(b) the absolute indeterminate error is assumed to be the standard deviation for the five data.

*(c) the absolute indeterminate error is assumed to be the 90% confidence interval for a single measurement.

15. Solutions containing A and B are known to obey Beer's law over an extensive concentration range. Molar absorptivity data for each are as follows:

	MOLAR ABSORPTIVITY, ϵ			MOLAR ABSORPTIVITY, ϵ	
λ, nm	A	B	λ, nm	A	B
400	893	0.00	560	440	622
420	940	0.00	580	297	880
440	955	0.00	600	167	1178
460	936	24.6	620	39.7	1492
480	874	102	640	3.45	1742
500	795	185	660	0.00	1806
520	691	289	680	0.00	1809
540	574	428	700	0.00	1757

*(a) A solution containing both solutes has an absorbance of 0.312 when measured in a 1.00-cm cell at 540 nm. What is the concentration of B in this solution if it is known that $[A] = 2.50 \times 10^{-4}$?

*(b) A solution containing both species has an absorbance of 0.696 both at 440 nm and at 600 nm when measured in a 1.00-cm cell. Calculate the concentrations of A and B.

(c) A solution containing both species has an absorbance in a 1.50-cm cell of 0.720 at 600 nm and of 0.699 at 480 nm. Calculate the concentrations of A and B.

(d) Construct an absorption spectrum for a solution that is (1) 5.00×10^{-4} M in A, (2) 4.00×10^{-4} M in B, and (3) 5.00×10^{-4} M in A and 4.00×10^{-4} M in B. For each spectrum assume a cell length of 1.08 cm.

(e) Construct transmittance spectra for the three solutions described in (d).

16. A. J. Mukhedkar and N. V. Deshpande [*Anal. Chem.*, **35**, 47(1963)] report on a simultaneous determination for cobalt and nickel based upon absorption by their respective 8-quinolinol complexes. Molar absorptivities corresponding to absorption maxima are the following:

	WAVELENGTH	
	365 nm	**700 nm**
ϵ_{Co}	3529	428.9
ϵ_{Ni}	3228	0.00

Calculate the formal concentrations of nickel and cobalt in each of the following solutions based upon the accompanying data.

	ABSORBANCE, 1.00-cm CELL	
SOLUTION	**365 nm**	**700 nm**
*a	0.816	0.0820
b	0.742	0.0543
c	0.680	0.0442

17. Solutions of P and of Q individually obey Beer's law over a large concentration range. Spectral data for these species, measured in 1.00-cm cells, are tabulated below.

	ABSORBANCE, A			ABSORBANCE, A	
λ, nm	8.55×10^{-5} F P	2.37×10^{-4} F Q	λ, nm	8.55×10^{-5} F P	2.37×10^{-4} F Q
400	0.078	0.550	560	0.126	0.255
420	0.087	0.592	580	0.170	0.170
440	0.096	0.599	600	0.264	0.100
460	0.102	0.590	620	0.326	0.055
480	0.106	0.564	640	0.359	0.030
500	0.110	0.515	660	0.373	0.030
520	0.113	0.433	680	0.370	0.035
540	0.116	0.343	700	0.346	0.063

(a) Plot an absorption spectrum (1.00-cm cells) for a solution that is 7.76×10^{-5} F with respect to P and 3.14×10^{-4} F with respect to Q.

*(b) Calculate the absorbance (1.00-cm cells) at 440 nm of a solution that is 2.00×10^{-5} F in P and 4.00×10^{-4} F in Q.

*(c) Calculate the absorbance (1.00-cm cells) at 620 nm for a solution that is 1.71×10^{-4} F in P and 7.62×10^{-4} F in Q.

18. Use the data in the previous problem to calculate the quantities sought (1.50-cm cells are used for all measurements).

	CONCENTRATION, MOLE /LITER		ABSORBANCE	
	P	Q	440 nm	620 nm
*(a)			0.862	0.450
(b)			0.313	0.799
*(c)			0.276	0.347
(d)			0.897	0.350
*(e)			0.495	0.816
(f)			0.202	0.297

19. The indicator HIn has an acid-dissociation constant of 5.20×10^{-6} at ordinary temperatures. The accompanying absorbance data are for 7.50×10^{-5} F solutions of the indicator measured in 1.00-cm cells in strongly acidic and strongly alkaline media.

λ, nm	ABSORBANCE		λ, nm	ABSORBANCE	
	pH = 1.00	pH = 13.00		pH = 1.00	pH = 13.00
420	0.535	0.050	550	0.119	0.324
445	0.657	0.068	570	0.068	0.352
450	0.658	0.076	585	0.044	0.360
455	0.656	0.085	595	0.032	0.361
470	0.614	0.116	610	0.019	0.355
510	0.353	0.223	650	0.014	0.284

Calculate the absorbance (1.00-cm cells) at 450 nm of a solution in which the total formal concentration of the indicator is 7.50×10^{-5} and the pH is

*(a) 4.670.

(b) 5.217.

*(c) 5.866.

20. What will be the absorbance at 595 nm (1.00-cm cells) of a solution that is 1.25×10^{-4} F with respect to the indicator in Problem 19 and whose pH is

*(a) 5.451.

(b) 6.213.

(c) 5.158.

21. The indicator described in Problem 19 was added to several buffered solutions; absorbance data (1.00-cm cells) at 450 nm and 595 nm were as follows:

	ABSORBANCE	
SOLUTION	450 nm	595 nm
*a	0.333	0.307
b	0.497	0.193
*c	0.666	0.121
d	0.210	0.375

Calculate the pH of each solution if the indicator concentration was 1.25×10^{-4} M.

*22. Absorptivity data for the cobalt and nickel complexes with 2,3-quinoxalinedithiol are as follows:

	WAVELENGTH	
	510 nm	656 nm
ϵ_{Co}	36,400	1,240
ϵ_{Ni}	5,520	17,500

A 0.376-g soil sample was dissolved and subsequently diluted to 50.0 ml. A 25.0-ml aliquot was treated to eliminate interferences; after addition of 2,3-quinoxalinedithiol, the volume was adjusted to 50.0 ml. This solution had an absorbance of 0.467 at 510 nm and 0.347 at 656 nm in a 1.00-cm cell. Calculate the percentages of cobalt and nickel in the soil.

17 atomic spectroscopy

Atomic spectroscopy, which has been extensively applied to qualitative and quantitative analysis, is based upon either absorption or emission of X-ray, ultraviolet, or visible radiation by atoms or monatomic ions. In order to obtain pure atomic spectra in the ultraviolet and visible regions, it is necessary to *atomize* the sample, a process whereby the constituent molecules are decomposed and converted to gaseous elementary particles.[1] Atomization is brought about by heating the sample to a temperature of several thousand degrees; a gaseous solution or *plasma* results, which is employed for emission or absorption analysis.

As shown in Table 17-1, atomic spectral methods are generally classified according to the method of atomization. All of these methods offer the advantages of high specificity, wide applicability, excellent sensitivity, speed, and convenience; they are among the most selective of all

---

[1] Atomic spectra in the X-ray region can be obtained without atomization of the sample because absorption and emission of X-rays involves interaction with the innermost nonbonding electrons. Thus these spectra are characteristic of an element, regardless of whether it is present in the elementary form or combined in a molecule.

TABLE 17-1

Classification of Atomic Spectral Methods

EMISSION METHODS			
Common Name	**Method of Atomization**	**Radiation Source**	**Usual Sample Treatment**
Arc spectroscopy	Electric arc	Sample in arc	Sample placed on electrode
Spark spectroscopy	Electric spark	Sample in spark	Sample placed on electrode
Flame emission or atomic emission	Flame	Sample in flame	Sample solution aspirated into a flame
Atomic fluorescence	Flame	Discharge lamp	Sample solution aspirated into a flame
X-ray fluorescence	None required	X-ray tube	Sample exposed to X-radiation
ABSORPTION METHODS			
Flame absorption or atomic absorption	Flame	Hollow cathode lamp	Sample solution aspirated into a flame
Flameless absorption	Heated surface	Hollow cathode lamp	Sample solution heated on a solid surface
X-ray absorption	None required	X-ray tube	Sample held in source beam

analytical procedures. Perhaps 70 elements can be determined. The sensitivities are typically in the parts-per-million to parts-per-billion range. An atomic spectral analysis can frequently be completed in a few minutes.

This chapter is concerned principally with ultraviolet and visible atomic emission and atomic absorption spectroscopy employing flames; occasional reference will be made to some of the other methods for producing atomic spectra.

Flame Spectroscopy

When an aqueous solution of an inorganic salt is aspirated into the hot flame of a burner, a substantial fraction of the metallic constituents is reduced to the elemental state; to a much lesser extent monatomic ions are also formed. Thus within the flame there is produced a gaseous solution of elementary particles.

TYPES OF FLAME SPECTROSCOPY

The temperature of a flame is sufficient to excite a small fraction of the monatomic particles to higher electronic energy states; their return to the ground state is accompanied by the production of emission lines, which serve as the basis for *flame emission spectroscopy.* Figure 17-1 shows one of the several emission lines produced when a potassium salt is atomized in a flame. Note that the potassium line is superimposed on a continuous background radiation arising from excitation of molecules such as carbon

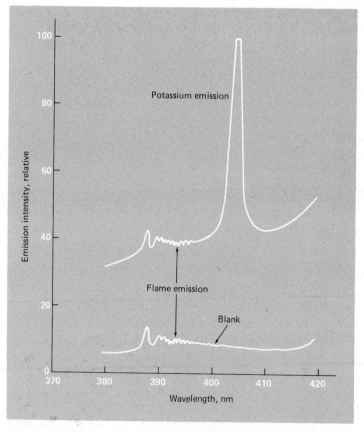

FIGURE 17-1 A portion of the emission spectrum for potassium in a hydrogen-oxygen flame. Note that the potassium band is superimposed upon the background emission from the flame. (The blank spectrum has been displaced downward.)

monoxide that are formed during the combustion process; cyanogen is a contributor to the background when air is used as the oxidant.

The wavelengths of emission lines are unique for each element; thus emission spectra offer a powerful method for qualitative analysis. Measurement of the intensity of these lines also permits quantitative analyses.

Generally the number of unexcited atoms in a flame at any instant far exceeds the number that are excited, and their presence can be employed for *atomic absorption analysis.* Here it is only necessary to interpose the flame plasma in the light path of a spectrophotometer similar to those discussed in Chapter 16. As with flame emission, the location of absorption lines serves to identify the components of a sample; the absorbance of a line, as in solution spectrophotometry, is generally proportional to the concentration of the absorbing species.

RELATIONSHIP BETWEEN FLAME ABSORPTION AND FLAME EMISSION SPECTROSCOPY

Atomic emission spectroscopy is based upon the intensity of radiation emitted by *excited* atoms, whereas the *unexcited* atoms serve as the basis for atomic absorption measurements. The unexcited species predominate by a large margin at the temperature of a flame. For example, at 2500°K only about 0.02% of the sodium atoms in a flame are excited to the $3p$ state at any instant; the percentages in higher electronic states are even less. At 3000°K the corresponding percentage is about 0.09.

This large ratio of unexcited to excited species in a flame is important in comparing absorption and emission procedures. Because atomic absorption methods are based upon a much larger population of particles, it might be expected to be the more sensitive procedure. This apparent advantage is offset, however, by the fact that an absorbance measurement involves a difference measurement ($A = \log P_0 - \log P$); when the two numbers are nearly alike, large relative errors in the difference result. As a consequence, the two procedures are often complementary in sensitivity, the one being advantageous for one group of elements and the other for a different group.

Another advantage of flame absorption spectroscopy is that the absolute number of unexcited particles is much less dependent upon temperature than is the number of excited species. For example, at about 2500°K a 20°K fluctuation causes a change of about 8% in the number of sodium atoms excited to the $3p$ state. Because so many more unexcited atoms exist at this temperature, however, their population changes by only about 0.02% for the same temperature variation.

It should be noted that temperature fluctuations do exert an indirect influence on atomic absorption by increasing the total number of atoms that are available for absorption. In addition, line broadening and a consequent decrease in peak height occur. Because of these indirect effects, a reasonable control of the flame temperature is required for quantitative atomic absorption measurements.

Atomic Absorption Spectroscopy

The radiation absorbed by a vaporized atom has exactly the same energy as that needed to cause excitation to a higher electronic state. Most of the atoms in the plasma are unexcited (p. 485); the most probable transitions will thus involve promotions from the ground state. Transitions from one excited state to another are far less likely, owing to the very small population of atoms in such states; absorption resulting from such a transition may be so weak as to be undetectable. Typically, then, an atomic absorption spectrum produced with a flame consists predominantly of *resonance* lines, which are the consequence of transitions from the ground state to upper levels. Note that *the wavelength of a resonance absorption line is identical to that of the emission line that corresponds to the same electronic transition.*

LINE WIDTHS IN ATOMIC ABSORPTION AND EMISSION SPECTRA

The natural line width of an atomic absorption or an atomic emission peak can be shown to be about 10^{-5} nm. Two effects, however, tend to cause observed widths to be between 0.002 and 0.005 nm. *Doppler broadening* arises from the rapid motion of the absorbing or emitting particles with respect to the detector. For those atoms traveling away from the detector the wavelength is effectively increased by the well-known Doppler effect; thus somewhat longer wavelengths are absorbed or emitted. The reverse is true of atoms moving toward the detector. *Pressure broadening* also occurs. Here collisions among atoms cause small changes in the ground-state energy levels and a consequent broadening of peaks. Note that both of these effects are enhanced as the temperature rises. Thus broader peaks are observed at more elevated temperatures.

MEASUREMENT OF ATOMIC ABSORPTION

Because atomic absorption lines are so very narrow, and because transition energies are unique for each element, analytical methods based on atomic absorption have the potential to be highly specific. On the other hand, the limited line widths create a measurement problem that is not encountered in solution absorption. No ordinary monochromator is capable of yielding a band of radiation that is as narrow as the peak width of an atomic absorption line (0.002 to 0.005 nm); thus the radiation isolated from a continuous source by a monochromator would necessarily have a large effective band width compared with that of the absorption line. Under these conditions Beer's law is not followed (p. 442). In addition, because the fraction absorbed from the beam would be small, the detector would receive a signal that was only slightly attenuated (that is, $P \to P_0$); the sensitivity of the measurement would, therefore, be poor.

This problem has been overcome by employing a source of radiation that emits a line of the same wavelength as the one to be used for the absorption analysis. For example, if the 589.6-nm absorption line of sodium is chosen for the analysis of that element, a sodium vapor lamp can be employed as a source. In such a lamp gaseous sodium atoms are excited by electrical discharge; the excited atoms then emit characteristic radiation as they return to lower energy levels. The emitted line will have a wavelength identical with the resonance absorption line. With a properly designed source (one that operates at a lower temperature than the flame to minimize Doppler broadening) the emission lines will have band widths that are significantly narrower than the absorption band widths. Thus the monochromator need have only the capability of isolating a suitable emission line for the absorption measurement (see Figure 17-2). The radiation employed in the analysis is then sufficiently limited in band width to permit measurements at the absorption peak; greater sensitivity and better adherence to Beer's law result.

Radiation Sources. The most common source for an atomic absorption measurement is the *hollow cathode lamp*, which consists of a tungsten anode and a cylindrical cathode. The cylinder is constructed of the metal whose spectrum is desired or serves to support a layer of that metal. The electrodes are housed in a glass tube that is filled with helium or argon at 1 to 2 mm pressure.

Ionization of the gas occurs when a potential is applied across the electrodes, and a current flows as the ions migrate to the electrodes. If the potential is sufficiently large, the gaseous cations acquire enough kinetic energy to dislodge some of the metal atoms from the cathode surface and produce an atomic cloud; this process is called *sputtering*. A portion of the sputtered metal atoms are in excited states and thus emit their characteristic radiation in the usual way. Eventually the metal atoms diffuse back to the cathode surface or the glass walls of the tube and are redeposited.

Hollow cathode lamps for various elements are sold commercially. Some have cathodes consisting of a mixture of several elements; lamps of this kind provide spectral lines for more than a single species.

Modulation. In the typical atomic absorption instrument it is necessary to eliminate interferences caused by *emission* of radiation by the flame. Most of the emitted radiation can be removed by locating the mono-

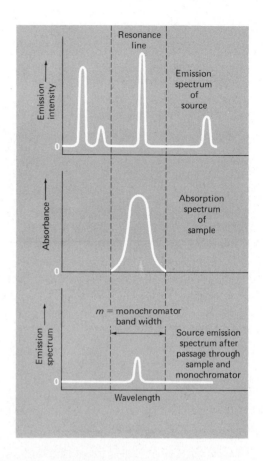

FIGURE 17-2 Absorption of a resonance line by atoms.

chromator between the flame and the detector. Nevertheless, this arrangement does not remove the radiation corresponding to the wavelength selected for the analysis; the flame will contain such radiation due to excitation and radiant emission by some atoms of the analyte. The difficulty is overcome by causing the intensity of the source to fluctuate at a constant frequency; this process is called *modulation*. The detector then receives two types of signal, an alternating one from the source and a continuous one from the flame. These signals are converted to the corresponding types of electrical response. A relatively simple electronic system is then employed to respond to and amplify only the ac part of the signal and to ignore the unmodulated dc signal.

A simple and entirely satisfactory way of modulating radiation from the source is to interpose a circular disk in the beam between the source and the flame. Alternate quadrants of this disk are removed to permit passage of light. Rotation of the disk at constant speed provides a beam that is chopped to the desired frequency. As an alternative, the power supply for the source can be designed for intermittent or ac operation.

INSTRUMENTS

An instrument for atomic absorption measurements has the same basic components as a spectrophotometer for measuring the absorption of solutions (see Figure 16-6); that is, it contains a source, a monochromator, a sample container (here, the flame or a hot surface), a detector, and an amplifier indicator. Both single-beam and double-beam instruments have been designed for atomic absorption measurements; the latter are the more common.

The major instrumental differences between atomic and solution absorption equipment are in the source and in the sample container. Features of these components require further discussion.

Burners. Two types of burners are employed in atomic absorption spectroscopy. In a *total consumption* burner the fuel and oxidizing gases are carried through separate passages to meet and mix with the solution of the analyte at an opening at the base of the flame. In a *premix* burner the sample is aspirated into a large chamber by a stream of the oxidant; here the fine mist of the analyte solution, the oxidant, and the fuel supply are mixed and then forced out the burner opening. Larger drops of sample collect in the bottom of the chamber and drain off. Figure 17-3 gives schematic diagrams of both burner designs.

Fuels used for flame production include natural gas, propane, butane, hydrogen, and acetylene. The last is perhaps most widely employed. The common oxidants are air, oxygen-enriched air, oxygen, and nitrous oxide. The acetylene–nitrous oxide mixture is advantageous when a hot flame is required.

Low-temperature flames (natural gas-air, for example) are used to advantage for elements that are readily converted to the atomic state, such as copper, lead, zinc, and cadmium. Elements such as the alkaline earths, on the other hand, form refractory oxides which require somewhat higher temperatures for decomposition; for these an acetylene-air mixture often produces the most sensitive results. Aluminum, beryllium, the rare earths, and certain other elements form unusually stable oxides; a reasonable concentration of their atoms can be obtained only at the high temperatures developed in an acetylene-oxygen or an acetylene–nitrous oxide flame.

Nonflame Atomizers. Several nonflame atomizers have been recently developed for the quantitative determination of small amounts of various elements. In a nonflame atomizer a few microliters of sample are evaporated and ashed at low temperatures on an electrically heated surface of carbon, tantalum, or other conducting material. The conductor can be a hollow tube, a strip or rod, a boat, or a trough. After ashing, a current of 100 A or more is passed through the conductor, which causes a rapid increase in temperature to perhaps 2000 to 3000°C; atomization of the sample occurs over a short period. Radiation from the source is passed

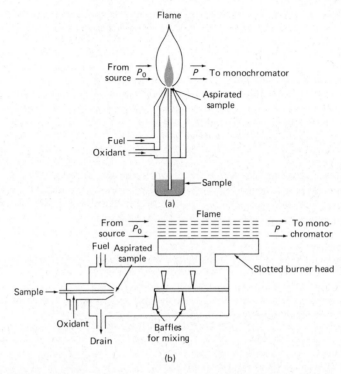

FIGURE 17-3 Burners for atomic absorption spectroscopy. (a) Total consumption burner. (b) Premix burner.

through the region immediately above the heated conductor. At a wavelength where absorption takes place, the absorbance is observed to rise to a maximum in a few seconds and then decay to zero, corresponding to the atomization and subsequent escape of the volatilized sample; analyses are based upon peak height.

Nonflame atomizers offer the advantage of unusually high sensitivity for small amounts of sample. Typically volumes between 0.5 and 10 μl are employed; under these circumstances absolute limits lie in the range of 10^{-10} to 10^{-13} g of the analyte.

The relative precision of nonflame methods is generally in the range of 5 to 10%; 1 to 2% can be expected for flame atomization.

Monochromators or Filters. An instrument for atomic absorption measurements must be capable of providing a sufficiently narrow band width to isolate the line chosen for the measurement from other lines that may interfere with or diminish the sensitivity of the analysis. For some of the alkali metals, which have only a few widely spaced resonance lines in the visible region, a glass filter suffices. An instrument employing readily interchangeable interference filters is available commercially. A separate filter (and light source) is used for each element. Satisfactory results for the analysis of 22 metals are claimed. Most instruments, however, incorporate a good-quality, ultraviolet and visible monochromator.[2]

Detectors and Indicators. The detector-indicator components for an atomic absorption spectrophotometer are fundamentally the same as for the typical ultraviolet and visible solution spectrophotometer (see Chapter 16). Generally photomultiplier tubes are used to convert the radiant energy received to an electrical signal. As we have pointed out, the electronic system must be capable of discriminating between the modulated signal from the source and the continuous signal from the flame. Both null-point and direct-reading meters, calibrated in terms of absorbance or transmittance, are used; some instruments provide a digital readout of analyte concentration.

APPLICATIONS

Atomic absorption spectroscopy provides a sensitive means for the determination of more than 60 elements; detection limits for some are provided in Table 17-2. Details concerning the quantitative determination of these and other elements can be found in several publications.[2]

INTERFERENCES

The absorption spectrum of each element is unique to that species. With a reasonably good monochromator it is possible to isolate an absorption peak that is free from interference by other elements. Unfortunately, however, this specificity does not extend to the chemical reactions occurring in the flame or above the heated surface of a nonflame atomizer; here interferences do occur as a consequence of interactions and chemical competitions that affect the number of atoms present in the light path under a specified set of conditions. These effects are not predictable from theory and must therefore be determined by experiment; they appear to be particularly troublesome with nonflame atomizers.

Cation Interferences. A few examples have been reported in which the absorption due to one cation is affected by the presence of a second. For example, the presence of aluminum is found to cause low results in the determination of magnesium; it has been shown that this interference results from the formation of a heat-stable, aluminum-magnesium

---

[2] For specifications see W. T. Elwell and J. A. F. Gidley, *Atomic-Absorption Spectrophotometry.* New York: Pergamon Press, 1966; R. G. Martinek, *Laboratory Management,* **6,** 24 (1968); J. Ramirez-Muñoz, *Atomic-Absorption Spectroscopy.* New York: Elsevier, 1968, pp. 182–202.

TABLE 17-2

Detection Limits for the Analysis of Selected Elements by Flame Absorption and Flame Emission Spectrometry

ELEMENT	WAVE-LENGTH, nm	DETECTION LIMIT, μg/ml		
		Flame Emission[a]	Flame Absorption[a]	Nonflame Absorption[b]
Aluminum	396.2	0.005 (N_2O)		0.03
	309.3		0.1 (N_2O)	
Calcium	422.7	0.005 (air)	0.002 (air)	0.0003
Cadmium	326.1	2 (N_2O)		0.0001
	228.8		0.005 (air)	
Chromium	425.4	0.005 (N_2O)		0.005
	357.9		0.005 (air)	
Iron	372.0	0.05 (N_2O)		0.003
	248.3		0.005 (air)	
Lithium	670.8	0.00003 (N_2O)	0.005 (air)	0.005
Magnesium	285.2	0.005 (N_2O)	0.0003 (air)	0.00006
Potassium	766.5	0.0005 (air)	0.005 (air)	0.0009
Sodium	589.0	0.0005 (air)	0.002 (air)	0.0001

[a] Taken from data compiled by E. E. Pickett and S. R. Koirtyohann, *Anal. Chem.*, **41** (14), 28A (1969). With permission of the American Chemical Society. Data are for an acetylene flame with the oxidant shown in parentheses.

[b] Data from J. W. Robinson and P. J. Slevin, *American Laboratory*, **4** (8), 14 (1972). With permission, International Scientific Communications, Inc.; wavelength not reported.

compound and a consequent decrease in the concentration of magnesium atoms in the flame. Beryllium, aluminum, and magnesium are reported to have a similar effect on calcium analyses. Fortunately this type of interference is relatively rare.

Anion Interferences. The absorption behavior for a metal may be influenced by the type and the concentration of anions present in the sample solution. This effect is not surprising inasmuch as the energy required to form atomic species from compounds must vary with the strength of the attraction between anion and cation. Such effects ordinarily become smaller with increases in flame temperature and may disappear entirely in some of the hotter flames.

Anion interference can sometimes be avoided by the addition of a complexing agent (EDTA, for example) to both the standards and the samples. In this way atom formation always results from decomposition of the complex. Alternatively, the standards can be made to approximate the anion composition of the sample to compensate for this effect.

ANALYTICAL TECHNIQUES

Both calibration curves and the standard addition method are suitable for atomic absorption spectroscopy.

Calibration Curves. Although absorbance should theoretically be proportional to concentration, deviations from linearity do occur. Thus empirical calibration curves must be prepared. In addition, there are sufficient uncontrollable variables in the production of an atomic vapor to warrant measuring the absorbance of at least one standard solution each time readings are taken. Any deviation of the standard from the original calibration curve can then be employed to correct the analytical results.

Standard Addition Method. The standard addition method is widely used in flame absorption spectroscopy. Here two or more aliquots of the sample are transferred to volumetric flasks. One is diluted to volume, and the absorbance of the solution is obtained. A known amount of analyte is added to the second, and its absorbance is measured after dilution to the same volume. Data for other additions may also be obtained. If a linear relationship between absorbance and concentration exists (and this should be established by several standard additions), the following relationships apply:

$$A_x = kC_x$$

$$A_T = k(C_s + C_x)$$

where C_x is the analyte concentration in the diluted sample, C_s is the contribution of the added standard to the concentration, and A_x and A_T are the two measured absorbances. Combination of the two equations yields

$$C_x = C_s \frac{A_x}{(A_T - A_x)} \qquad (17\text{-}1)$$

If several additions are made, A_T can be plotted against C_s. The resulting straight line can be extrapolated to $A_T = 0$. Substituting this value into Equation 17-1 reveals that at the intercept $C_x = -C_s$.

The standard addition method has the advantage that it often compensates for variations caused by physical and chemical interferences in the sample solution.

Accuracy. Under usual conditions the relative error associated with a flame absorption analysis is of the order of 1 to 2%. With special precautions this figure can be lowered to a few tenths of 1%.

Flame Emission Spectroscopy

Flame emission spectroscopy (also called flame photometry) has found widespread application to elemental analysis. The most important applications have been to the analysis of sodium, potassium, lithium, and calcium, particularly in biological fluids and tissue. For reasons of convenience, speed, and relative freedom from interference, flame emission spectroscopy has become the method of choice for these otherwise difficult-to-determine elements. The method has also been applied, with varying degrees of success, to the determination of perhaps half the elements in the periodic table. Thus flame emission spectroscopy must be considered to be one of the important tools for analysis.[3]

INSTRUMENTS FOR FLAME EMISSION SPECTROSCOPY

Both photometers and spectrophotometers are employed in flame emission methods. The latter find considerably more widespread application because of their greater selectivity and versatility. The typical flame spectrophotometer is similar in construction to the instruments discussed earlier in this chapter except that the flame now acts as the radiation source; the hollow cathode and chopper are not needed. Most modern instruments are adaptable to either emission or absorption analysis.

METALLIC SPECTRA IN FLAMES

An examination of the spectrum of a flame into which an aqueous solution of a metallic compound is aspirated (Figure 17-1) reveals the presence of both emission lines and bands. The line spectra are characteristic of the metallic atoms present. The band spectra, on the other hand, arise from the presence of molecules; here vibrational energy states are superimposed on electronic energy states and thus produce the closely spaced lines that make up the band.

Band Spectra. When hydrogen or hydrocarbon fuels are burned, band spectra due to such species as OH radicals, CN radicals, and C_2 molecules

---

[3] For a more complete discussion of the theory and applications of flame emission spectroscopy see J. A. Dean, *Flame Photometry*. New York: McGraw-Hill, 1960; B. L. Vallee and R. E. Thiers in *Treatise on Analytical Chemistry*, ed. I. M. Kolthoff and P. J. Elving, part I, vol. 6. New York: Interscience, 1965, Chapter 65.

are observed. Correction for this type of emission is necessary. In addition, some metallic elements form volatile oxides which produce analytically useful band spectra. This type of radiation is characteristic of the alkaline-earth and the rare-earth metals.

Effect of Temperature on Spectra. The temperature produced in a natural gas–air flame is so low (approximately 1800°C) that only the alkali- and alkaline-earth metals are excited. Generation of spectra for most other elements requires the use of oxygen as the oxidant and such fuels as acetylene, hydrogen, or cyanogen.

The optimum flame temperature for an analysis must be determined empirically and depends upon the excitation energy of the element, how it is combined in the sample, the sensitivity required, and what other elements are present. The high temperatures achieved with oxygen and nitrous oxide as oxidants are needed to excite many elements and may also be required for more easily excited elements when these occur as refractory compounds in the sample. Although high temperatures usually provide enhanced emission intensities, and thus higher sensitivity, there are notable instances where the use of a low-temperature flame is an advantage. For example, little enhancement in the characteristic emission by potassium is observed when the temperature is increased above 2000°C because of increases in the formation of potassium ions, which do not emit at the same wavelength as the element. At higher temperatures, moreover, there is an increased likelihood of interference from other elements. Thus the emission analysis for potassium (and the other alkali metals as well) is best performed with a low-temperature flame.

METHODS OF QUANTITATIVE ANALYSIS

Close control of many variables is essential for the acquisition of reliable flame photometric data. Whenever possible, the standards used for calibration should closely match the overall composition of the unknown solution. Ordinarily it is best to perform a calibration concurrently with the analysis. Even with these precautions measurements with a filter photometer can be expected to yield good results only where the sample solution has a relatively simple composition and the element being determined is a major constituent.

Several techniques have been suggested for the performance of a flame photometric analysis.[4] The use of a calibration curve or a standard addition is equally applicable here.

[4] For a summary of applications see J. A. Dean, *Flame Photometry*. New York: McGraw-Hill, 1960.

APPLICATIONS

Flame photometric methods have been applied to the analysis of numerous materials, including biological fluids, vegetable matter, cements, glasses, and natural waters.[4] The most important applications are for the determination of the alkali metals and calcium.

Table 17-2 compares the detection limits of flame photometry and flame absorption for several common elements. It is apparent that the relative sensitivity of the two procedures varies from element to element.

analytical separations

Few if any analytical measurements are truly specific for a single species. Were it not for this fact, chemical analysis would be a relatively simple undertaking. Properties that are useful for the determination of concentration are usually shared by several species. As a consequence, a separation step is more often the rule than the exception in a quantitative analysis.

Separation by Precipitation

Precipitation separations are based upon the solubility difference between the analyte and the undesired components. Solubility product considerations will generally provide guidance as to whether or not a given separation is theoretically feasible and will define the conditions required to achieve the separation. Unfortunately, however, other variables such as rate of precipitate formation and coprecipitation of unwanted species are crucial in determining the success or failure of a precipitation separation. These variables are not ordinarily susceptible to theoretical treatment within our present state of knowledge.

Numerous useful methods for the separation of inorganic species are based upon solubility differences. For example, ions such as tungsten(IV), tantalum(V), tin(IV), antimony(V), and silicon(IV) form oxides of limited solubility in strongly acidic media and can thus be isolated from most other cations. Hydrous oxides of iron, aluminum, and chromium are soluble in concentrated acids but precipitate from a slightly acidic medium; these ions can be separated from most divalent cations in acetate-buffered media.

Separations based on the formation of sulfides have found extensive use because the solubilities of metal sulfides differ greatly. Here control of the sulfide ion concentration is achieved by adjustment of pH (p. 101). The classical qualitative scheme is largely based upon sulfide separations. Quantitative applications are also extensive despite such difficulties as the toxic properties of the reagent and the likelihood of coprecipitation.

Chloride and sulfate ions are useful for separations because of their relatively specific behavior. The former can be used to separate silver from most other metals, whereas the latter is frequently employed to isolate a group of metals that includes lead, barium, strontium, and calcium.

Selected organic reagents used for the separation of inorganic ions were discussed in Chapter 6. Numerous other organic compounds have found similar applications.

Electrolytic precipitation constitutes a highly useful method for accomplishing separations (Chapter 15). In this process the more easily reduced species, be it the wanted or the unwanted component of the mixture, is isolated as a second phase. A mercury cathode is widely used for the electrolytic removal of many metal ions prior to the analysis of the residual solution. In general metals more easily reduced than zinc are conveniently deposited in the mercury; such ions as aluminum, beryllium, the alkaline earths, and the alkali metals remain in solution.

Extraction Methods

The distribution of a solute between two immiscible phases is an equilibrium process that can be treated by the law of mass action. Equilibrium constants for this process vary enormously among solutes; as a consequence, many useful separations are based on extraction.

THEORY

The partition of a solute between two immiscible solvents is governed by the *distribution law*. If we assume that the solute species A distributes itself between an aqueous and an organic phase, the resulting equilibrium may be written as

$$A_{aq} \rightleftharpoons A_{org}$$

where the subscripts "aq" and "org" refer to the aqueous and organic phases, respectively. Ideally the ratio of the activities of A in the two phases will be constant and independent of the total quantity of A. That is, at any given temperature

$$K = \frac{[A_{org}]}{[A_{aq}]} \tag{18-1}$$

where the equilibrium constant K is the *partition coefficient* or *distribution coefficient*. The terms in brackets are strictly the activities of A in the two solvents, but molar concentrations can frequently be substituted without serious error. Often K is approximately equal to the ratio of the solubility of A in each solvent.

The solute may exist in different states of aggregation in the two solvents. Then the equilibrium becomes

$$x(A_y)_{aq} \rightleftharpoons y(A_x)_{org}$$

and the partition coefficient takes the form

$$K = \frac{[(A_x)_{org}]^y}{[(A_y)_{aq}]^x}$$

Partition coefficients can be employed to establish the experimental conditions required to transfer a solute from one solvent to another. For example, consider a simple system that is adequately described by Equation 18-1.[1] Suppose further that we have V_{aq} ml of an aqueous solution containing a_0 mmole of A and that we propose to extract this with V_{org} ml of an immiscible organic solvent. At equilibrium a_1 mmole of A will remain in the aqueous layer, and we may write

$$[A_{aq}]_1 = \frac{a_1}{V_{aq}}$$

It follows, then, that

$$[A_{org}] = \frac{(a_0 - a_1)}{V_{org}}$$

Substitution of these quantities into Equation 18-1 and rearrangement give

$$a_1 = \left(\frac{V_{aq}}{V_{org}K + V_{aq}} \right) a_0 \tag{18-2}$$

The number of millimoles, a_2, remaining after a second extraction of the

---

[1] Suitable modification of this treatment can be made to take into account other equilibria; see H. A. Laitinen and W. E. Harris, *Chemical Analysis*, 2d ed. New York: McGraw-Hill, 1975, pp. 443–453.

water with an identical volume of solvent will, by the same reasoning, be

$$a_2 = \left(\frac{V_{aq}}{V_{org}K + V_{aq}}\right) a_1$$

When this expression is substituted into Equation 18-2, we obtain

$$a_2 = \left(\frac{V_{aq}}{V_{org}K + V_{aq}}\right)^2 a_0$$

After n extractions, the number of millimoles remaining is given by the expression

$$a_n = \left(\frac{V_{aq}}{V_{org}K + V_{aq}}\right)^n a_0 \tag{18-3}$$

Equation 18-3 can be rewritten in terms of the initial and final aqueous concentration of A by substituting the relationships

$$a_n = [A_{aq}]_n V_{aq} \quad \text{and} \quad a_0 = [A_{aq}]_0 V_{aq}$$

Thus

$$[A_{aq}]_n = \left(\frac{V_{aq}}{V_{org}K + V_{aq}}\right)^n [A_{aq}]_0 \tag{18-4}$$

EXAMPLE

The distribution coefficient of iodine between CCl_4 and water is 85. Calculate the number of millimoles of I_2 remaining in 100 ml of an aqueous solution, which was originally 1.00×10^{-3} M, after extraction with two 50-ml portions of CCl_4.

$$a_0 = 100 \times 1.00 \times 10^{-3} = 0.100 \text{ mmole}$$

$$a_2 = \left(\frac{100}{50 \times 85 + 100}\right)^2 0.100$$

$$= 5.28 \times 10^{-5} \text{ mmole}$$

Thus the two extractions should diminish the number of millimoles of I_2 in the aqueous solution from 0.1 to 5.28×10^{-5}.

The exponential nature of Equation 18-3 indicates that a more efficient extraction is achieved with several small volumes of solvent than a single large one. This effect is illustrated by comparing the results from the foregoing calculation with that which follows.

EXAMPLE

Calculate the number of millimoles of I_2 remaining if the aqueous solution in the preceding example had been extracted with a single 100-ml portion of CCl_4 rather than with two 50-ml portions.

$$a_1 = \frac{100}{100 \times 85 + 100} \times 0.100$$

$$= 1.16 \times 10^{-2} \times 0.100 = 1.16 \times 10^{-3} \text{ mmole}$$

Here the single extraction with the larger volume of CCl_4 leaves 1.16×10^{-3} mmole of I_2 in the aqueous layer; this is over 20 times greater than the remainder after the two extractions with half-volumes.

Figure 18-1 shows that the improved efficiency brought about by multiple extraction falls off rapidly as the number of subdivisions is increased. Clearly little is to be gained by dividing the extracting solvent into more than five or six portions.

APPLICATIONS

Extraction techniques are often more attractive than classical precipitation methods. The process of equilibration and separation of phases in a separatory funnel is inherently less tedious and time-consuming than precipitation, filtration, and washing. In addition, problems of coprecipitation and postprecipitation do not exist. Finally, and in contrast to the precipitation process, extraction procedures are ideally suited for isolating trace quantities of a species.

Extraction processes have found widespread use in the separation of organic compounds. In addition, a number of inorganic species can be separated by extraction with suitable solvents. For example, a single ether extraction of a 6-F hydrochloric acid solution will cause better than 50% of several ions to be transferred to the organic medium. Among these are iron(III), antimony(V), titanium(III), gold(III), molybdenum(VI), and tin(IV). Many other ions such as aluminum(III) and the divalent cations of cobalt, iron, lead, manganese, and nickel are unaffected by this extraction process.

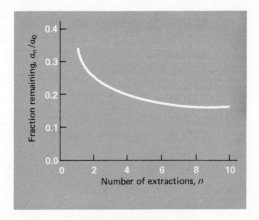

FIGURE 18-1 Plot of Equation 18-3, assuming that $K = 2$ and $V_{aq} = 100$. The total volume of the organic solvent is also assumed to be 100, so that $V_{org} = 100/n$.

Certain nitrate salts are selectively extracted by ether as well as other organic solvents. For example, uranium(VI) is conveniently separated from such elements as lead and thorium by ether extraction of an aqueous solution that is saturated with ammonium nitrate and has a nitric acid concentration of about 1.5 F. Bismuth and iron(III) nitrates are also extracted to some extent under these conditions.

Many of the organic reagents mentioned in Chapter 6, as well as other organic compounds, form chelates with various metal ions; these chelates are frequently soluble in such solvents as chloroform, carbon tetrachloride, benzene, and ether. Thus quantitative transfer of the metallic ions to the organic phase is possible.[2]

Chromatographic Separations[3]

Chromatography encompasses a diverse group of separation methods (see Table 18-1) that are of great importance to the analytical chemist, for they often permit the separation, isolation, and identification of components in mixtures that might otherwise be resolved with great difficulty if at all. The term "chromatography" is difficult to define rigorously, owing to the variety of systems and techniques to which it has been applied. In its broadest sense, however, chromatography refers to processes that are based on differences in rates at which the individual components of a mixture migrate through a stationary medium under the influence of a moving phase.

TYPES OF STATIONARY AND MOBILE PHASES

In chromatography the components of a mixture are carried through a porous *stationary phase* by means of a moving liquid or a gas called the *mobile phase*. The stationary phase may be a finely divided solid held in a narrow glass or metal tube; the mobile phase then percolates through the solid under the influence of gravity or as a result of pumping. Alternatively, the stationary phase may be a porous paper or a finely ground solid held on a glass plate; here the mobile phase moves through the solid either by capillary action or under the influence of gravity.

The stationary phase can be an immobilized liquid that is immiscible with the mobile phase. Several procedures are employed to fix the

---

[2] See E. B. Sandell, *Colorimetric Determination of Traces of Metals*, 3d ed. New York: Interscience, 1959.

[3] For more complete discussions of chromatography see E. Heftmann, *Chromatography*, 2d ed. New York: Reinhold, 1967; E. Lederer and M. Lederer, *Chromatography*, 2d ed. New York: American Elsevier, 1957; H. G. Cassidy, *Fundamentals of Chromatography*. New York: Interscience, 1957.

TABLE 18-1
Classification of Chromatographic Separations

NAME	TYPE OF MOBILE PHASE	TYPE OF STATIONARY PHASE	METHOD OF FIXING THE STATIONARY PHASE
Gas-liquid	Gas	Liquid	Adsorbed on a porous solid held in a tube or adsorbed on the inner surface of a capillary tube
Gas-solid	Gas	Solid	Held in a tubular column
Partition	Liquid	Liquid	Adsorbed on a porous solid held in a tubular column
Paper	Liquid	Liquid	Held in the pores of a thick paper
Thin layer	Liquid	Liquid or solid	A finely divided solid held on a glass plate; liquid may be adsorbed on particles
Gel	Liquid	Liquid	Held in the interstices of a polymeric solid
Ion exchange	Liquid	Solid	Finely divided ion-exchange resin held in a tubular column

stationary liquid in place. For example, a finely divided solid, coated with a thin layer of liquid, may be held in a glass or metal tube through which the mobile phase percolates. Ordinarily the solid plays no direct part in the separation, functioning only to hold the stationary liquid phase in place. In another technique a thin layer of liquid is held in place by adsorption on the inner walls of a capillary tube. A mobile gaseous phase is then caused to flow through the tube. A liquid phase can also be held in place on the fibers of paper or on the surface of finely ground particles held on a glass plate.

Common chromatographic procedures are classified in Table 18-1 according to the nature of the fixed and mobile phases.

THE PARTITION RATIO

All chromatographic separations are based upon differences in the extent to which solutes are partitioned between the mobile and the stationary phase. The equilibria involved can be described quantitatively by means of a temperature-dependent constant, the *partition ratio K*,

$$K = \frac{C_s}{C_m} \qquad (18\text{-}5)$$

where C_s is the total analytical concentration of a solute in the stationary phase and C_m is its concentration in the mobile phase.

The assumption that K in Equation 18-5 is constant is ordinarily satisfactory when the solute concentration is low. At high concentrations, however, marked departures are often observed, and conclusions that are based on the constancy of K must be modified accordingly. Fortunately the concentrations in a chromatographic column are frequently low. Chromatography carried out under conditions such that K is constant is called *linear chromatography*; we shall consider only this type.

A GENERAL DESCRIPTION OF THE CHROMATOGRAPHIC PROCESS

Chromatographic methods fall into three categories: (1) *elution analysis*, (2) *frontal analysis*, and (3) *displacement analysis*. We shall consider only elution analysis, which is the most widely encountered of the three.

In the elution method a single portion of the sample, dissolved in the mobile phase, is introduced at the head of the column (see Figure 18-2), whereupon the individual components distribute themselves between the two phases. Further additions of the mobile phase force the solvent containing a part of the sample down the column; partition again occurs between the mobile phase and fresh portions of the stationary phase. Simultaneously, partitioning between the fresh solvent and the solute held on the stationary phase occurs at the top of the column. With continued additions of solvent, solute molecules are carried down the column in a continuous series of transitions between the two phases. A solute molecule can move only when it is in the mobile phase; thus the average *rate* of movement depends upon the fraction of time it spends in that phase. This fraction is small for solutes with partition ratios that favor retention in the stationary phase, and it is large where retention in the mobile phase is more likely. The resulting differences in rates ideally cause a separation of the components in a mixture into bands or zones located along the length of the column (see Figure 18-2). Isolation can then be accomplished by passing a sufficient amount of the mobile phase through the column to cause these bands to pass out the end where they can be collected. Alternatively, the column packing can be removed and divided into portions containing the various components of the mixture.

The process whereby a solute is washed through the column by the addition of *fresh* solvent is called *elution*. If a detector that responds to the solutes is placed at the end of the column and its signal is plotted as a

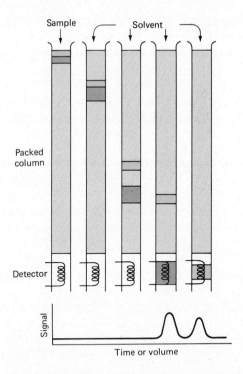

Sample — Solvent —

Packed
column

Detector

Signal

Time or volume

FIGURE 18-2 Schematic diagram of an elution chromatographic separation of a two-component mixture.

function of time (or of volume of the mobile phase), a series of symmetric peaks is obtained, as shown in the lower part of Figure 18-2. Such a plot, called a *chromatogram*, is useful for both qualitative and quantitative analysis. The positions of peaks may serve to identify the components of the sample. The areas under the peaks can be related to solute concentrations.

Two properties determine the effectiveness of a column in separating a pair of solutes. The first is the difference in rates at which the two species migrate through the column packing. Clearly, where this difference is large, the probability for a clean separation is high. The second is the breadth of the bands or zones of solutes as they exit from the column; narrow bands are desirable, of course, to minimize the chance of overlap.

The migration rate of a solute is determined largely by the chemical nature and the relative amounts of the stationary and mobile phases in the column. Band breadth, on the other hand, is influenced by such controllable variables as solvent flow rate, homogeneity of the packing, surface area of the stationary phase, and for immobilized liquids, the thickness of the liquid layer.

In the sections that follow we shall first consider factors that influence the rate at which a solute moves down a column. Consideration will then be given to variables that determine the width of solute zones.

MIGRATION RATE OF SOLUTES

If we define the *retention time t* as the time required for a solute to migrate through a column of length L, its rate of migration is given by (L/t). Similarly, if we let the time for the passage of a solvent molecule or a species that is not retained by the stationary phase be t_m, its rate of movement will be (L/t_m). A useful parameter in chromatography is the *retardation factor R_F*, which is the rate of movement of a solute molecule *relative* to the rate of movement of the solvent. That is,

$$R_F = \frac{L/t}{L/t_m} = \frac{t_m}{t} \tag{18-6}$$

It is important to note that R_F also corresponds to the fraction of time that an average solute molecule resides in the mobile phase. Because it can only migrate when it is in that phase, $(1 - R_F)$ must be the fraction of time in which the solute is fixed in place in the stationary phase. If it is assumed that dynamic equilibrium exists in a small cross section of a solute zone, the ratio of the fractional residence time in the two phases will be equal to the ratio of the amount of solute in the mobile phase to the amount in the stationary phase. That is,

$$\frac{R_F}{(1 - R_F)} = \frac{C_m V_m}{C_s V_s} \tag{18-7}$$

where C_m and C_s are molar concentrations of the solute in the two phases and V_m and V_s represent the volumes of the mobile and stationary phases, respectively, in the cross section. By substituting Equation 18-5 into Equation 18-7 and rearranging, we find that

$$R_F = \frac{V_m}{V_m + KV_s} \tag{18-8}$$

Clearly the rate at which the solute moves relative to that of the solvent decreases as its partition ratio increases.

COLUMN EFFICIENCY

In general terms the efficiency of a chromatographic column is a measure of its ability to separate two solutes. A highly efficient column will separate two similar solutes cleanly in a minimum column length.

Plate Height. The term used in chromatography to express column efficiency is the *height equivalent of a theoretical plate H*; the term is often shortened to *plate height*. The concept of plates and plate heights derives from the earliest theory of chromatography. According to *plate theory* a chromatographic column is envisioned as being composed of a series of discrete but contiguous, narrow, horizontal layers called plates. At each plate, equilibrium of the solute between the stationary and mobile phases is assumed to take place; movement of the solute and solvent is then viewed as a series of stepwise transfers from one plate to the next. The ability of a column to separate two solutes is directly dependent upon the number of equilibrations, that is, upon the number of plates. Column efficiency is measured by the *number of plates* N contained in each centimeter of column or, alternatively, in terms of H, the height of one of these plates in centimeters. These quantities are related by the equation

$$N = \frac{L}{H} \tag{18-9}$$

where L is the length of the column packing in centimeters. A highly efficient column is one that has a large number of plates per centimeter or a very small plate height.

Determination of Plate Height. From plate theory it is possible to show that the number of theoretical plates in a column, and thus the plate height, can be approximated from chromatograms. Figure 18-3 demonstrates the method.

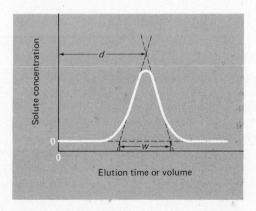

FIGURE 18-3 Theoretical plate determination from a chromatogram. $N = 16(d/w)^2$.

EXAMPLE

The elution peak for a solute was found to arrive at the column exit 9.34 min after sample injection; the width of the base of the peak (w in Figure 18-3) was 0.43 min. The column length was 18.3 cm. Calculate the column efficiency in terms of N and H.

$$N = 16(d/w)^2 = 16(9.34/0.43)^2 = 7.55 \times 10^3$$

$$H = 18.3 \text{ cm}/7.55 \times 10^3 \text{ plates} = 2.42 \times 10^{-3} \text{ cm/plate}$$

Knowledge of the plate height for a column packing makes it

possible to decide the length of column needed to separate two solutes with known partition ratios.

Although plate theory yields useful equations that account for the rate of solute migration and the symmetric shape of elution bands, it fails to provide a rationale for the effect of such variables as flow rate and packing characteristics on the *breadth* of elution bands and thus upon plate height. As a consequence, plate theory has been largely replaced by the *rate* or *kinetic theory of chromatography*. It should be noted, however, that the terms height equivalent of a theoretical plate and number of theoretical plates are retained as measures of column efficiency even though the rate theory does not picture a column as being made up of discrete sections or plates.

THE RATE THEORY OF CHROMATOGRAPHY[4]

The rate theory, which gives equations for zone shapes and migration rates that are analogous to plate theory, has the added advantage of accounting for the experimental variables that affect the breadth of a zone. Although the theory is mathematically complex, the principles that are involved can be described qualitatively.

Mass Transfer in a Chromatographic Column. As shown in Figures 18-2 and 18-3, the solute zones in linear chromatography are generally smooth and symmetric. This characteristic shape results from the random way in which individual solute particles migrate through the column in the mobile phase. Indeed, the elution curves approach the shape of the normal error (or Gaussian) curves shown in Figure 4-3 (p. 62). Recall that such curves also arise from the operation of random processes.

Let us first consider the behavior of an individual solute particle. During elution it undergoes many thousands of transfers between the stationary and mobile phases. The time it spends in a phase after a transfer is highly erratic, however, and depends upon its accidentally gaining sufficient thermal energy from its environment to accomplish a reverse transfer. Thus in some instances the residence time in a given phase may be transitory; in others the period may be relatively long. Movement can only occur when the particle is in the mobile phase. Thus the motion of an individual particle down the column is also highly erratic. Because of variability in the residence time, the average rate at which individual particles move relative to the mobile phase varies considerably. Some particles travel rapidly, owing to their accidental inclusion in the mobile phase for periods that exceed the average; others lag because their resi-

---

[4] For a more detailed presentation of the rate theory see J. C. Giddings, *J. Chem. Educ.*, **44**, 704 (1967); J. C. Giddings, *Dynamics of Chromatography*, part I. New York: Marcel Dekker, 1965.

dence in the mobile phase is less than average. The consequence of these random individual processes is a symmetric spread of velocities around the mean velocity that represents the most likely behavior of a particle.

As a zone moves down the column, migration occurs for a greater time and the zone profile becomes broader and lower. Thus the breadth of the zone is directly related to its residence time in the column and is inversely related to the flow rate for the mobile phase.

Longitudinal and Eddy Diffusion.   In addition to the distribution in the rates of mass transfer between phases, two other kinetic factors, *longitudinal diffusion* and *eddy diffusion*, also cause zone broadening. The rate theory takes these effects into account and leads to a general equation that describes most of the variables that influence the breadth of chromatographic peaks.[5] An approximate form of this equation is

$$H = A + \frac{B}{v} + Cv \qquad (18\text{-}10)$$

where H, the plate height, is employed as the broadening parameter and v is the velocity of flow of the mobile phase. The terms A, B, and C are constants for a given column.

The *eddy diffusion* term A is a measure of nonideal column behavior resulting from the multitude of pathways by which a molecule can find its way through the packing. These paths differ somewhat in length so that the residence times in the column for molecules of the same species are likewise variable. As a consequence, solute molecules that entered together reach the end of the column over a time interval. This spread tends to broaden the elution band and thus increase H. Eddy diffusion is *independent* of the flow velocity.

A second effect that leads to band broadening is *longitudinal diffusion* (B) in Equation 18-10. As solute molecules move through the column, molecular diffusion forces cause a migration from the concentrated center of the band toward the more dilute edges. The result is a broadening of the band and a lower column efficiency. As the flow rate increases, less time is available for this process to occur. Thus, as shown in Equation 18-10, longitudinal diffusion decreases as the velocity v becomes larger. Molecular diffusion is rapid in a gaseous phase; as a consequence, the longitudinal diffusion term is particularly important in gas-liquid chromatography.

We have already considered some of the effects responsible for the *mass-transfer term Cv* in Equation 18-10. In addition to differences in the rate of mass transfer, there is superimposed the influence of nonequilibrium conditions that result from rapid flow rates. As the flow rate

[5] See, for example, A. I. M. Keulemans, *Gas Chromatography*, 2d ed. New York: Reinhold, 1959, pp. 129–138.

becomes large, time is not available for attainment of true equilibrium. Thus the full efficiency of the column is not achieved, and an increase in H occurs.

Figure 18-4 shows the contribution of each of the three effects as a function of mobile-phase velocity (broken lines) as well as their net effect (solid line) on H of a column. Clearly, optimum efficiency is realized when the flow rate corresponds to the minimum in the curve.

Curves similar to the solid line in Figure 18-4 have been obtained experimentally and have provided information concerning the variables that influence A, B, and C in Equation 18-10. Band broadening due to eddy diffusion (A) is minimized by carefully packing the column with small, spherical particles. Particular care is needed to eliminate open channels.

The most important variable that affects the rate of approach to equilibrium (C) appears to be the thickness of the liquid layer that comprises the stationary phase. A marked improvement in column efficiency is obtained with very thin layers that enhance the probability of true equilibration. Equilibrium is also more likely at high temperatures and with low solvent viscosities.

CHROMATOGRAPHIC METHODS EMPLOYING A LIQUID MOBILE PHASE

In this section we shall consider the techniques and applications of the five common chromatographic methods that are based upon a liquid mobile phase. These include partition, adsorption, ion-exchange, paper, and thin-layer chromatography.

Partition Chromatography.　Liquid-liquid partition chromatography was developed by Martin and Synge, who showed that H for a properly

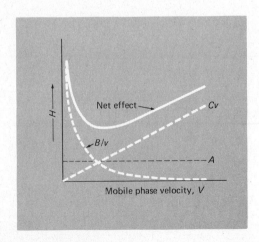

FIGURE 18-4　The effect of variables in Equation 18-10 on plate height.

prepared column can be as low as 0.002 cm. Thus a 10-cm column of this type may contain as many as 5000 theoretical plates. High separation efficiencies are to be expected even with relatively short columns.

The most widely used solid support for partition chromatography has been silicic acid or silica gel. This material adsorbs water strongly; the stationary phase is thus aqueous. For some separations the inclusion of a buffer or a strong acid (or base) in the water film has proved helpful. Polar solvents such as aliphatic alcohols, glycols, or nitromethane have also been employed as the stationary phase on silica gel. Other support media include diatomaceous earth, starch, cellulose, and powdered glass; water and a variety of organic liquids have been used to coat these solids.

The mobile phase may be a pure solvent or a mixture that is at least partially immiscible with the stationary phase. Better separations are sometimes realized if the composition of a mixed solvent is changed continuously as elution progresses (*gradient elution*). Other separations are improved by elution with a series of different solvents. The choice of mobile phase is largely empirical.

Partition chromatography has become a powerful tool for the separation of closely related substances. Typical examples include the resolution of the numerous amino acids formed in the hydrolysis of a protein, the separation and analysis of closely related aliphatic alcohols, and the separation of sugar derivatives. Figure 18-5 illustrates an application of the procedure to the separation of carboxylic acids. Here silica gel was employed as the support with 0.5-*M* sulfuric acid as the stationary

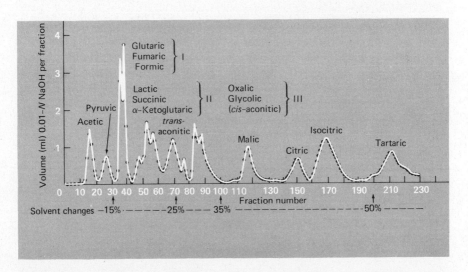

FIGURE 18-5 Partition chromatographic fractionation of acids on a silica gel column. Elution is by a chloroform solvent containing the indicated percentages of *n*-butyl alcohol. [From W. A. Bulen, J. E. Varner, and R. C. Burrell, *Anal. Chem.*, **24**, 187 (1952). With permission of the American Chemical Society.]

phase. The column was eluted with a series of butanol-chloroform mixtures that ranged from 15 to 50% in the alcohol. The eluted fractions were analyzed by titration with a standard solution of base. Where separation was incomplete, further treatment of the appropriate fractions on a longer column gave satisfactory isolation of the various components.

Adsorption Chromatography. In practice, adsorption chromatography is similar to the partition method. The type of equilibrium is fundamentally different, however, and depends upon the tendency of a finely divided solid surface to bind or adsorb certain types of compounds. Adsorbents include silica, alumina, calcium carbonate, sucrose, and others.

Separations based on adsorption chromatography rely on the equilibria that govern the distribution of the various solute species between the solvent and the surface of the solid. Large differences exist in the tendencies of compounds to be adsorbed. For example, a positive correlation can be discerned between adsorption properties and the number of hydroxyl groups in an organic molecule. A similar correlation exists with double bonds. Compounds containing certain functional groups are more strongly held than others. The tendency to be adsorbed decreases in the following order: acid > alcohol > carbonyl > ester > hydrocarbon. The nature of the adsorbent is also influential in determining the order of adsorption. Much of the available knowledge in this field is empirical; the choice of adsorbent and solvent for a given separation frequently must be made on a trial-and-error basis.

Ion-Exchange Chromatography. Ion exchange is a process involving an interchange of ions of like sign between a solution and an essentially insoluble solid in contact with the solution. Many substances, both natural and synthetic, act as ion exchangers. The ion-exchange properties of clays and zeolites have been recognized and studied for more than a century. Synthetic ion-exchange resins were first produced in 1935, and they have since found widespread laboratory and industrial application for water softening, water deionization, solution purification, and ion separation.

Synthetic ion-exchange resins are high molecular weight, polymeric materials containing large numbers of ionic functional groups per molecule. For cation exchange there is a choice between strong acid resins containing sulfonic acid groups (RSO_3H) or weak acid resins containing carboxylic acid ($RCOOH$) groups; the former have wider application. Anion-exchange resins contain basic functional groups, generally amines, attached to the polymer molecule. Strong base exchangers are quaternary amines ($RN(CH_3)_3{}^+OH^-$); weak base types contain secondary or tertiary amines.

A cation-exchange process is illustrated by the equilibrium

$$x\,RSO_3{}^-H^+ + M^{x+} \rightleftharpoons (RSO_3{}^-)_x M^{x+} + x\,H^+$$
$$\text{solid} \qquad \text{solution} \qquad \text{solid} \qquad \text{solution}$$

where M^{x+} represents a cation and R represents *a part* of a resin molecule. The analogous process involving a typical anion-exchange resin can be written

$$x\,RN(CH_3)_3{}^+OH^- + A^{x-} \rightleftharpoons (RN(CH_3)_3{}^+)_x\,A^{x-} + x\,OH^-$$

| solid | solution | solid | solution |

where A^{x-} is an anion.

Ion-exchange resins have been successfully employed as the stationary phase in elution chromatography. For example, Beukenkamp and Reiman separated sodium and potassium ions on a column packed with a sulfonic acid resin in its acidic form.[6] When the sample was introduced at the top of the column, the following exchange equilibrium was established for each of the alkali ions:

$$RH + B^+ \rightleftharpoons RB + H^+ \qquad \text{(18-11)}$$

where B^+ represents either Na^+ or K^+.

Equilibrium constants for these reactions take the form

$$K = \frac{[RB][H^+]}{[RH][B^+]} \qquad \text{(18-12)}$$

where $[RB]$ and $[RH]$ are the concentrations (strictly the activities) of the alkali and hydrogen ions in the solid resin phase. Equation 18-12 can be rewritten in the form

$$\frac{[RB]}{[B^+]} = \frac{K[RH]}{[H^+]} = K_D \qquad \text{(18-13)}$$

where K_D is defined as the distribution coefficient. Under the conditions extant during elution with hydrochloric acid, K_D is approximately constant because the hydrogen ion concentration of the eluate is large relative to the concentration of potassium and sodium ions. Also, the resin has an enormous number of exchange sites relative to the number of alkali metal ions in the sample. Thus the overall concentrations $[H^+]$ and $[RH]$ are not affected significantly by shifts in the equilibrium (Equation 18-11). Therefore under conditions where $[RH]$ and $[H^+]$ are large with respect to $[RB]$ and $[B^+]$, Equation 18-13 can be employed in the same way as Equation 18-5, and the general theory of chromatography, which we have already described, can be applied to a stationary phase consisting of an ion-exchange resin.

Note that K_D in Equation 18-13 represents the affinity of the resin for the ion B^+ *relative* to another ion (in this case H^+); where K_D is large, there is a strong tendency for the solid phase to retain ion B; where K_D is small, the reverse obtains. By selecting a common reference ion such as H^+, distribution ratios for different ions on a given type of resin can be compared. Such experiments reveal that polyvalent ions are much more strongly held than singly charged species. Within a given charge group,

[6] J. Beukenkamp and W. Reiman III, *Anal. Chem.*, **22**, 582 (1950).

however, differences appear that are related to the size of the hydrated ion as well as other properties. Thus for a typical sulfonated cation-exchange resin, values for K_D decrease in the order $Cs^+ > Rb^+ > K^+ > NH_4^+ > Na^+ > H^+ > Li^+$. For divalent cations the order is $Ba^{2+} > Pb^{2+} > Sr^{2+} > Ca^{2+} > Cd^{2+} > Cu^{2+} > Zn^{2+} > Mg^{2+}$.

The techniques for fractionation of ions with K_D values that are relatively close to one another are analogous to those described for adsorption and partition chromatography. For example, Beukenkamp and Reiman[6] showed that sodium and potassium ions in samples containing from about 200 to 300 mg of the alkali-metal chlorides could be cleanly separated on a 60-cm column of a sulfonic acid resin (approximately 60 g resin). Figure 18-6 shows the results of a typical separation.

A number of important separations are based upon ion exchange. Among these is the separation of the rare earths, primarily for preparative purposes. Here fractionation is enhanced by eluting with reagents that form complexes of differing stabilities with the various cations.[7] Ion-exchange resins have also been widely used to resolve mixtures of biological importance. For example, Figure 18-7 is a partial chromatogram of a synthetic mixture that simulates the composition of a protein hydrolysate. The amino acids histidine, lysine, and arginine appear with even higher volumes of eluent. Note that elutions were performed under varying

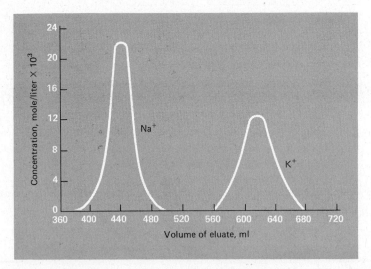

FIGURE 18-6 Elution of Na^+ and K^+ with HCl. Sample contained 1 mmole NaCl and 0.5 mmole KCl. [Adapted from J. Beukencamp and W. Reiman III, *Anal. Chem.*, **22**, 582 (1950). With permission of the American Chemical Society.]

[7] See a series of papers in *J. Amer. Chem. Soc.*, **69**, 2769–2881 (1947).

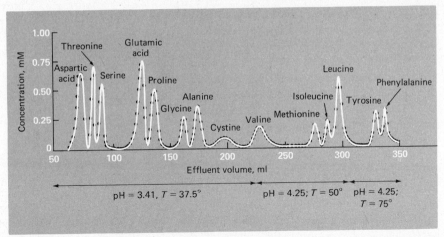

FIGURE 18-7 Separation of amino acids on a cation exchanger (Dowex 50, Na form; 100-cm column). Total sample ~ 6 mg. [Taken from S. Moore and W. H. Stein, *J. Biol. Chem.*, **192**, 663 (1951). With permission of the copyright owners.]

conditions for different parts of the chromatogram. The positions of the several peaks are reported to be reproducible to better than 5%.

Ion-exchange resins are also useful in the removal of interfering ions, particularly where these ions have a charge opposite to that of the species being determined. For example, iron(III), aluminum(III), and other cations cause difficulty in the determination of sulfate by virtue of their tendency to coprecipitate with barium sulfate. Passage of a solution to be analyzed through a column containing a cation-exchange resin results in retention of all cations and the liberation of a corresponding number of protons. The sulfate ion, on the other hand, passes freely through the column. The analysis can thus be performed on the effluent. An application of an anion-exchange resin to separate iron(III) from aluminum(III) is described in Chapter 20, Analysis 1–4. Here high concentrations of thiocyanate ion are employed to form anionic complexes selectively with the iron(III). These complexes are retained on the anion exchanger. The uncomplexed aluminum(III) passes directly through the column.

Yet another useful application of ion exchangers involves the concentration of an ion from a very dilute solution. Cation-exchange resins, for example, have been employed to collect traces of metallic elements from large volumes of natural water. The ions are then liberated from the resin by treatment with acid; the result is a considerably more concentrated solution for analysis.

An interesting application of ion-exchange resins is the determination of the total salt content of a sample. This analysis is accomplished by passing the sample through the acid form of a cation-exchange resin. Absorption of the cations causes the release of an equivalent quantity of hydrogen ion, which can be collected in the washings from the column and then titrated. Similarly, a standard acid solution can be prepared from a salt. For example, a cation-exchange column in the acid form can be treated with a weighed quantity of sodium chloride. The salt liberates an equivalent quantity of hydrochloric acid, which is then collected in the washings and diluted to a known volume.

Treatment of an anion-exchange resin with salts liberates the hydroxide ion by an analogous mechanism.

Paper Chromatography.[8]   Paper chromatography is a remarkably simple chromatographic method that has become an extremely valuable analytical tool to the organic chemist and the biochemist. Here a coarse paper serves in lieu of a packed column. A drop of solution containing the sample is introduced at some point on the paper. Migration then occurs as a result of flow by a mobile phase called the *developer*. Movement of the developer occurs by capillary action. In some applications the flow is in a downward direction (*descending development*). Here gravity also contributes to the motion. In *ascending development* the motion of the mobile phase is upward; in *radial development* it is outward from a central spot.

Usually paper chromatography appears to be a type of partition chromatography in which water absorbed on the hydrophilic surface of the paper acts as the stationary phase. An organic solvent then serves as the mobile phase. Suitable treatment of the paper makes possible the replacement of the water with a nonpolar, stationary liquid phase. Aqueous solutions can then be used as developers. In still another variation the paper is impregnated with anhydrous silica, alumina, or an ion-exchange resin; here partition occurs as a consequence of solid-liquid or ion-exchange equilibria.

The general principles developed earlier for column partition chromatography appear to apply as well to paper media. Thus the paper is treated as a column within which equilibration occurs according to Equation 18-5 (p. 504). It is customary to describe the movement of a particular solute in terms of its retardation factor R_F, which is defined as

$$R_F = \frac{\text{distance of solute motion}}{\text{distance of solvent motion}} \qquad \textbf{(18-14)}$$

To compensate for uncontrolled variables, the distance traveled by a solute frequently is compared with that for a standard substance under identical conditions. The ratio of these distances is designated as R_{std}.

Typical arrangements for paper chromatography are shown in Figure 18-8. The paper is usually 15 to 30 cm long and one to several centimeters wide. A drop of the sample solution is placed a short distance

---

[8] J. Sherma and G. Zweig, *Paper Chromatography and Electrophoresis*, vol. II. New York: Academic, 1971.

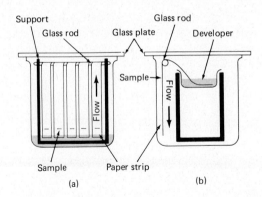

FIGURE 18-8 Apparatus for paper chromatography. (a) Ascending flow. (b) Descending flow.

from one end of the paper, and its position is marked with a pencil. The original solvent is allowed to evaporate. The end of the paper nearest the sample is then brought in contact with the developer. Both paper and developer are sealed in a container to prevent evaporation losses (see Figure 18-8). After the solvent has nearly traversed the length of the strip, the paper is removed and dried. The positions of the various components are then determined in any of a variety of ways. Reagents that form colored compounds in the presence of certain functional groups are available for this purpose. For example, a ninhydrin solution causes the development of blue or purple stains that reveal the location of amines and amino acids on the paper. Detection has also been based upon the absorption of ultraviolet radiation, fluorescence, and radioactivity.

Development can be carried out in two directions (*two-dimensional paper chromatography*). Here the sample is placed on one corner of a square sheet of paper. Development along one axis is performed in the usual way. After evaporation of the solvent, the paper is rotated 90 deg and is again developed, this time with a different solvent. Mixtures that cannot be resolved by a single solvent are often separated by this technique. Figure 18-9 is a typical two-dimensional chromatogram.

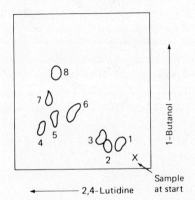

FIGURE 18-9 Two-dimensional chromatogram of an amino acid mixture. (1) Serine. (2) Glycine. (3) Alanine. (4) Tryptophan. (5) Methionine. (6) Valine. (7) Phenylalanine. (8) Leucine.

Both one- and two-dimensional paper chromatography can provide quantitative information that is accurate to perhaps $\pm 10\%$. In this instance standards are employed, and the analyses are based upon the size of spots relative to those for the standards.

Paper chromatography has been widely employed for the qualitative and semiquantitative analysis of inorganic, organic, and biochemical mixtures. Its greatest impact has been in biochemistry, where complicated mixtures of closely related compounds are so frequently encountered. One of the most important advantages of the method is its sensitivity; ordinarily a few micrograms of sample are sufficient for an analysis.

Thin-Layer Chromatography (TLC).[9] The techniques of thin-layer chromatography closely resemble those of paper chromatography. Here, however, partition occurs on a layer of finely divided adsorbent that is supported upon a glass plate.

Adsorbent materials for thin-layer chromatography include silica gel, alumina, diatomaceous earth, and powdered cellulose. An aqueous slurry of the finely ground adsorbent is spread over the surface of a glass plate or a microscope slide. The plate is then allowed to stand until the layer has set up. For many purposes, oven drying for several hours is useful.

The type of equilibrium involved in thin-layer chromatography depends upon the composition of the layer and the way it has been prepared. Thus the particles of a silica gel film deposited from aqueous solution and allowed to set up at room temperature undoubtedly remain coated with a thin film of water. If the sample is then distributed with an organic solvent, resolution will most likely be the result of liquid-liquid partition. On the other hand, if the silica film is dried by heating, partition may involve solid-liquid adsorption equilibria.

The sample is spotted at one end of the plate and then developed by the ascending technique used in paper chromatography. Development is carried out in a closed container saturated with developer vapor. The plate is then dried and sprayed with a reagent for detection of the components or, more commonly, exposed to iodine vapor. Brown spots indicate solute positions; identification is based upon R_F values.

Thin-layer chromatography is generally faster than paper chromatography and gives more reproducible R_F values. It is widely used for the identification of components in drugs, biochemical preparations, and natural products. As with paper chromatography, it can also provide semiquantitative information.

---

[9] See R. Maier and H. K. Mangold in *Advances in Analytical Chemistry and Instrumentation*, ed. C. N. Reilley, vol. 3. New York: Interscience, 1964, p. 369.

GAS-LIQUID CHROMATOGRAPHY (GLC)

Gas-liquid chromatography refers to the fractionation of components in a vaporized sample as a consequence of partition between a mobile gaseous phase and a stationary phase held in a column. The stationary phase is a liquid, supported upon an inert solid matrix. Here the position of gas-liquid equilibria determines the rate of migration of the solutes. Gas-liquid chromatography has become the most important and widely used of all of the column chromatographic methods and is probably the fractionation process used most often by all types of chemists.[10]

   In principle, gas-liquid and liquid-liquid partition chromatography differ only in that the mobile phase of the former is a gas rather than a liquid. The vaporized sample is introduced at the head of the column. Those components that have a finite solubility in the stationary liquid phase distribute themselves between this phase and the gas according to the equilibrium law. Elution is then accomplished by forcing an inert gas such as nitrogen or helium through the column. The rate at which the various components migrate along the column depends upon their tendency to dissolve in the stationary liquid phase. Ideally, bell-shaped elution curves are obtained. Qualitative identification of the components is based upon the time required for the peak to appear at the end of the column. Quantitative data are obtained from evaluation of the peak areas. A typical gas chromatogram is shown in Figure 18-10.

Definition of Some Terms. The time required for the maximum of a solute peak to reach the detector in a gas chromatographic column is called

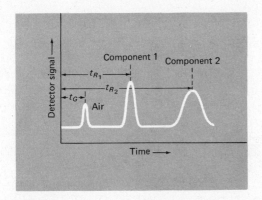

FIGURE 18-10 Gas chromatogram for a two-component system.

[10] For detailed discussions of the technique see L. S. Ettre and A. Zlatkis, *The Practice of Gas Chromatography*. New York: Interscience, 1967; R. A. Jones, *An Introduction to Gas-Liquid Chromatography*. New York: Academic, 1970; A. I. M. Keulemans, *Gas Chromatography*, 2d ed. New York: Reinhold, 1959; H. Purnell, ed., *New Developments in Gas Chromatography*. New York: Wiley, 1973.

the *retention time* (see Figure 18-10). If the sample contains a component whose solubility in the liquid phase is extremely low (oxygen or air, for example), its rate of movement will approach the rate of flow of the carrier gas, and its retention time t_G will be shorter than the retention time t_R of a component that partitions between the gas and liquid phases. Note, for example, the air peak in Figure 18-10.

Apparatus. The apparatus for gas-liquid chromatography can be relatively simple and inexpensive; several dozen models are offered by various instrument manufacturers. The essential components are illustrated in Figure 18-11. A brief description of each component follows.

Helium, nitrogen, carbon dioxide, and hydrogen from tank sources constitute the most commonly used carrier gases. Hydrogen has the obvious disadvantage of being explosive. Suitable flow-regulating valves are required, and some means of reproducing the flow rate is desirable.

Good column efficiency requires that the sample be of suitable size and be introduced as a "plug" of vapor. Slow injection and oversized samples cause band spreading and poor resolution. Liquid samples are injected into the column with a hypodermic syringe through a silicone rubber diaphragm. To prevent overloading of the column, the sample volumes must be small (of the order of 1 to 20 μliters). The injection port ordinarily is heated to a temperature above the boiling point of the sample so that vaporization is rapid. Special apparatus has been developed for introduction of solids and gases.

Two types of columns are employed in gas-liquid chromatography. The capillary type is fabricated from capillary tubing, the bore of which is coated with a very thin film of the liquid phase. Capillary columns have a very low pressure drop and can thus be of great length. Columns of several hundred thousand theoretical plates have been described. These columns, however, have very low sample capacities.

More convenient is the packed column, which consists of a glass or

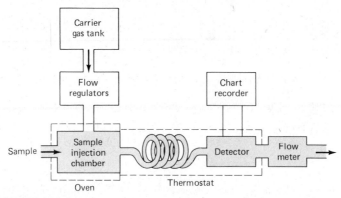

FIGURE 18-11   Block diagram of a gas-chromatographic apparatus.

metal tube roughly 10-mm in diameter, that ranges in length from 1 to 15 m. Ordinarily the tubes are folded or coiled so that they can be conveniently fitted into a thermostat. Such columns typically contain 100 to 1000 theoretical plates per foot. The best packed columns have a total of 30,000 to 60,000 theoretical plates.

The most widely used supports are made from diatomaceous earth and are marketed under such trade names as Celite, Dicalite, Chromosorb, C-22 Firebrick, and Sterchamol. Other supports have been fashioned from powdered Teflon, alumina, Carborundum, and microglass beads.

Desirable properties for the liquid phase in a gas-liquid chromatographic column include (1) *low volatility* (ideally its boiling point should be at least 200° higher than the maximum operating temperature for the column), (2) *thermal stability*, (3) *chemical inertness*, and (4) *solvent characteristics* such that K values for the solutes to be resolved fall within a suitable range.

For a reasonable residence time in the column a solute must show at least some degree of compatibility (solubility) with the liquid phase. Thus their polarities should be at least somewhat alike. A stationary liquid such as squalene (a high molecular weight saturated hydrocarbon) or dinonylphthalate might be chosen for the resolution of a nonpolar homologous series such as hydrocarbons, ethers, or esters. On the other hand, a more polar liquid such as polyethylene glycol would probably be more effective for the separation of alcohols or amines. Benzyldiphenyl might prove appropriate for aromatic hydrocarbons.

Among solutes of similar polarity the elution order usually follows the order of boiling points; where these differ sufficiently, clean separations are feasible. Solutes with nearly identical boiling points but different polarities frequently require a liquid phase that will selectively retain one (or more) of the components by dipole interaction or adduct formation.

Column temperature is an important variable that must be controlled to a few tenths of a degree for precise work. Circulating air baths, electrically heated metal blocks, and jackets fed with vapor from a constant boiling liquid have been used for temperature control.

In general, optimum resolution is associated with minimal temperature. The cost of lowered temperature, however, is an increase in elution time and therefore the time required to complete an analysis. Figure 18-12 illustrates this principle.

Detection devices for gas-liquid chromatography must respond rapidly and reproducibly to low concentrations of solutes emitted from the column. The solute concentration in a carrier gas at any instant is only a few parts in a thousand at most; frequently the detector is called upon to respond to concentrations that are smaller by one or two orders of magnitude (or more). In addition, the interval during which a peak passes the detector is usually a second or less. The detector must thus be capable of exhibiting its full response during this brief period.

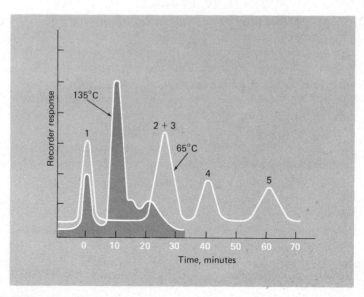

FIGURE 18-12 The effect of temperature on the separation of hexane isomers. (1) 2, 2-Dimethylbutane. (2) 2, 3-Dimethylbutane. (3) 2-Methylpentane. (4) 3-Methylpentane. (5) *n*-Hexane. [From C. E. Bennett, S. Dal Nogare, and L. W. Safranski, *Treatise on Analytical Chemistry*, ed. I. M. Kolthoff and P. J. Elving, part I, vol. 3. New York: Interscience, 1961, p. 1690. With permission.]

A relatively simple and broadly applicable detection system is based upon changes in the thermal conductivity of the gas stream. An instrument employed for this purpose is the *katharometer*. The sensing element of this device is an electrically heated source whose temperature at constant electrical power depends upon the thermal conductivity of the surrounding gas. The heated element may consist of a fine platinum or tungsten wire or, alternatively, a semiconducting thermistor. The resistance of the wire or thermistor gives a measure of the thermal conductivity of the gas. In contrast to the wire detector, the thermistor has a negative temperature coefficient.

In chromatographic applications a double detector is always employed, one element being placed in the gas stream *ahead* of the sample injection chamber and the other immediately beyond the column. In this way the thermal conductivity of the carrier gas is canceled, and the effects of variations in column temperature, pressure, and electrical power are minimized. The resistances of twin detectors are usually compared by incorporating them into two arms of a simple bridge circuit.

The thermal conductivities of hydrogen and helium are roughly 6 to 10 times greater than those of most organic compounds. Thus the presence of even small amounts of organic materials causes a relatively large decrease in thermal conductivity of the column effluent; the detector undergoes an increase in temperature as a result. The conductivities of nitrogen and carbon dioxide more closely resemble those of organic constituents; detection by thermal conductivity is less sensitive when these are used as carrier gases.

Thermal conductivity detectors are simple, rugged, inexpensive, nonselective, accurate, and nondestructive of the sample. They are not as sensitive, however, as other devices that have been developed.

Qualitative Analysis. Gas chromatograms are widely accepted as criteria of purity for organic compounds. Contaminants, if present, are revealed by the appearance of additional peaks. The areas under these peaks provide rough estimates of the extent of contamination. The technique is also useful for evaluating the effectiveness of purification procedures.

In theory retention-time data should be useful for the identification of components in mixtures. In fact, however, the number of variables that must be controlled in order to obtain reproducible retention times places limits on the applicability of the technique. On the other hand, gas-liquid chromatography provides an excellent means of confirming the presence of a suspected compound in a mixture, provided an authentic sample of the substance is available. No new peaks in the chromatogram of the mixture should appear upon addition of the pure compound, and enhancement of one of the existing peaks should be observed. The evidence is particularly convincing if the effect can be duplicated with different columns and at different temperatures.

Quantitative Analysis. The detector signal from a gas-liquid chromatographic column has had extensive quantitative and semiquantitative applications. Under carefully controlled conditions an accuracy of about 1% relative is attainable. As with most analytical tools, reliability is directly related to the complexity of the sample as well as the amount of effort spent in calibration and in control of variables.

Quantitative analysis can be based either on peak height or peak area. Peak heights, although more convenient to measure, are less satisfactory because of their greater dependence upon experimental variables. Flow rate is particularly critical because peaks tend to become broader and lower as the residence time of the solute on the column increases (p. 509). In addition, peak heights are more strongly affected by column temperatures, porosity of packing, and column length. Despite these limitations, peak height may be a better analytical parameter than peak area for solutes with low retention times. The peaks for such solutes are narrow and tall; accurate determination of their areas is thus difficult.

Peak areas generally increase linearly with the reciprocal of flow rate, and a satisfactory correction for changes in this variable can be applied. In contrast to peak heights, the *relative* areas of the peaks for two compounds remain constant and independent of flow rate, a valuable characteristic that permits use of internal standards. Temperature fluctuations have only a small effect on absolute peak areas and none whatsoever on relative ones.

The area A of a chromatographic peak can be evaluated in any of several ways.

1. Cut out the peak. Compare its mass with that for a portion of the recorder paper of known area.
2. Measure the area with a planimeter.
3. Multiply the height of the peak by its width at half-height; insofar as the curve has Gaussian characteristics, this product is equal to 0.84 A.
4. Form a triangle by drawing tangents through the inflection points on either side of the peak, and connect these with a third line along the recorder base line; the resulting triangle will have an area that is approximately 0.96 A.
5. Use a mechanical or electronic integrator. Such devices are included as accessories to many recorders.

Methods 4 and 5 appear to be the most widely used.

Calibration curves can be based upon weight, volume, or mole percent. A separate curve is needed for each solute to be determined because no detector responds in exactly the same way for each compound. The accuracy of the analysis depends upon how closely the standard samples approximate the composition of the unknown, the care with which they are prepared, and the degree to which the operational variables of the columns are controlled. The effect of column variables can be partially offset by adding an internal standard in known amount to both the calibration standards and the samples. The internal standard should have a retention time similar to that of the solute of interest. Its peak, however, must be separated from those of all of the components of the sample—a requirement often difficult to fulfill.

PROBLEMS

*1. The distribution coefficient for the species X between carbon tetrachloride and water is 23.5. Calculate the concentration of X remaining in the aqueous phase after 75.0 ml of 0.0500-M X are treated by extraction (assume that X neither associates nor dissociates in either phase) with
    (a) 60.0 ml of CCl_4.
    (b) two 30.0-ml portions of CCl_4.
    (c) six 10.0-ml portions of CCl_4.
    (d) twelve 5.00-ml portions of CCl_4.

2. The substance Y has a distribution coefficient of 13.2 between 2-hexanol and water. The coefficient is independent of concentration. What is the concentration of Y remaining in the aqueous phase if 40.0 ml of 0.0250-M Y are extracted with

   (a) two 20.0-ml portions of the alcohol?
   (b) four 10.0-ml portions of the alcohol?
   (c) ten 4.00-ml portions of the alcohol?

*3. What volume of alcohol would be required to reduce the concentration of Y (see Problem 2) in an aqueous solution to less than 1.00×10^{-4} M if 40.0 ml of 0.200-M Y were extracted with

   (a) 20.0-ml portions of alcohol?
   (b) 10.0-ml portions of alcohol?
   (c) 2.00-ml portions of alcohol?

4. The distribution coefficient of the metal complex MX_2 between water and chloroform was found to be 8.43. The formation constant for the complex in aqueous solution was 1.23×10^{12}. A solution that was 5.40×10^{-2} in M^{2+} and 0.250 F in X^- was prepared.

   (a) What total volume of $CHCl_3$ would be required to reduce the formal concentration of M(II) in a 25.0-ml aliquot of the aqueous solution to less than 1.00% of its original concentration if 20.0-ml portions of the organic solvent were to be used?
   (b) Repeat the calculation in (a) assuming that 10.0-ml portions of the $CHCl_3$ were to be employed.
   (c) Repeat the calculation in (b) assuming that it was desired to reduce the formal concentration of M(II) to 0.100% of the original.

*5. What would be the minimum partition coefficient that would permit removal of 99.00% of a solute from 50.0 ml of water with (a) two 20.0-ml extractions with benzene and (b) ten 5.00-ml extractions with benzene?

6. If 50.0 ml of water which is 0.0300 M in Q are to be extracted with four 10.0-ml portions of an immiscible organic solvent, what would be the minimum partition ratio that would allow transfer of all but the following percentages of the solute to the organic layer: (a) 1.00×10^{-1}, (b) 1.00×10^{-2}, and (c) 1.00×10^{-3}?

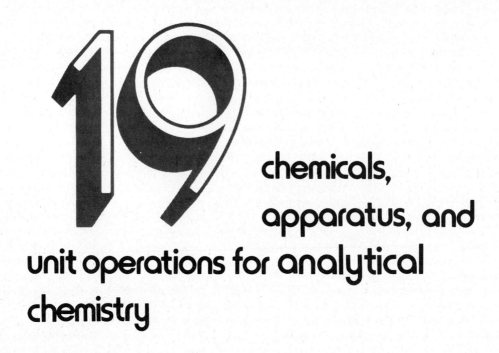

chemicals, apparatus, and unit operations for analytical chemistry

This chapter is concerned with practical aspects of the unit operations that are used in most analyses as well as the description of the apparatus and chemicals required for these operations.

Choosing and Handling Chemicals and Reagents

Of constant concern in analytical chemistry is the quality of reagents, inasmuch as the accuracy of an analysis is often affected by this factor.

CLASSIFICATION OF COMMERCIAL CHEMICALS

Technical or Commercial Grade. Chemicals labeled technical or commercial grade are of indeterminate quality and should be used only where high purity is not of paramount importance. Thus the potassium dichromate and the sulfuric acid used in the preparation of cleaning solution can be of this grade. In general, however, technical- or commercial-grade chemicals are not used in analytical work.

Chemically Pure, or CP Grade. The term *chemically pure* has little meaning. A chemical that is given this label is usually more refined than technical grade, but specifications to indicate the nature and extent of its impurities are not provided. In general, then, the quality of a CP reagent is too uncertain for analytical work. Any such reagent that must be used should be tested for contaminants that might affect the analysis; the performance of frequent reagent blanks may also be required.

USP Grade. USP chemicals have been found to conform to the tolerances set forth in the *United States Pharmacopoeia.*[1] The specifications are designed to limit contaminants that are dangerous to health; thus chemicals passing USP tests still may be quite heavily contaminated with impurities that are not physiological hazards.

Reagent Grade. Reagent-grade chemicals conform to the minimum specifications of the Reagent Chemical Committee of the American Chemical Society;[2] these are used, wherever possible, in analytical work. Some suppliers label their products with the maximum limits of impurity allowed by these specifications; others print the actual results of analyses for the various impurities.

Primary Standard Grade. *Primary standards* are substances that are obtainable in extraordinarily pure form (see Chapter 7). Primary standard-grade reagents, which are available commercially, have been carefully analyzed, and the assay value is printed on the label. An excellent source for primary standard chemicals is the National Bureau of Standards. This agency also supplies *reference standards*—complex mixtures that have been exhaustively analyzed.[3]

HANDLING REAGENTS AND SOLUTIONS

The availability of reagents and solutions with established purity is of prime importance for successful analytical work. A freshly opened bottle of a reagent-grade chemical can be used with confidence in most applications; whether the same confidence is justified when this bottle is half-full depends entirely upon the way it has been handled after being

[1]U.S. Pharmacopoeial Convention, *Pharmacopoeia of the United States of America*, 19th rev. Easton, Pa.: Mack, 1975.

[2]Committee on Analytical Reagents, *Reagent Chemicals, American Chemical Society Specifications*, 5th ed. Washington, D.C.: American Chemical Society, 1974.

[3]United States Department of Commerce, *Catalog and Price List of Standard Materials Issued by the National Bureau of Standards*, NBS Misc. Publ. 260 and its semiannual supplements. Washington, D.C.: Government Printing Office, 1968–.

opened. Only conscientious adherence to the rules given here will prevent contamination of reagents.

1. Select the best available grade of chemical for analytical work. If there is a choice, pick the smallest bottle that will supply the necessary quantity.

2. Replace the top of every container immediately after removal of reagent. Do not rely on someone else to do this.

3. Hold stoppers between the fingers; stoppers should never be set on the desk top.

4. Unless specifically directed to the contrary, *never return any excess reagent or solution to a bottle*. The minor saving represented by the return of an excess is likely to be a false economy in view of the risk of contaminating the entire bottle.

5. Again, unless specifically instructed otherwise, do not insert spoons, spatulas, or knives into a bottle containing a reagent chemical. Instead, shake the capped bottle vigorously to dislodge the contents. Then pour out the desired quantity.

6. Keep the reagent shelf and the laboratory balances clean. Be sure to clean up any spilled chemicals immediately.

Cleaning and Marking Laboratory Ware

Beakers and crucibles should be marked for identification purposes. The etched area on the sides of beakers and flasks can be marked semipermanently with a pencil. Special marking inks are available for porcelain surfaces; the marking is baked permanently into the glaze. A saturated solution of iron(III) chloride can also be used, although it is not as satisfactory as the commercial preparations.

Care must be taken to ensure that glass and porcelain ware is thoroughly clean before use. The apparatus should be washed with detergent, then rinsed first with copious amounts of tap water, and finally with several small portions of distilled water. A properly cleaned object will be coated with a uniform and unbroken film of water. It is seldom necessary to dry glassware before use; in fact this practice should be discouraged because it wastes time and can be a cause of contamination.

If a grease film persists after thorough cleaning with detergent, a cleaning solution consisting of sodium or potassium dichromate in concentrated sulfuric acid may be used. Extensive rinsing is required after using this solution to remove the last traces of dichromate ions, which adhere strongly to glass or porcelain surfaces. Cleaning solution is most effective when warmed to about 70°C; at this temperature it rapidly attacks plant and animal matter and is thus a potentially dangerous preparation. Any spillages should be diluted promptly with copious volumes of water.

PROCEDURE

Preparation of Cleaning Solution. In a 500-ml, heat-resistant conical flask mix 10 to 15 g of sodium or potassium dichromate with about 15 ml of water. Add concentrated sulfuric acid *slowly*; swirl the flask thoroughly between increments. The contents of the flask will become a semisolid red mass; add just enough sulfuric acid to dissolve this mass. Allow the solution to cool somewhat before attempting to transfer it to a storage bottle. The solution may be reused until it acquires the green color of chromium(III) ion, at which time it should be discarded.

Evaporation of Liquids

In the course of an analysis the chemist often finds it necessary to decrease the volume of a solution without loss of a nonvolatile solute. An arrangement such as illustrated in Figure 19-1 is generally satisfactory for most evaporations. A ribbed cover glass permits vapors to escape and protects the solution from accidental contamination. Less satisfactory is the use of glass hooks to provide space between the lip of the container and a conventional watch glass.

The evaporation process is occasionally difficult to control, owing to the tendency of some solutions to superheat locally. The bumping that results, if sufficiently violent, can cause a partial loss of the sample. This danger is minimized by careful and gentle heating. The introduction of glass beads, where permissible, is also helpful.

Unwanted constituents of a solution can frequently be eliminated by evaporation. For example, chloride and nitrate ions can be removed by adding sulfuric acid and evaporating until copious white fumes of sulfur trioxide are observed. Nitrate ion and nitrogen oxides can be eliminated by adding urea to an acidic solution, evaporating the solution to dryness, and gently igniting the residue. If large quantities of ammonium chloride must be removed, it is better to add concentrated nitric acid after evaporating

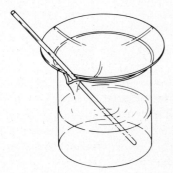

FIGURE 19-1 Arrangement for the evaporation of liquids.

the solution to a small volume. Rapid oxidation of the ammonium ion occurs upon heating; the solution is then evaporated to dryness.

Unwanted organic substances are frequently eliminated by adding sulfuric acid and evaporating the solution until sulfur trioxide fumes are observed. Nitric acid may be added at this point to hasten oxidation of the last traces of organic matter.

The Measurement of Mass

The measurement of mass is one of the commonest operations carried out by the chemist. Performance of many analyses will require the acquisition of highly reliable weighing data at one stage or another. For such measurements an *analytical balance*, which provides highly accurate information, is employed. Approximate weighing data are completely satisfactory for other purposes; these measurements are ordinarily obtained with a less precise but more rugged auxiliary *laboratory balance*.

THE DISTINCTION BETWEEN MASS AND WEIGHT

The reader should clearly recognize the difference between the concepts of mass and weight. The more fundamental of these is *mass*—an invariant measure of the quantity of matter in an object. The *weight* of an object, on the other hand, is the force of attraction exerted between the object and its surroundings, principally the earth. Because gravitational attraction is subject to slight geographical variation with altitude as well as latitude, the weight of an object is likewise a somewhat variable quantity. For example, the weight of a crucible would be less in Denver than in Atlantic City because the attractive force between it and the earth is less at the higher altitude. Similarly, it would weigh more in Seattle than in Panama because the earth is somewhat flattened at the poles and the force of attraction increases appreciably with latitude. The mass of this crucible, on the other hand, remains constant regardless of the location in which it is measured.

Weight and mass are simply related to each other through the familiar expression

$$W = Mg$$

where the weight, W, is given by the product of the mass, M, of the object and the acceleration due to gravity, g.

Chemical analyses are always based on mass in order to free the results from dependence on locality. The mass of an object is ordinarily obtained by means of a *balance* that permits comparison of the weight of the object with that of a known mass; because g affects both the known and unknown to the same extent, an equality in weight also indicates an equality in mass.

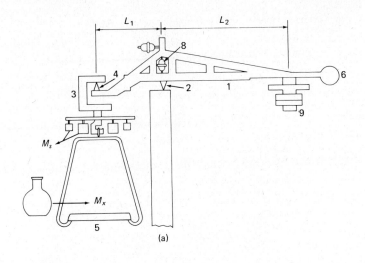

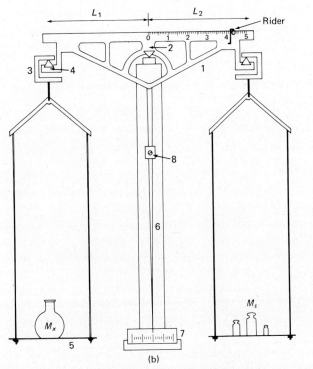

FIGURE 19-2 Schematic diagram of (a) a single-pan balance and (b) an equal-arm balance.

THE ANALYTICAL BALANCE[4]

An analytical balance is capable of detecting a weight difference that is about one one-millionth of its maximum load or better. For example, one widely used analytical balance can tolerate a load up to 160 g; the standard deviation of measurements with this balance is about ± 0.1 mg ($\pm 10^{-4}$ g).

Analytical balances are of two types, namely, *single-pan* and *equal-arm*. Both are widely used today. It seems probable, however, that the speed and convenience that characterize single-pan balances will ultimately cause the disappearance of the equal-arm balance from general laboratory use.

Analytical balances are also classified according to their sensitivities and load limits. The typical *macro analytical* balance has a maximum loading of 160 to 200 g, although oversize balances that can accommodate loads up to 2000 g are occasionally encountered. *Semimicro balances* are available that measure loads of 50 to 100 g with a standard deviation of about ± 0.01 mg. *Micro analytical* balances will tolerate loads of 10 to 20 g; a typical standard deviation with a micro balance is ± 0.001 mg.

The discussion that follows will deal exclusively with macro analytical balances, the type that is found in nearly every laboratory.

COMPONENTS OF AN ANALYTICAL BALANCE

Analytical balances differ considerably in appearance, design, and performance characteristics. Nevertheless, all contain common componen these will be considered in this section.

Beams. The heart of a balance is the beam, which is a lever that piv about a central support. Placement of the object to be weighed on one creates a moment of force that tends to rotate the beam about its pi This motion is offset by a counterforce, the magnitude of which measure of the mass of the object. Depending upon balance design, counterforce is either applied at the opposite end of the beam or else removed from the side upon which the object has been placed. counterforce is ordinarily gravitational and is derived from the mas standard weights. However, a measured electromagnetic force may al employed; more and more balances based on this design are appeari the market.

Figure 19-2 depicts examples of single-pan and equal-arm bal Common to both is the light-weight beam (1), which oscillates ab central prism-shaped knife-edge (2) as a result of the forces descri the preceding paragraph. The knife-edge and the plane surface upon

[4]For a description of commercially available balances see R. F. Hirsch, *J. Chei* **44**, A1023 (1967); **45**, A7 (1968); and G. W. Ewing, *J. Chem. Educ.*, **53**, A252, A292 (

it rests form a bearing having minimal friction. The performance of a balance is ultimately limited by the mechanical perfection of this bearing; thus both parts are precision formed and manufactured from unusually hard materials (agate or synthetic sapphire). Care must be taken in using any type of balance to avoid damage to the knife-edge and its bearing surface.

Stirrups and Pans. The single-pan balance shown in Figure 19-2a contains a second bearing around which motion can occur. Here the flat bearing surface is contained in the *stirrup* (3), which rests on an outer knife-edge (4). The stirrup serves to couple the pan (5), which holds the object to be weighed to the beam.

The equal-arm balance shown in Figure 19-2b requires two stirrups rather than one; thus the single-pan balance has one less moving part than its double-pan counterpart. Other things being equal, this smaller number of bearings is an advantage because it reduces the friction associated with beam motion.

Beam Deflection Detectors. Another necessary component of all balances is a device that detects and measures the displacement of the beam from its equilibrium position. The simplest detector is the mechanical pointer (6) and scale (7) of an equal-arm balance (Figure 19-2b). For small angles of deflection the horizontal displacement of the pointer tip along the scale is directly proportional to the difference in loading of the two pans. The *sensitivity* of the balance, which can be defined in terms of scale division displacement per unit of mass,[5] is directly proportional to the length of the pointer.

One method of increasing the sensitivity of a balance is to use a so-called "optical lever." Here a small mirror is mounted on the top of the beam so that it is turned through an angle by the deflection process. A light is reflected off the mirror and focused on a scale some distance away. In this way the sensitivity may be enhanced by as much as 100.

The typical single-pan balance employs an optical lever to detect beam deflections. A glass plate or reticle, scribed with a vertical scale, is attached to one end of the beam (see 6 in Figure 19-2a). Light from a small lamp is then passed through the screen; the resulting image is then enlarged and reflected onto a frosted glass surface.

Weights. An analytical balance is supplied with a combination of weights such that any desired loading up to the design maximum for the balance can be realized. As shown in Figure 19-2a, these weights (M_s) are attached to the beam of most single-pan balances; a mechanical means is provided

[5] A more general definition of sensitivity is the ratio of the change in response to the quantity measured. See *Anal. Chem.*, **26**, 1190 (1954).

for their individual removal. With equal-arm balances weights are usually added manually to one of the pans.

Dampers. *Dampers* are devices that shorten the time required for the beam of a balance to come to rest after it has been set in motion. A single-pan balance is ordinarily equipped with an *air damper*, consisting of a piston attached to the beam that moves within a concentric cylinder mounted to the balance case (see 9 and 10 in Figure 19-3). When the beam is set in motion, the enclosed air undergoes slight expansions and contractions because of the close spacing between piston and cylinder; the beam comes rapidly to rest as a result of this opposition to its motion. An equal-arm balance is more likely to be equipped with a *magnetic damper*, which consists of a metal plate (generally aluminum) secured to the end of the beam and positioned between the poles of a permanent magnet. The forces induced by the motion of the beam act to oppose further motion; the beam rapidly acquires its rest position as a result.

Other Components. Analytical balances are equipped with *beam arrests* and *pan arrests*. The beam arrest is a mechanical device that lifts the beam so that the central knife-edge is freed from contact with its bearing surface; in addition, the stirrups are simultaneously freed from contact with the outer knife-edge or edges. When engaged, pan arrests support most of the weight of the pans and thus prevent them from swinging. The purpose of the arrests is to prevent damage to the bearings of the balance when the objects are added to or removed from the pans. Whenever a balance is not in use, both arrests should be engaged.

To discriminate between small weight differences (<1 mg), it is necessary to protect the balance from air currents in the laboratory. Thus an analytical balance is enclosed in a case equipped with doors to permit the introduction and removal of objects.

STABILITY AND SENSITIVITY OF A BALANCE

An important property of a balance is *stability*; that is, when the beam is set in motion around the central knife-edge by the momentary application of a slight force, it must ultimately return to its original position. For a balance to be stable the center of gravity of the beam and its accouterments must lie *below the center knife-edge*; only then will there exist a restoring force (the weight of the beam) that will act to offset the displacement.

The sensitivity of a balance is inversely proportional to the distance between the central knife-edge and the center of gravity of the beam and pans (for a sensitive balance this distance is no more than a few thousandths of a centimeter). Balances are often equipped with a small weight that can be moved to adjust the sensitivity. These weights are labeled 8 in Figure 19-2.

METHODS OF WEIGHING

The classical method of weighing with an equal-arm balance involves determining the position of the deflection indicator in the absence of a load (this point is called the *zero point* of the balance). The object to be weighed is then placed on one pan and weights are added incrementally to the other side of the beam until the indicator is restored to the zero point.

The time required for a weighing of this type can be materially shortened by the addition of weights until a *rest point* for the indicator is reached that lies within a few scale divisions of the zero point. If the sensitivity of the balance is known—in, say, milligrams per scale division—the weight that must be added or subtracted to achieve the zero point can be readily calculated. It is important to note that the sensitivity of an equal-arm balance decreases with increased loading; thus it is necessary to determine the sensitivity under a load condition that approximates the weight on the pans.

Variations of sensitivity with load do not occur in *substitution weighing*. This technique is illustrated for the single-pan balance shown in Figure 19-2a. At the outset the side of the beam holding the pan is equipped with a full set of weights M_s; a fixed counter weight (9) provides the counterforce to level the beam. With an object M_x on the pan the beam is restored to its zero point or a rest point by the *removal of weights*. Note that the loading of this instrument, and thus its sensitivity, is constant and independent of the mass of M_x. Constant sensitivity offers the considerable advantage that the scale divisions of beam deflection can be made to read directly in weight. Single-pan analytical balances are calibrated in this way. The weight of an object can be obtained in a matter of seconds with a single-pan balance; the same operation may require minutes with an equal-arm instrument.

SINGLE-PAN ANALYTICAL BALANCE

Operational Features. The essential parts of a single-pan balance are shown in Figure 19-3. The pan (5) and a set of nonmagnetic, stainless steel weights (M_s) for loads that are larger than 100 mg are located on one side of the fulcrum; a fixed weight (9), which also serves as the piston of an air damper (10), is located on the other side. Restoration of the beam to its original position after an object has been placed on the pan is accomplished by removing weights (M_s) by means of knobs (C) in an amount equal to the mass of the object. In actual operation sufficient weights are removed to leave the system with a residual imbalance of 100 mg or less with the

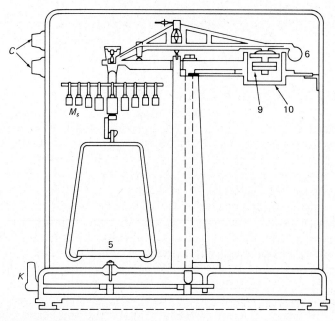

FIGURE 19-3 Schematic diagram of a single-pan balance. (Courtesy of Mettler Instrument Corporation)

object still heavy. The optical device (6) translates the resulting deflection of the beam into units of mass and displays the fractional weight to the operator as an image on a frosted-glass surface.

Weighing. The weight of an object is very simply obtained with a modern single-pan balance.

PROCEDURE

To zero the empty balance rotate the arrest knob (K in Figure 19-3) to the full release position. Then manipulate the zero-adjusting control until the illuminated scale indicates a reading of zero. Next arrest the balance, and place the object to be weighed on the pan. Turn the arrest knob to its partial release position. Rotate the dial controlling the heaviest likely weight for the object until the illuminated scale changes position or the notation "remove weight" appears; then turn the knob back one stop. Repeat this procedure with the other dials, working successively through the lighter weights. Then turn the arrest knob to its full release position, and allow the balance to achieve equilibrium. The weight of the object is found by taking the sum of the weights indicated on the dials and that which appears on the illuminated scale. A vernier is helpful in reading this scale to the nearest tenth of a milligram.

NOTE

Directions for operation of single-pan balances vary with make and model. Consult with the instructor for any modifications to this procedure that may be required.

EQUAL-ARM BALANCES

Weights. The weights for an equal-arm balance are manufactured from corrosion-resistant materials. Brass, plated with platinum, gold, or rhodium is preferred for denominations larger than 1 g, although highly polished, nonmagnetic stainless steel and lacquered brass weights also give good service.

Fractional weights (that is, weights smaller than 1 g) are fabricated from sheets of platinum, tantalum, or aluminum. The smallest increments are introduced by a rider or a chain; with the former, the beam is calibrated in units of 0.1 mg. The movement of a small wire rider along the beam has the effect of altering the loading from 0 to 5 or 10 mg depending on the design of the balance (see Figure 19-2b). The position of the rider is adjusted with a rodlike control that extends through the balance case.

Chain weights greatly facilitate the addition of fractional weights to an equal-arm balance. As illustrated in Figure 19-4, a small gold chain is attached some distance to the right of the fulcrum and also to an externally operated crank. The amount of weight supported by the beam can be varied by lengthening or shortening the chain. A scale arrangement and vernier make possible the direct reading of the weight that has been added. Because the chain generally allows the introduction of any weight within the range of 0 to 100 mg, the beam of such a balance will be notched for a dumbbell-shaped rider that permits addition of 100-mg increments from 0 to 1 g. Thus no weight smaller than 1 g needs to be manually introduced. A chain balance fitted with magnetic damping can provide accurate results with a minimum expenditure of time.

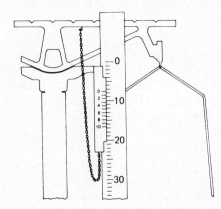

FIGURE 19-4 The chain balance. Note the arrangement of the chain and the vernier. The indicated loading is 7.6 mg. Note also that the beam is notched to accommodate the rider.

Weighing. Several techniques exist for determining the weight of an object with an equal-arm analytical balance. Regardless of method, however, the weighing operation consists of determining the weight needed to produce a rest point that is identical with the zero point. The beam of an undamped balance comes to rest very slowly; thus its equilibrium position is estimated by observing the swings of its pointer.

The deflection scale of the typical analytical balance generally does not have numerical values assigned to the division lines, these being left to the discretion of the operator. We shall employ the convention that assigns a value of 10 to the midline of the scale. The lines to the left, then, represent integers less than 10, whereas those to the right have values greater than 10.

PROCEDURE

The Method of Short Swings with an Undamped Balance. To obtain the zero point of the balance close the balance case, gently lower the beam support, and then carefully disengage the pan arrests to allow the beam to swing freely. The pointer should move from one to three divisions on either side of the mean. If the arc is greater than ±3 divisions, arrest the pans momentarily and then disengage the pan arrests again. If necessary, motion can be imparted to the beam by temporarily displacing the rider or chain from its zero position.

The zero point is obtained by computing the mean of the maximum consecutive left and right positions of the pointer. Several means may be averaged if desired.

The zero point will undergo slight changes with time, owing to changes in temperature and humidity. As a consequence, its value should be determined for each series of weighings. Pronounced shifts in the zero point are indicative of trouble; in such a case the instructor should be consulted.

After ascertaining that the pan arrests and beam supports have been reengaged, place the object to be weighed on the left pan of the balance. Make a rough estimate of its mass, and, using forceps, place on the right pan the single weight that is believed to have a slightly greater mass. Gently and partially disengage the beam support, and note the behavior of the pointer; it will swing away from the pan that has the heavier load. If no movement is seen, the loadings are within 1 to 2 g of each other. Under such circumstances cautiously disengage the pan arrests to determine which pan has the heavier load. Continue a systematic trial-and-error loading of weights until the closest approach to balance has been achieved where the object is still heavy. Then use the rider in a similar manner; the balance case should be closed henceforth. When balance apparently has been achieved, obtain a rest point in the manner described for determining the zero point, and compare it with the zero point. Subsequent corrections in the position of the rider can be checked in the same fashion until balance is attained.

Time can be saved by employing the balance sensitivity to calculate the number in the final decimal place. Here the weight of the object is determined to the nearest whole milligram, and a rest point (RP_1) is obtained. Then 1 mg is either added to or subtracted from the load on the balance, and a second rest point (RP_2) is taken. The difference between these rest points yields the sensitivity of the balance for the particular loading involved. The correction needed to shift either rest point into coincidence with the zero point can then be readily calculated.

EXAMPLE

Zero point (ZP)	10.2
Rest point with 10.422 g (RP_1)	11.1
Rest point with 10.423 g (RP_2)	8.7
Sensitivity $(11.1 - 8.7)$	2.4 divisions/mg
Difference between RP_1 and ZP	0.9 division

Because

$$\frac{1\text{ mg}}{2.4\text{ div}} = \frac{x\text{ mg}}{0.9\text{ div}}$$

$$x = 0.4\text{ mg}$$

This correction should be *added* to 10.422 g because the rest point corresponding to this loading (RP_1) lies to the *right* of the zero point. Thus the weight of the object is $10.422 + 0.0004 = 10.4224$ g. Note that the same result is obtained from RP_2 data:

$$\text{difference between ZP and } RP_2 = 1.5\text{ div}$$

Because

$$\frac{1\text{ mg}}{2.4\text{ div}} = \frac{x\text{ mg}}{1.5\text{ div}}$$

$$x = 0.6\text{ mg}$$

This correction should be *subtracted* from 10.423 g because the rest point corresponding to this loading lies to the *left* of the zero point. Thus the weight is again $10.423 - 0.0006 = 10.4224$ g.

The Method of Long Swings. With this method zero points and rest points are calculated from five consecutive pointer deflections, three to one side and two to the other. The average of the deflections to the left is added to the average to the right, and the sum is halved. Generally larger deflections are employed. The method of long swings provides the ultimate in precision for undamped balances; however, it is quite time-consuming.

Weighing with a Damped Balance.   A damped balance, particularly if it is also equipped with a chain, is well adapted for the direct determination of the weight of an object. Here the zero point is established by releasing the beam support and pan arrest and allowing time for the pointer to cease moving. Weighing is accomplished as before, the rider and chain being adjusted to give a rest point that is identical with the zero point.

SUMMARY OF RULES GOVERNING THE USE OF AN ANALYTICAL BALANCE

Continued good performance from an analytical balance depends entirely upon the treatment accorded it. Similarly, reliable data are obtained only when careful attention is paid to the details of the weighing operation. Because weighing data are required for virtually every quantitative analysis, it is desirable to summarize the rules and precautions that relate to their acquisition.

To Avoid Damage or Minimize Wear to the Balance and Weights.
1. Be certain that arresting mechanisms for the beam are engaged whenever the loading on the balance is being changed and when the balance is not in use.
2. Center the load on the pan as much as possible.
3. Protect the balance from corrosion. Only vitreous materials and nonreactive metal or plastic objects should be placed directly on the pans.
4. The weighing of volatile materials requires special precautions (p. 546).
5. Do not attempt to adjust the balance without the prior consent of the instructor.
6. The balance and case should be kept scrupulously clean. A camel's-hair brush is useful for the removal of spilled material or dust.
7. Free the beam of an equal-arm balance by releasing the beam support slowly, followed by the pan arrests. Reverse this order when arresting the beam; engage the pan arrests as the pointer passes through the center of the deflection scale.
8. Handle weights for an equal-arm balance gently and with special forceps. Weights should never be touched because moisture from the hands can initiate corrosion. Keep weights enclosed in their box when not in use.
9. If possible, avoid bringing weights for an equal-arm balance into the laboratory.

To Obtain Reliable Weighing Data.
10. Do not attempt to weigh an object until it has returned to room temperature.

11. Do not touch a dried object with bare hands; handle it with tongs or use finger pads to prevent the uptake of moisture.
12. Place the object on the left pan of an equal-arm balance, the weights on the right.
13. Recalibrate a weight that has been mishandled before using it again.
14. At the completion of the weighing, sum the weights at least twice to avoid arithmetic and reading errors. Some analysts first count the weights on the pan and then the empty spots in the weight box.
15. Perform all weighings for an analysis on the same balance, using the same set of weights.

BUOYANCY CORRECTION

When the density of the object weighed differs considerably from that of the weights, a *buoyancy* error is introduced. This discrepancy arises from the difference in buoyant force of the medium (air) as it acts on the object and the weights. A correction for this effect is seldom needed when solid objects are weighed because the density of most solids approaches that of the weights. With liquids, gases, or low-density solids, however, a correction must be applied. The corrected weight is calculated by means of the equation

$$W_1 = W_2 + W_2 \left(\frac{d_{air}}{d_1} - \frac{d_{air}}{d_2} \right) \qquad (19\text{-}1)$$

where W_1 is the corrected weight of the object and W_2 is the mass of the weights. The terms d_1 and d_2 represent the densities of the object and the weights, respectively, and d_{air} is the density of the air displaced by them; the density of air can be taken as 0.0012 g/ml. Table 19-1 gives the density d_2 of the various metals used in the construction of weights. The weights of a single-pan balance are ordinarily made of stainless steel.

TABLE 19-1
Density of Metals Used in the Manufacture of Weights

METAL	DENSITY, g/ml
Aluminum	2.7
Stainless steel	7.8
Brass	8.4
Tantalum	16.6
Gold	19.3
Platinum-iridium	21.5

EXAMPLE

A sample of an organic liquid having a density of 0.92 g/ml was weighed into a glass bottle. The weight of the empty bottle was 8.6500 g against stainless steel weights. After addition of the sample it was 9.8600 g. Correct the weight of the sample for buoyancy.

The apparent weight of the liquid is (9.8600 − 8.6500) or 1.2100 g. Inasmuch as the same buoyant force acted on the glass container during both weighings, we need to consider only the buoyant force acting on the 1.2100 g of liquid. Using the density of stainless steel from Table 19-1 and 0.0012 g/ml for the density of air in Equation 19-1, we obtain

$$W_1 = 1.2100 + 1.2100 \left(\frac{0.0012}{0.92} - \frac{0.0012}{7.8} \right)$$

$$= 1.2100 + 1.2100 \, (0.0013 - 0.00015)$$

$$= 1.2114 \text{ g}$$

OTHER SOURCES OF ERROR IN THE WEIGHING OPERATION

A serious weighing error may arise when the object to be weighed differs in temperature from its surroundings. The usual cause of this type of error is the failure by the chemist to allow sufficient time for a heated object to come to room temperature before it is weighed. Two sources are responsible for the effect. First, convection currents within the balance case exert a buoyant effect on the pan and object. Second, warm air entrapped within a closed container (such as a weighing bottle) weighs less than its surroundings. Both effects cause the apparent weight of the object to be low. For a typical weighing bottle or a glass crucible the error may be as large as 10 or 15 mg. To avoid this type of error heated objects should be cooled for 20 to 30 min before being weighed.

Occasionally a porcelain or glass object will acquire a static charge that is sufficient to cause the balance to perform erratically; this problem is particularly serious when the humidity is low. Spontaneous discharge frequently occurs after a short period. The use of a faintly damp chamois to wipe the object is a recommended preventative.

The optical scale of a single-pan balance should be checked regularly for accuracy, particularly under loading conditions that require essentially the full range of the scale; a 100-mg weight is useful in performing this check.

Many weighing errors are directly attributable to carelessness on the part of the analyst. To avoid such errors it is always good practice to repeat at least twice the process of reading weights.

AUXILIARY BALANCES

A variety of balances with lesser sensitivities than the analytical balance are employed in the laboratory. These offer the advantage of ruggedness, speed, large capacity, and convenience; such balances should be used in lieu of the analytical balance whenever maximum sensitivity is not required.

Auxiliary balances designed for top loading are particularly convenient. A sensitive top-loading balance permits weighing of loads as large as 150 to 200 g with a precision of about 1 mg—an order of magnitude less than the macro analytical balance. Balances of this type with maximum loads ranging up to as high as 25,000 (±0.05) g are available commercially. Some are fully automated and require no manual dialing or weight handling; often the restoring force is electromagnetic. A digital readout is typical.

Triple-beam and double-pan balances with sensitivities lower than most top-loading balances are also useful. The former is a single-pan balance in which three decades of weights can be added by moving sliding weights along three calibrated scales. Balances of this type ordinarily show precisions that are an order or two of magnitude less than that of a top-loading instrument. Their advantages include simplicity, ruggedness, and low cost.

Equipment and Manipulations Associated with Weighing

Most solids absorb atmospheric moisture and, as a consequence, change in composition. This effect assumes appreciable proportions when a large surface area is exposed, as with a sample or a reagent chemical that has been finely ground. It is ordinarily necessary to dry such solids before weighing to free the results from dependence upon the humidity of the surrounding atmosphere.

EQUIPMENT FOR DRYING AND WEIGHING

Weighing Bottles.   Solids are conveniently dried and stored in weighing bottles, two common varieties of which are illustrated in Figure 19-5. Ground-glass contacting surfaces ensure a snug fit between container and lid. In the newer design the lid acts as a cap; with this style of weighing bottle there is less possibility of the sample being entrained on, and subsequently lost from, a ground-glass surface. Weighing bottles usually have numbers from 1 to 100 etched on their sides for identification purposes. Plastic weighing bottles are available; ruggedness is the principal advantage possessed by these bottles over their glass counterparts.

FIGURE 19-5 Typical weighing bottles.

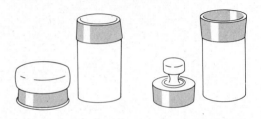

Desiccator, Desiccants. Oven drying is the most convenient method for removing absorbed moisture from a solid. The technique, of course, is not appropriate for samples that undergo decomposition at the ambient temperature of the oven. Furthermore, with some solids the temperatures attainable in ordinary drying ovens are insufficient to effect complete removal of the bound water.

While they cool, dried materials are stored in desiccators; these help prevent the uptake of moisture. As illustrated in Figure 19-6, the base section of a desiccator contains a quantity of a chemical drying agent. Samples are placed on a perforated plate that is supported by a constriction in the wall. Lightly greased ground-glass surfaces provide a tight seal between lid and base.

Several substances find use as drying agents in desiccators. Among these are anhydrous calcium chloride, calcium sulfate (Drierite[6]), anhy-

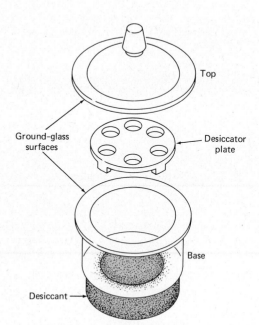

Top

Ground-glass
surfaces

Desiccator
plate

Base

Desiccant

FIGURE 19-6 Components of a typical desiccator.

[6]® W. A. Hammond Drierite Co.

drous magnesium perchlorate (Anhydrone[7] or Dehydrite[8]), and phosphorous pentoxide.

MANIPULATIONS ASSOCIATED WITH DRYING AND WEIGHING

Use of the Desiccator. Whether it is being replaced or removed, the lid of the desiccator is properly moved by a sliding, rather than a lifting, motion. An airtight seal is achieved by slight rotation and direct downward pressure upon the positioned lid.

When a heated object is placed in a desiccator, the increased pressure of the enclosed air is often sufficient to break the seal between the lid and base. If heating has caused the grease on the ground-glass surfaces to soften, there is the further danger that the lid may slide off and break. Upon cooling, the opposite effect is likely to occur, the interior of the desiccator now being under a partial vacuum. Both of these conditions can cause the contents of the desiccator to be physically lost or contaminated. Although it defeats the purpose of the desiccator somewhat, it is wise to allow some cooling to occur before finally sealing the lid. It also helps to break the seal several times during cooling in order to relieve any vacuum that may be developing. Finally, it is prudent to lock the lid in place with one's thumbs while moving the desiccator to ensure against accidental breakage.

Very hygroscopic materials should be stored in containers equipped with snugly fitting covers, the covers remaining in place during storage in the desiccator. Other substances may be stored with container covers removed.

Manipulation of Weighing Bottles. The moisture bound on the surface of many solid materials can be removed by heating at 105 to 110°C for about an hour. Figure 19-7 depicts the arrangement recommended for

FIGURE 19-7 Arrangement for the drying of samples.

[7]® J. T. Baker Chemical Company.

[8]® Arthur H. Thomas Company.

drying a sample in a weighing bottle. The bottle is contained in a beaker, which in turn is covered by a watch glass supported on glass hooks (or by a ribbed watch glass). The sample is thus protected from accidental contamination while the free access of air is maintained. This arrangement also satisfactorily accommodates crucibles containing precipitates that can be freed of moisture by simple drying. The beaker should be marked to permit identification.

Weighing data can be significantly affected by moisture picked up as a consequence of handling a dried weighing bottle with one's fingers. For this reason the bottle should be manipulated with tongs, chamois finger cots, clean cotton gloves, or strips of clean paper. The latter technique is illustrated in Figure 19-8, which shows the weighing of a sample by difference. The weight of the bottle and its contents is first obtained. Then the sample is transferred from the bottle to the container, the utmost care being taken to avoid losses during transfer. Gently tapping the weighing bottle with its top provides adequate control over the process; slight rotation of the bottle is also helpful. Finally, the bottle is again weighed.

Weighing of Hygroscopic Substances.   Hygroscopic substances often equilibrate rapidly, taking up moisture almost to capacity in a short time. Where this effect is pronounced, the approximate amount of each sample should be introduced to individual weighing bottles. After drying and cooling, the exact weight is determined by difference, care being taken to transfer the sample and replace the top of the weighing bottle as rapidly as possible.

Weighing of Liquids.   The weight of a liquid is always obtained by difference. Samples that are noncorrosive and relatively nonvolatile can be weighed into containers fitted with snugly fitted covers, such as weighing bottles; the mass of the container is subtracted from the total weight. If the sample is volatile or corrosive, it should be sealed in a weighed glass ampoule. The bulb of the ampoule is first heated. Then the neck is immersed in the sample. As cooling occurs, the liquid is drawn into the bulb. The neck is then sealed off with a small flame. After cooling, the bulb

FIGURE 19-8 Method for transferring the sample.

and contents are weighed, with any glass removed during the sealing also included in this weighing. The ampoule is then broken in the vessel where the sample is desired. A small volume correction for the glass in the ampoule is required for the most precise work.

Equipment and Manipulations for Filtration and Ignition

EQUIPMENT

Simple Crucibles. Simple crucibles serve as containers only and are of two general types. The weight of the more common variety remains constant within the limits of experimental error while in use. These crucibles are fabricated from porcelain, aluminum oxide, silica, or platinum and are employed to convert precipitates into suitable weighing forms. Nickel, iron, silver, or gold crucibles serve as containers for high-temperature fusions of difficultly soluble samples. Crucibles of this type may be appreciably attacked by the atmosphere and by their contents. Both effects will cause a change in weight; the latter will also contaminate the contents with species derived from the crucible. The analyst selects the crucible whose components will offer the least interference in subsequent steps of the analysis.

Filtering Crucibles. Filtering crucibles are also of two types. For most, the filtering medium is an integral part of the crucible. In addition, there is the *Gooch crucible*, which has a perforated bottom that supports a filter mat that is usually, but not always, asbestos. This mineral is of variable composition and structure, some forms being appreciably water soluble. Thus especially selected and prepared grades are required for use as a filtering medium. Long-fibered, acid-washed asbestos can be obtained from chemical suppliers.

Because asbestos is hygroscopic, care must be taken to bring the crucible and mat to constant weight under exactly the same conditions as will be required to convert the precipitate to a form suitable for weighing.

Small circles of glass matting are available commercially and can be used instead of asbestos in a Gooch crucible; they are used in pairs to protect against accidental disintegration while liquid is added. Glass mats can tolerate temperatures in excess of 500°C and are substantially less hygroscopic than asbestos.

Glass also finds application as a filtering medium in the form of fritted disks sealed permanently into filtering crucibles. These are known as *sintered glass crucibles* and are available in various porosities. With extreme care their use can be extended to temperatures as high as 500°C; normally, however, 150 to 200°C represents the upper practical limit.

Filtering crucibles made entirely of fused quartz may be taken to high temperatures and can be cooled rapidly without damage.

Unglazed porcelain is also useful as a filtering medium. Crucibles of this type have the versatility of temperature range possessed by the Gooch crucible and are not as costly as fused quartz. They require no preparation comparable to that for the Gooch crucible and are far less hygroscopic. Filtering crucibles made of aluminum oxide offer similar advantages.

A partial vacuum is used to draw the supernatant liquid through a filtering crucible; this procedure frequently shortens the time needed for filtration. Connection is made between the crucibles and a heavy-walled filtering flask with a rubber adapter (see Figure 19-9). A diagram for the complete filtration train is shown in Figure 19-14.

Filter Paper. Paper is an important filtering medium. Because it is appreciably hygroscopic, a paper filter is always destroyed by ignition if its contents are to be weighed. Ashless paper is manufactured from fibers that have been washed with hydrochloric and hydrofluoric acids and neutralized with ammonia. The residual ammonium salts in many papers are sufficient to affect analyses for amine nitrogen. Typically, 9- or 11-cm circles of such paper will leave an ash weighing less than 0.1 mg, an amount that is ordinarily negligible. Ashless paper is manufactured in various grades of porosity.

Gelatinous precipitates such as hydrous iron(III) oxide present special problems, owing to their tendency to clog the pores of the paper upon which they are being retained. This problem can be minimized by

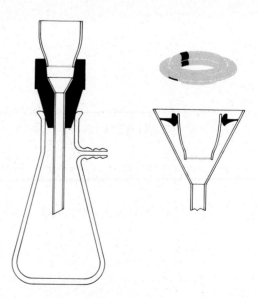

FIGURE 19-9 Adapters for filtering crucibles.

mixing a dispersion of ashless filter paper pulp with the precipitate prior to filtration. The pulp may be prepared by briefly treating a piece of ashless paper with concentrated hydrochloric acid and washing the disintegrated mass free of acid. Tablets of pulp are also commercially available.

Heating Equipment. Many precipitates may be weighed directly after the low-temperature removal of moisture; drying ovens, which are heated electrically and capable of maintaining uniform temperatures to within a degree or less, are convenient for this purpose. The maximum attainable temperature will range from 140 to 260°C, depending upon the make and model. For many precipitates 110°C is a satisfactory drying temperature.

The efficiency of a drying oven is increased significantly by the forced circulation of air. Further refinement is achieved by predrying the air to be circulated and by using vacuum ovens through which a small flow of predried air is maintained.

Ordinary heat lamps are useful for laboratory drying and provide temperatures capable of charring filter paper. A convenient method of treating precipitates collected on paper involves the use of heat lamps for initial drying and charring, followed by ignition at elevated temperatures in a muffle furnace.

Burners are convenient sources of intense heat. The maximum temperature attainable from a burner depends upon its design as well as the combustion properties of the gas used. Of the three common laboratory burners, the Meker provides the greatest heat, followed by the Tirrill and Bunsen types in that order.

Heavy-duty electric furnaces are capable of maintaining temperatures of 1100°C or higher with a control superior to that obtainable with a burner. Special long-handled tongs and asbestos gloves are required for protection while transferring objects to and from such furnaces.

A rough judgment of the temperature of an object can be gained from observation of its color; Table 19-2 will serve as a guide.

MANIPULATIONS ASSOCIATED WITH THE FILTRATION AND IGNITION PROCESSES

Preparation of Crucibles. A crucible used to convert a precipitate into a form suitable for weighing must maintain a substantially constant weight throughout the drying or ignition process. To demonstrate this property, the crucible first is cleaned thoroughly (filtering crucibles are conveniently cleaned by backwashing with suction) and then brought to constant weight, using the same heating and cooling cycle as will be required for the precipitate. Agreement within 0.2 mg between consecutive cycles can be considered as constant weight for most purposes.

TABLE 19-2
Estimation of Temperature by Color[a]

TEMPERATURE, °C	APPROXIMATE COLOR OF OBJECT AT THIS TEMPERATURE
700	Dull red
900	Cherry red
1100	Orange
1300	White

[a] Adapted from T. B. Smith, *Analytical Processes*, 2d ed. London: Edward Arnold, 1940, p. 431.

Filtration and Washing of Precipitates. The actual filtration process consists of three operations: decantation, washing, and transfer. Decantation involves gently pouring off the bulk of liquid phase while leaving the precipitated solid essentially undisturbed. The pores of any filtering medium become clogged with precipitate; hence the longer the transfer of solid can be delayed, the more rapid will be the overall filtration process. A stirring rod is employed to direct the flow of the decantate; see Figure 19-10b. Wash liquid is then added to the beaker and thoroughly mixed with the precipitate; after the solid has again settled, this liquid is also decanted through the filter. It can be seen that the principal washing of the precipitate is carried out *before* the solid is transferred; a more thoroughly washed precipitate and a more rapid filtration are the result.

The transfer process is illustrated in Figure 19-10d. The bulk of the precipitate is moved from beaker to filter by suitably directed streams of liquid. As always, a stirring rod is used to provide direction for the flow of liquid to the filtering medium.

The last traces of precipitate that cling to the walls of the beaker are dislodged with a *rubber policeman*, a small section of rubber tubing that has been crimped shut at one end; this device is fitted on the end of a stirring rod. A rubber policeman should be wetted with wash liquid before use. Any solid collected is combined with the main portion on the filter. Small pieces of ashless paper can be used to wipe the last traces of hydrous oxide precipitates from the wall of the beaker; these papers are ignited along with the cone containing the bulk of the precipitate.

Many precipitates have the exasperating property of *creeping* or spreading over wetted surfaces against the force of gravity. Filters are never filled to more than three-quarters of their capacity, owing to the possibility of losses of solid as a result of creeping.

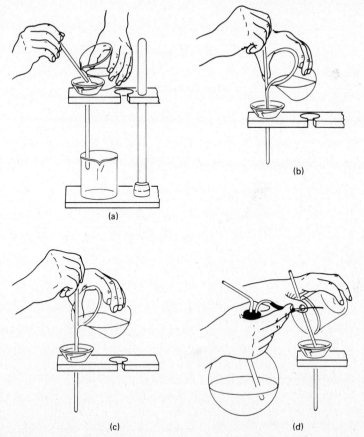

FIGURE 19-10   The filtering operation. Techniques for the decantation and transfer of precipitates are shown here.

A gelatinous precipitate should not be allowed to dry before the washing cycle is complete because the dried mass shrinks and develops cracks. Any liquid subsequently added merely passes through these cracks and accomplishes little or no washing.

DIRECTIONS FOR FILTRATION AND IGNITION WITH ASHLESS FILTER PAPER

Preparation of a Filter Paper.   Figure 19-11 illustrates the sequence followed in folding a filter paper and seating it in a 58-deg or 60-deg funnel. The paper is first folded exactly in half (a), firmly creased, and next loosely folded into quarters (b). A triangular portion is torn from one of the two single corners (c) parallel to the second fold. The paper is then opened to form a cone (d); after fitting in the funnel, the second fold is creased.

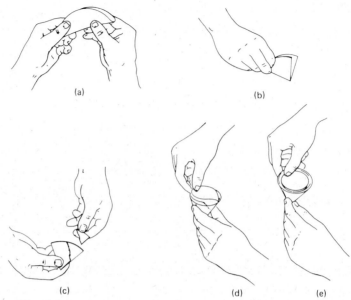

FIGURE 19-11 Technique for folding and seating a filter paper.

Seating (e) is completed by dampening the cone with water from a wash bottle and gentle *patting* with a finger. When the cone is properly seated, there will be no leakage of air between the paper and funnel, and the stem of the funnel will be filled with an unbroken column of liquid.

Transfer of Paper and Precipitate to Crucible. Upon completion of the filtration and washing steps the filter cone and its contents must be transferred from the funnel to a weighed crucible. Ashless paper has very low wet strength; considerable care must be exercised in performing this operation. The danger of tearing can be reduced considerably if the paper is allowed to dry partially prior to its removal from the funnel.

Figure 19-12 illustrates the preferred method of transfer. First, the triple-folded portion is drawn across the filter (a) to flatten the cone along its upper edge; the corners then are folded inward (b). Next, the top is folded over (c). Finally, the paper and contents are eased into the crucible (d) so that the bulk of the precipitate is near the bottom.

Ashing of a Filter Paper. If a heat lamp is to be used, the crucible is placed on a clean, nonreactive surface; an asbestos pad covered with a layer of aluminum foil is satisfactory. The lamp is then positioned about one-half inch from the top of the crucible and turned on. Charring of the paper will take place without further intervention; the process is considerably accelerated if the paper can be moistened with no more than one drop of strong ammonium nitrate solution. Removal of the remaining carbon is accomplished with a burner, as described in the following paragraphs.

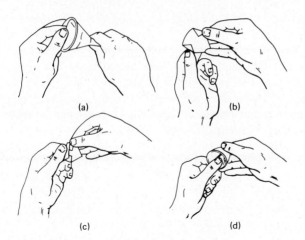

FIGURE 19-12 Method
for transferring a filter
paper and precipitate to
a crucible.

(a)   (b)

(c)   (d)

Considerably more attention is required when a burner is employed to ash a filter paper. Because the burner can produce much higher temperatures, there exists the danger of expelling moisture so rapidly in the initial stages of heating that mechanical loss of the precipitate occurs. A similar possibility arises if the paper is allowed to flame. Finally, so long as there is carbon present, there is also the possibility that the precipitate will be partially reduced; this is a serious problem where reoxidation following ashing of the paper is not convenient.

To minimize these difficulties the crucible is placed as illustrated in Figure 19-13; the tilted position allows for the ready access of air. A clean crucible cover should be located nearby, ready for use if necessary. Heating is then commenced with a small burner flame. The temperature is gradually increased as moisture is evolved and the paper begins to char. The smoke that is given off serves as a guide with respect to the intensity

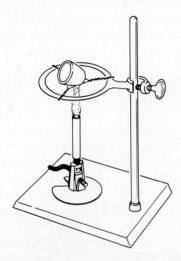

FIGURE 19-13 Ignition of the precipitate. Arrangement of the crucible for the preliminary charring of the paper is illustrated.

of heating that can be safely tolerated. Normally this will appear in thin wisps. An increase in the volume of smoke emitted indicates that the paper is about to flash; heating should be temporarily discontinued if this condition is observed. If, despite precautions, a flame does appear, it should be immediately snuffed out with the crucible cover. (The cover may become discolored, owing to the condensation of carbonaceous products. These must ultimately be removed by ignition to confirm the absence of entrained particles of precipitate.) Finally, when no further smoking can be detected, the residual carbon is removed by gradually increasing the flame. Strong heating, as necessary, can then be undertaken. Care must be exercised to avoid heating the crucible in the reducing portion of the flame.

The foregoing sequence will ordinarily precede the final ignition of the sample in a muffle furnace, in which a reducing atmosphere is equally undesirable.

DIRECTIONS FOR THE USE OF FILTERING CRUCIBLES

Preparation of a Gooch Crucible. A mat is prepared by arranging the Gooch crucible in a vacuum filtration train (see Figure 19-14). A few milliliters of the well-mixed and diluted asbestos suspension are poured into the crucible and allowed to stand for about a minute. In this interval the heavier filaments tend to settle and form a support for the finer fibers. Most of the liquid also drains off. Next, suction is applied to set the mat. The mat is then washed until no further loss of asbestos can be detected in the washings; several hundred milliliters may be required. Finally, the crucible is dried and ignited to constant weight.

A proper mat is one through which the pattern of holes in the crucible can be barely discerned when viewed against a strong light.

A vacuum must always be applied before attempting to transfer liquid to a Gooch crucible. Care must also be taken to avoid destroying the

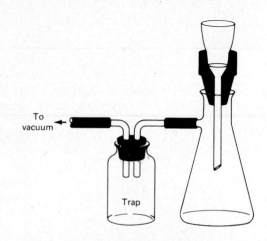

To vacuum

Trap

FIGURE 19-14 Train for vacuum filtration.

mat as liquids are introduced; the flow is always directed down the wall of the crucible with a stirring rod. The mat can be further protected with a small perforated disk called a *Witt plate*.

Ignitions with a Filtering Crucible. The use of a filtering crucible tends to shorten the time required for the filtration of a precipitate. The porous bottom of such a crucible, however, greatly increases the danger of reducing the precipitate with the burner flame. This difficulty is readily circumvented by placing the filtering crucible inside an ordinary crucible during the ignition step.

The reader is again cautioned against subjecting crucibles to unnecessarily abrupt changes in temperature.

RULES FOR THE MANIPULATION OF HEATED OBJECTS

1. A crucible that has been subjected to the full flame of a burner or to a muffle furnace should be allowed to cool momentarily on an asbestos plate before being moved to the desiccator.
2. Hot objects should not be placed directly on the desk top but should be set on an asbestos pad or a clean wire gauze.
3. Manipulations should be practiced first to assure that adequate control can be maintained with the implements to be used.
4. The tongs and forceps employed in handling heated objects should be kept scrupulously clean. The tips should not be allowed to come in contact with the desk top.

The Measurement of Volume

UNITS OF VOLUME

The fundamental unit of volume is the *liter*, defined as the volume occupied by 1 kg of water at the temperature of maximum density (3.98°C) and at 1 atm pressure.

The *milliliter* is one one-thousandth of a liter and is used when the liter represents an inconveniently large volume unit.

Another unit of volume is the *cubic centimeter*. Although this and the milliliter can be used interchangeably without effect in most situations, the two units are not strictly identical, the milliliter being equal to 1.000028 cm^3. It was originally intended that 1 kg of water should occupy exactly 1 dm^3. Owing to inadequacies of early experimental measurements, however, this relationship was not realized, and the small difference in units was the result. For volumetric analysis the liter (or the milliliter) is used.

EFFECT OF TEMPERATURE UPON VOLUME MEASUREMENTS

The volume occupied by a given mass of liquid varies with temperature. So also does the volume of the container that holds the liquid. The accurate measurement of volume may require taking both of these temperature effects into account.

Most volumetric measuring devices are constructed of glass, which fortunately has a small temperature coefficient. Thus, for example, a soft glass vessel will change in volume by about 0.003% per degree; with heat-resistant glass the change is about one-third of this value. Clearly variations in the volume of a container due to changes in temperature need be considered only for the most exacting work.

The coefficient of expansion for dilute aqueous solutions is approximately 0.025% per degree. The magnitude of this figure is such that a temperature variation of about 5°C will measurably affect the reliability of ordinary volumetric measurements.

EXAMPLE

A 40.00-ml sample is taken from a liquid refrigerated at 5°C. Calculate the volume this sample will occupy at 20°C.

$$V_{20°} = V_{5°} + 0.00025(20 - 5)(40.00) \qquad \textbf{(19-2)}$$

$$\doteq 40.00 + 0.15$$

$$= 40.15 \text{ ml}$$

Volumetric measurements must be referred to some standard temperature; to minimize the need for calculations such as these, 20.0°C has been chosen for this reference point. The ambient temperature in most laboratories is close to 20°C; thus the need seldom arises for a temperature correction in ordinary analytical work. The coefficient of cubic expansion for many organic liquids, however, is considerably greater than that for water or dilute aqueous solutions. Good precision in the measurement of these liquids may require corrections for temperature variations of a degree or less.

APPARATUS FOR THE PRECISE MEASUREMENT OF VOLUME

The reliable measurement of volume is performed with the *pipet*, the *buret*, and the *volumetric flask*. These are calibrated either to *deliver* or, alternatively, to *contain* a specified volume. Volumetric equipment is marked by the manufacturer to indicate not only the manner of calibration (usually with a TD for "to deliver" or a TC for "to contain") but also the temperature for which the calibration strictly refers. Pipets and burets are ordinarily designed and calibrated to deliver specified volumes, whereas volumetric flasks are calibrated on a "to contain" basis.

Pipets. Pipets are devices that permit the transfer of accurately known volumes from one container to another. Some of the common types are pictured in Figure 19-15; Table 19-3 provides further details on their use.

Volumetric or transfer pipets (Figure 19-15a) deliver a single fixed volume in the range of 0.5 to 200 ml; many are color coded by volume to facilitate sorting and identification. Measuring pipets are of two types, Mohr and serological (Figure 19-15b and c). Measuring pipets are calibrated in convenient units to allow delivery of any volume up to the maximum capacity; sizes from 0.1 to 25 ml exist. For situations that call for repeated delivery of a particular volume, a variety of automatic pipets that deliver volumes of 0.1 to 50 ml are available commercially. Hand-held syringe pipets (Figure 19-15d) that deliver volumes in the range between 1 and

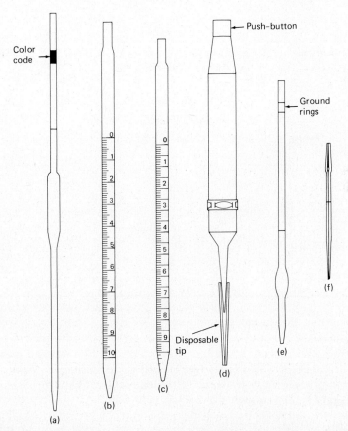

FIGURE 19-15 Typical pipets. (a) Volumetric. (b) Mohr. (c) Serological. (d) Syringe. (e) Ostwald-Folin. (f) Lambda.

TABLE 19-3
Pipets

NAME	TYPE OF CALIBRA-TION	FUNCTION	AVAILABLE CAPACITIES, ml	TYPE OF DRAINAGE
Volumetric	TD	Delivery of a fixed volume	1–200	Free drainage
Mohr	TD	Delivery of a variable volume	1–25	Drain to lower calibration line
Serological	TD	Same	0.1–10	Blow out last drop[a]
Serological	TD	Same	0.1–10	Drain to lower calibration line
Ostwald-Folin	TD	Delivery of a fixed volume	0.5–10	Blow out last drop[a]
Lambda	TC	To contain a fixed volume	0.001–2	Wash out with suitable solvent
Lambda	TD	Delivery of a fixed volume	0.001–2	Blow out last drop[a]
Syringe	TD	Delivery of a fixed volume	0.001–1	Tip emptied by syringe

[a] A frosted ring near the top of recently manufactured pipets indicates that the last drop is to be blown out.

1000 μliters (1 ml) are available; a useful feature of this type of pipet is that the liquid is contained in a disposable plastic tip. A measured volume of liquid is drawn into the tip by means of a spring operated piston activated by a push button at the top of the syringe. The liquid is then transferred by reversing the action of the piston by means of the push button. Remarkable precision is claimed for these devices (± 0.02 μliter for a 1 μliter measurement and ± 0.3 μliter at 1000 μliter).

Volumetric and measuring pipets are filled to an initial calibration mark at the outset; the manner in which the transfer is completed depends upon the particular type. Because of the attraction between most liquids and glass, a drop tends to remain in the tip of a drained transfer pipet. This drop is blown from some pipets but not from others. Table 19-3 and Figure 19-15 summarize the several varieties most likely to be encountered in an analytical laboratory.

Burets. Burets, like measuring pipets, enable the analyst to deliver any volume up to the maximum capacity. The precision attainable with a buret is appreciably better than that with a measuring pipet.

A buret consists of a calibrated tube containing the liquid and a valve arrangement by which the flow from a tip can be controlled. Principal differences among burets are to be found in the type of valve employed. The simplest consists of a closely fitting glass bead within a short length of rubber tubing. Only when the tubing is deformed can liquid flow past the bead. Burets equipped with glass stopcocks rely upon a lubricant between the ground-glass surfaces of stopcock and barrel for a liquid-tight seal. Some solutions, notably bases, will cause a stopcock to freeze upon long contact; thorough cleaning is necessary after each use.

Valves made of Teflon are commonly encountered; these are inert to attack by most common reagents and require no lubricant.

More elaborate burets are designed so that they are filled and zeroed automatically; these are of particular value in routine analysis.

Volumetric Flasks. Volumetric flasks are manufactured with capacities ranging from 5 ml to 5 liters and are usually calibrated to contain a specified volume when filled to the line etched on the neck. They are used in the preparation of standard solutions and the dilution of samples to known volumes prior to taking aliquot portions with a pipet. Some are also calibrated on a "to deliver" basis. These are readily distinguishable by two reference lines; if delivery of the stated volume is desired, the flask is filled to the upper of the two lines.

MANIPULATIONS ASSOCIATED WITH THE MEASUREMENT OF VOLUME

Only clean glass surfaces will support a uniform film of liquid; the presence of dirt or oil will tend to cause breaks in this film. The appearance of water breaks is a certain indication of an unclean surface; if the volume calibrations are to have meaning, the equipment must be kept scrupulously clean when in use.

As a general rule, the heating of calibrated glass equipment should be avoided. Rapid cooling can permanently distort the glass and cause a change in volume.

When a liquid is confined in a narrow tube such as a buret or a pipet, the surface exhibits a marked curvature called a *meniscus*. It is common practice to use the bottom of the meniscus as the point of reference in calibrating and using volumetric ware. This minimum can often be established more exactly if an opaque card or piece of paper is held behind the graduations (see Figure 19-16).

In judging volumes, the eye must be level with the liquid; otherwise the reading will be in error due to *parallax*. Thus, if one's eye level is above that of the liquid, it will appear that a smaller volume has been taken than is actually the case. An error in the opposite direction can be expected if the point of observation is too low (see Figure 19-16).

FIGURE 19-16 Method for reading a buret. The eye should be level with the meniscus. The black-white line on the card should be slightly below the meniscus. The reading shown is 34.38 ml. If viewed from 1, the reading will appear smaller than 34.38; from 2, it will appear larger.

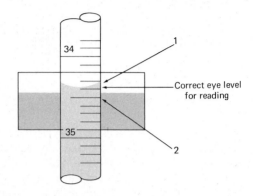

Directions for the Use of a Pipet. The following instructions pertain specifically to the manipulation of transfer pipets, but with minor modifications they may be used for other types as well. Liquids are usually drawn into pipets through the application of a slight vacuum. The mouth should not be used for suction because of the danger of accidentally ingesting liquids. The use of a rubber suction bulb or a rubber tube connected to an aspirator pump is strongly recommended (see Figure 19-17a).

PROCEDURE

Cleaning. Pipets may be cleaned with a warm solution of detergent or with tepid cleaning solution (see p. 529). Draw in sufficient liquid to fill the bulb to about one-third of its capacity. While holding it nearly horizontal, carefully rotate the pipet so that all interior surfaces are wetted. Drain and then rinse thoroughly with distilled water. Inspect for water breaks, and repeat the cleaning cycle if necessary.

Measurement of an Aliquot. As in cleaning, draw in a small quantity of the liquid to be sampled, and thoroughly rinse the interior surfaces. Repeat with at least two more portions. Then carefully fill the pipet somewhat past the graduation mark employing a suction bulb as in Figure 19-17a. Quickly place a *forefinger* over the upper end of the pipet to arrest the outflow of liquid (Figure 19-17b). Make certain that there are no bubbles in the bulk of the liquid or foam at the surface. Tilt the pipet slightly from the vertical, and wipe the exterior free of adhering liquid (Figure 19-17c). Touch the tip of the pipet to the wall of a glass vessel (*not the actual receiving vessel*), and slowly allow the liquid level to drop by partially releasing the forefinger (Note 1). Halt further flow as the bottom of the meniscus coincides exactly with the graduation mark. Then place the tip of the pipet well into the receiving vessel, and allow the sample to drain. When free flow ceases, rest the tip against an inner wall for a full 10 sec (Figure

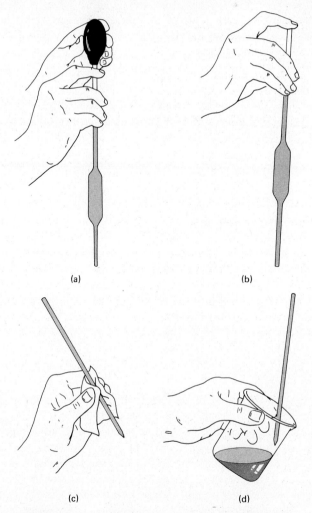

(a)   (b)

(c)   (d)

FIGURE 19-17   Technique for using a volumetric pipet.

19-17d). Finally, withdraw the pipet with a rotating motion to remove any droplet still adhering to the tip. *The small volume remaining inside the tip is not to be blown or rinsed into the receiving vessel.*

NOTES

1. The liquid can best be held at a constant level in the pipet if one's forefinger is slightly moist; too much moisture, however, makes control difficult.
2. It is good practice to avoid handling a pipet by the bulb to eliminate the possibility of changing the temperature of its contents.
3. Pipets should be thoroughly rinsed with distilled water after use.

Directions for the Use of a Buret. Before being placed in service, a buret must be scrupulously clean. In addition, it must be established that the valve is liquid-tight.

PROCEDURE

Cleaning. Thoroughly clean the tube with detergent and a long brush. If water breaks persist after rinsing, clamp the buret in an inverted position with the end dipped in a beaker of cleaning solution. Connect a hose from the buret tip to a vacuum line. Gently pull the cleaning solution into the buret, stopping well short of the stopcock (Note 1). Allow to stand for at least 15 min and then drain. Rinse thoroughly with distilled water, and again inspect for water breaks. Repeat the treatment if necessary.

Lubrication of a Stopcock Buret. Carefully remove all old grease from the stopcock and barrel with a paper towel, and dry both parts completely. Lightly grease the stopcock, taking care to avoid the area near the hole. Insert the stopcock into the barrel, and rotate it vigorously. When the proper amount of lubricant has been used, the area of contact between stopcock and barrel appears nearly transparent, the seal is liquid-tight, and no grease has worked its way in to the tip.

NOTES
1. Cleaning solution often disperses more stopcock lubricant than it oxidizes and leaves a buret with a heavier grease film than before treatment. For this reason cleaning solution should *not* be allowed to come in contact with lubricated stopcock assemblies.
2. Grease films unaffected by cleaning solution may yield to treatment with such organic solvents as acetone or benzene. Thorough washing with detergent should follow such treatment. The use of silicone lubricants is not recommended; contamination by such preparations is very difficult to remove.
3. So long as the flow of liquid is not impeded, fouling of the buret tip with lubricant is not a serious matter. Removal is best accomplished with organic solvents. A stoppage in the middle of a titration can be freed by *gently* warming the tip with a lighted match.
4. Before returning a buret to service after reassembly, it is advisable to test for leakage. Simply fill the buret with water, and establish that the volume reading does not change with time.

Filling. Make certain that the stopcock is closed. Add 5 to 10 ml of solution, and carefully rotate the buret to wet the interior completely. Allow the liquid to drain through the tip. Repeat this procedure two more times. Then fill the buret well above the zero mark. Free the tip of air bubbles by rapidly rotating the stopcock and allowing small quantities of solution to pass. Finally, lower the level of the solution to, or somewhat below, the zero marking. After allowing about a minute for drainage, take an initial volume reading.

Titration. Figure 19-18 illustrates the preferred method for the manipulation of a stopcock. Any tendency for lateral movement of the stopcock will be in the direction of firmer seatings.

With the tip well within the titration vessel, introduce the solution from the buret in increments of 1 ml or so. Swirl (or stir) the sample constantly to ensure efficient mixing. Decrease the volume of the additions as the titration progresses. In the immediate vicinity of the end point add reagent a drop at a time. When it is judged that only a few more drops are needed, rinse down the walls of the titration vessel before completing the titration. Allow 30 sec to elapse between the last addition of reagent and the reading of the buret.

NOTES

1. When unfamiliar with a particular titration, many analysts prepare an extra sample. No care is lavished on its titration since its functions are to reveal the nature of the end point and to provide a rough estimate of titrant requirements. This deliberate sacrifice of one sample often results in an overall saving of time.

2. Instead of being rinsed near the end of the titration, a flask can be carefully tipped and rotated so that the bulk of the liquid picks up any droplets adhering to the walls.

3. Volume increments smaller than a normal drop may be taken by allowing a small volume of liquid to form on the tip of the buret and then touching the tip to the wall of the flask. This droplet is then combined with the bulk of the solution as in Note 2.

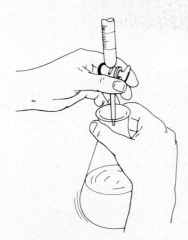

FIGURE 19-18 Recommended technique for the manipulation of a buret stopcock.

Directions for the Use of a Volumetric Flask.   Before use, volumetric flasks should be washed with detergent and, if necessary, cleaning solution. Only rarely need they be dried. If required, however, drying is best accomplished by clamping the flasks in an inverted position. The insertion of a glass tube connected to a vacuum line will hasten the process.

PROCEDURE

Weighing Directly into a Volumetric Flask.   The direct preparation of a standard solution requires that a known weight of solute be introduced into a volumetric flask. To minimize the possibility of loss during transfer, insert a powder funnel into the neck of the flask. The funnel must subsequently be washed free of solid.

Dilution to the Mark.   After transferring the solute, fill the flask about half-full, and swirl the contents to achieve solution. Add more solvent and again mix well. Bring the liquid level almost to the mark, and allow time for drainage; then use a medicine dropper to make such final additions of solvent as are necessary. Firmly stopper the flask, and invert repeatedly to assure uniform mixing. Finally, transfer the contents to a storage bottle that is either initially dry or has been thoroughly rinsed with several small portions of the solution from the flask.

> ### NOTE
> If, as sometimes happens, the liquid level accidently exceeds the calibration mark, the solution can be saved by correcting for the excess volume. Use a gummed label to mark the actual position of the meniscus. After the flask has been emptied, carefully refill to the etched graduation mark with water. Then with a buret determine the additional volume needed to duplicate the actual volume of the solution. This volume, of course, should be added to the nominal value for the flask when calculating the concentration of the solution.

CALIBRATION OF VOLUMETRIC WARE

The reliability of a volumetric analysis depends upon agreement between the volumes actually and purportedly contained (or delivered) by the apparatus. Calibration simply verifies this agreement if such already exists or provides the means for attaining agreement if it is lacking. The latter involves either the assignment of corrections to the existing volume markings or the striking of new markings that agree more closely with the nominal values.

In general a calibration consists of determining the mass of a liquid of known density contained (or delivered) by the apparatus. Although this appears to be a straightforward process, a number of important variables must be controlled. Principal among these is the temperature, which influences a calibration in two ways. First and most important, the volume occupied by a given mass of liquid varies with temperature. Second, the volume of the apparatus itself is variable, owing to the tendency of the glass to expand or contract with changes in temperature.

We noted earlier (p. 541) that the effect of buoyancy upon weighing data is most pronounced when the density of the object is significantly less than that of the weights. As a general rule, a buoyancy correction must be applied to data where water is the calibration fluid.

Finally, the liquid employed for calibration requires consideration. Water is the liquid of choice for most work. Mercury is also useful, particularly where small volumes are involved. Because mercury does not wet glass surfaces, the volume contained by the apparatus will be identical with that which is delivered. In addition, the convex meniscus of mercury gives rise to a small correction that must be applied to give the corresponding volume for a liquid forming a concave meniscus. The magnitude of this correction is dependent upon the diameter of the apparatus at the graduation mark.

The calculations associated with calibrations, while not difficult, are somewhat involved. First, the raw weighing data are corrected for buoyancy, using Equation 19-1. Next, the volume of the apparatus at the temperature (t) of calibration is obtained by dividing the density of the liquid at that temperature into the corrected weight. Finally, this volume is corrected to the standard temperature of 20°C by means of Equation 19-2.

Table 19-4 is provided to ease the computational burden of calibration. Corrections for buoyancy with respect to stainless steel and brass weights and for the volume change of the water, as well as its glass container, have been incorporated into these data. Multiplication by the appropriate factor from the table converts the mass of water measured at some other temperature to the volume it would occupy at 20°C.

EXAMPLE

A 25-ml pipet was found to deliver 24.976 g of water when calibrated against stainless steel weights at 25°C. Use the data in Table 19-4 to calculate the volume delivered by the pipet at this temperature and at 20°C.

$$\text{At } 25°C \qquad V = 24.976 \times 1.0040 = 25.08 \text{ ml}$$

$$\text{At } 20°C \qquad V = 24.976 \times 1.0037 = 25.07 \text{ ml}$$

General Directions for Calibration Work. All volumetric apparatus should be painstakingly freed of water breaks before being tested. Burets and pipets need not be dried. Volumetric flasks should be thoroughly drained.

TABLE 19-4

Volume Occupied by 1.000 g Water Weighed in Air against Stainless Steel Weights[a]

TEMPERATURE, t, °C	VOLUME, ml	
	At Temperature, t	Corrected to 20°C
10	1.0013	1.0016
11	1.0014	1.0016
12	1.0015	1.0017
13	1.0016	1.0018
14	1.0018	1.0019
15	1.0019	1.0020
16	1.0021	1.0022
17	1.0022	1.0023
18	1.0024	1.0025
19	1.0026	1.0026
20	1.0028	1.0028
21	1.0030	1.0030
22	1.0033	1.0032
23	1.0035	1.0034
24	1.0037	1.0036
25	1.0040	1.0037
26	1.0043	1.0041
27	1.0045	1.0043
28	1.0048	1.0046
29	1.0051	1.0048
30	1.0054	1.0052

[a]Corrections for buoyancy (stainless steel weights) and the change in volume of the container have been applied.

The water used for calibration should be drawn well in advance of use to permit it to reach thermal equilibrium with its surroundings. This condition is best assured by noting the temperature of the water at frequent intervals and waiting until no further changes are observed.

An analytical balance can be used for calibrations involving 50 ml or less. Weighings need never be more reliable than the nearest milligram;

this order of reproducibility is within the capabilities of most modern top-loading, single-pan balances. Weighing bottles or small, well-stoppered conical flasks are convenient receivers for small volumes.

PROCEDURE

Calibration of a Volumetric Pipet. Determine the empty weight of a stoppered receiver to the nearest milligram. Transfer a volume of water to the receiver with the pipet (p. 560), weigh the receiver and contents to the nearest milligram, and calculate the weight of water delivered from the difference in these weights. Repeat the calibration several times.

Calibration of a Buret. Fill the buret, and make certain that no bubbles are entrapped in the tip. Withdraw water until the level is at, or just below, the zero mark. Touch the tip to the wall of a beaker to remove any adhering drop. After allowing time for drainage, take an initial reading of the meniscus, estimating the volume to the nearest 0.01 ml. Allow the buret to stand for 5 min and recheck the reading; if the stopcock is tight, there should be no noticeable change. During this interval weigh (to the nearest milligram) a 125-ml conical flask fitted with a rubber stopper.

Once the tightness of the stopcock has been established, slowly (about 10 ml/min) transfer approximately 10 ml into the flask. Touch the tip to the wall of the flask. Wait 1 min, record the volume, and refill the buret. Weigh the flask and its contents to the nearest milligram; the difference between this and the initial weight gives the mass of water actually delivered. Convert this mass into the volume delivered, using Table 19-4. Compute the correction in this interval by subtracting the apparent volume from the true volume. The difference is the correction that must be applied to the *apparent volume* to give the *true volume*.

Starting again from the zero mark, repeat the calibration, using about 20 ml. Test the buret at 10-ml intervals over its entire volume. Prepare a plot of the correction to be applied as a function of the volume delivered.

NOTES
1. Any correction larger than 0.10 ml should be verified by duplicate determinations before being accepted.
2. Corrections associated with any interval may be determined from the plot.

Calibration of a Volumetric Flask. Weigh the clean, dry flask, placing it on the right-hand pan of a Harvard trip balance. Set a beaker on the left pan, and add lead shot until balance is achieved. Remove the flask, and in its place substitute known weights until the same point of balance is reached (Note 2). Carefully fill the flask with water of known temperature until the meniscus coincides with the graduation mark. Return the flask to the right-hand pan and the beaker to the left. Repeat the process of counterweighing with lead shot, followed by substituting weights for the flask. The difference between the two weighings gives the mass of water contained by the apparatus. Calculate the corresponding volume with the aid of Table 19-4.

NOTES

1. A glass tube with one end drawn to a tip is useful in making final adjustments of the liquid level.
2. The substitution method described here is employed to eliminate any error arising from unequal arm lengths of the balance. If a top-loading, single-pan balance is available, the substitution technique is unnecessary.

Calibration of a Volumetric Flask Relative to a Pipet.   The calibration of a flask relative to a pipet provides an excellent method for partitioning a sample into aliquots. The following directions pertain specifically to a 50-ml pipet and a 500-ml flask; other combinations are equally convenient.

With a 50-ml pipet carefully transfer 10 volumes to a 500-ml volumetric flask. Mark the location of the meniscus with a gummed label. Coat the label with paraffin to assure permanence. When a sample is diluted to this mark, the same 50-ml pipet will deliver a one-tenth aliquot of the total sample.

The Laboratory Notebook

RULES FOR KEEPING THE LABORATORY NOTEBOOK

1. The notebook should be permanently bound with consecutively numbered pages.
2. Entries should be legible and well spaced from one another. Most notebooks have more than ample room, so crowding of data is unnecessary.
3. The first few pages of the notebook should be reserved for a table of contents, which should be conscientiously kept up to date.
4. *All data should be entered in ink directly into the notebook.*

    a. Entries should be liberally identified with labels. If a series of weights refers to a set of empty crucibles, it should be labeled "Empty Crucible Weight" or something similar. The significance of an entry is obvious when it is recorded but may become unclear with the passage of time.

Gravimetric Determination of Chloride

The Chloride in a soluble sample was precipitated as AgCl and weighed as such

Sample weights	1	2	3
Wt bottle plus sample, g	27.6115	27.2185	26.8105
-less sample, g	27.2185	26.8105	26.4517
wt. sample, g	0.3930	0.4080	0.3588
Crucible weights, empty	~~20.7925~~	~~22.8311~~	~~21.2488~~
	20.7926	22.8311	~~21.2482~~
			21.2483
Crucible weights, with AgCl, g	~~21.4294~~	~~23.4920~~	~~21.8324~~
	~~21.4297~~	~~23.4914~~	21.8323
	21.4296	23.4915	
Weight of AgCl, g	0.6370	0.6604	0.5840
Percent Cl⁻	40.10	40.04	40.27
Average percent Cl⁻		40.12	
Relative standard deviation		3.0 parts per thousand	

Date Started   1-7-79

Date Completed   1-14-79

FIGURE 19-19   Sample summary page.

b. Each notebook page should be dated as it is used.

c. An erroneous entry should not be erased, nor should it be obliterated. Instead, it should be crossed out with a *single* horizontal line with the corrected entry located as nearby as possible. Numbers should never be written over; in time it may become impossible to decide what the correct number is.

d. Pages should not be removed from the notebook. It is sufficient to draw a single line diagonally across a page that is to be disregarded. A brief notation of the reason for striking out the page is useful.

SUGGESTED FORM

A satisfactory format involves the consecutive use of all pages for recording data. Upon completion of the analysis the next pair of facing pages is used to summarize the results. The right-hand page should contain the following:

1. The title of the experiment—for example, "The Gravimetric Determination of Chloride."
2. A brief statement of the principles upon which the analysis is based.
3. A summary of the data collected and the result calculated for each sample in the set.
4. A report of the best value for the set and a statement of the precision attained in the analysis.

A sample summary is shown in Figure 19-19. The left-hand page should show the following:

1. Equations for the principal reactions in the analysis.
2. An equation that shows the calculation employed in computing the results.
3. The calculations themselves.
4. A summary of observations—originally recorded in the notebook at the time of observation—that appear to bear upon a particular result of the analysis as a whole.

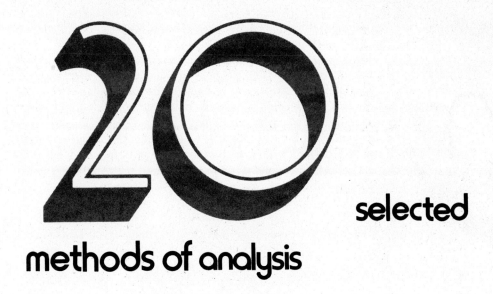

selected
methods of analysis

This chapter provides specific instructions for carrying out selected chemical analyses. The experiments are grouped as follows: 1, gravimetric methods; 2–11, volumetric methods; 12–15, electroanalytical methods; 16, spectral methods. Before starting experimental work, students should read and understand any description that precedes the directions as well as any notes that follow the directions; they should also refer to sections of earlier chapters that have been cited as sources for additional information.

1   Gravimetric Methods

ANALYSIS 1-1   Determination of Water in Barium Chloride Dihydrate

Discussion.   The water in a crystalline hydrate such as $BaCl_2 \cdot 2H_2O$ is readily determined gravimetrically. A sample is heated at a suitable temperature; its water content is taken as the difference in its weight before and after heating. Less frequently the evolved water is collected and weighed.

Procedure. Throughout this experiment perform all weighings to the nearest 0.1 mg.

Carefully clean two weighing bottles. Dry them for about 1 hr at 105 to 110°C; use a covered beaker to prevent accidental contamination (see Figure 19-7). Determine the weight of each bottle after it has cooled to room temperature in a desiccator. Repeat this cycle of heating, cooling, and weighing until successive weighings agree within 0.2 mg. Next introduce a quantity of unknown to each bottle (Note 1) and reweigh. Heat the samples for about 2 hr at 105 to 110°C; then cool and weigh as before. Repeat the heating cycle until constant weights for the bottles and their contents have been attained. Report the percentage of water in the sample.

NOTES

1. If the unknown consists of the pure dihydrate $BaCl_2 \cdot 2H_2O$, take samples weighing approximately 1 g. If the unknown is a mixture of the dihydrate and some anhydrous diluent, obtain the proper sample size from the instructor.

2. Barium chloride can be heated to elevated temperatures without danger of decomposition. If desired, the analysis can be performed in crucibles with a Bunsen flame as the source of heat.

3. Magnesium sulfate hepthahydrate can be used instead of barium chloride; heating to 140°C is needed to eliminate the last traces of moisture.

ANALYSIS 1-2 Determination of Chloride in a Soluble Sample

Discussion. The chloride content of a soluble salt can be determined by precipitation as silver chloride:

$$Ag^+ + Cl^- \rightarrow AgCl(s)$$

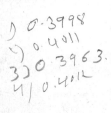

The precipitate is collected in a filtering crucible, washed, and brought to constant weight by drying at 105 to 110°C. Precipitation is carried out in acid solution to eliminate potential interference from anions of weak acids (for example, CO_3^{2-}) that form precipitates with silver in neutral media. A moderate excess of silver ion is required to diminish the solubility of the precipitate; a large excess will cause serious coprecipitation and should be avoided.

Silver chloride precipitates first as a colloid; it is coagulated with heat in the presence of a relatively high electrolyte concentration. A small quantity of nitric acid is added to the wash liquid to maintain the electrolyte concentration and prevent peptization during washing; the acid is volatilized during the subsequent heat treatment. See pages 125–129 for further information concerning the characteristics of colloidal precipitates.

Like other silver halides, silver chloride in suspension is susceptible to photodecomposition, the reaction being

$$AgCl(s) \rightarrow Ag(s) + \tfrac{1}{2}Cl_2(aq)$$

The precipitate acquires a violet color due to the accumulation of finely divided silver. If photochemical decomposition occurs in the presence of excess silver ion, the additional reaction,

$$3Cl_2(aq) + 3H_2O + 5Ag^+ \rightarrow 5AgCl(s) + ClO_3^- + 6H^+$$

will cause the analytical results to be high. In the absence of silver ion the results will be low. Dry silver chloride is virtually unaffected by exposure to light.

Unless elaborate precautions are taken, some photochemical decomposition of silver chloride is unavoidable; however, with reasonable care this effect will not induce an appreciable error in the analysis.

Iodide, bromide, and thiocyanate, if present in the sample, will be precipitated along with silver chloride and cause high results. In addition, the chlorides of tin and antimony are likely to precipitate as hydrous oxides under the conditions of the analysis.

Procedure. Clean (Note 1) and dry three fritted glass or porcelain filtering crucibles at 105 to 110°C; supply each with identifying marks, and bring them to constant weight during periods of waiting in the analysis.

Dry the unknown at 105 to 110°C for 1 to 2 hr in a weighing bottle (see Figure 19-7). Store in a desiccator while cooling. Weigh individual 0.4-g samples (to the nearest 0.1 mg) into 400-ml beakers (p. 545). To each add about 200 ml of distilled water and 3 to 5 ml of 6-F nitric acid. Slowly and with good stirring add 5% silver nitrate (w/v) to the cold solution until the precipitate is observed to coagulate (Note 2); then add an additional 3 to 5 ml. Heat almost to boiling, and digest the precipitate at this temperature for about 10 min. Check for completeness of precipitation by adding a few drops of silver nitrate to the supernatant liquid; should additional silver chloride appear, continue the addition of silver nitrate until precipitation is complete. Store in a dark place for at least 1 to 2 hr, preferably until the next laboratory period. Then decant (p. 550) the supernatant through a weighed filtering crucible. Wash the precipitate several times (while it is still in the beaker) with a cold solution consisting of 2 to 5 ml of 6-F nitric acid per liter of distilled water; decant these washings through the filter also. Finally, transfer the bulk of the precipitate to the crucible, using a rubber policeman to dislodge any particles that adhere to the walls of the beaker. Continue washing until the filtrate is found to be substantially free of silver (Note 3). Dry the precipitates at 105 to 110°C for about 1 hr. Store crucibles in a desiccator until they have cooled to room temperature. Determine the weight of the crucibles and their contents

(Note 4). Repeat the cycle of heating, cooling, and weighing until constant weight has been achieved. Report the percentage of chloride in the sample.

NOTES

1. Residual silver chloride can be removed from filtering crucibles by soaking them in a strong solution of sodium thiosulfate.
2. To determine an approximate amount of silver nitrate needed, calculate the volume that would be required if the sample were pure sodium chloride.
3. Washings are readily tested for their silver content by collecting a few milliliters in a test tube and adding a few drops of hydrochloric acid. Washing is judged complete when little or no turbidity is observed with this test.
4. Crucibles containing silver chloride should be weighed within 30 to 60 min after having been placed in the desiccator for cooling.

ANALYSIS 1-3 Determination of Iron in a Soluble Sample

Discussion. The analysis of iron in a soluble sample is based upon the precipitation of iron(III) as a hydrous oxide, followed by ignition in Fe_2O_3:

$$2Fe^{3+} + 6NH_3 + (x+3)H_2O \rightarrow Fe_2O_3 \cdot xH_2O(s) + 6NH_4^+$$

$$Fe_2O_3 \cdot xH_2O(s) \rightarrow Fe_2O_3(s) + xH_2O(g)$$

The hydrous oxide forms as a gelatinous mass that rapidly clogs the pores of most filtering media; for this reason a very coarse grade of ashless paper is employed. Even here, however, it is best to delay the transfer of the precipitate as long as possible and to wash by decantation.

The use of Fe_2O_3 as a weighing form requires that all iron present be in the +3 state, a condition that is readily achieved by treating the sample with nitric acid before precipitating the hydrous oxide:

$$3Fe^{2+} + NO_3^- + 4H^+ \rightarrow 3Fe^{3+} + NO + 2H_2O$$

The complex $FeSO_4 \cdot NO$ occasionally imparts a very dark color to the solution; it decomposes upon further heating.

Because it has negligible solubility, the precipitate can be safely washed with a hot solution of ammonium nitrate.

Procedure. Clean and supply identifying markings to three porcelain crucibles and covers; bring these to a constant weight by igniting them at 900 to 1000°C during periods of waiting throughout the analysis.

Double-precipitation method. Unless directed otherwise, do not dry the unknown sample. Consult the instructor for the proper sample size to be weighed into 600-ml beakers (p. 545). Dissolve the sample in 20 to 30 ml of distilled water to which about 5 ml of concentrated hydrochloric acid have been added. Then add 1 to 2 ml of concentrated nitric acid; heat gently to complete the oxidation, and remove any oxides of nitrogen. Dilute to 350 or 400 ml, and slowly add, with good stirring, freshly filtered aqueous ammonia until precipitation is complete (Notes 1 and 2). Digest briefly and then check completeness of precipitation with a few additional drops of ammonia. Decant the warm, clear supernatant liquid through ashless filter paper (Note 3), and wash the precipitate in the beaker with two 30-ml portions of hot 1% ammonium nitrate solution (see page 550 for information concerning the filtration and washing of precipitates).

Return the filter papers to their appropriate beakers. Add about 5 ml of concentrated hydrochloric acid to each, warm, then macerate the paper thoroughly with a stirring rod. Dilute to about 300 ml with distilled water, and reprecipitate the hydrous iron(III) oxide as before. Again decant the filtrate through ashless filter paper, and wash the precipitate repeatedly with hot 1% ammonium nitrate. When the filtrate gives little or no test for chloride ion (Note 4), transfer the precipitate quantitatively to the filter cone. Remove the last traces of precipitate adhering to the walls of the beakers by scrubbing with a small piece of ashless paper. Allow the precipitate to drain overnight, if possible. Then transfer the paper and contents (p. 552) to a porcelain crucible that has previously been ignited to constant weight. Char the paper at low temperature, taking care to allow free access of air (Note 5). Gradually increase the temperature until all carbon has been burned away. Finally, ignite to constant weight at 900 to 1000°C.

NOTES

1. Aqueous ammonia solutions attack the glass of their containers upon prolonged contact and thereby become contaminated with particles of silica. It is a wise precaution to filter the ammonia prior to use.

2. Aqueous ammonia should be added until its odor is unmistakable over the solution. The precipitate should appear reddish brown. If it is black (or nearly so), the presence of iron(II) is indicated; the sample is best discarded.

3. A porous grade of ashless paper is required for this gelatinous precipitate. Schleicher and Schuell No. 589 Black Ribbon or Whatman No. 41 is satisfactory. The time required for filtration is shortened by performing the principal washing of the precipitate while it is still in the beaker.

4. Before making a test for chloride in the washings, the sample of filtrate taken must first be acidified with dilute nitric acid.

5. Heat lamps (p. 552) are convenient for the initial charring of the filter papers.

Single-precipitation method. Proceed as directed in the first paragraph of the double-precipitation method, making the following changes:

(1) Instead of washing with only two 30-ml portions of 1% ammonium nitrate, wash repeatedly until the decantate is essentially free of chloride. (2) Then transfer the precipitate quantitatively to the filter and complete the analysis as before.

ANALYSIS 1-4 Determination of Aluminum by Homogeneous Precipitation

Discussion. Aluminum may be determined by precipitation as a hydrous oxide followed by ignition to Al_2O_3. Iron(III), if present, reacts analogously and causes the results of the analysis to be high. An effective scheme for separating these cations is based upon differences in their behavior with respect to thiocyanate ion. Iron(III) forms stable anionic complexes in the presence of high concentrations of thiocyanate while aluminum(III) does not. After being treated with a relatively large quantity of thiocyanate, the sample is passed through a column containing an anion-exchange resin (Chapter 18). Aluminum ion passes through while iron(III) is retained on the column as a thiocyanate complex.

The accompanying directions were adapted from data reported by Gordon, Salutsky, and Willard.[1] After separation from iron(III), aluminum is precipitated as a basic succinate with hydroxide ions homogeneously generated by the decomposition of urea (Chapter 6).

Procedure. To receive the sample, submit a clean, labeled, 500-ml volumetric flask. Dilute the sample to volume.

Mark three crucibles and covers. Bring these to constant weight by igniting them at 1000°C.

Prepare 1 liter of 0.30-F NH_4SCN wash solution, adjusting this to pH 1.0 ± 0.2 with dilute HCl or NH_3 (Note 3). *Do not continue past this point unless there is time to complete the steps in the next two paragraphs during the same day.*

Transfer three 50-ml aliquots of the sample solution to individual 250-ml beakers. Add 5 g of NH_4SCN to each sample, and adjust the pH to 1.0 with dilute HCl or NH_3 as needed (Note 3).

Wash the resin in an anion-exchange column with 50 ml of 0.30-F NH_4SCN (Note 2). Adjust the flow rate to 8 to 10 ml/min (no faster than 1 drop/sec); discard this wash liquid. Then pass the aliquot, prepared as in the previous paragraph, through the column at the same rate, collecting the eluate in a 600-ml beaker. Wash the column with 250 ml of 0.30-F NH_4SCN solution, which should be added in increments of about 15 ml, and collect the eluate in the same beaker as the sample. *Never allow the solution level*

[1]L. Gordon, M. L. Salutsky, and H. H. Willard, *Precipitation from Homogeneous Solution.* New York: Wiley, 1959.

to fall below the top level of the resin. Treat the other two samples identically. When finished with the column, leave the solution level at least 3 cm above the resin level.

For each sample dissolve 5 g of succinic acid, 10 g of NH_4Cl, and 4 g of urea in a minimum volume of water; filter, if necessary, to remove any undissolved solids. Then add this solution to the sample, bring the volume to about 400 ml with distilled water, and adjust the pH to between 3.1 and 3.4 with dilute HCl or NH_3 (Notes 3 and 4).

Provide each beaker with a special stirring rod (Note 5), and heat the solution to a gentle boil over a burner. Locate the flame directly beneath the stirring rod so that a stream of bubbles emanates from the end of the rod. Use considerable care while heating because the solution may tend to bump when the precipitate forms. Boil the solution gently for 2 hr *after* the appearance of the first opalescence that develops 15 to 30 min after heating has commenced. Keep the solution level constant during boiling by the periodic addition of distilled water.

Prepare 1 liter of 1% succinic acid (w/v), adding sufficient NH_3 to make this solution just alkaline to methyl red. The solution should be heated almost to boiling before being used in washing the precipitate.

After precipitation is complete, decant the supernatant liquid through ashless paper suitable for the collection of fine, crystalline solids (Note 6). Transfer the precipitate quantitatively to the paper, and wash with hot 1% succinic acid until it is essentially free of chloride ion.

To remove the last traces of precipitate from beaker and stirring rod, add about 10 ml of dilute HCl, cover the beaker with a watch glass, and gently reflux for about 5 min. Add NH_3 until the solution is faintly alkaline to methyl red, warm for a few minutes, and then filter through a fresh cone of ashless paper. Wash with hot 1% succinic acid as before.

Place the combined precipitates for each sample into individual porcelain crucibles that have been brought, with their covers, to constant weight. Gently dry and then burn off the papers. Finally ignite them at 1000°C. Weigh the ignited Al_2O_3 in the covered crucible. Repeat the ignition step until constant weight has been attained (Note 7).

Report the number of milligrams of Al(III) contained in the original sample.

NOTES

1. A stock solution for unknowns may be prepared by dissolving 50.00 g of reagent-grade aluminum and 0.15 to 0.2 g of $FeCl_3$ in 6-F HCl and diluting to 1.000 liter. Unknowns, consisting of 16 to 25 ml of this solution, are issued from a buret into 500-ml volumetric flasks. The stock solution is somewhat viscous. Several minutes should be allowed for drainage between delivery of the sample and reading of the buret.

2. A 30-cm column of 20- to 50-mesh Amberlite IRA 400 is satisfactory for the removal of iron(III). The column will give satisfactory service until

it is completely discolored with the iron(III)-thiocyanate complex. The resin can be regenerated with 2-F HCl; a substantial volume of acid is required.

3. A pH meter is convenient for adjusting the acidity. When using this instrument, remember that (a) the glass electrode is very fragile and (b) a few drops adhere to the electrodes when they are removed from solution. These drops should be washed into the proper beaker with distilled water.

4. When adjusting samples to pH 3.1 and 3.4, add NH$_3$ slowly and with good stirring. Be sure the solution is free of precipitate when the proper pH has been achieved; if any traces remain, dissolve with dilute HCl, and readjust the pH.

   If a pH meter is not available, the pH can be adjusted by adding NH$_3$ until the first permanent traces of precipitate appear and then discharging this turbidity with a minimum volume of 1-F HCl.

5. In addition to a cover glass, each beaker should be supplied with a special stirring rod to aid in the prevention of bumping. The end of an ordinary stirring rod is heated to the softening point and then firmly pressed upon the point of a thumbtack. The indentation produces a stream of bubbles that significantly reduces the danger of loss due to local overheating.

6. Whatman No. 40 or Schleicher and Schuell No. 589 White Ribbon is satisfactory for this filtration. See page 553 for folding instructions. The filtering process is greatly hastened through use of hot solutions.

7. Ignite the precipitates in the full flame of a Meker burner (or in a furnace at 1000°C) for at least 40 min. Subsequent heatings need not exceed 20 to 30 min.

   After each heating allow the crucibles to cool on an asbestos mat for 3 to 4 min before transferring them to a desiccator. Allow a uniform cooling time (25 to 30 min) in the desiccator before each weighing.

   Crucible covers should be in place during all cooling and weighing operations.

2 Precipitation Titrations with Silver Ion

Argentometric methods of analysis are discussed in Chapter 8.

Specific directions follow for the argentometric determination of chloride ion by the Mohr, the Volhard, and the Fajans methods. These may be used with little or no modifications for the titration of other anions; see Table 8-2.

PREPARATION 2-1 Direct Preparation Standard 0.1-F Silver Nitrate

Discussion. Silver nitrate may be obtained in primary standard purity. It has a high equivalent weight and is readily soluble in water. Both the solid and the aqueous solution must be scrupulously protected from dust and other organic materials and from sunlight. Metallic silver is produced by chemical reduction in the former instance and by photodecomposition in the latter. The reagent is expensive.

Silver nitrate crystals may be freed of surface moisture by drying at 110°C for about 1 hr. Some discoloration of the solid may result, but the amount of decomposition occurring in this time is ordinarily negligible.

Procedure. Use a laboratory balance to weigh approximately 17 g of silver nitrate into a clean, dry weighing bottle. Heat the bottle and contents (see Figure 19-7) for 1 hr, but not much longer, at 110°C; store in a desiccator. When cooled, determine the weight of the bottle and contents (to the nearest milligram) with an analytical balance. Carefully transfer the bulk of the solid into a powder funnel placed in the neck of a 1-liter volumetric flask. Reweigh the bottle to obtain the weight of the solid by difference. Wash the silver nitrate into the volumetric flask with copious amounts of distilled water. Remove and rinse the funnel; fill the flask half-full, and swirl it until the crystals are completely dissolved. Dilute to the mark and *mix the solution thoroughly* by upending the stoppered flask several times. Compute the normality of the solution.

NOTES

1. Weighing silver nitrate to the nearest milligram will incur a maximum error of only 1 part in 17,000 in the value of the normality; because uncertainties in the subsequent analyses commonly exceed this figure, more accurate weighing is of no value.
2. If desired, silver nitrate solutions prepared to approximately the desired strength can be standardized against sodium chloride.
3. Once prepared, and when not actually in use, $AgNO_3$ solutions should be stored in a dark place.
4. Do not discard unused silver nitrate solutions before consulting with the instructor.

PREPARATION 2-2 Preparation and Standardization of 0.1-*F* Potassium Thiocyanate

Discussion. Standard solutions of KSCN as well as $AgNO_3$ are required for the Volhard procedure. The former is standardized against the latter (Preparation 2-1).

Procedure. Dissolve approximately 9.8 g of KSCN in about 1 liter of water. *Mix well.*

Measure 25- to 30-ml samples (to the nearest 0.01 ml) of standard $AgNO_3$ solution from a buret into conical flasks, and dilute to approximately 100 ml. Add about 5 ml of 6-F HNO_3, followed by 5 ml of iron(III) ammonium sulfate indicator (Note 3). Titrate with the KSCN solution, swirling the flask vigorously until the red-brown color of $FeSCN^{2+}$ is permanent for 1 min. Calculate the concentration of the KSCN solution. Results from duplicate standardizations should agree within 2 to 3 ppt; if this precision has not been attained, perform further standardization titrations.

NOTES

1. Potassium thiocynate is usually somewhat moist; the direct preparation of a standard solution is not ordinarily attempted. However, Kolthoff and Lingane[2] report that gentle fusion followed by storage over calcium chloride yields a product that can be used for the direct preparation of standard solutions.

2. A potassium thiocyanate solution retains its titer over extended periods of time.

3. The indicator is readily prepared by dissolving about 10 g of iron(III) ammonium sulfate, $NH_4Fe(SO_4)_2 \cdot 12H_2O$, in 100 ml of freshly boiled 6-F nitric acid. The acid prevents the formation of hydrous ferric oxide.

ANALYSIS 2-1   Determination of Chloride by the Mohr Method

Discussion.   The Mohr method is discussed on page 176.

Procedure.   Dry the unknown for 1 hr at 100 to 110°C. Carefully weigh 0.25- to 0.35-g samples (to the nearest 0.1 mg) into 250-ml flasks; dissolve each in about 100 ml of water. Add a pinch of calcium carbonate or sodium hydrogen carbonate, making further additions if necessary until effervescence ceases. Then introduce 1 to 2 ml of 5% potassium chromate, and titrate with standard silver nitrate solution to the first permanent appearance of a buff color due to silver chromate. Determine an indicator blank by suspending a small quantity of $CaCO_3$ in about 100 ml of water containing 1 to 2 ml of 5% K_2CrO_4; use the color developed in this determination to judge the end point in the actual titrations.

NOTE

Mohr titrations should be performed at room temperature. Elevated temperatures significantly increase the solubility of silver chromate; its sensitivity as an indicator for this titration undergoes a corresponding decrease.

[2] I. M. Kolthoff and J. J. Lingane, *J. Amer. Chem. Soc.*, **57**, 2126 (1935).

ANALYSIS 2-2 Determination of Chloride by the Volhard Method

Discussion. The Volhard method is discussed on page 177. As applied to the analysis of chloride ion, a measured excess of standard silver nitrate is added to the sample, following which the excess is back-titrated with a standard potassium thiocyanate solution.

As indicated in Table 8-2, the Volhard method may be employed for the determination of numerous substances. Where the solubility of the silver salt formed is increased by the presence of strong acid, a filtration is mandatory prior to back-titration. The use of nitrobenzene (or filtration) can be omitted if the salts produced are less soluble than silver thiocyanate in acid media; silver bromide and silver iodide are the only common examples.

The Volhard method cannot be employed in the presence of oxidizing agents because of the susceptibility of thiocyanate to attack. It is often possible to eliminate this source of interference by prior treatment of the sample with a reducing agent. The method also fails in the presence of cations that form slightly soluble thiocyanates, notably palladium and mercury. Again, preliminary treatment may eliminate these sources of interference.

Procedure. Dry the unknown at 100 to 110°C for 1 hr. Weigh several 0.25- to 0.35-g samples (to the nearest 0.1 mg) into 250-ml conical flasks. Dissolve each sample in 100 ml of distilled water, and acidify with 5 ml of 6-F HNO$_3$. Introduce an excess of standard silver nitrate, and be sure to note the volume taken. Add 5 ml of iron(III) indicator and 5 ml of chloride-free nitrobenzene (Note 4). Shake vigorously. Titrate the excess silver with standard thiocyanate until the color of FeSCN^{2+} is permanent for 1 min.

NOTES
1. With the concurrence of the instructor a larger quantity of the unknown can be weighed into a volumetric flask and diluted to a known volume. The determination can then be made upon aliquot portions of this solution.
2. To approximate the volume of standard AgNO$_3$ that constitutes an excess, calculate the amount that would be required for one of the samples, assuming that it is 100% NaCl. When actually adding the silver solution to the sample, swirl the flask vigorously, and add 3 or 4 ml in excess of the volume required to cause the AgCl to coagulate.
3. Nitric acid is introduced to improve observation of the end point. Because the lower oxides of nitrogen tend to attack thiocyanate, the acid should be freshly boiled.
4. Nitrobenzene is a hazardous chemical and must be handled with respect. Poisoning can result not only from prolonged breathing of its vapors but also from absorption of the liquid through the skin. In the

event of spillage on one's person, the affected areas should be promptly and thoroughly washed with soap and warm water. Clothing soaked with the liquid should be removed immediately and laundered.

The instructor may wish to dispense with the use of nitrobenzene. In this case the solution should be heated to coagulate the AgCl after the addition of excess standard $AgNO_3$. The precipitate is filtered and washed a few times with dilute nitric acid; the filtrate and washings are then titrated with the KSCN solution after the addition of indicator.

5. At the outset of the back-titration, an appreciable quantity of silver ion is adsorbed on the surface of the precipitate. As a result, there is a tendency for a premature appearance of the end-point color. Because success of the method depends upon an accounting for all excess silver ions, thorough and vigorous agitation is essential to bring about desorption of this ion from the precipitate. A magnetic stirrer is helpful for this purpose.

ANALYSIS 2-3 Determination of Chloride by the Fajans Method

Discussion. The Fajans method is discussed on page 179. This analysis employs a direct titration with dichlorofluorescein as the indicator; dextrin is added to maintain the silver chloride in the colloidal state. Silver nitrate is the only standard solution that is required.

Procedure. Dry the unknown for 1 hr at 100 to 110°C. Weigh individual 0.20- to 0.29-g (Note 1) samples (to the nearest 0.1 mg) into 500-ml flasks. Dissolve each in 175 to 200 ml of distilled water. Introduce 10 drops of dichlorofluorescein solution (Note 3) and about 0.1 g of dextrin; immediately titrate with standard silver nitrate to the first permanent appearance of the pink color of silver dichlorofluorescinate (Note 2).

NOTES
1. Kolthoff[3] states that the chloride concentration should be within the range of 0.025 to 0.005 M. These directions yield solutions approaching the maximum concentration only when the sample taken is pure sodium chloride at the upper weight limit. If the approximate percentage of chloride in the sample is known, a corresponding decrease in volume or increase in sample size can be tolerated.
2. Silver chloride is particularly sensitive to photodecomposition in the presence of the indicator; the titration will fail if attempted in direct sunlight. Where this problem exists, the approximate equivalence point should first be ascertained by a trial titration, this value being used to calculate the volume of silver nitrate required for the other samples.

[3] I. M. Kolthoff, W. M. Lauer, and C. J. Sunde, *J. Amer. Chem. Soc.*, **51**, 3273 (1929).

The addition of the indicator and dextrin should be delayed until the bulk of the silver nitrate has been added to subsequent samples, after which the titration should be completed without delay.

3. The indicator may be prepared as a 0.1% solution of dichlorofluorescein in 70% alcohol or as a 0.1% aqueous solution of the sodium salt.

3 Neutralization Titrations

Neutralization titrations are performed with standard solutions of strong acids or strong bases. In principle a single solution (acid or base) should suffice; in practice, however, it is convenient to have both a standard acid and a standard base available to locate end points more exactly. One solution is standardized against a primary standard; the normality of the other is then found by determining the acid-base ratio (that is, the volume of acid required to neutralize 1.000 ml of base).

The instructions that follow include methods for the standardization of both an acid or a base against appropriate primary standards. Directions for determining the acid-base ratio are also provided.

PREPARATION 3-1 Preparation of Indicator Solutions

Discussion. Acid-base indicators exist for virtually any pH range between 1 and 13.[4] The theory of indicator behavior is discussed in Chapter 9. Directions follow for the preparation of indicator solutions that will permit the performance of most common analyses.

Procedure. Stock solutions of indicators generally contain 0.5 to 1.0 g of indicator per liter.

Methyl orange and methyl red. Dissolve the sodium salt directly in distilled water.

Phenolphthalein and thymolphthalein. Dissolve the solid indicator in a solution that is 80% ethyl alcohol by volume.

Sulfonphthaleins. Dissolve the sulfonphthaleins in water by adding sufficient NaOH to react with the sulfonic acid group of the indicator. To prepare stock solutions, triturate 100 mg of the solid indicator with the specified volume of 0.1-N NaOH; dilute to 100 ml with distilled water. The volumes of base in milliliters required are as follows: *bromocresol green,* 1.45; *bromothymol blue,* 1.6; *bromophenol blue,* 1.5; *thymol blue,* 2.15; *cresol red,* 2.65; *phenol red,* 2.85.

The sodium salts of several sulfonphthaleins are available commercially. These substances can be dissolved directly in water.

[4]See, for example, L. Meites, *Handbook of Analytical Chemistry.* New York: McGraw-Hill, 1963, p. 3-35.

PREPARATION 3-2   Preparation of 0.1-*N* Hydrochloric Acid

Discussion.   See page 228.

Procedure.   Add about 8 ml of concentrated HCl to approximately 1 liter of distilled water. Mix thoroughly and store in a glass-stoppered bottle.

> **NOTE**
> For very dilute HCl solutions it is advisable to eliminate CO_2 from the water by a preliminary boiling.

PREPARATION 3-3   Preparation of Carbonate-Free 0.1-*N* Sodium Hydroxide

Discussion.   See pages 229 to 232.

Procedure.   If directed by the instructor, prepare a bottle for protected storage as in Figure 10-1. Boil approximately 1 liter of distilled water; after it has cooled, transfer the water to the storage bottle. Decant 4 to 5 ml of commercial 50% NaOH, add to the water, and *mix thoroughly*. Protect the solution from unnecessary contact with the atmosphere.

STANDARDIZATION 3-1   Determination of the Acid-Base Ratio

Discussion.   See page 229.

Procedure.   If both acid and base solutions have been prepared, it is convenient to determine the volume ratio between them before the standardization of one.

Rinse the burets with three or four portions of the solutions they are to contain. Then fill each buret, and remove the air bubble in the tip by opening the stopcock briefly. Cover the top of the buret that contains the base with a test tube. Record the initial buret readings. Deliver a 35- to 40-ml portion of the acid into a 250-ml conical flask. Touch the tip of the buret to the inside wall of the flask, and rinse down with a little distilled water. Add 2 drops of phenolphthalein, and then introduce NaOH until the solution is definitely pink. Now add HCl dropwise until the solution is again colorless, rinse the inside surface of the flask with water, and again add NaOH until the solution acquires a light pink hue that persists for 30 sec. Drops that are smaller than normal can be delivered by touching the buret tip to the wall of the flask and rinsing down with water. Record the final buret readings. Repeat the experiment, and calculate the volume ratio between the acid and the base. Duplicate titrations should yield values that lie within 1 to 2 ppt of the mean.

NOTES

1. The end point slowly fades as CO_2 is absorbed from the atmosphere.
2. The volume ratio can also be obtained with an indicator that has an acidic transition range such as bromocresol green. With this particular indicator, titrate the solution to the faintest tinge of green. If appreciable carbonate is present, the ratios obtained with phenolphthalein will differ significantly from those with bromocresol green.
3. Use the acid-base ratio and the normality of the solution that is standardized to calculate the concentration of the other.

STANDARDIZATION 3-2 Standardization of Hydrochloric Acid against Sodium Carbonate

Discussion. See page 228.

Procedure. Dry a quantity of primary standard sodium carbonate for 2 hr at 110°C, and cool in a desiccator. Weigh 0.2- to 0.25-g samples (to the nearest 0.1 mg) into 250-ml conical flasks, and dissolve in about 50 ml of distilled water. Introduce 3 drops of bromocresol green, and titrate until the solution just changes from blue to green. Boil the solution for 2 to 3 min, cool to room temperature, and complete the titration. If a dilute base solution has also been prepared, the end point can be established with more certainty by adding small increments of the base and acid until a minimal volume of acid causes the color to change from blue to green. The volume of base used must, of course, be recorded; this volume is multiplied by the acid-base ratio and then subtracted from the total amount of acid to establish the volume needed for the titration of the sodium carbonate.

Determine an indicator correction by titrating approximately 100 ml of 0.05-F sodium chloride and 3 drops of indicator. Subtract the volume of the blank from the titration data.

NOTE

The indicator should change from green to blue as a result of the heating step. If it does not, an excess of acid was added originally. This excess can be back-titrated with base, provided its combining ratio with the acid (p. 584) has been established; otherwise the sample must be discarded.

STANDARDIZATION 3-3 Standardization of Sodium Hydroxide against Potassium Hydrogen Phthalate

Discussion. See page 232.

Procedure. Dry a quantity of primary standard potassium hydrogen phthalate for 2 hr at 110°C and cool in a desiccator.

Weigh 0.7- to 0.9-g samples (to the nearest 0.2 mg) into 250-ml conical flasks, and dissolve in 50 to 75 ml of freshly boiled and cooled distilled water. Add 2 drops of phenolphthalein, and titrate with base until the pink color of the indicator persists for 30 sec. If an HCl solution has also been prepared, the end point can be established with more certainty by adding small increments of acid and base until a minimal volume of NaOH causes development of the pink color. The volume of acid must, of course, be recorded; a correction to the volume of base can then be calculated from the acid-base ratio (p. 584).

(handwritten: ① Exp # 3-1 p 586.)

ANALYSIS 3-1 Determination of Potassium Hydrogen Phthalate *(handwritten: (KHP))*

Discussion. The sample consists of potassium hydrogen phthalate, $KHC_8H_4O_4$, mixed with neutral, soluble salts.

Procedure. Dry the unknown for 2 hr at 110°C. Weigh samples that will require 30- to 40-ml titrations into 250-ml conical flasks, and dissolve in 50 to 75 ml of freshly boiled and cooled distilled water. Introduce 2 drops of phenolphthalein, and titrate with standard 0.1-N NaOH. Consider the first pink that persists for 30 sec as the end point. A standard acid solution, if available, may be employed to establish the end point more exactly.

ANALYSIS 3-2 Determination of the Equivalent Weight of a Weak Acid

Discussion. The equivalent weight of an acid is an aid in establishing its identity (p. 240). This quantity is readily determined by titrating a weighed quantity of the purified acid with standard sodium hydroxide.

Procedure. Weigh 0.3-g samples (to the nearest 0.1 mg) of the purified acid into 250-ml conical flasks, and dissolve in 50 to 75 ml of freshly boiled water (Note). Add 2 drops of phenolphthalein, and titrate with standard base to the first persistent pink color. Calculate the equivalent weight of the acid, assuming 100.0% purity for the sample.

> **NOTE**
> Acids that have limited solubility in water can be dissolved in ethanol or an ethanol-water mixture. As received, the alcohol may be measurably acidic and should be rendered faintly alkaline to phenolphthalein before being used as a solvent. As an alternative, a sparingly soluble acid can be dissolved in a known volume of standard base, the excess being determined by back-titration with standard acid.

ANALYSIS 3-3 Determination of the Acid Content of Vinegar

Discussion. The total acid content of vinegar is conveniently determined by titration with standard base. Even though other acids are also present, the results of the analysis are customarily reported in terms of acetic acid, the principal acidic constituent. Vinegars assay approximately 5% acid, expressed as acetic acid.

Procedure. Pipet 25 ml of vinegar into a 250-ml volumetric flask, and dilute to the mark with boiled and cooled distilled water. Mix thoroughly and pipet 50-ml aliquots into 250-ml flasks. Add 50 ml of water, 2 drops of phenolphthalein, and titrate with standard 0.1-N NaOH to the first permanent pink color.

Calculate the total acidity as grams of acetic acid per 100 ml of sample.

ANALYSIS 3-4 Determination of Sodium Carbonate in an Impure Sample

Discussion. The titration of sodium carbonate is discussed in connection with its use as a primary standard (p. 228). This analysis is conveniently performed concurrently with the acid standardization.

Procedure. Dry the unknown for 2 hr at 110°C, and then cool in a desiccator. Weigh samples that will require 30- to 40-ml titrations (see instructor) into 250-ml flasks. Dissolve in 50 to 75 ml of boiled water, and add 2 drops of bromocresol green. Titrate with acid to a color change from blue to green. Boil the solution for 2 to 3 min, cool, and complete the titration. If additional acid is not required after boiling, either discard the sample or back-titrate the excess acid with standard base.

Calculate the percentage of sodium carbonate in the sample.

ANALYSIS 3-5 Determination of Sodium Carbonate and Sodium Hydrogen Carbonate in a Mixture

Discussion. See pages 237 to 239.

Procedure. If the sample is a solid, weigh dried portions into 250-ml volumetric flasks, dissolve in 50 to 75 ml of water, and dilute to the mark with boiled and cooled water.

If the sample is a solution, pipet suitable aliquots into 250-ml volumetric flasks, and dilute to volume. Mix thoroughly.

To determine the total number of milliequivalents of the two components, transfer 25-ml aliquots to 250-ml conical flasks, and titrate with standard 0.1-N HCl, following the directions given for Analysis 3-4.

To determine the bicarbonate content pipet additional 25-ml portions into flasks; *treat each aliquot individually from here.* Add a carefully measured excess of standard 0.1-N NaOH to an aliquot (conveniently 50.00 ml). Immediately add 10 ml of 10% $BaCl_2$ and 2 drops of phenolphthalein. Titrate the excess NaOH at once with standard 0.1-N HCl to the disappearance of the pink color.

Titrate a blank consisting of 25 ml of water, 10 ml of the $BaCl_2$ solution, and *exactly* the same volume of NaOH as used with the samples. The difference between the volume of HCl needed for the blank and the sample corresponds to the $NaHCO_3$ present.

Calculate the percentage of $NaHCO_3$ and of Na_2CO_3 in the sample.

ANALYSIS 3-6 Determination of Sodium Carbonate and Sodium Hydroxide in a Mixture

Discussion. See pages 237 to 239.

Procedure. Transfer the sample to a 250-ml volumetric flask immediately upon receipt, and dilute to the mark with boiled and cooled distilled water. Mix thoroughly. Keep the flask tightly stoppered to avoid absorption of CO_2.

Pipet 25-ml aliquots of the sample solution into 250-ml conical flasks, add 2 drops of bromocresol green, and titrate with 0.1-N HCl, following the procedure for Analysis 3-4. This titration will give the total number of milliequivalents of NaOH and Na_2CO_3.

To determine the quantity of NaOH present transfer a 25-ml aliquot of the sample solution into a 250-ml conical flask. Slowly add 10 ml of a neutral 10% $BaCl_2$ solution and 2 drops of phenolphthalein. Titrate immediately with standard 0.1-N HCl. To avoid contamination of the sample by atmospheric CO_2, complete this part of the analysis as rapidly as possible. Repeat with additional aliquots of the sample.

Calculate the weight of NaOH and Na_2CO_3 in the original sample.

ANALYSIS 3-7 Determination of Ammonium Ion

Discussion. A method for the isolation of ammonium ion as ammonia by distillation is discussed on page 234 in connection with the Kjeldahl method.

Procedure. Transfer the sample, which should contain 2 to 4 meq of ammonium salt, to a 500-ml Kjeldahl flask, and add enough water to give a total volume of about 200 ml.

Arrange a distillation apparatus similar to that shown in Figure 10-2. Use a buret or pipet to measure precisely 50.00 ml of standard 0.1-N HCl into the receiver flask. Clamp the flask so that the tip of the adapter extends just below the surface of the standard acid. Circulate water through the jacket of the condenser.

For each sample prepare a solution containing approximately 45 g of NaOH in about 75 ml of water. Cool this solution to room temperature before use. With the Kjeldahl flask tilted, slowly pour the caustic down the side of the container to minimize mixing with the solution in the flask (Note 1). Add several pieces of granulated zinc (Note 2) and a small piece of litmus paper. *Immediately* connect the flask to the spray trap. Very *cautiously* mix the solution by gentle swirling. After mixing is complete, the litmus paper should indicate that the solution is basic.

Immediately bring the solution to a boil, and distill at a steady rate until one-half to one-third of the original solution remains. Control the rate of heating during this period to prevent the receiver acid from being drawn back into the distillation flask. After the distillation is judged complete, lower the receiver flask until the tip of the adapter is well clear of the standard acid. Then discontinue heating, disconnect the apparatus, and rinse the inside of the condenser with a small amount of water. Disconnect the adapter, and rinse it thoroughly. Add 2 drops of bromocresol green or methyl red, and titrate the distillate with standard 0.1-N NaOH to the color change of the indicator.

A modification of this procedure makes use of about 50 ml of 4% boric acid solution in place of the standard HCl in the receiver flask (p. 236). The distillation is then carried out in an identical fashion. When complete, the ammonium borate produced is titrated with a standard 0.1-N HCl solution, using 2 to 3 drops of bromocresol green as indicator.

Calculate the percent $(NH_4)_2SO_4$ in the sample.

NOTES
1. The dense caustic solution should form a second layer on the bottom of the flask. Mixing must be avoided at this point in order to prevent loss of the volatile ammonia. Manipulations should be carried out as rapidly as possible.
2. Granulated zinc is added to minimize bumping during the distillation. It reacts slowly with the alkali to give small bubbles of hydrogen that minimize superheating of the liquid.

ANALYSIS 3-8 Determination of Nitrogen in an Organic Mixture; the Kjeldahl Method

Discussion. These directions are suitable for the analysis of protein in samples such as blood meal, wheat flour, macaroni, dry cereals, and pet foods. Prereduction is not required for such materials. A simple modification will permit analysis of samples that contain more highly oxidized forms of nitrogen.[5] The Kjeldahl analysis is discussed on page 234.

[5]See *Official Methods of Analysis*, 11th ed. Washington, D.C.: Association of Official Analytical Chemists, 1975, p. 15.

Procedure. Weigh three samples (see instructor for size) onto individual 9-cm filter papers; drop each into a 500-ml Kjeldahl flask (the paper wrapping will prevent the sample from clinging to the neck of the flask). Add 25 ml of concentrated sulfuric acid and 10 g of powdered K_2SO_4. Add catalyst (Note 1) and clamp the flask in an inclined position in a hood. Heat the mixture carefully until the H_2SO_4 is boiling; discontinue heating briefly if foaming becomes excessive. Continue the heating until the solution becomes colorless or light yellow; this may take as long as 2 to 3 hr. If necessary, *cautiously* replace the acid lost by evaporation.

Remove the flame, and allow the flask to cool; swirl the liquid if it begins to solidify. Cautiously dilute with 250 ml of water, and cool to room temperature under a water tap. If mercury was used as the catalyst, introduce 25 ml of a 4% sodium sulfide solution. Then complete the analysis as described in the third and fourth paragraphs of the procedure for Analysis 3-7.

Calculate the percent nitrogen in the samples.

NOTES

1. As a catalyst, one may use any of the following: a drop of mercury, 0.5 g of HgO, a crystal of $CuSO_4$, 0.1 g of Se, or 0.2 g $CuSeO_3$. Alternatively, the catalyst may be omitted.
2. It is recommended that a blank, which includes the filter paper, be carried through all steps of the analysis.

4 Complex Formation Titrations

PREPARATION 4-1 Indicators

Procedure. *Eriochrome Black T.* Dissolve 200 mg of the solid in a solution consisting of 15 ml of triethanolamine and 5 ml of absolute ethanol. Solutions should be freshly prepared every two weeks; refrigeration slows their deterioration.

Calmagite. Prepare a 0.1% (w/v) aqueous solution of the indicator.

PREPARATION 4-2 Direct Preparation of Standard 0.01-F EDTA

Procedure. Dry the purified dihydrate ($Na_2H_2Y \cdot 2H_2O$) at 80°C to remove superficial moisture. After cooling, weigh about 3.8 g (to the nearest milligram) into a 1-liter volumetric flask, using a powder funnel; dilute to the mark with distilled water.

NOTES

1. W. J. Blaedel and H. T. Knight [*Anal. Chem.*, **26**, 741 (1954)] give specific instructions for the purification of commercial preparations of the disodium salt.

2. The direct preparation of standard solutions requires the total exclusion of polyvalent cations. If any doubt exists regarding the quality of the distilled water, pretreatment by passing it through a cation-exchange resin is recommended.

3. If desired, the anhydrous salt can be employed instead of the dihydrate. The weight taken should be adjusted accordingly.

4. If desired, an EDTA solution that is approximately $0.01\ F$ can be prepared and standardized against a Mg^{2+} solution of known strength.

PREPARATION 4-3 Preparation of Standard $0.10\text{-}F$ Magnesium Complex of EDTA

Discussion. A standard solution of the magnesium complex of EDTA is required for the displacement titration procedure described on page 592.

Procedure. To 3.722 g of $Na_2H_2Y \cdot 2H_2O$ in 50 ml of distilled water add an equivalent quantity (2.465 g) of $MgSO_4 \cdot 7H_2O$. Introduce a few drops of phenolphthalein followed by sufficient sodium hydroxide to turn the solution faintly pink. Dilute the solution to 100 ml. When properly prepared, portions of this solution should assume a dull violet color when treated with pH-10 buffer and a few drops of Eriochrome black T (Erio T) or Calmagite indicator. Furthermore, a single drop of $0.01\text{-}F\ Na_2H_2Y$ should cause a color change to blue, whereas an equal quantity of $0.01\text{-}F\ Mg^{2+}$ should cause a change to red. The composition of the solution should be adjusted by adding Mg^{2+} or Na_2H_2Y until these criteria are met.

ANALYSIS 4-1 Determination of Magnesium by Direct Titration

Procedure. The sample will be issued as an aqueous solution; transfer it to a clean 500-ml volumetric flask, and dilute to the mark. Mix thoroughly. Take 50.00-ml aliquots, adding 1 to 2 ml of pH-10 buffer (Note 1) and 2 to 4 drops of Erio T or Calmagite indicator solution to each. Titrate with $0.01\text{-}F\ Na_2H_2Y$ until a color change from red to pure blue occurs. Express the results of the analysis in terms of milligrams of Mg^{2+} per liter of the diluted solution.

NOTES

1. A pH-10.0 buffer is readily prepared by diluting 570 ml of aqueous NH_3 (sp gr 0.90) and 70 g NH_4Cl to about 1 liter.

2. The color change of the indicator is slow in the vicinity of the end point. Care must be taken to avoid overtitration.

3. Other alkaline earths, if present, will also be titrated along with magnesium and should be removed prior to the analysis; $(NH_4)_2CO_3$ is a suitable reagent. Most polyvalent cations also interfere and should be precipitated as hydroxides.

4. Because this reaction may be expressed as

$$Mg^{2+} + H_2Y^{2-} \rightleftharpoons MgY^{2-} + 2H^+$$

the normality of the chelating solution is twice its formal concentration. For a 40-ml titration each aliquot should contain about 10 mg of magnesium ion.

ANALYSIS 4-2   Determination of Calcium by Displacement Titration

Discussion.   See page 262.

Procedure.   Weigh the sample (Note 1) into a 500-ml volumetric flask, and dissolve in a minimum quantity of dilute HCl. Neutralize the solution with sodium hydroxide (Note 2), and dilute to the mark. Take 50.00-ml aliquots for titration, treating each as follows: add approximately 2 ml of pH-10 buffer, 1 ml of the magnesium chelate solution, and 2 to 4 drops of Erio T or Calmagite indicator. Titrate with standard 0.01-F Na_2H_2Y until a color change from red to blue occurs. Report the percentage of calcium oxide in the sample.

NOTES

1. The sample taken should contain about 150 to 160 mg of Ca^{2+}.

2. To neutralize the solution introduce a few drops of methyl red, and add base until the red color is discharged.

3. The amount of MgY^{2-} solution added is not critical. Its presence is required because the indicator does not exhibit a satisfactory color change when Ca^{2+} is present alone. Because the calcium chelate is more stable, the reaction

$$Ca^{2+} + MgY^{2-} \rightleftharpoons CaY^{2-} + Mg^{2+}$$

takes place. The introduction of titrant leads to the preferential formation of the calcium chelate for the same reason. At the equivalence point the reaction is actually between H_2Y^{2-} and Mg^{2+}, for which Erio T is an excellent indicator.

4. Interferences with this method are substantially the same as with the direct determination of magnesium and are eliminated in the same way.

ANALYSIS 4-3   Determination of Calcium by Back-Titration

Procedure.   Prepare the sample as directed for the substitution analysis of calcium. To each 50.00-ml aliquot add about 2 ml of pH-10 buffer and 2 to 4 drops of Erio T or Calmagite indicator. Run in an excess of 0.01-F Na_2H_2Y solution from a buret, and record the volume taken. Titrate the excess chelating agent with standard 0.01-F $MgSO_4$ solution until a color change from blue to red occurs.

> **NOTE**
>
> The magnesium sulfate solution is conveniently prepared by dissolving about 2.5 g of the heptahydrate in 1 liter of distilled water. The equivalence of this solution with respect to the Na_2H_2Y solution should be determined by a direct titration.

ANALYSIS 4-4   Determination of Water Hardness

Discussion.   See page 263.

Procedure.   Acidify 100-ml aliquots with a few drops of HCl, and boil gently for a few minutes to remove CO_2. Cool, add a few drops of methyl red, and neutralize the solution with NaOH. Introduce 2 ml of pH-10 buffer, 2 to 4 drops of Erio T or Calmagite indicator, and titrate with standard 0.01-F Na_2H_2Y until the color changes from red to pure blue. Report the results of the analysis in terms of milligrams of $CaCO_3$ per liter of water.

> **NOTE**
>
> If the color change of the indicator is sluggish, the absence of magnesium is indicated. In this event add 1 to 2 ml of standard 0.1-F MgY^{2-} solution.

5 Titrations with Potassium Permanganate

This section contains methods for the analysis of iron and calcium by titration with a standard solution of potassium permanganate. The former requires that iron(III) be converted completely to iron(II) prior to titration. Two procedures are given; one involves the use of a Jones reductor and the other calls for the use of a tin(II) solution.

The Jones reductor provides a general means of reducing several ions to their lower oxidation state before titration. A detailed description of its preparation is therefore given at the outset.

PREPARATION 5-1   Preparation and Use of a Jones Reductor

Discussion. See pages 314 to 315.

Procedure. The reductor (Figure 13-1) should be thoroughly cleaned and fitted with a porcelain disk to support a mat of glass wool or asbestos at its lower end. This mat must be sufficiently thick to prevent the passage of zinc granules when the reductor is in use.

The zinc should be 20 to 30 mesh and free of impurities such as iron. Cover a suitable quantity of the metal with 1-F HCl for about 1 min. Decant the liquid, and cover the zinc with a 0.25-F solution of $Hg(NO_3)_2$ or $HgCl_2$. Stir the mixture vigorously for 3 min, decant, and wash two or three times with distilled water. Fill the reductor tube with water, and slowly add the amalgam until a column of about 30 cm is achieved. Wash with 500 ml of water, being sure that the packing is always covered by liquid. Keep the reductor filled with water during storage.

NOTE

Solutions that are 0.5 to 5 N in HCl or H_2SO_4 may be used in the reductor; warm solutions do not harm the packing.

PREPARATION 5-2 Preparation of Approximately 0.1-N Potassium Permanganate

Discussion. The precautions that must be observed in the preparation and storage of permanganate solutions are discussed on pages 317 to 320.

Procedure. Dissolve 3.2 g of $KMnO_4$ in about 1 liter of distilled water. Heat to boiling, and keep hot for about 1 hr. Cover and let stand overnight. Filter the solution through a fine-porosity sintered glass funnel or crucible (Note) or through a Gooch crucible with an asbestos mat (p. 547). Store the solution in a clean, glass-stoppered bottle, and keep in the dark when not in use.

NOTE

After filtration the fritted plate can be cleaned with dilute H_2SO_4 containing a few milliliters of 3% H_2O_2.

STANDARDIZATION 5-1 Standardization of Potassium Permanganate against Sodium Oxalate

Discussion. Directions are provided for standardization by the McBride method and the method of Fowler and Bright; see page 320 for a discussion of these methods as well as other primary standards for permanganate solutions.

Procedure. *Method of McBride.* Weigh 0.2- to 0.3-g samples (to the nearest 0.1 mg) of dried $Na_2C_2O_4$ into 400-ml beakers, and dissolve in approximately 250 ml of 0.90-F H_2SO_4. Heat to 80 to 90°C and titrate with the $KMnO_4$, stirring vigorously with a thermometer. The reagent should be introduced slowly so that the pink color is discharged before further additions are made (Note 4). If the solution temperature drops below 60°C, reheat. The end point is the first persistent pink color. Determine an end point blank by titrating an equal volume of the water and sulfuric acid.

Method of Fowler and Bright. Dry primary standard-grade $Na_2C_2O_4$ for 1 hr at 110 to 120°C. Cool in a desiccator, and weigh (to the nearest 0.1 mg) suitable portions (0.2 to 0.3 g for 0.1-N $KMnO_4$) into 400-ml beakers. Add 250 ml of 0.90-F sulfuric acid that has been boiled for 10 to 15 min and cooled to room temperature. Stir until dissolved. A thermometer is convenient for this purpose because the temperature must be measured later. Introduce from a buret sufficient permanganate to consume 90 to 95% of the oxalate (about 40 ml of 0.1-N $KMnO_4$ for a 0.3-g sample; a preliminary titration by the McBride method will provide the approximate volume required). Let stand until the solution is decolorized. Warm to 55 to 60°C, and complete the titration, taking the first pale pink color that persists for 30 sec as the end point. Determine an end point correction by titrating 250 ml of 0.90-F sulfuric acid at this same temperature. Correct for the blank, and calculate the normality.

NOTES

1. To measure the volume of $KMnO_4$ take the surface of the liquid as a point of reference. Alternatively, provide sufficient backlighting with a flashlight or match to permit reading of the meniscus in the conventional manner.

2. Permanganate solutions should not be allowed to stand in burets any longer than necessary because decomposition to MnO_2 may occur. Freshly formed MnO_2 can be removed from burets and glassware with a solution of 1-F H_2SO_4 containing a small amount of 3% H_2O_2.

3. Any $KMnO_4$ spattered on the sides of the titration vessel should be washed down immediately with a stream of water.

4. If the addition of $KMnO_4$ is too rapid, some MnO_2 will be produced in addition to Mn^{2+}; evidence for this is a faint brown discoloration of the solution. This is not a serious problem so long as sufficient oxalate remains to reduce the MnO_2 to Mn^{2+}; the titration is temporarily discontinued until the solution clears. The solution must be free of MnO_2 at the equivalence point.

ANALYSIS 5-1 Determination of Iron in an Ore after Prereduction with the Jones Reductor

Procedure. *Sample preparation.* Dry the unknown and weigh portions of suitable size into 250-ml beakers. Add 10 ml of concentrated HCl, and heat until the sample is decomposed. If solution of the sample is incomplete, filtration and fusion of the dark residue with Na_2CO_3 (Note 1) may be necessary.

Heat a blank consisting of 10 ml of HCl for the same time as the samples.

Evaporation of HCl. Add 15.0 ml of 1:1 H_2SO_4 to the dissolved samples and blank, and heat until fumes of SO_3 are observed (Hood). Cool and then slowly add 100 ml of water. Swirl the solution to dissolve any iron salts that may have been deposited on the side of the beaker. The first few milliliters of water should be added a drop at a time down the side of the beaker.

Reduction of iron. Drain the solution from the Jones reductor to within 3 cm of the top of the packing (Note). Pass 200 ml of 1-F H_2SO_4 through the column, again draining to about 3 cm from the top of the zinc. Disconnect the receiver, discard its contents, and rinse it with distilled water. Again attach the receiver, and pass 50 ml of 1-F H_2SO_4, followed by the sample or blank, through the reductor at a rate of about 50 to 100 ml/min. Wash the beaker with five 10-ml portions of 1-F H_2SO_4, passing each through the reductor. Finally, pass 100 ml of water through the column. Leave the reductor filled with water. Throughout the entire process do not permit the amalgam to become exposed to the atmosphere.

Titration. Disconnect the receiver at once, rinse the tip of the reductor into the flask, and add 5 ml of 85% H_3PO_4. Titrate with standard 0.10-N $KMnO_4$.

The blank should be treated in the same way as the sample, the volume of $KMnO_4$ consumed being subtracted from that required for the sample.

NOTES

1. If dark particles persist after treatment with acid, filter the solution through ashless paper, wash with 5 to 10 ml of 6-F HCl, and retain the filtrate and washings. Place the paper in a small platinum crucible and ignite. Mix 0.5 to 0.7 g of finely ground anhydrous Na_2CO_3 with the residue, and heat until a clear liquid melt is obtained. Cool, add 5 ml of water, followed by the cautious addition of an equal volume of 6-F HCl. Warm the crucible and combine the contents with the original filtrate. Evaporate the solution to about 15 ml, and proceed with step labeled "Evaporation of HCl."

2. A Jones reductor that has not been in constant use may be partially clogged with various oxides. The first acid passed through may consume some oxidant. Therefore test the reductor before use by passing about 200 ml of 1-F H_2SO_4 and 100 ml of water. The emergent solution should

require no more than 0.6 ml of 0.1-N KMnO₄. If necessary, repeat the
operation until a blank of the required size is obtained.

ANALYSIS 5-2 Determination of Iron in an Ore after Prereduction with
Tin(II)

Discussion. The chemistry of the tin(II) reduction of iron(III) is discussed on page 323.

Procedure. *Special solutions for the reduction of iron.* Approximately
100 titrations can be performed with the following solutions:

1. Tin(II) chloride, 0.25 F. Dissolve 60 g of iron-free $SnCl_2 \cdot 2H_2O$ in
   100 ml of concentrated HCl; warm if necessary. After solution is
   complete, dilute to about 1 liter, and store in a well-stoppered bottle. A
   few pieces of mossy Sn in the bottle will prevent air oxidation of the
   Sn(II).
2. Mercury(II) chloride. Dissolve 50 g of $HgCl_2$ in about 1 liter of water.
3. Zimmermann-Reinhardt reagent. Dissolve 210 g of $MnSO_4 \cdot 4H_2O$ in
   about 1 liter of water. Cautiously add 375 ml of concentrated H_2SO_4
   and 375 ml of 85% phosphoric acid. Dilute to about 3 liters.

Sample preparation. Dry the ore at 105 to 110°C, and weigh
individual samples into 500-ml conical flasks. A sample of optimum size
will require 25 to 40 ml of the standard KMnO₄. Add 10 ml of concentrated
HCl, 3 ml of SnCl₂ solution, and heat at just below boiling until the sample
is decomposed as indicated by the disappearance of all of the dark particles
(see Note 1 of the preceding directions). A pure white residue may remain.
A blank consisting of 10 ml of HCl and 3 ml of SnCl₂ should be heated for
the same length of time. If the solutions become yellow during the heating,
add another milliliter or two of SnCl₂. After the decomposition is complete,
add approximately 0.2-F KMnO₄ dropwise until the solution is just yellow.
Dilute to about 15 ml. Add KMnO₄ to the blank until the solution just turns
pink. Then just decolorize with SnCl₂. *Carry samples individually through
subsequent steps.*

Reduction of iron. Heat the solution containing the sample nearly
to boiling, and add SnCl₂ drop by drop until the yellow color disappears.
Add 2 drops in excess (Note). Cool to room temperature, and *rapidly* add
10 ml of the HgCl₂ solution. A small quantity of white precipitate should
appear. If no precipitate forms or if the precipitate is gray, the sample
should be discarded. The blank solution should also be treated with 10 ml
of the HgCl₂ solution.

Titration. After 2 to 3 min add 25 ml of Zimmermann-Reinhardt
reagent and 300 ml of water. Titrate *immediately* with the KMnO₄ to the

first faint pink that persists for 15 to 20 sec. Do not titrate rapidly at any time. Correct the volume of $KMnO_4$ for the blank titration.

NOTE

The solution may not become entirely colorless but may instead acquire a pale yellow-green hue. Further additions of $SnCl_2$ will not alter this color. If too much $SnCl_2$ is inadvertently introduced, add 0.2-F $KMnO_4$ until the yellow color is restored, and repeat the reduction.

ANALYSIS 5-3 Determination of Calcium in an Impure Calcium Carbonate

Discussion. The precipitation of calcium as the oxalate, followed by the titration of the oxalic acid liberated when the solid is dissolved, is discussed on page 323 and is the basis for the directions that follow.

Procedure. Dry the unknown at 110°C. Weigh samples large enough to contain about 80 mg of calcium into 600-ml beakers. Cover each with a watch glass. Add 10 ml of water, followed by 10 ml of 6-F HCl from a pipet; the acid should be added slowly to prevent losses by spattering. Heat the solution to drive off the CO_2. Wash the watch glass and sides of the beaker with water, and dilute the solution to about 150 ml. Heat to 60 to 80°C, and add 50 ml of a warm solution containing about 3 g of $(NH_4)_2C_2O_4 \cdot H_2O$ [if the $(NH_4)_2C_2O_4$ solution is not clear, filter before use]. Add 3 to 4 drops of methyl red. Then introduce 1:1 NH_3 dropwise from a pipet until the color changes from red to yellow. Allow the solution to stand for about 30 min (but no longer than 1 hr if magnesium is present) without further heating.

Filter the solution through a medium-porosity glass or porcelain filtering crucible or a Gooch crucible fitted with a mat of asbestos (p. 547) or glass fiber. Wash the beaker and precipitate with 10- to 20-ml portions of chilled distilled water until the washings show only a faint cloudiness when tested with an acidified $AgNO_3$ solution (Note). A quantitative transfer of the precipitate is unnecessary.

Rinse the outside of the crucible with water, and return it to the beaker in which the precipitate was formed. Add 150 ml of water and 50 ml of 3-F H_2SO_4. Heat to 80 to 90°C to dissolve the precipitate. Titrate with 0.1-N $KMnO_4$ with the crucible still in the beaker. The temperature of the solution should not be allowed to drop below 60°C. Take as an end point the first pink color that persists for 15 to 20 sec.

Report the percentage of $CaCO_3$ in the sample.

NOTE

If the sample contains large concentrations of sodium or magnesium ions, more accurate results can be obtained by reprecipitation of the calcium oxalate. To do this, filter the precipitate through paper, and wash it four or five times with 0.1% $(NH_4)_2C_2O_4$ solution. Pour 50 ml of hot 1:4 HCl through the paper, collecting the washings in the beaker in which the precipitation was made. Wash the paper several times with hot 1:100 HCl, and dilute all of the washings to about 200 ml. Reprecipitate the calcium oxalate as before, this time collecting the precipitate in a Gooch or glass crucible and washing it with cold water. Proceed with the analysis as before.

ANALYSIS 5-4 Determination of Calcium in a Limestone

Discussion. Limestones are composed principally of calcium carbonate; dolomitic limestones contain large concentrations of magnesium carbonate in addition. Also present in smaller amounts are calcium and magnesium silicates as well as carbonates and silicates of iron, aluminum, manganese, titanium, the alkalies, and other metals.

Hydrochloric acid will often decompose limestones completely; only silica remains undissolved. Some limestones are more readily decomposed if first ignited; a few will yield only to a carbonate fusion.

The following method is remarkably effective for the analysis of calcium in most limestones. Iron and aluminum, in amounts equivalent to the calcium, do not interfere. Small amounts of titanium and manganese can be tolerated.[6]

Procedure. Dry the unknown for 1 to 2 hr at 110°C. If the material can be readily decomposed with acid, weigh (to the nearest 0.1 mg) 0.25- to 0.3-g samples into 250-ml beakers and cover each beaker with a watch glass. Add 10 ml of water and 10 ml of concentrated HCl, taking care to avoid loss by spattering. Proceed with the analysis as described in the next paragraph. If the limestone is not completely decomposed by acid, weigh the sample into a small porcelain crucible and ignite. Raise the temperature slowly to 800 to 900°C, and maintain this temperature for 30 min. After it cools, place the crucible and contents in a 250-ml beaker, add 5 ml of water, and cover with a watch glass. Carefully add 10 ml of concentrated HCl, and heat to boiling. Remove the crucible with a stirring rod, rinsing thoroughly with water.

Add 5 drops of saturated bromine water to oxidize any iron present, and boil for 5 min to remove the excess bromine. Dilute to 50 ml, heat to boiling, and add 100 ml of hot, filtered 5% $(NH_4)_2C_2O_4$ solution. Add 3 to 4 drops of methyl red, and precipitate the calcium oxalate by the dropwise addition of 1:1 NH_3. The rate of addition should be 1 drop every 3 or 4 sec until the solution turns to the intermediate orange-yellow color of the

[6]For further details of the method see J. J. Lingane, *Ind. Eng. Chem.*, *Anal. Ed.*, **17**, 39 (1945).

indicator (pH 4.5 to 5.5). Allow the solution to stand for 30 min but no longer and filter (see Note); a Gooch crucible or a medium-porosity filtering crucible can be used. Wash the precipitate with several 10-ml portions of cold water. Rinse the outside of the crucible, and return it to the beaker in which the calcium oxalate was originally formed. Add 150 ml of water and 50 ml of 3-F H_2SO_4.

Heat the solution to 80 to 90°C. If a Gooch crucible was used, stir to break up the asbestos pad. Then titrate with 0.1-N $KMnO_4$. The solution should be kept above 60°C throughout the titration.

Report the percentage of CaO in the sample.

NOTE
If Mg^{2+} is absent, the period of standing need not be limited to 30 min.

6 Titrations with Cerium(IV)

PREPARATION 6-1   Preparation of Orthophenanthroline Indicator

Discussion. The orthophenanthroline complex of iron(II) is a useful indicator for titrations with cerium(IV). Its properties are listed in Table 12-2.

Procedure. Dissolve 1.485 g of orthophenanthroline in 100 ml of water that contains 0.695 g of $FeSO_4 \cdot 7H_2O$. Indicator solutions are available commercially.

PREPARATION 6-2   Preparation of Approximately 0.1-N Cerium(IV)

Procedure. Carefully add 50 ml of concentrated H_2SO_4 to 500 ml of water, and then add 63 g of $Ce(SO_4)_2 \cdot 2(NH_4)_2SO_4 \cdot 2H_2O$ with continual stirring. Cool, filter if the solution is not clear, and dilute to about 1 liter.

If $Ce(NO_3)_4 \cdot 2NH_4NO_3$ is used, weigh about 55 g into a 1-liter beaker. Add about 60 ml of 95% H_2SO_4, and stir for 2 min. *Cautiously* introduce 100 ml of water, and again stir for 2 min. Repeat the operations of adding water and stirring until all of the salt has dissolved. Then dilute to about 1 liter.

STANDARDIZATION 6-1 Standardization of Cerium(IV) against Arsenic(III)

Procedure. Dry primary standard As_2O_3 for 1 hr at 110°C, and weigh 0.2-g portions (to the nearest 0.1 mg) into 250-ml flasks. Dissolve the samples in 15 ml of 2-F NaOH. After solution is complete, promptly acidify with 25 ml of 3-F H_2SO_4 (Note 1). Dilute to about 100 ml, add 3 drops of 0.01-F osmium tetroxide (Note 2) and 1 drop of iron(II)-ortho-phenanthroline indicator. Titrate to a color change from red to colorless (or very pale blue).

NOTES

1. Arsenic(III) will undergo air oxidation in an alkaline environment. Arsenite solutions should be neutralized as soon as possible.
2. The catalyst solution may be purchased from the G. Frederick Smith Chemical Company, Columbus, Ohio; it should be about 0.01 F in OsO_4 and 0.1 F in H_2SO_4.

STANDARDIZATION 6-2 Standardization of Cerium(IV) against Sodium Oxalate

Procedure. Weigh 0.25- to 0.3-g portions of dried sodium oxalate (to the nearest 0.1 mg) into 250-ml beakers, and dissolve in 75 ml of water. Add 20 ml of concentrated HCl and 1.5 ml of 0.017-F ICl (Note 1). Heat to 50°C, and add 2 to 3 drops of iron(II)-orthophenanthroline indicator. Titrate with cerium(IV) until the solution turns colorless or pale blue and the pink does not return within 1 min. The temperature should be between 45 and 50°C throughout (Note 2).

NOTES

1. Iodine monochloride can be prepared as follows: mix 25 ml of 0.04-F KI, 40 ml of concentrated HCl, and 20 ml of 0.025-F KIO_3. Add 5 to 10 ml of CCl_4 and shake thoroughly. Titrate with either KI or KIO_3 until the CCl_4 layer is barely pink after shaking. The former should be added if the CCl_4 is colorless and the latter if it is too pink.
2. Do not permit the temperature to exceed 50°C because the indicator may be destroyed.

ANALYSIS 6-1 Determination of Iron in an Ore

Discussion. Quadrivalent cerium oxidizes iron(II) smoothly and rapidly at room temperature. Orthophenanthroline is an excellent indicator for the titration. In contrast to the analysis based upon oxidation with perman-ganate, consumption of the reagent by chloride ion is of no concern.

The problems associated with the solution of the sample and the prereduction of the iron are the same as in the permanganate method (pp. 595–598), the only major difference being that here there is no need for the Zimmermann-Reinhardt solution.

Procedure. Follow the directions for sample preparation and reduction of iron given in the procedure for Analysis 5-1 (p. 595) if the Jones reductor is to be employed; use the directions given in the procedure for Analysis 5-2 (p. 597) if the reduction is to be carried out by tin(II).

If the Jones reductor is employed, complete the analysis by titration with cerium(IV), using 1 drop of orthophenanthroline indicator.

If the tin(II) chloride is used, allow about 2 to 3 min after the addition of the $HgCl_2$. Then add 300 ml of 1-F HCl, a drop or two of orthophenanthroline, and titrate to the color change of the indicator. For accurate work a blank should be carried through the entire procedure.

7 Titration with Potassium Dichromate

PREPARATION 7-1 Preparation of Sodium Diphenylamine Sulfonate Solution

Procedure. Dissolve 0.2 g of sodium diphenylamine sulfonate in 100 ml of water.

PREPARATION 7-2 Preparation of 0.1-N Potassium Dichromate

Procedure. Dry primary standard $K_2Cr_2O_7$ for 2 hr at 150 to 200°C. After it cools, weigh 4.9 g of the solid (to the nearest milligram) into a 1-liter volumetric flask, dissolve, and dilute to the mark with distilled water; mix well.

If the purity of the salt is suspect, recrystallize three times from water before drying. Alternatively, prepare a solution of approximate normality by dissolving about 5 g of $K_2Cr_2O_7$ in a liter of water, and standardize the solution against weighed samples of electrolytic iron wire employing the procedure that follows. Dissolve 0.20- to 0.25-g wire samples (weighed to the nearest 0.1 mg) in a minimum amount of dilute HCl, prereduce the iron with Sn(II) (p. 597) or a Jones reductor (p. 595), and proceed as in Analysis 7-1.

ANALYSIS 7-1 Determination of Iron in an Ore

Procedure. Follow the directions for sample preparation given in the procedure for Analysis 5-1 (p. 595) or 5-2 (p. 597). No Zimmermann-Reinhardt reagent is required after the Sn(II) reduction.

Introduce about 10 ml of concentrated H_2SO_4 and 15 ml of syrupy H_3PO_4 to the prereduced solution. Add water if necessary to bring the volume to about 250 ml. Cool, add 8 drops of diphenylamine sulfonate indicator, and titrate with dichromate to the violet-blue end point.

8 Iodimetric Titrations

The oxidizing properties of iodine, the composition and stability of triiodide solutions, and the application of this reagent to volumetric analysis
are discussed on pages 331 to 335.

PREPARATION 8-1   Preparation of Starch Indicator

Procedure.   Make a paste by rubbing about 2 g of soluble starch and
10 mg of HgI_2 in about 30 ml of water. Pour this into 1 liter of boiling water,
and heat until a clear solution results. Cool and store in stoppered bottles.
For most titrations 3 to 5 ml of this solution should be sufficient.

PREPARATION 8-2   Preparation of Approximately 0.1-N Triiodide
Solution

Procedure.   Weigh about 40 g of KI into a 100-ml beaker. Add 12.7 g of I_2
and 10 ml of water. Stir until solution is complete. Filter the solution through
a Gooch or a sintered glass crucible; then dilute to about 1 liter. Store it in a
glass-stoppered bottle. If possible, allow the solution to stand for two to three
days before standardizing.

PREPARATION 8-3   Direct Preparation of Standard 0.1-N Arsenic(III)
Solution

Discussion.   A slightly acidic solution of arsenic(III) is stable indefinitely
and can be conveniently employed for the periodic standardization of the
less stable triiodide solutions. See page 334 for the properties of
arsenic(III) solutions.

Procedure.   Dry a quantity of primary standard As_2O_3 for about 1 hr at
110°C. Cool, then weigh about 5 g (to the nearest 0.2 mg) into a 1-liter
volumetric flask, and dissolve in about 100 ml of 1-F NaOH. After solution
is complete, *immediately* (Note) neutralize the base with 1-F HCl (a small
piece of litmus paper in the solution is adequate for an indicator). Dilute to
the mark, and mix well.

NOTE
Alkaline solutions of As(III) are readily air oxidized. Thus the neutralization should be performed promptly.

STANDARDIZATION 8-1   Standardization of a Triiodide Solution against
Primary Standard Arsenic(III) Oxide

Procedure. Dry a quantity of primary standard As_2O_3 for about 1 hr at 110°C. For standardization of 0.1-N I_2 solutions weigh 0.2-g samples (to the nearest 0.1 mg), and dissolve in 10 ml of 1-F NaOH. When solution is complete, dilute with about 75 ml of water, and add 2 drops of phenolphthalein. Promptly introduce 6-F HCl until the red color just disappears (see Note for Preparation 8-3). Then add about 1 ml of acid in excess. Carefully add 3 to 4 g of solid $NaHCO_3$, in small portions at first to avoid losses of solution due to effervescence of the CO_2. Add 5 ml of starch indicator, and titrate the solution to the first faint purple or blue color that persists for at least 30 sec.

STANDARDIZATION 8-2 Standardization of Triiodide Solution against Standard Arsenic(III) Solution

Procedure. Withdraw convenient aliquots of the arsenic(III) solution, add solid $NaHCO_3$, and proceed as in Standardization 8-1.

ANALYSIS 8-1 Determination of Arsenic(III) Oxide in an Impure Sample

Procedure. Dry the unknown for about 1 hr at 110°C. Weigh suitable samples into a beaker; the amount taken should be sufficient to require a 35- to 40-ml titration with 0.1-N I_3^-. Dissolve in base, acidify promptly with HCl (see Note in Preparation 8-3), buffer with $NaHCO_3$, and titrate with the standard triiodide solution following the procedure given for Standardization 8-1.

ANALYSIS 8-2 Determination of Antimony in Stibnite

Discussion. See page 334.

Procedure. Dry the unknown for 1 hr at 110°C. After it cools, weigh sufficient quantities of the ore to consume 30 to 40 ml of 0.1-N I_3^- into 500-ml conical flasks. Add about 0.3 g of KCl and 10 ml of concentrated HCl. Heat the mixture to just below boiling (Hood), and maintain this temperature until only a white or slightly gray residue of silica remains.

Add 3 g of solid tartaric acid to the solution; heat for another 10 to 15 min. While swirling the solution, slowly add water from a pipet until the volume is about 100 ml. The addition of water should be slow enough to prevent formation of white SbOCl. If reddish Sb_2S_3 forms, stop the addition of water, and heat further to remove H_2S; add more acid if necessary.

Add 3 drops of phenolphthalein to the solution and 6-F NaOH until the first pink color is observed. Add 6-F HCl dropwise until the solution is decolorized and then 1 ml in excess. Add 4 to 5 g of $NaHCO_3$, taking care to avoid losses of solution during the addition. Add 5 ml of starch, and titrate to the first blue color that persists for 30 sec or longer.

9 Iodometric Methods of Analysis

Numerous methods are based upon the reducing properties of iodide ion; the reaction product, iodine, is ordinarily titrated with a standard thiosulfate solution. Iodometric methods are discussed on pages 336 to 343.

PREPARATION 9-1 Preparation of Approximately 0.1-N Sodium Thiosulfate

Procedure. Boil about 1 liter of distilled water for at least 5 min. Cool and add about 25 g of $Na_2S_2O_3 \cdot 5H_2O$ and 0.1 g of Na_2CO_3. Stir until solution is complete, then transfer to a clean stoppered bottle (glass or plastic). Store in the dark.

STANDARDIZATION 9-1: Standardization of Sodium Thiosulfate against Potassium Iodate

Procedure. Weigh (to the nearest 0.1 mg) about 0.12-g samples of dried, primary standard KIO_3 into 250-ml conical flasks. Dissolve in 75 ml of water, and add about 2 g of iodate-free KI. *From here treat each sample individually.* After the iodide has dissolved, add 10 ml of 1.0-N HCl, and titrate immediately with thiosulfate until the color of the solution becomes pale yellow. Add 5 ml of starch, and titrate to the disappearance of the blue color.

To minimize the weighing error, this procedure can be modified by weighing a 0.6-g sample into a 250-ml volumetric flask, dissolving in water, and diluting to the mark. A 50-ml aliquot of this solution can then be used for each standardization.

STANDARDIZATION 9-2 Standardization of Sodium Thiosulfate against Potassium Dichromate

Procedure. Dry primary standard $K_2Cr_2O_7$ for 1 to 2 hr at 150 to 200°C, and weigh (to the nearest 0.1 mg) 0.20- to 0.23-g portions into 500-ml flasks. Dissolve in 50 ml of water. Then add a freshly prepared solution containing 3 g of KI and 5 ml of 6-N HCl in 50 ml of water. Swirl gently, cover the flask with a watch glass, and let it stand in a dark place for 5 min. Wash down the sides of the flask, add 200 ml of water, and titrate with the thiosulfate solution. When the yellow color of the iodine becomes faint, add 5 ml of starch. Continue the titration until a color change from the blue starch–iodine complex to the green color of the Cr(III) ion is observed.

ANALYSIS 9-1 Determination of Copper in an Ore

Discussion. The analysis for copper is a typical iodometric method. The directions that follow will permit the determination of copper in samples of ore. Additional information concerning this analysis is to be found on pages 340 to 342.

Procedure. Weigh appropriate-sized samples (150- to 250-mg Cu) of the finely ground and dried ore into 150-ml beakers, and add 20 ml of concentrated HNO_3. Heat (Hood) until all of the Cu is in solution. If the volume becomes less than 5 ml, add more HNO_3. Continue the heating until only a white or slightly gray siliceous residue remains (Note 1). Evaporate to about 5 ml.

Add 25 ml of distilled water, and boil to bring all soluble salts into solution. If the residue is small and nearly colorless, no filtration is necessary. Otherwise, filter the suspension, collecting the filtrate in a 250-ml conical flask. Wash with several small portions of hot 1:100 HNO_3; discard the paper. Evaporate the filtrate and washings to about 25 ml, cool, and slowly add concentrated NH_3 to the first appearance of the deep blue tetraamine copper(II) complex. A faint odor of NH_3 should be detectable over the solution. If it is not, add another drop of NH_3 and repeat the test. Avoid an excess (Note 2).

From this point on treat each sample individually. Add 2.0 ± 0.1 g NH_4HF_2 (Caution! Note 3), and swirl until completely dissolved. Then add 3 g of KI, and titrate immediately with 0.1-N $Na_2S_2O_3$. When the color of the iodine is nearly discharged, add 2 g of KSCN and 3 ml of starch. Continue the titration until the blue starch–iodine color is decolorized and does not return for several minutes.

NOTES

1. If the ore is not readily decomposed by the HNO_3, add 5 ml of concentrated HCl and heat (Hood) until only a small white or gray residue remains. Do not evaporate to dryness. Cool, add 10 ml of concentrated H_2SO_4, and evaporate until copious white fumes of SO_3 are observed (Hood). Cool and carefully add 15 ml of water and 10 ml of saturated bromine water. Boil the solution vigorously in a hood until all of the bromine has been removed. Cool and proceed with the filtration step in the second paragraph.

2. If too much NH_3 is added, neutralize the excess with 3-F H_2SO_4.

3. Ammonium hydrogen fluoride is a toxic and corrosive chemical. Avoid contact with the skin. If exposure does occur, immediately rinse the affected area with copious amounts of water.

ANALYSIS 9-2   Determination of Copper in a Brass

Discussion.   See pages 340 to 342.

Procedure.   Weigh (to the nearest 0.1 mg) 0.3-g samples of the clean, dry metal into 250-ml conical flasks, and add 5 ml of 6-F HNO_3. Warm the solution (Hood) until decomposition is complete. Then add 10 ml of concentrated H_2SO_4, and evaporate to copious white fumes of SO_3 (Hood). Allow the mixture to cool. Carefully add 20 ml of water; boil for 1 to 2 min and cool.

Add concentrated NH_3 dropwise and with thorough mixing until the first dark-blue color of the tetraamine copper(II) complex appears. The solution should smell faintly of NH_3. Add 3-F H_2SO_4 dropwise until the color of the complex just disappears. Then add 2.0 ml of syrupy phosphoric acid. Cool to room temperature.

From this point on treat each sample individually. Dissolve 4.0 g of KI in 10 ml of water, and add this to the sample. Titrate immediately with standard $Na_2S_2O_3$ until the iodine color is no longer distinct. Add 5 ml of starch solution, and titrate until the blue begins to fade. Add 2 g of KSCN, and complete the titration.

ANALYSIS 9-3   Determination of Dissolved Oxygen by the Winkler Method

Discussion.   For a discussion of the chemistry of this analysis see page 342.

Success of the method is critically dependent upon the manner in which the sample is manipulated; at all stages every effort must be made to assure that oxygen is neither introduced to nor lost from the sample. Biological Oxygen Demand (BOD) bottles are designed to minimize the entrapment of air.

The sample should be free of any solutes that will oxidize iodide or reduce iodine. Numerous modifications have been developed to permit the use of the Winkler method in the presence of such species.

Procedure.   The following special solutions are required:

1. *Manganese(II) sulfate.* Dissolve 48 g of $MnSO_4 \cdot 4H_2O$ in sufficient water to give 100 ml of solution.
2. *Potassium iodide–sodium hydroxide.* Dissolve 15 g of KI in about 25 ml of water, add 66 ml of 50% NaOH, and dilute to 100 ml.
3. *Sodium thiosulfate*, 0.025 N. See Preparation and Standardization 9-1 (p. 605); approximately 6.2 g of $Na_2S_2O_3 \cdot 5H_2O$ are needed.

Transfer the sample to a BOD bottle, taking care to minimize exposure to air. Use a tube to introduce the water to the bottom of the bottle; remove the tube slowly while the bottle is overflowing.

Add 1 ml of $MnSO_4$ with a dropper; discharge the reagent well below the surface (some overflow will occur). Similarly, introduce 1 ml of the KI–NaOH solution. Place the stopper in the bottle; be sure that no air becomes entrapped. Invert the bottle to distribute the precipitate uniformly.

When the precipitate has settled at least 3 cm below the stopper, introduce 1 ml of concentrated (18 F) H_2SO_4 well below the surface (Note 1). Replace the stopper and mix until the precipitate dissolves (Note 2). Measure 200 ml of the acidified sample into a 500-ml conical flask. Titrate with 0.025-N $Na_2S_2O_3$ until the iodine color becomes faint. Then introduce 5 ml of starch indicator and complete the titration.

Report the milliliters of O_2 (STP) dissolved in each liter of sample.

NOTES

1. Care should be taken to avoid exposure to the overflow, as the solution is quite alkaline.
2. A magnetic stirrer is effective in bringing about the solution of the precipitate.

10   Titrations with Potassium Bromate

Applications of potassium bromate to organic analyses are described on pages 327 to 330. Directions follow for a typical application, namely, the determination of phenol in water.

PREPARATION 10-1   Direct Preparation of Standard 0.12-N Potassium Bromate

Procedure.   Dry reagent-grade $KBrO_3$ for 1 hr at 100 to 110°C and cool. Weigh approximately 3.3 g (to the nearest milligram) into a 1-liter volumetric flask, dilute to the mark, and mix thoroughly.

PREPARATION 10-2   Preparation of Starch Indicator (see Preparation 8-1)

PREPARATION 10-3   Preparation of 0.1-N Sodium Thiosulfate (see Preparation 9-1)

STANDARDIZATION   10-1   Standardization   of   Sodium   Thiosulfate against Potassium Bromate

Discussion. The standard bromate solution is employed to generate a known amount of I_2.

$$BrO_3^- + 6I^- + 6H^+ \rightarrow Br^- + 3I_2 + 3H_2O$$

The iodine is then titrated with the sodium thiosulfate solution.

Procedure. Pipet 25-ml aliquots of the $KBrO_3$ solution into conical flasks, add about 2 to 3 g of KI and 5 ml of 6-N H_2SO_4. Titrate the liberated I_2 until the solution becomes faintly yellow. Add 5 ml of starch solution, and continue the titration until the blue color disappears. Calculate the normality of the thiosulfate solution.

ANALYSIS 10-1 Determination of Phenol

Procedure. Transfer a sample containing between 3 and 4 mmole of phenol to a 250-ml volumetric flask, and dilute to the mark. Pipet 25-ml aliquots of this solution into 250-ml glass-stoppered conical flasks, and add exactly 25 ml of the standard bromate reagent. Add about 0.5 g of KBr to each flask and about 5 ml of 3-F H_2SO_4. Stopper each flask *immediately* after addition of the acid to prevent loss of Br_2. Mix and let stand for at least 10 min. Weigh 2 to 3 g of KI (not accurately) for each sample. Add the KI rapidly to each flask, restoppering them immediately thereafter. Swirl the solution until the KI is dissolved. Then titrate with standard $Na_2S_2O_3$ until the solution is faintly yellow. Add 5 ml of starch, and complete the titration.

See the instructor for the method of reporting the results.

11 Titrations with Potassium Iodate

The oxidizing properties and applications of iodate ion are described on pages 330 to 331.

PREPARATION 11-1 Direct Preparation of Standard 0.025-F Potassium Iodate

Discussion. The KIO_3 will be employed for two purposes, namely, (1) the standardization of a sodium thiosulfate solution and (2) the oxidation of I^- and I_2 in the sample to ICl_2^-. For the first application a measured quantity of the reagent reacts with an excess of iodide to form a known quantity of I_3^-, which is then titrated with the thiosulfate solution. Here

$$IO_3^- + 8I^- + 6H^+ \rightleftharpoons 3I_3^- + 3H_2O$$

In its reaction with thiosulfate each I_3^- undergoes a 2-electron reduction so that the equivalent weight of potassium iodate is gfw $KIO_3/6$; the normality of the solution is six times its formality.

The reaction of KIO_3 with the I^- and I_3^- in the sample is described by the equations

$$2I^- + IO_3^- + 6Cl^- + 6H^+ \rightarrow 3ICl_2^- + 3H_2O$$

$$2I_2 + IO_3^- + 10Cl^- + 6H^+ \rightarrow 5ICl_2^- + 3H_2O$$

Here each IO_3^- acquires 4 electrons; thus the equivalent weight is gfw $KIO_3/4$, and the normality of the solution is four times its formality.

Procedure. Dry reagent-grade KIO_3 for 1 hr at 100 to 110°C and cool. Weigh about 2.68 g (to the nearest milligram) into a 500-ml volumetric flask, dilute to the mark, and mix thoroughly.

STANDARDIZATION 11-1 Standardization of Sodium Thiosulfate against 0.025-F Potassium Iodate

Procedure. Prepare about 1 liter of approximately 0.1-N $Na_2S_2O_3$ following the instruction in Preparation 9-1, page 605.

Pipet 25-ml aliquots of the 0.025-F KIO_3 into 250-ml conical flasks; add about 50 ml of water and 2 g of iodate-free KI to each. *From here on treat each aliquot individually.* Add about 10 ml of 1.0-N HCl and titrate with the thiosulfate until the color becomes pale yellow. Add 5 ml of starch indicator (p. 603), and titrate until the blue color disappears.

ANALYSIS 11-1 Determination of Iodine and Iodide Ion in an Aqueous Mixture

Discussion. The iodine in the mixture is determined by direct titration of an aliquot of the sample with standard thiosulfate solution. The concentration of I_2 plus I^- is then determined by the direct titration of another aliquot with the standard iodate solution. Here the solution contains a high concentration of Cl^-, and the reaction product is exclusively ICl_2^- (see equation in Preparation 11-1). The end point is observed as the disappearance of the iodine color in a small amount of chloroform $CHCl_3$ (see p. 331).

Procedure. Obtain the unknown in a clean, 500-ml volumetric flask. Dilute to the mark and mix thoroughly.

To determine the iodine in the unknown titrate 50.0-ml aliquots of the sample with standard sodium thiosulfate employing starch as an indicator.

To determine the combined concentration of I_2 and I^- in the sample, fill a buret with the unknown solution and run about 25.0 ml into a 250-ml conical flask. Add about 40 ml of concentrated HCl and 5 ml of CHCl₃. Titrate with KIO₃ until the purple color disappears from the chloroform layer. As the end point is approached, make dropwise additions of reagent, and swirl vigorously between additions. To eliminate the likelihood of overtitration introduce enough of the sample solution to reestablish the iodine color in the nonaqueous phase. Titrate as before with iodate. Repeat, as necessary, to establish the end point.

Report the grams I_2 and KI in the sample.

12   Potentiometric Titrations

The uses of potential measurements in analytical chemistry are discussed in Chapter 14. In this section a general procedure for carrying out potentiometric titrations is first given; directions for typical applications follow.

GENERAL INSTRUCTIONS FOR PERFORMING POTENTIOMETRIC TITRATIONS

1. Dissolve the sample in 50 to 250 ml of water. Rinse the electrodes with distilled water, and immerse them in the sample solution. Allow for magnetic (or mechanical) stirring. Position the buret so that the reagent can be delivered without splashing.
2. Connect the electrodes to the potentiometer, commence stirring, measure, and record the initial potential.
3. Measure and record the potential after each addition of reagent. Introduce fairly large volumes (say, 5 ml) at the outset. Withold each succeeding addition until the potential remains constant within 1 to 2 mV (or 0.05 pH unit) for 30 sec. A stirring motor will occasionally cause erratic potential readings; it may be advisable to turn off the motor during the actual measuring process. Judge the volume of reagent to be added by calculating an approximate value of $\Delta E/\Delta V$ after each addition. In the immediate vicinity of the equivalence point introduce the reagent in exact 0.1-ml increments. Continue the titration 2 to 3 ml past the equivalence point. Increase the volumes added as $\Delta E/\Delta V$ once again acquires small values.
4. Locate the end point by one of the methods described on pages 378 to 379.

ANALYSIS 12-1   Determination of Chloride and Iodide in a Mixture

Discussion.   The potentiometric titration of halide mixtures is discussed on page 380. The indicator electrode can be simply a polished silver wire or a commercial billet-type variety. A calomel electrode can be used as reference, although diffusion of Cl^- from the salt bridge will cause results to be slightly high. An alternative is to place the calomel electrode in a saturated KNO_3 solution and make contact with the solution to be titrated with a KNO_3 bridge. If desired, the solution can be made acidic with several drops of nitric acid; a glass electrode can then be employed as the reference because the pH of the solution will remain substantially constant during the titration.

Procedure. Prepare a standard 0.1-N solution of $AgNO_3$ according to Preparation 2-1 (p. 578).

The sample will be issued as a solution. Take aliquots containing a total of 2 to 4 mfw of Cl^- and I^-; dilute to 100 ± 10 ml with water. Acidify with HNO_3 and titrate with the silver nitrate as described in "General Instructions." Use small increments of titrant near the two end points. Plot the data and establish the end point for each ion (p. 378). Plot a theoretical titration curve, assuming the measured concentrations of the two constituents to be correct. Report the number of milligrams of I^- and Cl^- in the sample or as instructed.

ANALYSIS 12-2 Titration of a Weak Acid

Procedure. Prepare and standardize a 0.1-N solution of carbonate-free sodium hydroxide according to Preparation 3-3 (p. 584).

Dissolve between 1 and 4 meq of the acid in about 100 ml of water. Some of the less soluble organic acids may require a larger volume of water. Alternatively, the acid may be dissolved in 5 to 10 ml of ethyl alcohol and then diluted to 50 to 100 ml (see also p. 586).

Calibrate the system by rinsing the electrodes with water and immersing them in a buffer solution of known pH. Follow the instrument instructions for adjusting the meter to this pH.

Rinse the electrodes thoroughly and immerse them in the sample solution. Add 1 to 2 drops of phenolphthalein. Titrate the solution as directed in "General Instructions" (p. 611). Some samples will contain more than one replaceable hydrogen; be alert for more than one break in the titration curve. Note the volume at which the indicator changes color.

Plot the titration data, and determine the end point or points. Compare these with the phenolphthalein end point. The derivative method for end point determination may also be used.

Calculate the number of milliequivalents of H^+ present in the sample. Evaluate the dissociation constant(s) for the species titrated.

ANALYSIS 12-3 Determination of Carbonate and Bicarbonate in a Mixture

Discussion. See pages 237 to 239.

Procedure. Prepare and standardize a 0.1-N solution of HCl according to Preparation 3-2 and Standardization 3-2.

Dissolve the sample, which contains a total of 2 to 3 mfw of the two solutes, in about 200 ml of water. Add 2 drops of phenolphthalein, and titrate as directed in "General Instructions" (p. 611). After the phenolphthalein indicator has become colorless, add 2 drops of methyl orange or bromocresol green. Carry the titration 3 to 5 ml beyond the second end point. Note the titrant volumes where the two indicators change color.

Plot the data and determine the end points. Compare the potentiometric and indicator end points. Estimate the two dissociation constants for carbonic acid from the curve, and compare these with literature values.

Calculate the percent Na_2CO_3 and $NaHCO_3$ in the sample.

ANALYSIS 12-4 Determination of the Species Present in a Phosphate Mixture

Discussion. The sample will contain one or two of the following components: HCl, H_3PO_4, NaH_2PO_4, Na_2HPO_4, Na_3PO_4, and NaOH. The object is to determine which compatible species are present as well as their amounts.

A study of pages 212 to 220 may assist in interpreting the data.

Procedure. Depending upon its composition, the analysis may require only standard HCl or standard NaOH. For other compositions separate aliquots will have to be titrated, one with acid and the other with base. The appropriate titrant(s) can be determined from the initial pH of the sample and the titration curve for H_3PO_4 (p. 216). Instructions for the preparation and standardization of 0.1-N NaOH and HCl are found on pages 584 to 586.

The sample will be issued as a solution, which should be diluted to volume in a 250-ml volumetric flask. Mix well.

Pipet exactly 50.0 ml of the sample into a beaker, and determine its pH. Titrate with either standard HCl or NaOH until one or two end points have been passed; follow "General Instructions" (p. 611). If necessary, titrate a second aliquot with the reagent not employed in the first titration. On the basis of the titration curves, choose acid-base indicators that would be suitable for the titrations, and perform the duplicate analyses with these.

Identify and report the total number of millimoles of the component(s) in the unknown solution. Also, calculate and report approximate values for any dissociation constants for H_3PO_4 that can be obtained from the titration data.

13 Electrogravimetric Methods

ANALYSIS 13-1 Determination of Copper in an Aqueous Solution

Discussion. See pages 394 to 397.

Procedure. The sample to be analyzed should be free of chloride ion (Note 1) and contain from 0.2 to 0.3 g of Cu in 100 ml of water. Transfer the solution to a 150-ml electrolytic beaker. Add 3 ml of concentrated H_2SO_4 and 2 ml of freshly boiled and cooled 6-F HNO_3.

Prepare the electrodes by immersing them in hot 6-F HNO_3 that contains about 1 g of KNO_2 (Note 2). Wash thoroughly with distilled water, rinse several times with small portions of ethyl alcohol or acetone, and dry in an oven at 110°C for 2 to 3 min. Cool and weigh the cathode carefully on an analytical balance (Note 3).

Attach the cathode to the negative terminal of the electrolytic apparatus and the anode to the positive terminal. Elevate the beaker containing the solution so that all but a few millimeters of the cathode is covered. Start the stirring motor, and adjust the potential so that a current of about 2 A passes through the cell (Note 4). When the blue copper color has entirely disappeared from the solution, add sufficient water to raise the liquid level by a detectable amount, and continue the electrolysis with a current of about 0.5 A. If no further copper deposit appears on the newly covered portion of the cathode within 15 min, the electrolysis is complete. If additional copper does deposit, continue testing for completeness of deposition from time to time as directed above.

When no more copper is deposited after a 15-min period, stop the stirring and slowly lower the beaker while continuously playing a stream of wash water on the electrodes. Rinse the electrodes thoroughly with a fine stream of water. *Maintain the applied potential until rinsing is complete* (Note 5). Disconnect the cathode, and immerse it in a beaker of distilled water. Then rinse with several portions of alcohol or acetone. Dry the cathode in an oven for 2 to 3 min at 110°C, cool, and weigh.

NOTES

1. The presence of chloride results in attack on the platinum anode. This is not only destructive but will also lead to errors because the dissolved platinum will codeposit with the copper at the cathode.
2. Grease and organic material can be removed by bringing the electrode to red heat in a flame.
3. The cathode surface should not be touched with the fingers because grease and oil cause nonadherent deposits.
4. Alternatively, the electrolysis may be carried out without stirring. Here the current should be kept below 0.5 A. Several hours will be required to complete the analysis.

5. It is important to maintain the application of a potential until the electrodes have been removed from the solution and washed free of acid. If this precaution is not observed, some copper may redissolve.

14 Coulometric Titrations

ANALYSIS 14-1 Titration of Arsenic(III) with Coulometrically Generated Iodine

Discussion. See pages 397 to 407 for a discussion of coulometric titrations. A typical application of constant current coulometry is the electrolytic generation of iodine. This reagent is used to titrate arsenic(III) in the accompanying experiment.

The apparatus required is a constant current supply (5 to 30 mA), a stopclock, and a potentiometer (see Figure 15-3). The current is determined by measuring the potential drop across a precision standard resistor (50 to 100 Ω) with a potentiometer. The titration vessel can be a 150- to 250-ml beaker provided with magnetic stirring. The electrode system consists of a platinum wire cathode and a platinum foil anode (at least 2 cm^2 in area). Iodine is generated at the anode and hydrogen at the cathode. The latter does not interfere with the analysis. Reaction conditions are similar to those described in Analysis 8-1 for the iodimetric determination of As(III).

Procedure. Weigh a sample (to the nearest 0.1 mg) that will contain about 0.1 g As$_2$O$_3$, and dissolve in 30 ml of 1-F NaOH. Neutralize promptly with 1-F HCl (litmus paper); add 1 or 2 drops excess HCl, and dilute to exactly 1 liter in a volumetric flask.

See page 603 for the preparation of a starch suspension.

Turn on the constant current generator. Set and measure the current with the potentiometer. Place about 50 ml of water containing 3 to 4 g of NaHCO$_3$ in the titration vessel. Add about 5 ml of 1-F KI and 5 ml of the starch indicator.

Immerse the electrodes in the solution and start the stirrer. Switch on the electrolysis and timer unit, and titrate until the solution develops a permanent faint blue color. Stop the electrolysis and reset the timer to zero. Pipet 10 ml of the sample into the cell, and titrate to the same faint blue color. Record the elapsed time.

Additional titrations can be performed by adding fresh aliquots of the sample to the solution just titrated.

Calculate the percentage of As$_2$O$_3$ in the sample.

15 Voltammetry

This section contains directions for the performance of selected polarographic analyses and amperometric titrations. Chapter 15 should be consulted for a discussion of voltammetry.

Enormous variation exists among equipment for polarographic work; the reader should consult instructional literature for the particular instrument that is available. The characteristics of many commercial instruments have been summarized by Meites.[7]

EXPERIMENT 15-1 Polarographic Behavior of Cadmium(II)

Procedure. Into 50-ml volumetric flasks transfer 5.0 ml of 0.1% gelatin, 10.0 ml of 0.5-F KNO_3, and, respectively, 1.00, 3.00, 10.0, and 30.0 ml of 5.0×10^{-3} F Cd^{2+}; dilute each solution to the mark and mix well. Prepare an additional solution with 10.0 ml of Cd^{2+} and 10.0 ml of KNO_3, but no gelatin.

Rinse the polarographic cell with the solution containing no Cd^{2+} ion, and equilibrate in a constant temperature waterbath (about 25°C). Obtain a polarogram from 0 to −2.0 V. Bubble nitrogen through the solution for 15 min, and obtain the polarogram once again.

Obtain polarograms at 25°C for each of the Cd^{2+}-containing solutions after the removal of O_2. If a manual polarograph is used, establish a complete curve for the sample containing 10.0 ml of Cd^{2+}; take only enough points to determine i_d for the remainder.

1. Calculate half-wave potentials for the two oxygen waves.
2. Compare the cadmium waves obtained with and without the addition of gelatin.
3. Calculate $E_{1/2}$ for each determination; compare the average value with published data.
4. Plot E versus $\log i/(i_d - i)$ for the solution containing 10.0 ml of Cd^{2+} solution and gelatin; see Equation 16-6. Evaluate n from this plot.

EXPERIMENT 15-2 Polarographic Behavior of a Mixture of Ions

Procedure. Prepare 50 ml of a solution that contains 5 ml of the gelatin, 10 ml of 0.5-F KNO_3, 10 ml of the standard cadmium solution, and 10 ml of 5.0×10^{-3} F Zn^{2+}. After the removal of oxygen, obtain the polarogram.

Determine i_d/C and $E_{1/2}$ for each wave.

[7]L. Meites, *Polarographic Techniques*, 2d ed. New York: Interscience, 1965, pp. 36–55.

EXPERIMENT 15-3 Amperometric Titrations with a Dropping Mercury Electrode

Discussion. Amperometric titrations are discussed on pages 421 to 426.

Procedure. *Titration of lead.* Prepare 0.100-F Pb(NO$_3$)$_2$ and 0.050-F K$_2$Cr$_2$O$_7$ solutions and a pH-4.2 acetate buffer (add sufficient 0.1-F NaOAc to a solution that is 0.2 F in HOAc and 0.2 F in KNO$_3$ until the desired pH is obtained).

Transfer 5.00 ml of the Pb(NO$_3$)$_2$ solution to the titration vessel; add 5 ml of 0.1% gelatin and 40 ml of the acetate buffer. Insert a dropping electrode and a saturated calomel electrode. Bubble nitrogen through the solution for 15 min; then measure the current at -1.0 V after the gas flow has been stopped. Add exactly 1.00-ml increments of K$_2$Cr$_2$O$_7$ until a total of 10 ml have been added. Correct the currents for dilution, and plot the titration curve. Determine the end point graphically.

Repeat the titration at zero applied potential. For this titration oxygen removal is unnecessary. Contrast the curves and explain the differences.

Titration of sulfate ion. Transfer 25.0 ml of 0.0200-F K$_2$SO$_4$ to the titration vessel, and add 25 ml of a 40% (by volume) ethyl alcohol solution containing a drop of methyl red. Deaerate the solution, and titrate it with the 0.1-F Pb(NO$_3$)$_2$, following the directions given in the preceding section. Determine the end point graphically.

EXPERIMENT 15-4 Amperometric Titrations with a Rotating Platinum Electrode

Discussion. This experiment illustrates the use of the amperometric method to locate the end point in the titration of As(III) with standard KBrO$_3$. The reactions are

$$BrO_3^- + 5Br^- + 6H^+ \rightarrow 3Br_2 + 3H_2O$$

$$Br_2 + H_3AsO_3 + H_2O \rightarrow 2Br^- + H_3AsO_4 + 2H^+$$

A rotating platinum electrode, a saturated calomel electrode, and a 100-ml titration vessel are required.

Procedure. Prepare a 0.0075-F arsenious acid solution (p. 603) and a 0.0100-F potassium bromate solution (p. 608).

Transfer 10 ml of the arsenious acid solution to the titration vessel, and add 40 ml of a solution that is about 2 F in HCl and 0.1 F in KBr. Fill a 10-ml buret with the bromate solution. Adjust the electrode so that it rotates at about 600 rpm. Apply +0.2 V to the rotating electrode, and titrate as described in the foregoing section. Locate the end points graphically.

16   Methods Based upon Absorption of Radiation

Typical analytical applications of absorption measurements have been described in Chapter 16. Directions follow for (1) a determination of iron that requires the generation of a calibration curve, (2) a method for the analysis of manganese in steel that makes use of a standard addition, and (3) an experiment involving the resolution of a mixture that contains two species with mutually overlapping absorption spectra.

ANALYSIS 16-1   Determination of Iron in Water

Discussion.   An excellent and sensitive method for the determination of iron is based upon the formation of the orange-red iron(II)-orthophenanthroline complex. Orthophenanthroline is a weak base; in acidic solution the principal species is the phenanthrolium ion PhH^+. Thus the complex-formation reaction is best described by the equation

$$Fe^{2+} + 3PhH^+ \rightleftharpoons Fe(Ph)_3^{2+} + 3H^+$$

The equilibrium constant for this reaction is 2.5×10^6 at 25°C. Quantitative formation of the complex is observed in the pH region between 2 to 9. Ordinarily a pH of about 3.5 is recommended to prevent the precipitation of various iron salts such as phosphates. Careful control of pH is not required.

When orthophenanthroline is used for the analysis of iron, an excess of reducing agent is added to keep the iron in the +2 state; either hydroxylamine hydrochloride or hydroquinone is convenient for this purpose. Once formed, the color of the complex is stable for long periods of time.

The experiment can be performed with a spectrophotometer at 508 nm or with a photometer equipped with a green filter. Consult the manufacturer's instructions for operating details of the instrument to be used.

Procedure.   *Special solutions.*
1. Hydroxylamine hydrochloride, 10%. Dissolve 10 g of $H_2NOH \cdot HCl$ in about 100 ml of water. Introduce a sufficient amount of sodium citrate to bring the pH to 4.5.
2. Sodium citrate solution, 250 g per liter of solution.
3. o-Phenanthroline, 0.3% of the monohydrate in water. Store in a dark place. Discard the reagent when it becomes colored.
4. Standard iron solution, 0.1 mg Fe/ml. Weigh 0.702 g of analytical reagent grade $FeSO_4 \cdot (NH_4)_2SO_4 \cdot 6H_2O$ into a 1-liter volumetric flask; dissolve in 50 ml of water containing 1 ml of concentrated H_2SO_4. Dilute to the mark; mix well.

Alternatively, dissolve 0.1000 g of electrolytic iron wire in 6 to 10 ml of 6-F HCl, and dilute to exactly 1 liter.

Preparation of calibration curve. Transfer a 5-ml aliquot of the standard iron solution into a beaker, and add a drop of bromophenol blue indicator. Introduce sodium citrate solution from a pipet until the intermediate color of the indicator is achieved. Note the volume of citrate required, and discard the solution. Now measure a second 5-ml aliquot of the iron standard into a 100-ml volumetric flask; add 1 ml of the hydroxylamine and 3 ml of the orthophenanthroline solution. Introduce the same quantity of citrate solution as was required for the preliminary titration, and allow the mixture to stand 5 min. Dilute to the mark.

Clean the cells for the instrument. Rinse one of these with the solution of the complex and then fill. Rinse and fill the second cell with a blank containing all reagents except the iron solution. Carefully wipe the windows of the cells with tissue, and place them in the instrument. Measure the absorbance of the standard against the blank.

Prepare at least three other standards so that an absorbance range between about 0.1 to 1.0 will be covered. Construct a calibration curve for the instrument.

Analysis of sample. Transfer a 5-ml aliquot of the sample into a beaker, and add 1 drop of bromophenol blue. Adjust the pH of the solution with sodium citrate solution or 0.1-N H$_2$SO$_4$ from a pipet until the intermediate color of the indicator is achieved. Note the volume of reagent required and then discard the solution. Transfer a fresh 5-ml aliquot to a 100-ml volumetric flask; add 1 ml of the hydroxylamine and 3 ml of the orthophenanthroline solution. Introduce the same quantity of citrate solution or sulfuric acid as was required to adjust the pH. After 5 min dilute to the mark, and measure the absorbance. Repeat the analysis if necessary, using quantities of sample that will give an absorbance in the range of the calibration curve.

Calculate the milligrams of iron per liter of the sample solution.

ANALYSIS 16-2 Determination of Manganese in Steel

Discussion. Small quantities of manganese are readily determined colorimetrically by oxidation to the highly colored permanganate ion. Potassium periodate is an effective oxidizing reagent:

$$5IO_4^- + 2Mn^{2+} + 3H_2O \rightarrow 2MnO_4^- + 5IO_3^- + 6H^+$$

Permanganate solutions containing an excess of periodate are relatively stable.

Interferences to this procedure are few. The presence of colored ions can be compensated for by employing as a blank an aliquot of the sample that has not been oxidized by the periodate. This method of correction is not effective in the presence of appreciable quantities of cerium(III) or chromium(III) ions; both are oxidized by the periodate to a greater or lesser extent, and their reaction products absorb in the region commonly employed for the permanganate.

The accompanying method is applicable to most steels except those containing large amounts of chromium. The sample is dissolved in nitric acid; any carbon present is removed by oxidation with peroxodisulfate. Phosphoric acid is added to complex the iron(III) and prevent the color of this species from interfering with the analysis. The standard addition method (p. 470) is used to establish the relationship between absorbance and concentration.

A spectrophotometer set at 525 nm or a photometer with a green filter may be used for absorbance measurements. Consult the instruction manual of the instrument to be used for specific operating instructions.

Procedure. Prepare a standard manganese(II) solution as follows. Dissolve approximately 0.100 g of Mn (weighed to the nearest 0.1 mg) in about 10 ml of HNO_3. Boil the solution gently to eliminate oxides of nitrogen; cool, then transfer the solution quantitatively to a 1-liter volumetric flask, and dilute to the mark.

Weigh duplicate 1.0-g samples of the steel (to the nearest milligram), and dissolve in 50 ml of 4-F HNO_3 with gentle boiling; heating for 5 min should suffice. Cautiously add about 1 g of ammonium peroxodisulfate, and boil gently for 10 to 15 min. If the solution is pink or contains a brown oxide of manganese, add approximately 0.1 g of sodium hydrogen sulfite or ammonium hydrogen sulfite, and heat for another 5 min. Cool and dilute the solution to exactly 100 ml in a volumetric flask.

Pipet three 20.0-ml aliquots of each sample into small beakers. Treat as follows:

	H_3PO_4, ml	Std Mn Soln, ml	KIO_4, g
Aliquot 1	5	0.00	0.4
Aliquot 2	5	5.00	0.4
Aliquot 3	5	0.00	0.0

Boil each solution gently for 5 min; cool and dilute to 50.0 ml in volumetric flasks. Determine the absorbance of aliquots 1 and 2, with aliquot 3 serving as blank.

Report the average percentage of manganese in the sample.

ANALYSIS 16-3 Determination of the pH of a Buffer Mixture

Discussion. In this analysis the absorptivities of the acid and base forms of an indicator are determined at each of two wavelengths. The concentration of each form in an unknown buffer is then determined by spectral measurement at the two wavelengths (p. 407); the hydronium ion concentration can then be calculated from these data.

The accompanying directions make use of methyl red as the indicator; bromocresol green is equally satisfactory.

The behavior of methyl red in aqueous solution can be described by the reaction

$$InH^+ \ + \ H_2O \ \rightleftharpoons \ In \ + \ H_3O^+$$

$$\underset{red}{} \qquad\qquad \underset{yellow}{}$$

and

$$K_a = 1.00 \times 10^{-5} = \frac{[In][H_3O^+]}{[InH^+]}$$

Spectrophotometric determination of [In] and [InH$^+$] permits the calculation of [H$_3$O$^+$].

Procedure. Dissolve approximately 40 mg (weighed to the nearest 0.1 mg) of methyl red in a minimum amount of dilute NaOH, and dilute to 1 liter in a volumetric flask.

Determination of individual absorption spectra. Transfer 25.0-ml aliquots of stock methyl red solution to each of two 100-ml volumetric flasks. To one add 25.0 ml of 0.4-F HCl; to the other add 25.0 ml of 0.4-F NaOH. Dilute each to the mark; mix thoroughly. Obtain the absorption spectra for the acid and conjugate-base forms of the indicator between 400 and 600 nm, using water as a blank; record absorbance values at 10-nm intervals routinely and at smaller intervals as necessary to define the curves. Evaluate the absorptivity for each species at wavelengths that correspond to their respective absorption maxima.

Determination of the pH of an unknown buffer. Transfer a 25.0 ml aliquot of the stock methyl red solution to a 100-ml volumetric flask. Add 50.0 ml of the unknown buffer, dilute to the mark, and measure the absorbance of the diluted solution at the wavelengths for which absorptivity data were calculated. Report the pH of the buffer.

answers
to problems

Chapter 2

1. (a) 59.4 mfw   (b) 5.48×10^{-4} mfw   (c) 65.0 mfw
    (d) 10.0 mfw   (e) 1.55×10^{-2} mfw

3. (a) 7.7×10^{5} mg   (b) 118 mg   (c) 1.96×10^{6} mg
    (d) 428 mg   (e) 1.78×10^{7} mg

5. (a) $1.60 \times 10^{-3}\,F$ (b) $3.20 \times 10^{-3}\,M$ (c) $4.80 \times 10^{-3}\,F$ (d) 0.0546%
    (e) 0.0800 mmole   (f) 86.4 ppm   (g) 2.495   (h) 2.319

7. (a) $0.470\,M_{Na^+}$; $9.9 \times 10^{-3}\,M_{K^+}$ (b) pNa = 0.328; pK = 2.004

9. (a) pNa = 1.301; pSO_4 = 1.699; pOH = 2.000
    (c) pH = −0.176; pNO_3 = −0.279; pZn = 0.699
    (e) pH = 4.135; pBa = 4.000; $pClO_4$ = 3.564

10. (a) $7.6 \times 10^{-13}\,M$ (c) $3.7 \times 10^{-7}\,M$ (e) $1.8 \times 10^{-4}\,M$ (g) $1.5 \times 10^{2}\,M$

11. $F_{(NH_4)_2SO_4}$ = 1.70; $M_{NH_4^+}$ = 3.40

13. (a) Dissolve 22.5 g ethylene glycol in water, and dilute to 250 ml.
    (b) Mix 22.5 g ethylene glycol with 228 g water.
    (c) Dissolve 22.5 ml ethylene glycol in water, and dilute to 250 ml.

15. (a) Dissolve 2.54 g I_2 in CCl_4, and dilute to 1.00 liter.
    (b) Dissolve 2.06 g $(NH_4)_2SO_4$ in water, and dilute to 250 ml.
    (c) Dilute 9.17 ml of the reagent to 100 ml.
    (d) Dilute 29.4 ml of the reagent to 250 ml.
    (e) Dissolve 0.103 g $BaCl_2 \cdot 2H_2O$, and dilute to 3.00 liters.

17. (a) Dissolve 49.6 g $K_2Cr_2O_7$ in water, and dilute to 750 ml.
    (b) Dissolve 1.74×10^{3} g K_2SO_4 in water, and dilute to 50.0 liters.
    (c) Dilute 19.6 ml of the reagent to 250 ml.
    (d) Dilute 556 ml of the reagent to 20.0 liters.

19. (a) 7.3×10^2 g (b) Dilute 32 ml of the reagent to 2.00 liters.

21. Dilute about 409 ml of the reagent to 1.00 liter.

23. (a) 7.21 g (b) 1.93×10^3 g (c) 0.845 g (d) 2.04 g (e) 0.443 g

25. (a) 1.35 g (b) 23.3 ml (c) 81.9 ml (d) 90.2 ml (e) 2.8×10^2 ml

Chapter 3

1. (a) $K_{sp} = [Cu^+][CN^-]$ (c) $K_{sp} = [Ag^+]^2[CO_3^{2-}]$ (e) $K_{sp} = [Al^{3+}][OH^-]^3$

2. (a) $s = (K_{sp})^{1/2}$ (c) $s = (K_{sp}/4)^{1/3}$ (e) $s = (K_{sp}/27)^{1/4}$

3. (a) $K_b = \dfrac{[C_2H_5NH_3^+][OH^-]}{[C_2H_5NH_2]} = 4.28 \times 10^{-4}$

   (c) $K_a = \dfrac{[C_5H_5N][H_3O^+]}{[C_5H_5NH^+]} = \dfrac{K_w}{K_b} = 5.9 \times 10^{-6}$

4. (a) $K = [AgCl_2^-]/[Cl^-]$ (c) $\beta_4 = \dfrac{[Cd(NH_3)_4^{2+}]}{[Cd^{2+}][NH_3]^4}$

   (e) $K_1K_2K_3 = \dfrac{[H_3O^+]^3[AsO_4^{3-}]}{[H_3AsO_4]}$

5. (a) 6.2×10^{-8} (c) 7.4×10^{-8} (e) 3.2×10^{-13} (g) 2.0×10^{-20}

6. (a) 0.44 g/100 ml (b) 1.8×10^{-3} g/100 ml (c) 7.9×10^{-2} g/100 ml

8. $0.98\ M$

10. (a) $7.9 \times 10^{-6}\ M$

11. (a) $0.0250\ M$ (b) $6.9 \times 10^{-4}\ M$ (c) $1.5 \times 10^{-8}\ M$ (d) $1.7 \times 10^{-2}\ M$

13.

	$[H_3O^+]$	$[OH^-]$
(a)	3.5×10^{-5}	2.9×10^{-10}
(c)	2.26×10^{-12}	4.42×10^{-3}
(e)	3.84×10^{-9}	2.60×10^{-6}
(g)	4.32×10^{-4}	2.31×10^{-11}

15. (a) $6.61 \times 10^{-2}\ M$ (b) $6.73 \times 10^{-9}\ M$ (e) $2.2 \times 10^{-5}\ M$

16. (a) $3.37 \times 10^{-6}\ M$

17.

	$[Cl^-]$	$[Hg^{2+}]$	$[HgCl_2]$
(a)	2.0×10^{-7}	6.0×10^{-3}	4.0×10^{-3}
(b)	1.1×10^{-5}	5.3×10^{-6}	1.0×10^{-2}
(c)	1.3×10^{-2}	3.5×10^{-12}	1.0×10^{-2}

18. (a) $4.8 \times 10^{-7}\ M$ (b) $4.8 \times 10^{-6}\ M$ (c) $4.8 \times 10^{-5}\ M$

Chapter 4

1. **(a)** $249.7 = 250$ mg **(b)** 250 mg **(c)** 12 mg **(d)** 3.3 mg $= 3$ mg or 13 ppt
   (e) -3 mg or -12 ppt **(f)** -5 mg or 21 ppt

3.

	STUDENT A	STUDENT B
Ave. abs dev	$0.018 = 0.02\%$ H_2O	$0.018 = 0.02\%$ H_2O
Ave. rel dev	10 ppt	2 ppt
Abs error	-0.03% H_2O	-0.05% H_2O
Rel error	-19 ppt	-7 ppt

5. **(a)** 1.3×10^2 ppt **(b)** 25 ppt **(c)** 13 ppt **(d)** 2.5 ppt **(e)** 1.2 ppt

7. $s = 4.3 = 4$ mg; $(s)_r = 17$ ppt

9. Student A, $s = 0.025\%$ H_2O or 15 ppt; student B, $s = 0.025\%$ H_2O or 3.4 ppt

11. $s = 0.033 = 0.03\%$ S

13. $s = 3.1 = 3$ ppb

15.

	(a)	**(b)**	**(c)**
90% C.I.	16.2 ± 2.6	16.2 ± 1.9	16.2 ± 1.3
95% C.I.	16.2 ± 4.1	16.2 ± 2.9	16.2 ± 2.1

17. $N = 6.9 = 7$ measurements; $N = 17.04 = 17$ measurements

19. **(a)** 6.21 ± 0.34 **(b)** 6.21 ± 0.26

21. $s = 0.0085\%$ S; 95% C.L. $= \pm 0.021\%$ S

23. $s = 0.059$; 95% C.I. $=$ **(a)** 2.88 ± 0.15 and **(b)** 2.88 ± 0.07

25.

	1	**2**	**3**	**4**	**5**	**6**
(a) $s = $	0.045	0.053	0.036	0.028	0.020	0.034

   (b) $s \rightarrow \sigma = 0.038$
   (c) yes
   (d) 3.55 ± 0.030 and 3.55 ± 0.026
   (e) 3.86 ± 0.25 and 3.86 ± 0.053

26. **(a)** $N = 2.39 = 3$ readings **(c)** $N = 1.4 = 2$ readings

27. 90% C.L. $= \pm 0.20$; 95% C.L. $= \pm 0.24$

28. **(a)** $2.96 = 3$ replications

30. **(a)** 3 **(b)** 5 **(c)** 2 **(d)** 7 **(e)** 5 **(f)** 3

32.

	s	$(s)_r$, ppt	METHODS FOR REPORTING Y	
			1	2
(a)	0.36	85	$4.2(\pm0.4)$	42
(b)	0.072	15	$4.82(\pm0.07)$	4.8
(c)	2.4×10^{-7}	14	$1.78(\pm0.02) \times 10^{-5}$	1.78×10^{-5}
(d)	0.63	18	$34.6(\pm0.6)$	35
(e)	7.8×10^{-18}	4.1	$1.880(\pm0.008) \times 10^{-15}$	1.88×10^{-15}
(f)	0.0051	440	$1.2(\pm0.5) \times 10^{-2}$	1.2×10^{-2}
(g)	5.9×10^{-4}	110	$-5.3(\pm0.6) \times 10^{-3}$	-5×10^{-3}
(h)	3.5×10^{-5}	16	$2.20(\pm0.04) \times 10^{-3}$	2.20×10^{-3}

34. **(a)** 16.473 **(b)** 2.5 **(c)** 1.090 **(d)** 1×10^{14} **(e)** 10^{-12} **(f)** 6×10^{-35}

36.

	Q_{exp}	Q_{crit}	DECISION
(a)	0.51	0.76	retain
(b)	0.92	0.76	reject
(c)	0.44	0.76	retain
(d)	0.86	0.76	reject
(e)	0.80	0.94	retain

Chapter 5

1. **(a)** 0.55 g **(b)** 0.10 g **(c)** 0.0026 g

3. **(a)** 3.6×10^{-5} and 1.9×10^{-9} F **(b)** 1.2×10^{-17} and 1.2×10^{-27} F

5.

		TlI	AgI	PbI$_2$	BiI$_3$
(a)	Solubility	2.6×10^{-4} F	9.1×10^{-9} F	1.2×10^{-3} F	1.3×10^{-5} F
	Order of solubility	2	4	1	3
(b)	Solubility	6.5×10^{-7} F	8.3×10^{-16} F	7.1×10^{-7} F	8.1×10^{-16} F
	Order of solubility	2	3	1	4
(c)	Solubility	6.5×10^{-7} F	8.3×10^{-16} F	1.3×10^{-4} F	6.7×10^{-7} F
	Order of solubility	3	4	1	2

7. **(a)** Ni^{2+} **(b)** 2.6×10^{-6} M **(c)** Not feasible

8. **(b)** Feasible; $[Ag^+] = 8.3 \times 10^{-11}$ to 1.8×10^{-9} **(d)** Not feasible
(f) Feasible; $[OH^-] = 3.4 \times 10^{-11}$ to 1.3×10^{-5}

9. **(a)** Feasible; $[H_3O^+] = 3.4 \times 10^{-4}$ to 1.2×10^{-3}
(c) Feasible; $[H_3O^+] = 1.6 \times 10^{-10}$ to 1.1×10^{-6}
(e) Not feasible

10. **(a)** 4.6×10^{-8} **(b)** 3.8×10^{-17} **(c)** 2.0×10^{-23} **(d)** 2.5×10^{-37}

12. **(a)** 1.2×10^{-8} and $9.1 \times 10^{-9}\,F$ **(b)** 2.7×10^{-2} and $1.6 \times 10^{-2}\,F$
   (c) 3.5×10^{-4} and $1.3 \times 10^{-4}\,F$ **(d)** 1.4×10^{-5} and $2.0 \times 10^{-6}\,F$

14. **(a)** $1.3 \times 10^{-4}\,F$ **(b)** $3.9 \times 10^{-4}\,F$

15.

	$[H_3O^+],\,M$	$[Ag^+],\,M$	SOLUBILITY, F
(a)	1.0×10^{-6}	1.80×10^{-4}	6×10^{-5}
(b)	1.0×10^{-4}	1.76×10^{-3}	6×10^{-4}
(c)	1.0×10^{-2}	2.24×10^{-2}	7×10^{-3}

17. **(a)** $8.4 \times 10^{-5}\,F$ **(b)** $1.63 \times 10^{-5}\,F$ **(c)** $1.55 \times 10^{-5}\,F$
The last two should be rounded to 1.6×10^{-5}.

18. **(a)** $4.5 \times 10^{-3}\,F$

19. **(a)** 2.6×10^{-3} and $2.6 \times 10^{-5}\,F$ **(c)** 3.3×10^{-2} and $1.5 \times 10^{-3}\,F$

20. **(a)** $4.0 \times 10^{-5}\,F$ **(b)** $4.8 \times 10^{-6}\,F$ **(c)** $2.2 \times 10^{-7}\,F$ **(d)** $2.0 \times 10^{-7}\,F$

22. $3.1 \times 10^{-3}\,F$

Chapter 6

1. **(a)** $\dfrac{2\ \text{gfw CHF}_3}{3\ \text{gfw CaF}_2}$ **(d)** $\dfrac{\text{gfw (C}_6\text{H}_5)_6\text{Si}_2}{36\ \text{gfw BaCO}_3}$

  (b) $\dfrac{2\ \text{gfw Fe}_3\text{Al}_2\text{Si}_3\text{O}_{12}}{3\ \text{gfw Fe}_2\text{O}_3}$ **(e)** $\dfrac{2\ \text{gfw Na}_3\text{PO}_4}{\text{gfw P}_2\text{O}_5 \cdot 24\text{MoO}_3}$

  (c) $\dfrac{\text{gfw (C}_6\text{H}_5)_6\text{Si}_2}{2\ \text{gfw SiO}_2}$ **(f)** $\dfrac{24\ \text{gfw Mo}}{\text{gfw P}_2\text{O}_5 \cdot 24\text{MoO}_3}$

3. **(a)** $1.21\,g$ **(c)** $0.576\,g$

4. 2.23%

6. $0.0366\,g$

8. $30\,ml$

10. **(a)** $9.0\,ml$ **(b)** $8.4\,ml$

12. 26.9%

14. 46.4%

16. 20.3%

18. $377\,mg/tablet$

20. 22.8%

22. 33.3%; two tablets per dose

24. $17.6\%\ TeO_2$ and $82.4\%\ In_2O_3$

Chapter 7

1. (a) $pK = pI = 1.699$; $pH = pOH = 7.000$
   (c) $pBa = 2.58$; $pH = 11.72$; $pOH = 2.28$
   (e) $pCa = 3.34$; $pCd = 3.28$; $pCl = 3.04$; $pNO_3 = 2.98$; $pH = pOH = 7.00$

2. (a) $pPb = 4.37$; $pIO_3 = 4.07$ (c) $pPb = 1.602$; $pIO_3 = 5.45$ (e) $pPb = -0.08$

3. (a) $[H_3O^+] = 4.8 \times 10^{-9}$ (c) $[Br^-] = 0.940$ (e) $[Li^+] = 1.85$
   (g) $[Mn^{2+}] = 9.917 \times 10^{-1}$

4. (a) Complex formation reaction:

$$\frac{\text{gfw Fe(OH)}_2\text{Cl}}{3}; \frac{\text{gfw Fe}_2\text{O}_3}{2 \times 3}; 2 \times \text{gfw NaBr}; \frac{\text{gfw C}_6\text{H}_2\text{Br}_4}{2}$$

   (b) Oxidation-reduction reaction:

$$\frac{\text{gfw I}_2}{1 \times 2}; \frac{\text{gfw H}_2\text{S}}{2}; \frac{\text{gfw As}_2\text{S}_5}{5 \times 2}; \frac{\text{gfw KI}}{1}$$

   (c) Acid-base reaction:

$$\frac{\text{gfw H}_2\text{SO}_4}{2}; \frac{\text{gfw Mg(OH)}_2}{2}; \frac{\text{gfw Mg}_2\text{P}_2\text{O}_7}{2 \times 2}; \frac{\text{gfw MgO}}{2}$$

   (d) Oxidation-reduction reaction:

$$\frac{\text{gfw Ce}^{3+}}{1}; \frac{\text{gfw NaH}_2\text{AsO}_3}{2}; \frac{\text{gfw As}_2\text{O}_3}{2 \times 2}; \frac{\text{gfw Ce}_2(\text{SO}_4)_3}{2 \times 1}$$

6. (a) Precipitation reaction:

$$\frac{\text{gfw Ce}_2(\text{SO}_4)_3}{2 \times 3}; \frac{\text{gfw NaIO}_3}{1}; \frac{\text{gfw I}_2\text{O}_5}{2 \times 1}; \frac{\text{gfw AlOH(IO}_3)_2}{2 \times 1}$$

   (b) Acid-base reaction:

$$\frac{\text{gfw As}_2\text{O}_3}{2 \times 2}; \frac{\text{gfw H}_3\text{AsO}_4}{2}; \frac{\text{gfw KOH}}{1}; \frac{\text{gfw As}}{2}$$

   (c) Oxidation-reduction reaction:

$$\frac{\text{gfw K}_2\text{Cr}_2\text{O}_7}{2 \times 3}; \frac{\text{gfw Fe}_3\text{O}_4}{3 \times 1}; \frac{\text{gfw Na}_2\text{CrO}_4}{3}; \frac{\text{gfw Fe}}{1}$$

   (d) Complex formation reaction:

$$2 \times \text{gfw NH}_3; \frac{\text{gfw Ag}_2\text{SO}_4}{2}; \frac{\text{gfw (NH}_4)_2\text{SO}_4}{1}; \frac{\text{gfw N}_2}{1}$$

8. (a) $0.0248\ F$ (b) $0.0248\ N$ (c) $0.0496\ N$ (d) $0.0496\ N$
   (e) $0.0744\ N$ (f) 8.04 mg/ml

10. (a) $0.160\ N$ (b) $0.800\ N$ (c) $0.480\ N$ (d) $0.640\ N$

11. **(a)** $9.66 \times 10^{-3} F$ **(b)** $9.66 \times 10^{-3} N$
   (c) $2.90 \times 10^{-2} N$ **(d)** 2.21 mg $Zn_2P_2O_7/ml$

13. **(a)** 0.250 meq **(b)** 0.500 meq **(c)** 0.250 meq **(d)** 0.250 meq
   (e) 0.125 meq **(f)** 1.50 meq

15. **(a)** 0.0680 g **(b)** 9.2 **(c)** 0.884 **(d)** 9.74 **(e)** 0.385 **(f)** 0.770

17. Dilute the following quantities to 2.00 liters:
   (a) 13.3 ml **(b)** 71.4 g **(c)** 360 ml **(d)** 2.2 ml

19. Dissolve the following quantities, and dilute to 500 ml:
   (a) 7.29 g **(b)** 3.71 g **(c)** 18.1 g **(d)** 5.49 g **(e)** 3.75 ml

21. 0.0810 N

23. 0.105 N

25. $N_{H_2SO_4} = 0.290$; $N_{NaOH} = 0.305$

27. 20.7%

29. 4.86 mg/liter

31. 7.04 g/100 ml

Chapter 8

1. **(a)** 0.111 F **(b)** 0.111 N **(c)** 7.77 mg $CrOHCl_2/ml$ $AgNO_3$
   (d) 14.6 mg CHI_3/ml $AgNO_3$

3. **(a)** 0.132 N **(b)** 0.0663 N **(c)** 0.133 N **(d)** $6.71 \times 10^{-3} N$

5. **(a)** Dissolve 12.4 g and dilute to 1.50 liters.
   (b) Dissolve 6.01 g and dilute to 500 ml.
   (c) Dissolve 5.89 g and dilute to 800 ml.
   (d) Dissolve 3.89 g and dilute to 2.00 liters.

7. **(a)** 0.215 N **(b)** 16.4 mg/ml

9. 29.4% F and 61.7% Na_2SeF_6

11. 17.5%

13. Only one Cl in the compound is titrated.

15. 21.3 ppm

17. 3.82 mg/ml

19. 42.5% $BaCl_2$ and 57.5% KBr

20.

REAGENT VOL FROM EQ POINT, ml	(a) pAg	(c) pAg	(e) pCd
−10.00	2.42	6.53	2.24
−1.00	3.47	6.01	3.30
−0.10	4.48	5.51	4.30
0.00	5.98	4.80	5.40
+0.10	7.48	3.70	5.95
+1.00	8.48	2.70	6.44
+10.00	9.43	1.74	6.92

Chapter 9

1.

	APPROXIMATE METHOD $[H_3O^+]$	pH	QUADRATIC METHOD $[H_3O^+]$	pH
(a)	3.85×10^{-3}	2.41	3.77×10^{-3}	2.42
(b)	1.38×10^{-3}	2.86	1.37×10^{-3}	2.86
(c)	4.21×10^{-3}	2.38	4.12×10^{-3}	2.38
(d)	1.30×10^{-1}	0.88	7.06×10^{-2}	1.15
(e)	2.26×10^{-1}	0.65	8.56×10^{-2}	1.07
(f)	7.54×10^{-6}	5.12	7.54×10^{-6}	5.12

3.

	APPROXIMATE METHOD $[OH^-]$	pH	QUADRATIC METHOD $[OH^-]$	pH
(a)	4.20×10^{-4}	10.62	4.11×10^{-4}	10.61
(b)	5.64×10^{-4}	10.75	5.48×10^{-4}	10.74
(c)	4.12×10^{-6}	8.62	4.12×10^{-6}	8.62
(d)	1.98×10^{-6}	8.30	1.98×10^{-6}	8.30
(e)	2.18×10^{-4}	10.34	2.16×10^{-4}	10.33
(f)	8.54×10^{-7}	7.93	8.54×10^{-7}	7.93

5. (a) 2.00   (b) 2.30   (c) 7.00   (d) 11.70   (e) 12.00   (f) 2.00

7. (a) 9.84   (b) 11.70   (c) 7.22   (d) 4.67   (e) 2.30   (f) 7.52

9. (a) 5.82   (b) 1.70   (c) 11.23   (d) 11.82   (e) 12.48   (f) 10.41

11.

	APPROXIMATE METHOD $[H_3O^+]$	pH	QUADRATIC METHOD $[H_3O^+]$	pH
(a)	1.49×10^{-4}	3.83	1.49×10^{-4}	3.83
(b)	1.73×10^{-2}	1.76	1.46×10^{-2}	1.84
(c)	7.48×10^{-3}	2.13	6.94×10^{-3}	2.16
(d)	4.47×10^{-3}	2.35	4.28×10^{-3}	2.37
(e)	1.88×10^{-2}	1.72	1.56×10^{-2}	1.81
(f)	5.48×10^{-6}	5.26	5.48×10^{-6}	5.26

13.

	APPROXIMATE METHOD		QUADRATIC METHOD	
	$[OH^-]$	pH	$[OH^-]$	pH
(a)	4.80×10^{-5}	9.68	4.80×10^{-5}	9.68
(b)	7.07×10^{-1}	13.85	5.96×10^{-2}	12.78
(c)	1.24×10^{-5}	9.09	1.24×10^{-5}	9.09
(d)	1.41×10^{-2}	12.15	1.26×10^{-2}	12.10
(e)	3.73×10^{-6}	8.57	3.73×10^{-6}	8.57
(f)	3.46×10^{-4}	10.54	3.45×10^{-4}	10.54

15.

	APPROXIMATE METHOD		EXACT METHOD	
	$[H_3O^+]$	pH	$[H_3O^+]$	pH
(a)	2.51×10^{-5}	4.60	2.2×10^{-5}	4.66
(b)	5.6×10^{-10}	9.25	6.0×10^{-10}	9.22
(c)	1.70×10^{-3}	2.77	8.88×10^{-4}	3.05
(d)	1.6×10^{-10}	9.79	2.4×10^{-10}	9.62
(e)	4.2×10^{-12}	11.37	1.8×10^{-11}	10.75
(f)	6.62×10^{-5}	4.18	6.44×10^{-5}	4.19

17. Q = quadratic solution
(a) 9.12 (b) 10.18 (c) 2.03, Q (d) 10.77, Q

19. (a) 4.83 (b) 4.68 (c) 4.44

21. (a) $\dfrac{[NH_4^+]}{[NH_3]} = 2.22$ (b) $\dfrac{[CH_3NH_3^+]}{[CH_3NH_2]} = 60$ (c) $\dfrac{[CN^-]}{[HCN]} = 1.7$ (d) $\dfrac{[OCl^-]}{[HOCl]} = 24$

23. (a) 1.29 (b) 0.82 (c) 0.31

25. (a) −0.09 (b) −0.21 (c) −0.02

27. (a) 0.48 (b) 0.29 (c) 0.04

29. 0.816 g

31. 185 ml

33. The first answer is by the approximate method and the second by the quadratic equation.
(a) 3.05, 3.08 (b) 8.10, 8.10 (c) 6.90, 6.90 (d) 12.08, 11.89 (e) 12.68, 12.12

35. The first answer is by the approximate method and the second by the quadratic equation.
(a) 2.62, 2.67 (b) 2.93, 2.95 (c) 7.20, 7.20 (d) 2.32, 2.35
(e) 6.60, 6.60 (f) 7.80, 7.80

37. (a) 146 ml (b) 232 ml

39.

VOL HCl, ml	pH (a)	pH (c)
0.00	11.79	9.07
10.00	11.16	5.71
20.00	10.68	5.23
35.00	9.84	4.38
39.00	9.09	3.64
40.00	6.02	3.29
41.00	2.96	2.96
45.00	2.28	2.28
50.00	2.00	2.00

40.

VOL NaOH, ml	pH (a)	pH (d)
0.00	2.61	2.80
5.00	3.61	3.99
12.50	4.21	4.60
20.00	4.81	5.20
24.00	5.59	5.98
25.00	8.52	8.71
26.00	11.42	11.42
30.00	12.10	12.10

41. (a) 3.83 (b) 8.34 (c) 11.75 (d) 6.83 (e) 9.85

43. (a) 2.02 (b) 4.62 (c) 9.25 (d) 2.50 (e) 6.70

45. (a) 1.70 (b) 1.10 (quadratic method)

47. (a) 2.33 (b) 7.38 (c) 2.05 (d) 3.05 (e) 6.50

49.

VOL NaOH, ml	pH (a)	pH (b)
0.00	1.23	11.51
10.00	1.45	10.50
20.00	1.69	9.73
24.00	1.80	8.95
25.00	1.83	8.34
26.00	1.87	7.23
35.00	2.16	6.53
45.00	2.74	5.75
49.00	3.48	4.97
50.00	7.16	3.98
51.00	11.00	3.00
60.00	11.96	2.04

50. (a)

pH	D	α_0	α_1	α_2
1.00	1.536×10^{-2}	0.651	3.49×10^{-1}	1.89×10^{-4}
3.00	5.751×10^{-5}	1.74×10^{-2}	9.32×10^{-1}	5.05×10^{-2}
5.00	3.441×10^{-6}	2.91×10^{-5}	1.56×10^{-1}	8.44×10^{-1}
7.00	2.910×10^{-6}	3.44×10^{-9}	1.84×10^{-3}	9.98×10^{-1}
9.00	2.905×10^{-6}	3.44×10^{-13}	1.84×10^{-5}	1.000
11.00	2.905×10^{-6}	3.44×10^{-17}	1.84×10^{-7}	1.000
13.00	2.905×10^{-6}	3.44×10^{-21}	1.84×10^{-9}	1.000

(b)

pH	D	α_0	α_1	α_2
1.00	1.000×10^{-2}	1.00	5.70×10^{-7}	6.84×10^{-21}
3.00	1.000×10^{-6}	1.00	5.70×10^{-5}	6.84×10^{-17}
5.00	1.006×10^{-10}	9.94×10^{-1}	5.67×10^{-3}	6.80×10^{-13}
7.00	1.570×10^{-14}	6.37×10^{-1}	3.63×10^{-1}	4.36×10^{-9}
9.00	5.800×10^{-17}	1.72×10^{-2}	9.83×10^{-1}	1.18×10^{-6}
11.00	5.702×10^{-19}	1.75×10^{-4}	1.00	1.20×10^{-4}
13.00	5.768×10^{-21}	1.73×10^{-6}	9.88×10^{-1}	1.19×10^{-2}

51.

	D	α_0	α_1	α_2	α_3	PRINCIPAL SPECIES
(a)	5.219×10^{-18}	1.92×10^{-7}	0.136	0.864	3.63×10^{-5}	HPO_4^{2-} ($\sim 86\%$) $H_2PO_4^-$ ($\sim 14\%$)
(b)	8.28×10^{-6}	1.21×10^{-3}	0.648	0.351		Ox^{2-} ($\sim 35\%$) HOx^- ($\sim 65\%$)
(c)	1.83×10^{-9}	5.46×10^{-5}	0.940	0.060		HSO_3^- ($\sim 94\%$) SO_3^{2-} ($\sim 6\%$)
(d)	8.11×10^{-9}	0.123	0.877	5.6×10^{-5}	2.3×10^{-14}	$H_2PO_4^-$ ($\sim 88\%$) H_3PO_4 ($\sim 12\%$)

52. (a) 5.18×10^{-2} **(b)** 2.42×10^{-4} **(c)** 3.02×10^{-2} **(d)** 3.08×10^{-2}

Chapter 10

1. **(a)** gfw $H_3PO_4/2$ **(c)** gfw HCl/1 **(e)** gfw $C_6H_4(COOCH_3)_2/2$

2. **(a)** gfw NaOH/1 **(c)** gfw $Na_2B_4O_7/2$ **(e)** gfw NaOH/1

3. **(a)** Dissolve 12 g NaOH, and dilute to about 2.0 liters.
   (b) Dilute exactly 113 g HCl to 5.00 liters.
   (c) Dilute 18 ml of H_3PO_4 to 2.5 liters.
   (d) Dissolve exactly 3.06 g, and dilute to 500 ml.

5. **(a)** $\bar{x} = 0.06466\ N$ **(b)** $(s)_r = 0.9$ ppt **(c)** 95% C.I. = 0.06466 ± 0.00009

7. **(a)** 2.7 to 3.5 g **(b)** 0.11 to 0.14 g **(c)** 1.03 to 1.32 g

8. 0.133 g/100 ml

10. 355 g/eq

12. 7.79%

14. 8.82% protein

16. 863 ppm

18. 0.0150 N

20. 28.6% $(NH_4)_2SO_4$ and 39.0% NH_4NO_3

22. 45.3%

24. **(a)** 12.5% **(b)** 75.0%

26. 26.1% $NaHC_2O_4$ and 16.7% $H_2C_2O_4 \cdot 2H_2O$

27. 78.8% KOH, 17.1% K_2CO_3, and 4.1% H_2O

29. **(a)** 18.4 ml **(b)** 36.7 ml **(c)** 73.8 ml **(d)** 32.1 ml

31. **(a)** 4.86 mg NaOH/ml
    (b) 7.82 mg Na_2CO_3/ml and 5.92 mg $NaHCO_3$/ml
    (c) 4.00 mg Na_2CO_3/ml and 4.81 mg NaOH/ml
    (d) 8.07 mg Na_2CO_3/ml
    (e) 11.3 mg $NaHCO_3$/ml

33. **(a)** 2.41 mg HCl/ml
    (b) 2.13 mg H_3PO_4/ml and 2.91 mg HCl/ml
    (c) 8.33 mg H_3PO_4/ml and 6.18 mg NaH_2PO_4/ml
    (d) 6.88 mg NaH_2PO_4/ml
    (e) 3.41 mg HCl/ml

Chapter 11

1. **(a)** 0.0496 F **(b)** 0.0992 N **(c)** 9.64 mg/ml **(d)** 12.5 mg/ml

3. **(a)** Dissolve 21.9 g H_4Y, and dilute to 1.50 liters.
   (b) 0.0500 F **(c)** 5.58 mg/ml

5. **(a)** Dissolve 2.800 g in water, and dilute to 750.0 ml.
   (b) 1.03 mg/ml

7. **(a)** 0.0289 N **(b)** 2.32%

9. 3.97×10^3 mg Ca/2.00 liters and 6.81×10^2 mg Mg/2.00 liters

11. 62.3%

13. 13.0% NaBr and 65.2% $NaBrO_3$

15. 0.0839 F

17. 5.08% Pb, 6.27% Mg, and 8.76% Zn

19. 7.0% Pb, 23.6% Zn, 66.4% Cu, and 2.96% Sn

20. (a) 3.0×10^{10} (b) 3.2×10^{12} (c) 5.3×10^{13}

23.

VOL, ml	pSr	VOL, ml	pSr
0.00	1.82	25.00	5.09
10.00	2.12	25.10	5.78
24.00	3.39	26.00	6.78
24.90	4.40	30.00	7.48

Chapter 12

1. (a) $\underline{MnO_2(s)} + HNO_2 + H^+ \rightleftharpoons Mn^{2+} + NO_3^- + H_2O$
   (b) $\underline{2VO^{2+}} + 3I^- + 4H^+ \rightleftharpoons 2V^{3+} + I_3^- + 2H_2O$
   (c) $2Fe^{2+} + \underline{SeO_4^{2-}} + 4H^+ \rightleftharpoons 2Fe^{3+} + H_2SeO_3 + H_2O$
   (d) $3Mn^{2+} + \underline{2MnO_4^-} + 2H_2O \rightleftharpoons 5MnO_2(s) + 4H^+$
   (e) $\underline{O_2(g)} + 6I^- + 4H^+ \rightleftharpoons 2H_2O + 2I_3^-$
   (f) $Sn^{2+} + \underline{2AgI(s)} \rightleftharpoons Sn^{4+} + 2Ag(s) + 2I^-$

3. (a) 0.319 V   (b) -0.096 V   (c) -0.159 V

5. (a) 1.26 V   (b) 1.02 V   (c) 1.064 V   (d) 1.26 V   (e) 0.190 V

7. (a) 0.274 V   (b) -1.144 V

9. (a) 0.008 V, cathode   (b) -0.110 V, anode   (c) 0.31 V, cathode
   (d) -0.03 V, anode   (e) 0.06 V, cathode   (f) -1.5 V, anode

11. (a) 0.612 V, galvanic   (b) 1.702 V, galvanic   (c) 0.221 V, galvanic
   (d) -0.745 V, electrolytic   (e) 0.051 V, galvanic   (f) 1.62 V, galvanic

13. (a) 6×10^9 (b) 1.20×10^{-6} (c) 7×10^{12} (d) 2×10^{47} (e) 8×10^{46}
   (f) 4.8×10^{-11}

15. -0.587 V

17. 0.377 V

19. 3.5×10^{-11}

21. 6×10^{18}

23. 2.9×10^{-11}

25. 1.1×10^8

27. 3.7×10^{-8}

29. **(a)** −0.03 V   **(b)** 1.01 V   **(c)** 0.322 V   **(d)** 1.07 V

31.

		E, V	
VOL, ml	**(a)**	**(c)**	**(e)**
10.00	−0.265	0.35	0.211
24.00	−0.173	0.44	0.257
24.90	−0.113	0.50	0.287
25.00	0.018	0.95	1.07
25.10	0.083	1.18	1.39
26.00	0.113	1.21	1.40
30.00	0.133	1.23	1.41

Chapter 13

1. **(a)** $S_2O_8^{2-} + 2Ce^{3+} \rightleftharpoons 2SO_4^{2-} + 2Ce^{4+}$
 (b) $2Mn^{2+} + 5NaBiO_3(s) + 4H^+ \rightleftharpoons 2MnO_4^- + 5BiO^+ + 2H_2O + 5Na^+$
 (c) $H_2MoO_4 + 2H^+ + Ag(s) + Cl^- \rightleftharpoons MoO_2^+ + AgCl(s) + 2H_2O$
 (d) $H_2O_2 + V^{3+} + 2H_2O \rightleftharpoons 2H^+ + V(OH)_4^+$
 (e) $SCN^- + 3Br_2 + 4H_2O \rightleftharpoons SO_4^{2-} + 6Br^- + 8H^+ + CN^-$
 (f) $3H_2C_2O_4 + BrO_3^- \rightleftharpoons 6CO_2 + Br^- + 3H_2O$
 (g) $2Ce^{4+} + H_2O_2 \rightleftharpoons 2Ce^{3+} + O_2 + 2H^+$
 (h) $3HClO + I^- \rightleftharpoons 3Cl^- + IO_3^- + 3H^+$

3. **(a)** Dissolve 24.5 g, and dilute to 2.00 liters.
 (b) Dissolve 2.09 g, and dilute to 500 ml.
 (c) Dissolve 38.1 g, and dilute to 2.50 liters.

5. 2.35 mg C_6H_5OH/ml $KBrO_3$

7. 0.515 mg $K_2Cr_2O_7$/ml BaS_2O_3

9. 0.0408 *N*

11. 39.8%

13. **(a)** 10.5% Sb   **(b)** 14.6% Sb_2S_3

15. 0.594%

17. 0.359 g

19. **(a)** 12.6% $FeO \cdot Cr_2O_3$ **(b)** 5.84% Cr

21. **(a)** $MnO_4^- + 4Mn^{2+} + 8H^+ + 20F^- \rightleftharpoons 5MnF_4^- + 4H_2O$
 (b) 0.120 *N*   **(c)** 52.4%

23. 67.0% Fe and 16.5% Cr

25. 36.8%

27. 1.94%

29. 0.167 mg H_2S and 0.465 mg SO_2/liter

31. 0.354 mg CO/liter

33. 18.2%

35. 8.14 mg I_2 and 3.96 mg KI/ml

37. 97.6%

39. 34.0%

Chapter 14

1. (a) 0.031 V
   (b) $Cu|CuBr(\text{sat'd}),Br^-(xM)\|SCE$
   (c) $pBr = (E_{SCE} - E_{cell} - 0.031)/0.0591 = (0.211 - E_{cell})/0.0591$
   (d) 3.27

3. (a) $SCE\|Hg_2Br_2(\text{sat'd}), Br^-(xM)|Hg$
   $pBr = (E_{cell} + 0.110)/0.0591$
   (b) $SCE\|Ag_2C_2O_4(\text{sat'd}),C_2O_4{}^{2-}(xM)|Ag$
   $pC_2O_4 = 2(E_{cell} - 0.248)/0.0591$
   (c) $SCE\|Sn^{4+}(1.00 \times 10^{-4} M),Sn^{2+}(xM)|Pt$
   $pSn(II) = 2(E_{cell} + 0.206)/0.0591$

5. 5.14

7. -1.71 V

9. (a) 0.146 V   (b) 0.087 V   (c) 0.028 V

11. (a) 2.76   (b) 7.24

13. 1.6×10^{-12}

15. (a)

VOL Ce^{4+}, ml	E vs. SCE, V	VOL Ce^{4+}, ml	E vs. SCE, V
5.00	0.064	49.00	0.142
10.00	0.074	50.00	0.46
15.00	0.081	51.00	1.10
25.00	0.092	55.00	1.14
40.00	0.110	60.00	1.16

16.

VOL NaCl, ml	E vs. SCE, V	VOL NaCl, ml	E vs. SCE, V
5.00	0.089	39.00	0.193
15.00	0.102	40.00	0.365
25.00	0.119	41.00	0.451
30.00	0.131	45.00	0.471
35.00	0.150	50.00	0.479

18. (a) 0.406 V (b) 0.111 V

19. (a) pH = 10.679 and $[H_3O^+] = 2.096 \times 10^{-11}$
(b) $[H_3O^+] = 2.016 \times 10^{-11}$ to 2.180×10^{-11}
(c) -4% to $+4\%$

21. 3.497

Chapter 15

1. (a) Cd (b) -1.945 V

2. -0.066 V

4. 21.9%

6. 0.689 mg/ml

8. 23.2%

10. 16.8 ppm

12. 0.665 g/liter

13. 0.587%

14. 3.92%

16. 0.0525 mg Cr/cm^2

18. 2.62 ppm

19. 11.7% Cd and 4.10% Zn

21. 30.8%

23. 2.36% CCl_4 and 2.85% $CHCl_3$

25. 7.77×10^{-3} M

27.

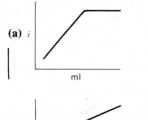

(a) *i* — ml

(c) *i* — ml

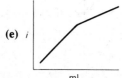

(e) *i* — ml   or   *i* — ml

Chapter 16

1. (a) $\epsilon = 7.28 \times 10^3$; $T = 0.336$ or 33.6%

(b) $A = 0.0536$; $\epsilon = 630$

(c) concn $= 3.15 \times 10^{-6}\ M$; $T = 0.748$ or 74.8%

(d) $A = 0.506$; concn $= 1.26 \times 10^{-5}\ M$ or 4.11 ppm

(e) $T = 0.113$ or 11.3%; $b = 1.86$ cm

(f) $A = 1.016$; $b = 1.02$ cm

(g) concn $= 5.56 \times 10^{-5}\ M$; $A = 1.33$; $T = 0.047$ or 4.7%

(h) $A = 0.763$; $T = 0.173$ or 17.3%

3. concn $= 4.28 \times 10^{-5}\ M$; $\epsilon = 2.32 \times 10^4$

5. *a* can be expressed in any of a number of concentration terms. Two would be $3.37 \times 10^{-2}\ \mathrm{cm}^{-1}\ \mathrm{mg}^{-1}$ liter and $33.7\ \mathrm{cm}^{-1}\ \mathrm{g}^{-1}$ liter.

7. (a) 0.127   (b) 0.746 or 74.6%   (c) 0.428

9. (a) 16.9%   (b) 2.87%   (c) 6.17×10^3

10. $6.49 \times 10^{-3}\%$

12.

	F_{HIn}	$[In^-]$	$[HIn]$	A_{430}	A_{600}
(a)	10.0×10^{-4}	8.54×10^{-4}	1.46×10^{-4}	0.286	1.67
(c)	2.50×10^{-4}	2.39×10^{-4}	1.14×10^{-5}	0.020	0.468
(e)	0.500×10^{-4}	4.95×10^{-5}	4.90×10^{-7}	0.001	0.0970

13. (a) $\pm 7.6\%$ (b) $\pm 1.4\%$ (c) $\pm 1.7\%$

14. (a) 26, 17, and 63 ppt   (c) 33, 22, and 81 ppt

15. (a) $3.94 \times 10^{-4}\ M$ B   (b) $7.29 \times 10^{-4}\ M$ A and $4.88 \times 10^{-4}\ M$ B

16. (a) $1.91 \times 10^{-4}\ F$ Co and $4.4 \times 10^{-5}\ F$ Ni

17. (b) 1.033   (c) 0.829

18.   [P]   [Q]
   (a) 6.67×10^{-5} 1.98×10^{-4}
   (c) 5.78×10^{-5} 4.71×10^{-5}
   (e) 1.38×10^{-4} 6.91×10^{-5}

19. (a) 0.544   (c) 0.197

20. (a) 0.380

21. (a) 5.58   (c) 4.69

22. 1.56×10^{-2}% Co and 2.98×10^{-2}% Ni

Chapter 18

1. (a) $2.53 \times 10^{-3}\,M$ (b) $4.62 \times 10^{-4}\,M$ (c) $1.00 \times 10^{-5}\,M$ (d) $6.12 \times 10^{-7}\,M$

3. (a) 80.0 ml   (b) 60.0 ml   (c) 30.0 ml

5. (a) 22   (b) 5.8

APPENDIXES

1 use of exponential numbers

Scientists frequently find it useful or necessary to employ exponential numbers to express quantitative data. This appendix describes exponential notation and the manipulation of exponential numbers.

EXPONENTIAL NOTATION

An exponent is used to describe the process of repeated multiplication or division. For example, 3^5 means $3 \times 3 \times 3 \times 3 \times 3 = 243$. The 5 is the exponent of the number or base 3; in obtaining 243 we are raising or taking 3 to the fifth power.

A negative exponent implies repeated division. Thus 3^{-5} means

$$\frac{1}{3} \times \frac{1}{3} \times \frac{1}{3} \times \frac{1}{3} \times \frac{1}{3} = \frac{1}{3^5} = 3^{-5} = 0.00412$$

Note that the *reciprocal* of an exponential number is obtained by changing the sign of the exponent.

It is important to note that a number raised to the first power is the number itself; any number raised to the zero power has a value of 1. For example,

$$4^1 = 4$$

$$4^0 = 1$$

$$67^0 = 1$$

Fractional Exponents. The inverse of raising a number to a power is extracting or taking the root of a number. This process is symbolized by a fractional exponent. Thus

$$(243)^{1/5} = 3$$

Here we have found the fifth root of 243. Other examples are

$$4^{1/2} = 2 \quad \text{and} \quad 27^{1/3} = 3$$

Also

$$4^{-1/2} = \frac{1}{4^{1/2}} = \frac{1}{2}$$

Combining Exponents in Multiplication and Division. Two exponential numbers with the same base can be multiplied or divided by adding or subtracting exponents. Thus

$$3^3 \times 3^2 = (3 \times 3 \times 3)(3 \times 3) = 3^{(3+2)} = 3^5 = 243$$

$$3^4 \times 3^{-2} \times 3^0 = (3 \times 3 \times 3 \times 3)(\tfrac{1}{3} \times \tfrac{1}{3}) \times 1 = 3^{(4-2+0)} = 3^2 = 9$$

$$\frac{4^4}{4^2} = \frac{4 \times 4 \times 4 \times 4}{4 \times 4} = 4^{(4-2)} = 4^2 = 16$$

$$\frac{2^3}{2^{-1}} = \frac{2 \times 2 \times 2}{1/2} = 2^4 = 16$$

Note that in the last example the exponent is given by the relationship

$$3 - (-1) = 3 + 1 = 4$$

USE OF EXPONENTS IN SCIENTIFIC NOTATION

Scientists and engineers often deal with very large or very small numbers, which makes ordinary decimal notation awkward or impossible. For example, to express the number of atoms in a mole in decimal notation would require that the number 602 be followed by 21 zeros.

   In scientific notation a number is written as a multiple of two numbers, one of which is in decimal notation and the other of which is expressed as a power of 10. Thus the number of particles in the mole can be conveniently written in scientific notation as 6.02×10^{23}. Other examples are

$$4.32 \times 10^3 = 4.32 \times 10 \times 10 \times 10 = 4320$$

$$4.32 \times 10^{-3} = 4.32 \times \tfrac{1}{10} \times \tfrac{1}{10} \times \tfrac{1}{10} = 0.0432$$

$$0.002002 = 2.002 \times \tfrac{1}{10} \times \tfrac{1}{10} \times \tfrac{1}{10} = 2.002 \times 10^{-3}$$

$$375 = 3.75 \times 10 \times 10 = 3.75 \times 10^2$$

It should be clear from these examples that the conversion of a number from scientific to purely decimal notation requires that the decimal point be shifted a number of places that is equal to the exponent. The shift is to the right if the exponent is positive and to the left if it is negative. The process is reversed when numbers are converted from decimal to scientific notation.

   It should be noted that several equivalent scientific forms exist for any given number. Thus

$$4.32 \times 10^3 = 43.2 \times 10^2 = 432 \times 10^1$$

$$= 0.432 \times 10^4 = 0.0432 \times 10^5$$

ARITHMETIC OPERATIONS WITH SCIENTIFIC NOTATION

The use of scientific notation is often helpful in preventing decimal errors in arithmetic calculations. Some examples follow.

Multiplication. Here the decimal parts of the number are multiplied and the exponents added. Thus

$$420,000 \times 0.0300 = 4.20 \times 10^5 \times 3.00 \times 10^{-2}$$
$$= 12.60 \times 10^3 = 1.260 \times 10^4$$
$$0.0060 \times 0.000020 = 6.0 \times 10^{-3} \times 2.0 \times 10^{-5}$$
$$= 12 \times 10^{-8} = 1.2 \times 10^{-7}$$

Division. Here the decimal numbers are divided and the exponent of the denominator subtracted from that of the numerator. For example,

$$\frac{0.015}{5000} = \frac{15 \times 10^{-3}}{5.0 \times 10^3} = 3.0 \times 10^{-6}$$

Addition and Subtraction. To add or subtract numbers expressed in scientific notation it is necessary that all be expressed to a common power of 10. The decimal parts of the number are then added or subtracted. Thus

$$2.00 \times 10^{-11} + 4.00 \times 10^{-12} - 3.00 \times 10^{-10}$$
$$2.00 \times 10^{-11} + 0.40 \times 10^{-11} - 30.0 \times 10^{-11} = -27.6 \times 10^{-11} = -2.8 \times 10^{-10}$$

Raising to a Power. Both parts of an exponential are individually raised to the desired power; for example,

$$(2.0 \times 10^{-3})^4 = (2.0)^4 \times (10^{-3})^4 = 16 \times 10^{-12} = 1.6 \times 10^{-11}$$

Extraction of a Root. Here the number should be expressed in such a way that the exponent of 10 is exactly divisible by the desired root. The root of each part is then extracted. For example,

$$\sqrt[3]{4.0 \times 10^{-5}} = \sqrt[3]{40 \times 10^{-6}} = \sqrt[3]{40} \times \sqrt[3]{10^{-6}}$$
$$= 3.4 \times 10^{-2}$$

2 logarithms

A logarithm of a number is the power to which some base number must be raised in order to give that number. The following are some simple examples of base 10 logarithms:

$$\log 1000 = \log 10^3 \ = \ 3$$
$$\log 100 \ = \log 10^2 \ = \ 2$$
$$\log 10 \ \ = \log 10^1 \ = \ 1$$
$$\log 1 \ \ \ = \log 10^0 \ = \ 0$$
$$\log 0.1 \ \ = \log 10^{-1} = -1$$
$$\log 0.01 \ = \log 10^{-2} = -2$$

Before describing how logarithms are obtained for numbers that are not integer multiples of 10, it is worthwhile to consider the nature of the logarithm of a product, quotient, power, or root.

LOGARITHMS OF PRODUCTS, QUOTIENTS, POWERS, AND ROOTS

In Appendix 1 it was shown that the exponent of the product of two numbers with the same base is the sum of the exponents. Thus

$$10^3 \times 10^2 = 10^5$$

Similarly, the exponent for a quotient is the difference between the exponents of the numerator and denominator.

$$\frac{10^3}{10^2} = 10^1$$

An exponential number is raised to a power by multiplication of exponents; the extraction of a root involves division. Thus

$$(10^3)^4 = 10^{(4 \times 3)} = 10^{12}$$
$$(10^8)^{1/2} = 10^{8/2} = 10^4$$

We have noted that a logarithm to the base 10 is simply the exponent of 10 needed to express the desired number. Thus the following statements can be made:

1. The logarithm of a product is the sum of the logarithms of the individual numbers.
2. The logarithm of a quotient is the difference between the logarithms of the individual numbers.
3. The logarithm of a base number raised to some power is the logarithm of the base number multiplied by that power.
4. The logarithm of the root of a number is the logarithm of that number divided by the root. For example,

$$\log (100 \times 1000) = \log 10^2 + \log 10^3$$
$$= 2 + 3 = 5$$

$$\log \frac{100}{1000} = \log 10^2 - \log 10^3$$
$$= 2 - 3 = -1$$

$$\log (1000)^2 = 2 \log 10^3 = 2 \times 3 = 6$$

$$\log (0.01)^6 = 6 \log 10^{-2} = 6 \times (-2) = -12$$

$$\log (1000)^{1/3} = \tfrac{1}{3} \log 10^3 = \tfrac{1}{3} \times 3 = 1$$

DETERMINATION OF A LOGARITHM FROM A TABLE OF LOGARITHMS

It is now necessary to consider how the logarithm of a number such as 4.32 can be derived. We have seen that the logarithm of 1 is 0 and of 10 is 1; thus log 4.32 must lie between these two numbers. Its exact magnitude can be found in a table of logarithms such as that found inside the back cover. The first column of this table, labeled n, gives the first two digits of the number whose log is being sought; the row at the top of the table gives the third digit. Note that log tables ordinarily do not show decimal points; however, the table does in fact cover a range of n from 0.00 to 9.99.

To find log 4.32 we move down the first column to 43 and then over to the column labeled 2; we find

$$\log 4.32 = 0.6355$$

As a second example let us derive the logarithm of 6.643. Proceeding as before we find

$$\log 6.64 = 0.8222$$

$$\log 6.65 = 0.8228$$

The logarithm of the desired number is obtained by *interpolation* between the two numbers; that is, we *assume* that the logarithm of 6.643 lies 0.3 of the way between 0.8222 and 0.8228. Thus

$$(0.8228 - 0.8222)0.3 = 0.00018 = 0.0002$$

and

$$\log 6.643 = 0.8222 + 0.0002 = 0.8224$$

To extend a log table to numbers lying outside the range of 0 to 9.99, advantage is taken of the fact that the logarithm of a product is the sum of the logarithms of the individual numbers. For example, to find the logarithm of 664.3 we write the number in scientific notation. Then

$$\log 6.643 \times 10^2 = \log 6.643 + \log 10^2$$
$$= \quad 0.8224 \ + \quad 2$$
$$= 2.8224$$

As another example,

$$\log 0.00247 = \log 2.47 \times 10^{-3}$$
$$= \log 2.47 + \log 10^{-3}$$
$$= \ 0.3927 \ - \quad 3$$
$$= -2.6073$$

For some purposes it is useful to dispense with the subtraction step and report the log as a *negative* integer and a *positive* decimal number. That is,

$$\log 0.00247 = \bar{3}.3927$$

Another common method for expressing negative logarithms involves adding and subtracting 10 from the log. Thus

$$\log 2.47 \times 10^{-3} = \bar{3}.3927$$
$$+10 \quad -10$$
$$\log 2.47 \times 10^{-3} = \overline{7.3927 - 10}$$

These examples demonstrate that the logarithm of a number is the sum of two parts, a *characteristic* which lies to the left of the decimal point and a *mantissa* which lies to the right. The characteristic is the logarithm of 10 raised to a power; it indicates the location of the decimal point in the original number when it is expressed in decimal notation. The mantissa, on the other hand, is the logarithm of a number which lies between 0 and $9.999\cdots$.

CALCULATION OF THE ANTILOG

Conversion of the logarithm to the original number is called taking the *antilog*.

EXAMPLE

Find x if log x is equal to (1) 6.4320, (2) $\bar{4}.375$, and (3) -11.7734.

1. Let us first separate the logarithm into its characteristic and mantissa and find the antilog of each. Thus

$$\log x = 6.4320 = 6 + 0.4320$$
$$\text{antilog } 6 = 10^6$$

To find antilog 0.4320 we search the table of logarithms for the two numbers that bracket this number and find log 2.70 = 0.4314 and log 2.71 = 0.4330. The difference between the two is 0.0016; the difference between the number of interest and the lower of these is $(0.4320 - 0.4314) = 0.0006$. Therefore log 0.4320 must lie $0.0006/0.0016 = 0.4$ of the way between 2.70 and 2.71. Thus

$$\text{antilog } 0.4320 = 2.70 + 0.004 = 2.704$$

From our earlier discussion, however, it is evident that the antilog of a sum is the product of the individual antilogs. Therefore

$$\log x = 6 + 0.4320$$

$$\text{antilog } x = (\text{antilog } 6)(\text{antilog } 0.4320)$$

$$= \quad 10^6 \quad \times \quad 2.704 \quad = 2.704 \times 10^6$$

2. $\quad \log x = \bar{4}.375 = -4 + 0.375$

$$\text{antilog } x = (\text{antilog } -4)(\text{antilog } 0.375)$$

$$= \quad 10^{-4} \quad \times \quad 2.37 \quad = 2.37 \times 10^{-4}$$

3. $\quad \log x = -11.7734$

Because a table of logarithms applies to positive numbers only, a first requirement here is to provide this logarithm with a positive mantissa. To accomplish the conversion we add and subtract the next largest integral number to and from the logarithm. Thus

$$\begin{array}{cc} +12 & -12 \end{array}$$

$$\log x = -11.7734$$

$$= \quad 0.2266 - 12$$

$$\text{antilog } x = (\text{antilog } 0.2266)(\text{antilog } -12)$$

$$1.685 \quad \times \quad 10^{-12}$$

USE OF LOGARITHMS FOR NUMERICAL CALCULATIONS

The advent of the inexpensive electronic calculator has largely displaced logarithm tables as a means of carrying out otherwise tedious arithmetic calculations. The following example will, however, demonstrate how logarithms can be employed for this purpose.

EXAMPLE

Evaluate

$$x = \frac{38.26 \times 0.1020 \times 0.0864 \times 100}{1.675}$$

The several multiplications are accomplished in a single step by adding the logarithms:

$$
\begin{array}{ll}
\log 38.26 & 1.5827 \\
\log 0.1020 & 9.0086 - 10 \\
\log 0.0864 & 8.9365 - 10 \\
\log 100 & \underline{2.0000} \\
& 21.5278 - 20
\end{array}
$$

After eliminating 20 from the characteristic and remainder, we subtract log 1.675

$$
\begin{array}{ll}
& 1.5278 \\
\log 1.675 - & \underline{0.2240} \\
\log x & 1.3038
\end{array}
$$

Thus

$$x = 20.13$$

3 the quadratic equation

Problems dealing with chemical equilibrium frequently require the solution of a quadratic equation. The expression in question is first written in the form $ax^2 + bx + c = 0$, where a, b, and c are numerical constants. These are then substituted into the equation

$$x = \frac{-b \pm \sqrt{b^2 - 4ac}}{2a} \tag{1}$$

EXAMPLE

The hydronium ion concentration in a system is given by the equation

$$2 \times 10^{-2} = \frac{4 \times 10^{-3}}{2[H_3O^+]} - [H_3O^+]$$

To solve for $[H_3O^+]$ the expression is rearranged to

$$2[H_3O^+]^2 + 4 \times 10^{-2}[H_3O^+] - 4 \times 10^{-3} = 0$$

Substituting into Equation 1,

$$[H_3O^+] = \frac{-4 \times 10^{-2} \pm \sqrt{(4 \times 10^{-2})^2 - (-4 \times 2 \times 4 \times 10^{-3})}}{2 \times 2}$$

$$= \frac{-4 \times 10^{-2} \pm \sqrt{0.16 \times 10^{-2} + 3.2 \times 10^{-2}}}{4}$$

$$= \frac{-4 \times 10^{-2} \pm 1.83 \times 10^{-1}}{4}$$

Of the two solutions to this equation only the positive value possesses significance in the present context. We therefore conclude that

$$[H_3O^+] = \frac{-4 \times 10^{-2} + 18.3 \times 10^{-2}}{4} = 3.6 \times 10^{-2}$$

4 solution of higher order equations

Although methods for the rigorous solution of equations containing terms raised to powers greater than 2 exist, it is usually simpler to proceed by the systematic approximation procedure shown below.

EXAMPLE

The solubility s of a precipitate is given by the expression

$$s^2(1.0 \times 10^{-3} + s) = 2.0 \times 10^{-6}$$

In solving for s, all terms are first collected on one side of the equation:

$$s^3 + 1.0 \times 10^{-3}s^2 - 2.0 \times 10^{-6} = 0$$

Values for s are then systematically tried. The one that causes the left side of the equation to approach zero most closely represents the desired solution.

Because s represents a solubility, we need concern ourselves only with a positive root. If we assume a value of 0.0 for s, the equation becomes

$$-2.0 \times 10^{-6} = 0$$

Assuming $s = 1.0$, we find that

$$1.0 + 1.0 \times 10^{-3} - 2.0 \times 10^{-6} = 0$$

$$1.001 = 0$$

Clearly s will have a value between 0 and 1. Using $s = 0.10$,

$$1.0 \times 10^{-3} + 1.0 \times 10^{-5} - 2.0 \times 10^{-6} = 0$$

$$0.001008 = 0$$

With $s = 0.010$,

$$1.0 \times 10^{-6} + 1.0 \times 10^{-7} - 2.0 \times 10^{-6} = 0$$

$$-0.9 \times 10^{-6} = 0$$

Because this gives a slightly negative number, s must lie between 0.10 and 0.010. Further, it appears that the latter represents the closer estimate of the two.

Setting $s = 0.012$,

$$1.73 \times 10^{-6} + 1.44 \times 10^{-7} - 2.0 \times 10^{-6} = 0$$

$$1.87 \times 10^{-6} - 2.0 \times 10^{-6} = 0$$

This estimate still gives a slightly negative value for s. Use of a slightly larger value, $s = 0.013$, gives a positive result:

$$2.20 \times 10^{-6} + 1.69 \times 10^{-7} - 2.0 \times 10^{-6} = 0$$

$$2.37 \times 10^{-6} - 2.0 \times 10^{-6} = 0$$

Thus s lies between 0.012 and 0.013. If it were desired to define this quantity more closely, values between these limits could be tested in the same way.

With a simple electronic calculator these calculations can be performed quite rapidly.

5 simplification of equations by neglect of terms

Terms that consist of sums or differences can frequently be simplified without serious error, provided the quantities making up the term differ sufficiently in magnitude. Thus in the term $(0.001 + x)$ two possibilities for simplification exist:

$$(0.001 + x) = x \qquad \text{provided } x \gg 0.001$$

or

$$(0.001 + x) = 0.001 \qquad \text{provided } x \ll 0.001$$

A knowledge of the physical significance of x will frequently provide the basis for choice between these assumptions. It is clear, however, that the validity of either will ultimately depend upon the numerical value of x and the accuracy with which the answer must be known. *Whenever a simplifying assumption has been made, the answer obtained must be considered provisional until it is established that the assumption was indeed justified.*

EXAMPLE

Considering again the example shown in Appendix 4,

$$s^2(1.0 \times 10^{-3} + s) = 2.0 \times 10^{-6}$$

If we assume that s is very small with respect to 1.0×10^{-3}, the equation simplifies to

$$1.0 \times 10^{-3} s^2 = 2.0 \times 10^{-6}$$

$$s = 4.5 \times 10^{-2}$$

Here the value for s is larger instead of smaller than 1.0×10^{-3}. Clearly the original assumption was not justified.

If we assume that s is large with respect to 1.0×10^{-3}, the equation becomes

$$s^3 = 2.0 \times 10^{-6}$$

and

$$s = 1.26 \times 10^{-2}$$

This assumption provides a value for s that is about 13 times greater than 1.0×10^{-3}. Whether the approximation is proper or not will depend upon the magnitude of error we can tolerate in the answer. If an uncertainty of something less than 1 part in 10 (10%) is satisfactory for our purposes, we may use this value of s as the solution to the equation. If, on the other hand, a greater accuracy is required, we must proceed with the method of successive approximations shown in Appendix 4. It is noteworthy that the approximate value for s obtained here (1.26×10^{-2}) provides an excellent starting point for a more rigorous solution by the technique shown in Appendix 4.

6

some standard and formal electrode potentials[a]

HALF-REACTION	E^0, V	FORMAL POTENTIAL, V
$Ag^+ + e \rightleftharpoons Ag(s)$	+0.799	0.228, 1-F HCl; 0.792, 1-F HClO$_4$; 0.77, 1-F H$_2$SO$_4$
$AgBr(s) + e \rightleftharpoons Ag(s) + Br^-$	+0.095	
$AgCl(s) + e \rightleftharpoons Ag(s) + Cl^-$	+0.222	0.228, 1-F KCl
$Ag(CN)_2^- + e \rightleftharpoons Ag(s) + 2CN^-$	−0.31	
$Ag_2CrO_4(s) + 2e \rightleftharpoons 2Ag(s) + CrO_4^{2-}$	+0.446	
$AgI(s) + e \rightleftharpoons Ag(s) + I^-$	−0.151	
$Ag(S_2O_3)_2^{3-} + e \rightleftharpoons Ag(s) + 2S_2O_3^{2-}$	+0.01	
$Al^{3+} + 3e \rightleftharpoons Al(s)$	−1.66	
$H_3AsO_4 + 2H^+ + 2e \rightleftharpoons H_3AsO_3 + H_2O$	+0.559	0.577, 1-F HCl, HClO$_4$
$Ba^{2+} + 2e \rightleftharpoons Ba(s)$	−2.90	
$BiO^+ + 2H^+ + 3e \rightleftharpoons Bi(s) + H_2O$	+0.32	
$BiCl_4^- + 3e \rightleftharpoons Bi(s) + 4Cl^-$	+0.16	
$Br_2(l) + 2e \rightleftharpoons 2Br^-$	+1.065	1.05, 4-F HCl

[a] Sources for E^0 values: W. M. Latimer, *The Oxidation States of the Elements and Their Potentials in Aqueous Solutions*, 2d ed. Englewood Cliffs, N.J.: Prentice-Hall, 1952; A. J. deBethune and N. A. S. Loud, *Standard Aqueous Electrode Potentials and Temperature Coefficients at 25°C*. Skokie, Ill.: Clifford A. Hampel, 1964. Source of formal potentials: E. H. Swift and E. A. Butler, *Quantitative Measurements and Chemical Equilibria*. San Francisco: Freeman, 1972.

HALF-REACTION	E^0, V	FORMAL POTENTIAL, V
$Br_2(aq) + 2e \rightleftharpoons 2Br^-$	$+1.087^b$	
$BrO_3^- + 6H^+ + 5e \rightleftharpoons \frac{1}{2}Br_2(l) + 3H_2O$	$+1.52$	
$Ca^{2+} + 2e \rightleftharpoons Ca(s)$	-2.87	
$C_6H_4O_2(quinone) + 2H^+ + 2e$ $\rightleftharpoons C_6H_4(OH)_2$	$+0.699$	0.696, 1-F HCl, HClO$_4$, H$_2$SO$_4$
$2CO_2(g) + 2H^+ + 2e \rightleftharpoons H_2C_2O_4$	-0.49	
$Cd^{2+} + 2e \rightleftharpoons Cd(s)$	-0.403	
$Ce^{4+} + e \rightleftharpoons Ce^{3+}$		1.70, 1-F HClO$_4$; 1.61, 1-F HNO$_3$; 1.44, 1-F H$_2$SO$_4$; 1.28, 1-F HCl
$Cl_2(g) + 2e \rightleftharpoons 2Cl^-$	$+1.359$	
$HClO + H^+ + e \rightleftharpoons \frac{1}{2}Cl_2(g) + H_2O$	$+1.63$	
$ClO_3^- + 6H^+ + 5e \rightleftharpoons \frac{1}{2}Cl_2(g) + 3H_2O$	$+1.47$	
$Co^{2+} + 2e \rightleftharpoons Co(s)$	-0.277	
$Co^{3+} + e \rightleftharpoons Co^{2+}$	$+1.842$	
$Cr^{3+} + e \rightleftharpoons Cr^{2+}$	-0.41	
$Cr^{3+} + 3e \rightleftharpoons Cr(s)$	-0.74	
$Cr_2O_7^{2-} + 14H^+ + 6e \rightleftharpoons 2Cr^{3+} + 7H_2O$	$+1.33$	
$Cu^{2+} + 2e \rightleftharpoons Cu(s)$	$+0.337$	
$Cu^{2+} + e \rightleftharpoons Cu^+$	$+0.153$	
$Cu^+ + e \rightleftharpoons Cu(s)$	$+0.521$	
$Cu^{2+} + I^- + e \rightleftharpoons CuI(s)$	$+0.86$	
$CuI(s) + e \rightleftharpoons Cu(s) + I^-$	-0.185	
$F_2(g) + 2H^+ + 2e \rightleftharpoons 2HF(aq)$	$+3.06$	
$Fe^{2+} + 2e \rightleftharpoons Fe(s)$	-0.440	
$Fe^{3+} + e \rightleftharpoons Fe^{2+}$	$+0.771$	0.700, 1-F HCl; 0.732, 1-F HClO$_4$; 0.68, 1-F H$_2$SO$_4$
$Fe(CN)_6^{3-} + e \rightleftharpoons Fe(CN)_6^{4-}$	$+0.36$	0.71, 1-F HCl; 0.72, 1-F HClO$_4$, H$_2$SO$_4$

[b] These potentials are hypothetical because they correspond to solutions that are 1.00 M in Br$_2$ or I$_2$. The solubilities of these two compounds at 25°C are 0.21 M and 0.0133 M, respectively. In saturated solutions containing an excess of Br$_2$(l) or I$_2$(s) the standard potentials for the half-reactions $Br_2(l) + 2e \rightleftharpoons 2Br^-$ or $I_2(s) + 2e \rightleftharpoons 2I^-$ should be used. On the other hand, at Br$_2$ and I$_2$ concentrations less than saturation these hypothetical electrode potentials should be employed.

HALF-REACTION	E^0, V	FORMAL POTENTIAL, V
$2H^+ + 2e \rightleftharpoons H_2(g)$	0.000	-0.005, 1-F HCl, HClO$_4$
$Hg_2^{2+} + 2e \rightleftharpoons 2Hg(l)$	$+0.789$	0.274, 1-F HCl; 0.776, 1-F HClO$_4$; 0.674, 1-F H$_2$SO$_4$
$2Hg^{2+} + 2e \rightleftharpoons Hg_2^{2+}$	$+0.920$	0.907, 1-F HClO$_4$
$Hg^{2+} + 2e \rightleftharpoons Hg(l)$	$+0.854$	
$Hg_2Cl_2(s) + 2e \rightleftharpoons 2Hg(l) + 2Cl^-$	$+0.268$	0.242, sat'd KCl; 0.282, 1-F KCl; 0.334, 0.1-F KCl
$Hg_2SO_4(s) + 2e \rightleftharpoons 2Hg(l) + SO_4^{2-}$	$+0.615$	
$HO_2^- + H_2O + 2e \rightleftharpoons 3OH^-$	$+0.88$	
$I_2(s) + 2e \rightleftharpoons 2I^-$	$+0.5355$	
$I_2(aq) + 2e \rightleftharpoons 2I^-$	$+0.620^b$	
$I_3^- + 2e \rightleftharpoons 3I^-$	$+0.536$	
$ICl_2^- + e \rightleftharpoons \frac{1}{2}I_2(s) + 2Cl^-$	$+1.06$	
$IO_3^- + 6H^+ + 5e \rightleftharpoons \frac{1}{2}I_2(s) + 3H_2O$	$+1.195$	
$IO_3^- + 6H^+ + 5e \rightleftharpoons \frac{1}{2}I_2(aq) + 3H_2O$	$+1.178^b$	
$IO_3^- + 2Cl^- + 6H^+ + 4e \rightleftharpoons ICl_2^- + 3H_2O$	$+1.24$	
$H_5IO_6 + H^+ + 2e \rightleftharpoons IO_3^- + 3H_2O$	$+1.60$	
$K^+ + e \rightleftharpoons K(s)$	-2.925	
$Li^+ + e \rightleftharpoons Li(s)$	-3.045	
$Mg^{2+} + 2e \rightleftharpoons Mg(s)$	-2.37	
$Mn^{2+} + 2e \rightleftharpoons Mn(s)$	-1.18	
$Mn^{3+} + e \rightleftharpoons Mn^{2+}$		1.51, 7.5-F H$_2$SO$_4$
$MnO_2(s) + 4H^+ + 2e \rightleftharpoons Mn^{2+} + 2H_2O$	$+1.23$	1.24, 1-F HClO$_4$
$MnO_4^- + 8H^+ + 5e \rightleftharpoons Mn^{2+} + 4H_2O$	$+1.51$	
$MnO_4^- + 4H^+ + 3e \rightleftharpoons MnO_2(s) + 2H_2O$	$+1.695$	
$MnO_4^- + e \rightleftharpoons MnO_4^{2-}$	$+0.564$	
$N_2(g) + 5H^+ + 4e \rightleftharpoons N_2H_5^+$	-0.23	
$HNO_2 + H^+ + e \rightleftharpoons NO(g) + H_2O$	$+1.00$	
$NO_3^- + 3H^+ + 2e \rightleftharpoons HNO_2 + H_2O$	$+0.94$	0.92, 1-F HNO$_3$
$Na^+ + e \rightleftharpoons Na(s)$	-2.714	
$Ni^{2+} + 2e \rightleftharpoons Ni(s)$	-0.250	
$H_2O_2 + 2H^+ + 2e \rightleftharpoons 2H_2O$	$+1.776$	
$O_2(g) + 4H^+ + 4e \rightleftharpoons 2H_2O$	$+1.229$	

HALF-REACTION	E^0, V	FORMAL POTENTIAL, V
$O_2(g) + 2H^+ + 2e \rightleftharpoons H_2O_2$	+0.682	
$O_3(g) + 2H^+ + 2e \rightleftharpoons O_2(g) + H_2O$	+2.07	
$Pb^{2+} + 2e \rightleftharpoons Pb(s)$	−0.126	−0.14, 1-F HClO$_4$; −0.29, 1-F H$_2$SO$_4$
$PbO_2(s) + 4H^+ + 2e \rightleftharpoons Pb^{2+} + 2H_2O$	+1.455	
$PbSO_4(s) + 2e \rightleftharpoons Pb(s) + SO_4^{2-}$	−0.356	
$PtCl_4^{2-} + 2e \rightleftharpoons Pt(s) + 4Cl^-$	+0.73	
$PtCl_6^{2-} + 2e \rightleftharpoons PtCl_4^{2-} + 2Cl^-$	+0.68	
$Pd^{2+} + 2e \rightleftharpoons Pd(s)$	+0.987	
$S(s) + 2H^+ + 2e \rightleftharpoons H_2S(g)$	+0.141	
$H_2SO_3 + 4H^+ + 4e \rightleftharpoons S(s) + 3H_2O$	+0.45	
$S_4O_6^{2-} + 2e \rightleftharpoons 2S_2O_3^{2-}$	+0.08	
$SO_4^{2-} + 4H^+ + 2e \rightleftharpoons H_2SO_3 + H_2O$	+0.17	
$S_2O_8^{2-} + 2e \rightleftharpoons 2SO_4^{2-}$	+2.01	
$Sb_2O_5(s) + 6H^+ + 4e \rightleftharpoons 2SbO^+ + 3H_2O$	+0.581	
$H_2SeO_3 + 4H^+ + 4e \rightleftharpoons Se(s) + 3H_2O$	+0.740	
$SeO_4^{2-} + 4H^+ + 2e \rightleftharpoons H_2SeO_3 + H_2O$	+1.15	
$Sn^{2+} + 2e \rightleftharpoons Sn(s)$	−0.136	−0.16, 1-F HClO$_4$
$Sn^{4+} + 2e \rightleftharpoons Sn^{2+}$	+0.154	0.14, 1-F HCl
$Ti^{3+} + e \rightleftharpoons Ti^{2+}$	−0.37	
$TiO^{2+} + 2H^+ + e \rightleftharpoons Ti^{3+} + H_2O$	+0.1	0.04, 1-F H$_2$SO$_4$
$Tl^+ + e \rightleftharpoons Tl(s)$	−0.336	−0.551, 1-F HCl; −0.33, 1-F HClO$_4$, H$_2$SO$_4$
$Tl^{3+} + 2e \rightleftharpoons Tl^+$	+1.25	0.77, 1-F HCl
$UO_2^{2+} + 4H^+ + 2e \rightleftharpoons U^{4+} + 2H_2O$	+0.334	
$V^{3+} + e \rightleftharpoons V^{2+}$	−0.255	−0.21, 1-F HClO$_4$
$VO^{2+} + 2H^+ + e \rightleftharpoons V^{3+} + H_2O$	+0.361	
$V(OH)_4^+ + 2H^+ + e \rightleftharpoons VO^{2+} + 3H_2O$	+1.00	1.02, 1-F HCl, HClO$_4$
$Zn^{2+} + 2e \rightleftharpoons Zn(s)$	−0.763	

7 solubility product constants[a]

Ksp

SUBSTANCE	FORMULA	gfw	K_{sp}
Aluminum hydroxide	$Al(OH)_3$	78.00	2×10^{-32}
Barium carbonate	$BaCO_3$	197.35	5.1×10^{-9}
Barium chromate	$BaCrO_4$	253.33	1.2×10^{-10}
Barium iodate	$Ba(IO_3)_2$	487.15	1.57×10^{-9}
Barium manganate	$BaMnO_4$	256.28	2.5×10^{-10}
Barium oxalate	BaC_2O_4	225.36	2.3×10^{-8}
Barium sulfate	$BaSO_4$	233.40	1.3×10^{-10}
Bismuth oxide chloride	$BiOCl$	260.43	7×10^{-9}
Bismuth oxide hydroxide	$BiOOH$	241.99	4×10^{-10}
Cadmium carbonate	$CdCO_3$	172.41	2.5×10^{-14}
Cadmium hydroxide	$Cd(OH)_2$	146.41	5.9×10^{-15}
Cadmium oxalate	CdC_2O_4	200.42	9×10^{-8}
Cadmium sulfide	CdS	144.46	2×10^{-28}
Calcium carbonate	$CaCO_3$	100.09	4.8×10^{-9}
Calcium fluoride	CaF_2	78.08	4.9×10^{-11}
Calcium oxalate	CaC_2O_4	128.10	2.3×10^{-9}
Calcium sulfate	$CaSO_4$	136.14	1.2×10^{-6}
Copper(I) bromide	$CuBr$	143.45	5.2×10^{-9}

[a]Taken from L. Meites, *Handbook of Analytical Chemistry*. New York: McGraw-Hill, 1963, p. 1-13.

SUBSTANCE	FORMULA	gfw	K_{sp}
Copper(I) chloride	$CuCl$	98.99	1.2×10^{-6}
Copper(I) iodide	CuI	190.44	1.1×10^{-12}
Copper(I) thiocyanate	$CuSCN$	121.62	4.8×10^{-15}
Copper(II) hydroxide	$Cu(OH)_2$	97.55	1.6×10^{-19}
Copper(II) sulfide	CuS	95.60	6×10^{-36}
Iron(II) hydroxide	$Fe(OH)_2$	89.86	8×10^{-16}
Iron(II) sulfide	FeS	87.91	6×10^{-18}
Iron(III) hydroxide	$Fe(OH)_3$	106.87	4×10^{-38}
Lanthanum iodate	$La(IO_3)_3$	663.62	6.2×10^{-12}
Lead carbonate	$PbCO_3$	267.20	3.3×10^{-14}
Lead chloride	$PbCl_2$	278.10	1.6×10^{-5}
Lead chromate	$PbCrO_4$	323.18	1.8×10^{-14}
Lead hydroxide	$Pb(OH)_2$	241.20	2.5×10^{-16}
Lead iodide	PbI_2	461.00	7.1×10^{-9}
Lead oxalate	PbC_2O_4	295.21	4.8×10^{-10}
Lead sulfate	$PbSO_4$	303.25	1.6×10^{-8}
Lead sulfide	PbS	239.25	7×10^{-28}
Magnesium ammonium phosphate	$MgNH_4PO_4$	137.32	3×10^{-13}
Magnesium carbonate	$MgCO_3$	84.32	1×10^{-5}
Magnesium hydroxide	$Mg(OH)_2$	58.33	1.8×10^{-11}
Magnesium oxalate	MgC_2O_4	112.33	8.6×10^{-5}
Manganese(II) hydroxide	$Mn(OH)_2$	88.95	1.9×10^{-13}
Manganese(II) sulfide	MnS	87.00	3×10^{-13}
Mercury(I) bromide	Hg_2Br_2	561.00	5.8×10^{-23}
Mercury(I) chloride	Hg_2Cl_2	472.09	1.3×10^{-18}
Mercury(I) iodide	Hg_2I_2	654.99	4.5×10^{-29}
Silver arsenate	Ag_3AsO_4	462.53	1×10^{-22}
Silver bromide	$AgBr$	187.78	5.2×10^{-13}
Silver carbonate	Ag_2CO_3	275.75	8.1×10^{-12}
Silver chloride	$AgCl$	143.32	1.82×10^{-10}
Silver chromate	Ag_2CrO_4	331.73	1.1×10^{-12}
Silver cyanide	$AgCN$	133.89	7.2×10^{-11}
Silver iodate	$AgIO_3$	282.77	3.0×10^{-8}

SUBSTANCE	FORMULA	gfw	K_{sp}
Silver iodide	AgI	234.77	8.3×10^{-17}
Silver oxalate	$Ag_2C_2O_4$	303.76	3.5×10^{-11}
Silver sulfide	Ag_2S	247.80	6×10^{-50}
Silver thiocyanate	AgSCN	165.95	1.1×10^{-12}
Strontium oxalate	SrC_2O_4	175.64	5.6×10^{-8}
Strontium sulfate	$SrSO_4$	183.68	3.2×10^{-7}
Thallium(I) chloride	TlCl	239.82	1.7×10^{-4}
Thallium(I) sulfide	Tl_2S	440.80	1×10^{-22}
Zinc hydroxide	$Zn(OH)_2$	99.38	1.2×10^{-17}
Zinc oxalate	ZnC_2O_4	153.39	7.5×10^{-9}
Zinc sulfide	ZnS	97.43	4.5×10^{-24}

8 dissociation constants for acids[a]

NAME	FORMULA	DISSOCIATION CONSTANT, 25°C		
		K_1	K_2	K_3
Acetic	CH_3COOH	1.75×10^{-5}		
Arsenic	H_3AsO_4	6.0×10^{-3}	1.05×10^{-7}	3.0×10^{-12}
Arsenious	H_3AsO_3	6.0×10^{-10}	3.0×10^{-14}	
Benzoic	C_6H_5COOH	6.14×10^{-5}		
Boric	H_3BO_3	5.83×10^{-10}		
1-Butanoic	$CH_3CH_2CH_2COOH$	1.51×10^{-5}		
Carbonic	H_2CO_3	4.45×10^{-7}	4.7×10^{-11}	
Chloroacetic	$ClCH_2COOH$	1.36×10^{-3}		
Citric	$HOOC(OH)C(CH_2COOH)_2$	7.45×10^{-4}	1.73×10^{-5}	4.02×10^{-7}
Ethylenediamine-tetraacetic	$C_6H_{12}N_2(COOH)_4$	1.0×10^{-2}	2.1×10^{-3} $K_4 = 5.5 \times 10^{-11}$	6.9×10^{-7}
Formic	$HCOOH$	1.77×10^{-4}		
Fumaric	$trans\text{-}HOOCCH:CHCOOH$	9.6×10^{-4}	4.1×10^{-5}	
Glycolic	$HOCH_2COOH$	1.48×10^{-4}		
Hydrazoic	HN_3	1.9×10^{-5}		
Hydrogen cyanide	HCN	2.1×10^{-9}		
Hydrogen fluoride	H_2F_2	7.2×10^{-4}		

[a] Taken from L. Meites, *Handbook of Analytical Chemistry.* New York: McGraw-Hill, 1963, p. 1-21.

NAME	FORMULA	DISSOCIATION CONSTANT, 25°C		
		K_1	K_2	K_3
Hydrogen peroxide	H_2O_2	2.7×10^{-12}		
Hydrogen sulfide	H_2S	5.7×10^{-8}	1.2×10^{-15}	
Hypochlorous	HOCl	3.0×10^{-8}		
Iodic	HIO_3	1.7×10^{-1}		
Lactic	$CH_3CHOHCOOH$	1.37×10^{-4}		
Maleic	cis-$HOOCCH:CHCOOH$	1.20×10^{-2}	5.96×10^{-7}	
Malic	$HOOCCHOHCH_2COOH$	4.0×10^{-4}	8.9×10^{-6}	
Malonic	$HOOCCH_2COOH$	1.40×10^{-3}	2.01×10^{-6}	
Mandelic	$C_6H_5CHOHCOOH$	3.88×10^{-4}		
Nitrous	HNO_2	5.1×10^{-4}		
Oxalic	$HOOCCOOH$	5.36×10^{-2}	5.42×10^{-5}	
Periodic	H_5IO_6	2.4×10^{-2}	5.0×10^{-9}	
Phenol	C_6H_5OH	1.00×10^{-10}		
Phosphoric	H_3PO_4	7.11×10^{-3}	6.34×10^{-8}	4.2×10^{-13}
Phosphorous	H_3PO_3	1.00×10^{-2}	2.6×10^{-7}	
o-Phthalic	$C_6H_4(COOH)_2$	1.12×10^{-3}	3.91×10^{-6}	
Picric	$(NO_2)_3C_6H_2OH$	5.1×10^{-1}		
Propanoic	CH_3CH_2COOH	1.34×10^{-5}		
Pyruvic	$CH_3COCOOH$	3.24×10^{-3}		
Salicylic	$C_6H_4(OH)COOH$	1.05×10^{-3}		
Succinic	$HOOCCH_2CH_2COOH$	6.21×10^{-5}	2.32×10^{-6}	
Sulfamic	H_2NSO_3H	1.03×10^{-1}		
Sulfuric	H_2SO_4	strong	1.20×10^{-2}	
Sulfurous	H_2SO_3	1.72×10^{-2}	6.43×10^{-8}	
Tartaric	$HOOC(CHOH)_2COOH$	9.20×10^{-4}	4.31×10^{-5}	
Trichloroacetic	Cl_3CCOOH	1.29×10^{-1}		

dissociation constants for bases[a]

NAME	FORMULA	gfw	DISSOCIATION CONSTANT, K, 25°C
Ammonia	NH_3	17.03	1.76×10^{-5}
Aniline	$C_6H_5NH_2$	93.13	3.94×10^{-10}
1-Butylamine	$CH_3(CH_2)_2CH_2NH_2$	73.14	4.0×10^{-4}
Dimethylamine	$(CH_3)_2NH$	45.08	5.9×10^{-4}
Ethanolamine	$HOC_2H_4NH_2$	61.08	3.18×10^{-5}
Ethylamine	$CH_3CH_2NH_2$	45.08	4.28×10^{-4}
Ethylenediamine	$NH_2C_2H_4NH_2$	60.10	$K_1 = 8.5 \times 10^{-5}$
			$K_2 = 7.1 \times 10^{-8}$
Hydrazine	H_2NNH_2	32.05	1.3×10^{-6}
Hydroxylamine	$HONH_2$	33.03	1.07×10^{-8}
Methylamine	CH_3NH_2	31.06	4.8×10^{-4}
Piperidine	$C_5H_{11}N$	85.15	1.3×10^{-3}
Pyridine	C_5H_5N	79.10	1.7×10^{-9}
Trimethylamine	$(CH_3)_3N$	59.11	6.25×10^{-5}

[a]Taken from L. Meites, *Handbook of Analytical Chemistry*. New York: McGraw-Hill, 1963, p. 1-21.

INDEX

Entries in boldface refer to specific laboratory instructions; *t* refers to a table.

Absolute error, E, 49

Absorbance, A, 439, 440*t*; measurement of, 443 (*see also* Absorption analysis, Beer's law); variables affecting, 469

Absorptiometric method(s), 433 (*see also* Absorption analysis, Atomic absorption spectroscopy, Photometric analysis, Spectrophotometric analysis)

Absorption, atomic, 461 (*see also* Atomic absorption); charge-transfer, 466

Absorption analysis, calibration curves for, 470 (*see also* Beer's law); errors in, 471–475; instruments for, 454 (*see also* Photometers, Spectrophotometers); of mixtures, 470, 620; standard addition method, 470; wavelength selection for, 454

Absorption of electromagnetic radiation, 460–466; analytical applications, 466–471, **618–621**; detectors for, 451; instruments for measurement of, 454–460; in a magnetic field, 463; measurement of, 438; species responsible for, 464–466; terminology and symbols for, 440*t*

Absorption filter(s), 446 (*see also* Filters)

Absorption spectra, 442, 460–464; effect of slit width upon, 466

Absorptivity, a, 440*t*

Accuracy, 49; in absorption analysis, 471; of atomic absorption methods, 494; of potentiometers, 352; of quantitative data, gas-liquid chromatography, 523

Acetic acid, apparent dissociation constant for, 104*t*; determination of, in vinegar, **586**

Acid(s), Brønsted-Lowry definition, 5; preparation and standardization of, 228, **584**; primary standards for 228, **585**; strength of, 7; titration of, 186 (*see also* Neutralization titrations), coulometric, 401, 403*t*

Acid-base behavior, Brønsted-Lowry theory, 5

Acid-base equilibria, 32–39, 186–220; effect of ionic strength on, 201

Acid-base indicator(s), 186–190; preparation of solutions, **583**; selection of, 192, 205;

transition ranges for, 187, 189*t*; variables affecting behavior of, 190

Acid-base ratio, 583, **584**

Acid-base titration(s), 186, **583** (*see also* Neutralization titrations)

Acid-dissociation constant(s), K_a, 32, 663*t* (*see also* Dissociation constants); evaluation of, 382

Acid error, glass electrode, 364

Acid mixture(s), titration curves for, 206

Acid salt(s), calculation of pH for solutions of, 209

Acid strength, 7

Activity, a, 106; direct potentiometric measurement of, 372; effect, on chemical equilibria, 106–111, on electrode potentials, 279, on Nernst equation, 294, on potentiometric titration curves, 382

Activity coefficient(s), f, 106–111, 109*t*

Addition, propagation of errors in, 74

Addition reaction(s), 329; amperometric end point for, 424*t*

Adsorbent(s), for thin-layer chromatography, 518

Adsorption, coprecipitation by, 125

Adsorption chromatography, 512

Adsorption indicators, for precipitation titrations, 179 (*see also* Fajans method)

Air damper(s), 534

Air oxidation, of arsenic (III), 334; of iodide ion, 332 (*see also* Iodide ion)

Aliquot, measurement of, **560**

Alizarin yellow, transition range for, 189*t*

Alkalimetric titration(s), involving EDTA, 262

Alkaline error, glass electrode, 364

Alkene(s), absorption of electromagnetic radiation by, 464, 465*t*; bromination of, 329; polarographic behavior of, 421

Alkyne(s), absorption of electromagnetic radiation by, 465*t*

α-Value(s), for EDTA solutions, 251, 254, 255*t*; for polyfunctional acids, 218–220

Alumina, as solid chromatographic support, 518

Aluminum, as auxiliary reducing reagent, 314; gravimetric determination of, **576–578**

Aluminum oxide filtering crucible(s), 548

American Society for Testing Materials, recommendations for absorptiometric nomenclature, 440

Amido group(s), absorption of electromagnetic radiation by, 465*t*

Amine group(s), titration of, 240

Amine oxide(s), polarographic behavior of, 420

Aminopolycarboxylic acid(s), titrations with, 250 (*see also* EDTA)

Ammonia, distillation of, 235 (*see also* Kjeldahl method); gas-sensing electrode for, 370

Ammonium ion, determination of, **588**; elimination by evaporation, 529

Ammonium peroxodisulfate, as auxiliary oxidizing reagent, 317

Ammonium salt(s), analysis for, 236, **588**

Amperometric titration(s), applications, 424, **617**; with one microelectrode, 421; with two microelectrodes, 425

Amphiprotic solute(s), 5; calculation of pH for solutions of, 209

Amplitude, *a*, of electromagnetic radiation, 434

Ampoule(s), for weighing of liquids, 546

Analyte, 1

Analytical balance(s), 531 (*see also* Laboratory balances, Single-pan balances, Equal-arm balances); rules for use of, **540**

Analytical separations, 497–524 (*see also* Precipitation, Chromatographic separations)

Ångström unit, Å, 434

Anhydrone, 545

Anion(s), interference by, in atomic spectroscopy, 492; metal indicator electrodes for, 358

Anion exchange resins, types of, 512

Anode, definition of, 272

Antimony, iodimetric determination of, 334, **604**

Approximate equilibrium constant expression(s), 26

Arc spectroscopy, 483*t*

Argentometric method(s), 180, 181*t*, **578–583**; amperometric end points for, 425

Arithmetic mean, 47 (*see also* Mean)

Arnd's alloy, 237

Aromatic hydrocarbon(s), absorption of electromagnetic radiation by, 465*t*

Arsenic(III), amperometric titration of, 426, **617**; iodimetric determination of, 333, **604**; as volumetric reducing reagent, 336*t*

Arsenic(III) oxide, analysis of impure, **604**; as primary standard, for cerium(IV) solutions, 326, **600**, for iodine solutions, 334, **603**, for

permanganate solutions, 321

Arsenic(III) solution(s), air oxidation of, 334; preparation of, **603**

Arsenious oxide, 334 [*see also* Arsenic(III) oxide]

Asbestos, as filtering medium, 547

Ascarite, 230

Ascending development, paper chromatography, 516

Ashing, of filter paper, **552**

Ashless filter paper, 548; preparation of, **551**

Asymmetry potential, glass electrode, 363

Atom(s), excitation of, in flames, 483

Atomic absorption, 461

Atomic absorption spectroscopy, 483*t*, 485–493; applications, 491, 492*t*; instruments for, 489

Atomic emission spectroscopy, 483*t* (*see also* Flame emission spectroscopy)

Atomic fluorescence spectrometry, 483*t*

Atomic plasma, at high temperatures, 482

Atomic spectra, in flames, 494

Atomic spectroscopy, 482 (*see also* Atomic absorption spectroscopy, Flame emission spectroscopy); classification of methods, 483*t*

Atomization of samples, at high temperatures, 482

Attenuation, of electromagnetic radiation, 438 (*see also* Absorption of electromagnetic radiation)

Autocatalysis, by manganese dioxide, in decomposition of permanganate solutions, 319

Automatic pipet(s), 557, 558*t*

Automatic titrators, 384

Autoprotolysis, of solvents, 6

Auxiliary balance(s), 543

Auxiliary reagent(s), for oxidation-reduction reactions, 313–317

Average, of results, 47 (*see also* Mean)

Azo group(s), absorption of electromagnetic radiation by, 465*t*; polarographic behavior of, 420

Back titration, 143; with EDTA, 261, **593**

Bacteria, attack on thiosulfate by, 338

Balance(s), 530 (*see also* specific types)

Band spectra, in flames, 494

Band width, effective, 446

Barium chloride dihydrate, determination of water in, **571**

Barium thiosulfate, as primary standard for iodine solutions, 334

Barrier-layer cell(s), 451

Base(s), Brønsted-Lowry definition, 5; effect on

ground-glass fittings, 232; preparation and standardization of, 229, **584**, **585**; primary standards for, 232; strength of, 7; titration of, 233 (*see also* Neutralization titrations), coulometric, 402, 403*t*

Base-dissociation constant(s), K_b, 32, 666

Base mixture(s), titration curves for, 206

Bausch and Lomb Spectronic 20 spectrophotometer, 459

Beam, of analytical balance, 531

Beam arrest(s), 534

Beam deflection detector(s), for analytical balances, 533

Beckman DU-2 spectrophotometer, 458

Beer's law, 438; and atomic absorption, 487; experimental verification of, 470; limitations to, 441; variables affecting, 469

Benzoic acid, as primary standard for bases, 232

Benzyldiphenyl, as stationary phase, gas-liquid chromatography, 521

β, overall equilibrium constant, 248

Bias, in estimate of standard deviation, 64; in observation, 53

Bicarbonate ion, titration of, 237

Biological Oxygen Demand (BOD) bottle(s), 607

Blank determinations, for detection of determinate error, 55

Boric acid, collection of ammonia in solutions of, 236 (*see also* Kjeldahl method)

Bottle(s), Biological Oxygen Demand (BOD), 607; weighing, 543

Brass, iodometric determination of copper in, 342, **607**

Broadening, of atomic line spectra, 486

Bromate ion, iodometric determination of, 340*t*; as volumetric reagent, 327 (*see also* Potassium bromate)

Bromide ion, coulometric titration of, 403*t*

Bromination, analytical applications of, 328; of phenol, **609**

Bromine, analytical applications of, 328–330; elemental analysis for, 237*t*; generation of, coulometric, 403, 404*t*, from potassium bromate, 328; iodometric determination of, 340*t*; titration of arsenic(III) with, **617**

Bromocresol green, preparation of solution, **583**; transition range for, 189*t*

Bromophenol blue, preparation of solution, **583**

Bromothymol blue, preparation of solution, **583**; transition range for, 189*t*

Brønsted-Lowry concept, acid-base behavior, 5

Buffer capacity, 200

Buffer solution(s), 197–201; calculation of pH for, 194; spectrophotometric determination of pH in, **620**

Bunsen burner(s), 549

Bunsen monochromator(s), 447

Buoyancy, effect of, on weighing data, 541

Buret(s), 558; calibration of, **567**; directions for use of, **562**

Burner(s), for atomic absorption spectroscopy, 489

Cadmium, as auxiliary reducing reagent, 314

Cadmium (II), polarographic analysis for, **616**

Calcium chloride, as desiccant, 544

Calcium ion, coulometric titration of, 403*t*; liquid membrane electrode for, 366; volumetric analysis for, with EDTA, **592**, with permanganate, 323, **598–600**

Calcium oxalate, as primary standard for acids, 229

Calcium sulfate, as desiccant, 544

Calculation(s), of electrochemical cell potentials, 283; equilibrium, 24 (*see also* under specific types); for gravimetric analysis, 117; neglect of terms in, 654; propagation of errors in, 74; stoichiometric, 15; for volumetric analysis, 148–160

Caldwell and Moyer modification, Volhard method, 179, **581**

Calibration, for elimination of instrumental error, 54; of volumetric ware, 564, **567**

Calibration curve(s), for absorption analysis, 470, 493, **619** (*see also* Beer's law); for flame emission spectroscopy, 495; for direct potentiometry, 374; for polarographic analysis, 420, **616**

Calmagite, indicator for EDTA titrations, 261; preparation of solution, **590**

Calomel electrode(s), 355

Capillary column(s), for gas-liquid chromatography, 520

Carbon, elemental analysis for, 237*t*

Carbonate(s), analysis for, 237*t*

Carbonate(s), analysis for, 237–239, **588**

Carbonate error, 230, 402

Carbonate-free base solutions. preparation and storage of, 231, **584**

Carbonate-hydrogen carbonate mixture(s), analysis of, 237–239, **587**, potentiometric, **612**

Carbonate-hydroxide mixture(s), analysis of, 237–239, **588**

Carbonate ion, homogeneous generation of, 132*t*

Carbon-carbon double bond(s), as chromophore, 464; polarographic behavior of, 421

Carbon dioxide, as carrier gas, gas-liquid

chromatography, 520; effect on standard base solutions, 230

Carbonyl group(s), as chromophore, 465*t*; polarographic behavior of, 420

Carboxylic acid group(s), as chromophore, 465*t*; polarographic behavior of, 420; titration of, 239

Carrier gas(es), for gas-liquid chromatography, 520

Catalyst(s), for Kjeldahl digestions, 235, 590

Cathode, definition of, 272

Cathode depolarizer(s), 396

Cation(s), EDTA complexes with, 253*t*; metal electrodes for, 357 (*see also* Indicator electrodes)

Cation exchange resin(s), types of, 512

Cation interference(s), atomic absorption spectroscopy, 491

Cell(s), for absorption analysis, 450, errors due to positioning of, 473; for coulometric analysis, 402; electrochemical, 270 (*see also* Electrochemical cells); for electrogravimetric analysis, 395; photovoltaic, 451, for polarographic analysis, 409; Weston standard, 352

Cell potential(s), calculation of, 287–289; changes in, during discharge, 289

Cerium(IV), coulometric generation of, 399, 404*t*; as volumetric oxidizing reagent, 318*t*, 324–326

Cerium(IV) salts, 325*t*

Cerium(IV) solution(s), analytical applications, 326, **601**; preparation and standardization of, 325, **600**; stability of, 326

Chain weight, for equal-arm balance, 537

Characteristic, of logarithm, 646

Charge-balance equation(s), 89

Charge-transfer absorption, 466

Charring, of filter papers, 549, **552**

Chelate(s), 247; extraction of, 502

Chelating reagent(s), in gravimetric analysis, 137; in volumetric analysis, 250

Chemical(s), grades of, 526

Chemical coulometer(s), 405

Chemical deviation(s), to Beer's law, 441

Chemical equilibrium, 24 (*see also* Equilibrium)

Chemical formula(s), 8

Chemical indicator(s), acid-base, 186; oxidation-reduction, 306

Chemically pure, CP, designation for chemicals, 527

Chlorate ion, iodometric determination of, 340*t*

Chloride(s), extraction of, with ether, 501

Chloride ion, coulometric titration of, 403*t*; elimination by evaporation, 529; gravimetric

determination of, **572**; oxidation of, by permanganate, 323; volumetric determination of, 176–180, **580–583**, **611**

Chlorine, coulometric generation of, 404*t*; elemental analysis for, 237*t*; iodometric determination of, 340*t*

Chlorophenol red, transition range for, 189*t*

Chopping, of electromagnetic radiation, 454, 488

Chromate ion, as indicator for Mohr method, 176 (*see also* Mohr method); as titrant for lead(II), 422

Chromatogram(s), 505

Chromatographic analysis, column efficiency for, 506

Chromatographic column(s), migration of solutes in, 504, 508

Chromatographic separation(s), 502–524 (*see also* under specific types); classification of, 503*t*

Chromatography, 502

Chromium(VI), titration of lead with, **617**; as volumetric oxidizing reagent, 326 (*see also* Potassium dichromate)

Chromophore(s), 464, 465*t*

Clark and Lubs buffers, 201

Cleaning solution, preparation of, **529**

Coagulation of colloids, variables affecting, 125–127

Colloidal precipitate(s), 125–129; coprecipitation on, 128; peptization of, 128

Colloidal suspension(s), 123

Colorimeter(s), 444, 454, 455

Column(s), for gas-liquid chromatography, 520

Column efficiency, elution chromatography, 506, 520

Common ion effect, 25; on solubility equilibria, 30

Competing equilibria, calculations involving, 89–98; effect of, on electrode potentials, 294, on oxidation-reduction titrations, 294, on solubility of precipitates, 88–89

Completeness of reaction, effect on end points, 171, 205, 257, 305

Complex formation, absorption analysis based on, 468; effect, on electrode potentials, 285, on half-wave potential, 415; equilibrium constants for, 39 (*see also* Formation constants)

Complex formation titrations, 247–263, 250*t*, **590**; amperometric end point for, 424*t*; coulometric, 402, 403*t*; curves for, 254; equivalent weight for, 151; potentiometric, 381

Complexing reagent(s), application to

iodimetry, 333; effect, on EDTA titration curves, 258, on electrolytic deposits, 396, upon solubility of precipitates, 95–99; use in atomic absorption spectroscopy, 493

Complexometric titration(s), 247 (*see also* Complex formation titrations)

Computation(s), propagation of errors in, 74–77

Concentration, determination of, from absorption measurements, 470, **618** (*see also* Beer's law), from polarographic measurements, 420, **616**, by standard addition method, 374 (*see also* Standard addition method); effect, on electrode potentials, 278, on particle size of precipitates, 124, on titration curves, 170, 192, 204, 305; methods of expressing, 10–15, 17*t*, 154; relation to potential, direct potentiometric measurements, 372

Concentration equilibrium constant(s), K', K'_a, 110, 201 (*see also* Conditional constants)

Concentration gradient(s), in electrochemical cells, 392

Concentration polarization, 391; of polarographic microelectrode, 414

Conditional constant(s), for EDTA titrations, 255, 259

Conduction, in glass membrane electrode, 363

Confidence interval(s), 65–72

Confidence level(s), 68*t*

Conjugate acid-base pairs, 6; equilibrium constants for, 33 (*see also* Buffer solutions)

Conjugation, effect on absorption of electromagnetic radiation, 464, 465*t*

Constant boiling hydrochloric acid, concentration of, 228

Constant current coulometric analysis, 398 (*see also* Coulometric titrations)

Constant error(s), 53

Constant weight, 549

Continuous spectrum, 437

Controlled-potential coulometry, 398, 405 (*see also* Coulometric titrations)

Convection, material transport in solution by, 393

Coordination compound(s), 136

Coordination number, 247

Copper, as auxiliary reducing reagent, 314; as catalyst for Kjeldahl oxidation, 235; electrogravimetric analysis for, **614**; as indicator electrode, 357; iodometric determination of, 340, **606**; as primary standard for thiosulfate solutions, 339

Copper(I), coulometric generation of, 403, 404*t*

Copper(II), coulometric titration of, 403*t*

Coprecipitation, 123; effect of, on colloidal precipitates, 128, on crystalline precipitates, 130; errors resulting from, 131

Corning 015 glass, 362

Coulomb C, 397

Coulometer(s), 405

Coulometric titration(s), 398–407; of arsenic (III), **615**

Counter-ion layer, 127

Creeping, of precipitates, 550

Cresol purple, transition range for, 189*t*

Cresol red, preparation of solution, **583**

Crucible(s), 547; preparation for use, 549

Crystalline precipitate(s), 129–131

Crystalline suspension(s), 123

Cubic centimeter, 555

Current(s), dark, 451; diffusion, 412; effect on electrochemical cells, 389 (*see also* Electrogravimetric analysis, Coulometric titrations, Voltammetry); limiting, 412; transport of, through electrochemical cells, 271, 389; variations at dropping mercury electrode, 416

Current density, effect, on electrolytic deposits, 396, on kinetic polarization, 393

Current efficiency, 398

Current maxima, in polarographic wave, 419

Current-voltage curve(s), polarographic analysis, 408 (*see also* Polarographic analysis)

Cut-off filter(s), 447

Cuvette(s), for absorption analysis, 450

Cyanide ion, as masking agent, 262; as a reagent, complex formation titrations, 250*t*

Damped balance(s), weighing with, 540

Damper(s), for analytical balance, 534

Dark current(s), 451; as source of indeterminate error, absorption measurements, 472*t*

Data, evaluation of, 46–79; statistical treatment of, 60–74

Dead-stop end point, 425

Debye-Hückel equation, for activity coefficients, 108

Decantation, of filtrates, 550

Decinormal calomel electrode, potential of, 355*t*

Decomposition potential, 412 (*see also* Polarographic waves)

Degrees of freedom, in data, 64

Dehydrite, 545

Density, of solutions, 13

Depolarizer(s), cathode, 396

Descending development, paper chromatography, 516

Desiccant(s), 544

Desiccator(s), 544, **545**

Detection limit(s), for atomic absorption and flame emission spectroscopy, 492t

Detector(s), for atomic absorption spectrophotometers, 491; for electromagnetic radiation, 451; errors due to, 472t; for gas-liquid chromatography, 521

Determinate error(s), 51–56

Deuterium lamp(s), as source for ultraviolet radiation, 445

Devarda's alloy, 236

Developer(s), paper chromatography, 516

Deviation, 48; standard, 61 (*see also* Standard deviation)

Diatomaceous earth, as solid chromatographic support, 511, 518, 521

Dichlorofluorescein, preparation of solution, **583**

Dichromate ion, as volumetric reagent, 326 (*see also* Potassium dichromate)

Difference(s), propagation of errors in, 74

Differential titration(s), 383

Diffraction grating(s), 449

Diffusion, material transport in solution by, 392; in polarographic analysis, 413; of solutes, in chromatographic columns, 509

Diffusion current, i_d, polarographic, 412, 414; evaluation of, 419

Digestion, of colloidal precipitates, 129; of crystalline precipitates, 130

Dilution, effect on buffer solutions, 197

Dimethylglyoxime, as precipitating reagent, 137

Dinonylphthalate, as stationary phase, gas-liquid chromatography, 521

Diphenylamine, transition range for, 307t

2,3'-Diphenylaminedicarboxylic acid, transition range for, 307t

Diphenylamine sulfonic acid, as indicator, 327, for permanganate titrations, 318; preparation of solutions, **602**; transition range for, 307t

Direct potentiometric measurements(s), 371–376

Direct titration(s), with EDTA, 261, **591**

Dispersion of electromagnetic radiation, by monochromators, 448, 449

Displacement analysis, 504

Displacement titration(s), involving EDTA, 262, **592**

Dissociation constant(s), for complex ions, 40 (*see also* Formation constant); for conjugate acid-base pairs, 33; evaluation, from electrochemical measurements, 292, from potentiometric titration curves, 382; for weak acids, K_a, 32, 663; for weak bases, K_b, 37, 666

Distribution coefficient, 41, 499; ion-exchange chromatography, 513

Distribution law, 498

Division, propagation of errors in, 75

Doppler broadening, of atomic spectra, 486

Double-beam balance(s), 543

Double-beam instrument(s) for absorption analysis, 454, 456

Double bond(s), absorption of electromagnetic radiation by, 464 (*see also* Alkenes)

Double monochromator(s), 450

Drierite, 544

Dropping mercury electrode, DME, 408, 416; amperometric titrations with, 422, **617**

Drying, determination of water by, **571**; of precipitates, 133; of samples, 544

Drying oven(s), 549

Dumas method, for nitrogen, 233

Dynode(s), 453

Eddy diffusion term, A, elution chromatography, 509

EDTA, 250–263; α_4 values for, 255t (*see also* α-Values); coulometric release of, 402

EDTA complex(es), 252; formation constants for, 253t

EDTA solution(s), composition as a function of pH, 251; preparation and standardization of, 251, **590**

EDTA titration(s), 250; curves for, 254; end points for, 260; factors affecting, 254, 258; mercury electrode for, 358; potentiometric, 381; scope and applications, 261–263, **590–593**

Effective band width, 446; of grating monochromators, 449; of prism monochromators, 448

Effective equilibrium constant(s), K', EDTA titrations, 255 (*see also* Conditional constants)

Efficiency, of chromatographic column, 506; current, 398

Electrical double layer, around colloidal particles, 127

Electricity, quantity of, q, 397

Electrochemical cell(s), 270–272; change in, during passage of current, 389–394 (*see also* Electrogravimetric analysis, Coulometric analysis, Polarographic analysis); mechanisms for current passage through, 271; polarization of, 390 (*see also* Concentration polarization, Kinetic polarization); potential of, 287 (*see also* Cell potentials); reversibility of, 271; schematic representation of, 286

Electrochemical methods of analysis, 350, 389

(*see also* under specific types)

Electrode(s), for amperometric titrations, 422; calomel, 305; for differential titrations, 383; dropping mercury, 408; for electrochemical cells, 271; for electrogravimetric analysis, 394; gas-sensing, 368; generator, 399 (*see also* Generator electrode); indicator, 357 (*see also* Indicator electrodes); membrane, 359 (*see also* Glass electrode); polarization of, 391; for polarographic analysis, 408; for potentiometric titrations, 350 (*see also* Indicator electrodes, Reference electrodes); reference, 354 (*see also* Reference electrodes); rotating platinum, 422; sign convention for, 272; silver-silver chloride, 356

Electrode calibration method, direct potentiometric measurements, 372

Electrode composition, effect on kinetic polarization, 393

Electrode dimensions, effect on concentration polarization, 393

Electrodeposition, 394 (*see also* Electrogravimetric analysis)

Electrode potential, E, 273; calculation of, 283; changes in, during passage of current, 289; effect, of complex-forming and precipitating reagents upon, 285, of concentration upon, 278–286; measurement of, 275; sign convention for, 276; standard, E^0, 279 (*see also* Standard electrode potential)

Electrode potential for the system, 296

Electrode process(es), 271

Electrode sign(s), 272

Electrogravimetric analysis, 394–397; of copper, **614**; separations by, 498

Electrolysis, 394 (*see also* Coulometric analysis, Electrogravimetric analysis)

Electrolyte(s), classification of, 4, 5*t*; supporting, 412 (*see also* Polarographic waves)

Electrolyte concentration, effect, on chemical equilibrium, 103 (*see also* Activity), on concentration polarization, 393, on stability of colloids, 127, 129

Electrolytic cell(s), 271

Electromagnetic radiation, 433; absorption of, 437 (*see also* Absorption of electromagnetic radiation, Beer's law); detectors for, 451; dispersion of, 448; energy range for, 436; generation of, 437; monochromatic, 438; properties of, 433–437; sources for, 444

Electromagnetic spectrum, 436

Electronic state(s), of atoms and molecules, 437

Electron-transfer reaction(s), 268 (*see also* Oxidation-reduction reactions)

Electrostatic force(s), material transport in solution by, 393

Elemental analysis, volumetric, 233, 237*t*

Elution analysis, 504

Elution chromatography, gradient, 511; with liquid mobile phase, 510 (*see also* under specific types); rate theory of, 508; zone shapes in, 508

Emission spectroscopy, classification of methods, 483*t*

Empirical formula(s), 8

End point(s), 144, 146 (*see also* under specific reaction types, reagents); factors affecting, 170; potentiometric, 308 (*see also* Potentiometric titrations)

Energy, absorption of, by atoms and molecules, 433, 436

Epoxide(s), polarographic behavior of, 420

Equal-arm balance(s), 531, 537; weighing with, **538**

Equilibrium, 24; acid base, 32, 190; complex formation, 39, 254; distribution, 41, 498; ion-exchange, 512; oxidation-reduction, 41, 268; solubility, 29, 87, 167; stepwise, 41 (*see also* specific types)

Equilibrium constant(s), 24–42 (*see also* specific types); calculation of, from electrode potential data, 289–294; concentration, K'_a, 110, 201 overall, β, 248

Equivalence point(s), 144; potentiometric determination of, 378

Equivalence point potential(s), evaluation of, 297–300

Equivalent, eq, 16*t*, 156

Equivalent weight, 16*t*, 148–154; of weak acids, determination of, **586**

Eriochrome Black T, for EDTA titrations, 260; preparation of solution, **590**

Erioglaucin A, transition range for, 307*t*

Error(s), 49; in absorption analysis, 471–475; classification of, 51 (*see also* Determinate error, Indeterminate error); constant, 53; due to coprecipitation, 130; in electrode calibration methods, 373; in measurement of pH, 364, 376; parallax, 559; in polarographic analysis, 418; propagation of, in computation, 74–77; proportional, 54; titration, 144, 189; in weighing operation, 541

Ester(s), titration of, 240

Ether, extraction of metal chlorides with, 501

p-Ethoxychrysoidine, transition range for, 307*t*

Ethylenediaminetetraacetic acid, 250 (*see also* EDTA)

Evaporation, of liquids, 529

Excitation, of atoms and molecules, 437, 483
Exponential number(s), manipulation of, 643
Extinction, E, 440t
Extinction coefficient, k, 440t
Extraction(s), separations based on, 498–502

Fajans method, 179, 181t; determination of chloride, **582**
Faraday, F, 279, 397
Ferroin(s), 325 [see also 1,10-Phenanthroline iron(II) complex]
Ferrous ammonium sulfate, as a reducing reagent, 336 [see also Iron(II)]
Filter(s), for photometric analysis, 446; selection of, 457
Filterability, of precipitates, 122
Filtering crucible(s), 547; use of, **554**
Filter paper, 548 (see also Ashless filter paper)
First-order electrode(s), 357
Flame(s), atomic excitation in, 483
Flame absorption spectroscopy, 483t (see also Atomic absorption spectroscopy)
Flame emission spectroscopy, 483–486, 492t, 494–496
Flame photometry, 483 (see also Flame emission spectroscopy)
Flameless absorption spectroscopy, 483t
Flask(s), Kjeldahl, 235; volumetric, 559 (see also Volumetric flasks)
Fluorescein, as adsorption indicator, 179
Fluorescence, relaxation by, 438
Fluoride, as masking agent for iron(III), 341; solid-state electrode for, 368
Fluorine, elemental analysis for, 237t
Formality, F, 10, 17t
Formal potential(s), 294, 656
Formation constants(s), K_f, 40, 248 (see also Complex formation); for EDTA complexes, 253t; evaluation of, from electrochemical measurements, 292, from polarographic data, 415
Formula weight, fw, 8, 16t
Fowler and Bright method, for standardization of permanganate solutions, 320, **595**
Frequency, ν, 434
Frontal analysis, 504
Fuel mixture(s), for atomic absorption spectroscopy, 489; for flame emission spectroscopy, 495

Galvanic cell(s), 271
Galvanometer(s), for polarographic analysis, 417

Gas(es), as auxiliary reducing reagents, 315
Gas electrode(s), 273
Gas-liquid chromatography, GLC, 503t, 519–524
Gas-sensing electrode(s), 359, 368
Gas-solid chromatography, 503t
Gaussian distribution, 60; of solutes in mobile phase, elution chromatography, 508
Gelatin, as maximum suppressor, 419
Gel chromatography, 503t
Generator electrode(s), coulometric analysis, 399, 401, 402, 403
Glass, as dispersing agent for electromagnetic radiation, 448; as filtering medium, 547; temperature coefficient for, 556
Glass electrode(s), 360–365; alkaline and acid errors, 364; direct measurement of pH with, 375; origin of potential in, 360, 363; for univalent cations, 365
Globar, 445
Gold, as indicator electrode, 359
Gooch crucible, 547; preparation of, **554**
Gradient elution, 511
Gram equivalent weight, 16t (see also Equivalent weight)
Gram formula weight, gfw, 8, 16t
Gram molecular weight, gmw, 9, 16t
Graphical end-point detection, for amperometric titrations, 421; for potentiometric titrations, 378
Grating monochromator(s), 449
Gravimetric analysis, 117–139; of aluminum, **576**; calculations for, 117–122; of chloride, **572**; for copper, **614**; coprecipitation errors affecting, 131; for hydrate water, **571**; of iron, **574**; precipitating reagents for, 134–139; precipitation from homogeneous solution, 131, **576**
Gravimetric factor(s), 118
Ground state, of atoms and molecules, 437
Gunning modification, to Kjeldahl analysis, 234 (see also Kjeldahl method)

H$_4$Y, 250 (see also EDTA)
Half-cell(s), 271; potential of, 273 (see also Electrode potential)
Half-reaction(s), 271; potential of, 273 (see also Electrode potential)
Half-reaction(s), 269
Half-wave potential, $E_{1/2}$, 414 (see also Polarographic waves)
Halide(s), titration of, 167–180; coulometric, 403t; Fajans method, 179, **582**; Mohr method,

176, **580**; potentiometric, 380, **611**; Volhard method, 177, **581**

Halogen group(s), organic, polarographic behavior of, 421

Hardness, of water, analysis for, 263, **593**

Heated object(s), manipulation of, **555**

Heat lamp(s), use of, 549, **552**

Height equivalent of a theoretical plate, 506

Helium, as mobile phase, gas-liquid chromatography, 519

Henderson-Hasselbalch equation, 195

Hertz, Hz, 434

Heyrovsky, Jaroslav, 408

Higher order equation(s), solution of, 652

Hollow cathode lamp(s), atomic absorption spectroscopy, 487

Homogeneous precipitation, 131 (see also Precipitation from homogeneous solution)

Hydrate water, determination of, **571**

Hydration, of glass membrane electrode, 362

Hydrochloric acid solution(s), preparation and standardization, 228, **584, 585**

Hydrogen, codeposition of, during electrolysis, 396; overvoltage effects with, 394

Hydrogen carbonate-carbonate mixture(s), titration of, 237–239, **587, 612**

Hydrogen electrode(s), 273

Hydrogen lamp(s), for ultraviolet radiation, 445

Hydrogen-oxygen coulometer, 405

Hydrogen peroxide, as auxiliary oxidizing reagent, 317

Hydrogen sulfide, as auxiliary reducing reagent, 315

Hydronium ion, 6; coulometric generation of, 402

Hydronium-ion concentration, calculation of, in solutions of strong acids or strong bases, 28, 190, in solutions of weak acids or bases, 33–39, 193–205; measurement of, 360

Hydrogen lamp(s), as source for ultraviolet radiation, 445

Hydroquinone(s), polarographic behavior of, 421

Hydrous oxide(s), determination of iron(III) as, **574**; separations based on, 100, 498

Hydroxide(s), solubility of, 94 (see also Hydrous oxides)

Hydroxide-carbonate mixture(s), analysis of, 237–239 (see also Winkler method)

Hydroxide ion(s), coulometric generation of, 401; homogeneous generation of, 131, 132t; **576**

Hydroxide-ion concentration, calculation of, in solutions of strong bases, 28, of weak bases, 39, 193–206

8-Hydroxyquinoline, bromination of, 329; homogeneous generation of, 132t; as precipitating reagent, 137, as titrant for magnesium ion, 421

Hygroscopic substance(s), storage of, 545; weighing of, 546

Hypoiodite ion, interference by, in iodimetry, 333, in iodometry, 337

Ignition, or precipitates, 133, 551

Inclusion(s), coprecipitation by, 130

Independent analysis, for detection of determinate error, 55

Interdeterminate error(s), 51, 56–60; in absorption measurements, 471–475; accumulation in computations, 74–77

Indicator(s), for bromate titrations, 328; for cerium(IV) titrations, 325; for dichromate titrations, 327; for EDTA titrations, 260; for neutralization titrations, 186, 189t (see also Acid-base indicators); oxidation-reduction, 306 (see also Oxidation-reduction indicators); for precipitation titrations, 174–180; for volumetric analysis, 144 (see also under specific reaction types, reagents)

Indicator blank(s), 176

Indicator electrode(s), 350, 357–370

Indigo tetrasulfonate, transition range for, 307t

Infrared radiation, sources for, 445

Infrared region, absorption in, 462; quantitative absorption analysis in, 469

Inorganic precipitating reagents, for gravimetric analysis, 134, 135t

Instability constant(s), 40 (see also Formation constant)

Instrumental deviation(s), to Beer's law, 442, 471–475

Instrumental error(s), 52

Intensity, I, of electromagnetic radiation, 435

Interference(s), elimination by ion exchange, 515

Interference filter(s), 447

Internal standard(s), for gas-liquid chromatography, 524

Iodate ion, as volumetric reagent, 330 (see also Potassium iodate)

Iodide ion, air oxidation of, 332, 339; coulometric titration of, 403t; as a reducing reagent, 336 (see also Iodometric methods of analysis)

Iodimetry, 331 (see also Iodine)

Iodine, amperometric end point for, 426; applications of, **604**; generation of, coulometric, 404t, **615**, from potassium iodate, 330, **610**; as oxidizing reagent, 318*t*, 331–335, **604**

Iodine-iodide mixture(s), titration of, 330, **610**

Iodine monochloride, preparation of, **601**

Iodine solution(s), indicators for, 333, preparation and standardization of, 332, 334, **603**; stability of, 332

Iodemetric methods of analysis, 331, 336–343; determination of copper, 340, **606**

Ion exchange, in membrane indicator electrodes, 363, 365; separation of iron(III) from aluminum(III) by, **576**

Ion-exchange chromatography, 503*t*, 512–516

Ionic strength, μ, 105; effect upon acid-base equilibria, 201

Ion product constant for water, K_w, 27

IR drop, 389

Iron, analysis for, gravimetric, **574**, photometric, **618**, volumetric, 321, 327, **595**, **601**, **602**; as primary standard for permanganate solutions, 321

Iron(II), analysis for, with cerium(IV), **601**, with dichromate, 327, **602**, with permanganate, 321, **595**; charge-transfer complexes involving, 466; coulometric generation of, 404*t*; as volumetric reducing reagent, 336

Iron(III), charge-transfer complexes involving, 466; as indicator for thiocyanate ion, 177, **580**; masking of, by fluoride ion, 341; reduction of, with reductors, 321, **595**, with tin(II) chloride, 323, **597**; separation from aluminum by ion exchange, 515, **576**

Irreversible reaction(s), polarograms for, 416

Irreversability, electrochemical, 271

Isomorphous inclusion(s), 130

IUPAC convention, for electrode potentials, 276

Jones reductor, 314, 316*t*; preparation and use of, **593**; reduction of iron(III) with, **595**

Junction potential, 287 (*see also* Liquid junction potential)

Katharometer, 522

Kinetic polarization, 393; in amperometric titration, 426

Kjeldahl method, for amine nitrogen, 234, **589**

Knife edge(s), for analytical balance, 531

Laboratory balance(s), 530

Laboratory notebook, rules for keeping, **568**

Laboratory ware, cleaning of, 528

Lambda pipet(s), 558*t*

Lanthanide element(s), absorption of electromagnetic radiation by, 465

Lead, as auxiliary reducing reagent, 314

Lead(II), amperometric titration of, 421, **617**; coulometric titration of, 403*t*

Le Châtelier principle, 25

Ligand(s), 247

Limestone, determination of calcium in, **599**

Limiting current, polarographic wave, 412 (*see also* Polarographic waves)

Linear chromatography, 504

Line spectrum, 437

Line width(s), of atomic absorption and atomic emission spectra, 486

Liquid(s), evaporation of, 529; weighing of, 546

Liquid junction potential, 287, 371, 390

Liquid-liquid partition chromatography, 510*t* (*see also* Partition chromatography)

Liquid-membrane electrode(s), 359, 365

Liter, 555

Littrow prism(s), 448

Logarithm(s), manipulations involving, 646; significant figures in, 79

Longitudinal diffusion term, *B*, elution chromatography, 509

Long swings, method of weighing, **539**

Macro analytical balance(s), 531 (*see also* Single-pan, Equal-arm balances)

Magnesium-EDTA solution, preparation of, **591**

Magnesium ion, titration of, amperometric, 421; with EDTA, **591**

Magnesium perchlorate, as desiccant, 545

Magnesium sulfate, preparation of solution(s), **593**

Magnetic damper(s), 534

Manganese, photometric determination of, in steel, **619**

Manganese(II), as catalyst in the permanganate oxidation of oxalate, 319

Manganese(III), coulometric generation of, 403, 404*t*

Manganese dioxide, effect on permanganate solutions, 319

Mantissa, of logarithm, 646

Masking agent(s), 262; for iron, 341

Mass, chemical units of, 8, 16*t*; measurement of, 530–543

Mass-action effect, 25 (*see also* Common-ion effect)

Mass-balance equation(s), 89

Mass transfer, in chromatographic column, 508

Mass transfer term, C_v, elution chromatography, 509

Material transport, during operation of cells, 392; in solution, 413

Maximum suppressor(s), polarographic analysis, 419

McBride method, for standardization of permanganate solutions, 320, **595**

McIlvane buffers, 201

Mean, of results, 47; statistical tests involving, 65–74

Mean activity coefficient(s), 108

Measuring pipet(s), 557

Median, or results, 47

Meker burner(s), 549

Membrane indicator electrode(s), 359–370 (*see also* Glass electrode, Liquid-membrane electrodes, Precipitate electrodes, Gas-sensing electrodes)

Meniscus, 559

Mercaptan(s), coulometric titration of, 403t; polarographic behavior of, 421

Mercury, as auxiliary reducing reagent, 314; as catalyst for Kjeldahl oxidation, 235

Mercury electrode(s), for EDTA titrations, 358, 381; for polarographic analysis, 410 (*see also* Polarographic analysis)

Mercury(II), as reagent, for complex formation titrations, 250t; for oxidation of tin(II), 323, **597**

Mercury(II)-EDTA complex, use, in coulometric titrations, 402

Mercury(II) oxide, as primary standard for acids, 229

Metal(s), as auxiliary reducing reagents, 314; as indicator electrodes, 357

Metal-ion indicator(s), for EDTA titrations, 260

Metastannic acid, 342

Method error(s), 52

Methylene blue, transition range for, 307t

Methyl orange, preparation of solution, **583**; transition range for, 189t

Methyl red, as maximum suppressor, 419; preparation of solution, **583**; transition range for, 189t

Methyl yellow, transition range for, 189t

Micro analytical balance(s), capacity of, 531

Microelectrode(s), for voltammetry, 408 (*see also* Dropping mercury electrode, Rotating platinum electrode)

Micron, μ, 434

Microorganism(s), effect on stability of thiosulfate solutions, 338

Microwave region, absorption in, 462

Migration, of solutes through chromatographic columns, 508 (*see also* Chromatographic columns)

Milliequivalent, meq, 16t

Milliequivalent weight, 148 (*see also* Equivalent weight)

Milliformula weight, mfw, 9, 16t

Milliliter, 555

Millimicrom, $m\mu$, 434

Millimole, mmole, 9, 16t

Mixture(s), absorption of electromagnetic radiation by, 439; polarographic waves for, 416, **616**; potentiometric titration curves for, 380, **611, 613**; spectrophotometric analysis of, 470, **620**

Mobile phase(s), 502; for gas-liquid chromatography, 519; for paper chromatography, 516; for partition chromatography, 511; types of, 503t

Modulation, of output from hollow cathode lamp, 487

Mohr method, 176, 181t; determination of chloride, **580**

Mohr pipet(s), 558t

Mohr's salt, 336 [*see also* Iron(II)]

Molar absorptivity, ϵ, 439, 440t

Molar extinction coefficient, 440t

Molarity, M, 10, 17t

Mole, 9, 16t

Molecular absorption, 462

Molecular weight, 8

Monochromatic radiation, 438; and Beer's law, 442

Monochromator(s), 446, 447–450; for atomic absorption spectroscopy, 487, 490

Muffle furnace(s), 549

Multiple equilibria, steps in solving problems involving, 89–98

Multiplication, propagation of errors in, 75

Nanometer, nm, 434

National Bureau of Standards, as source, for reference standards, 527, for standard samples, 55

Neglect of terms, in equations, 654

Nernst equation, 279; limitations to use of, 294

Nernst glower, 445

Nessler tube(s), 455

Neutralization titration(s), 227–240, 401, 403t, **583–590**; curves for, 146–148 (*see also* Titration curves); equivalent weight for, 149;

indicators for, 186 (*see also* Acid-base indicators); potentiometric, 382, **612**; theory of, 186–220

Neutral red, transition range for, 189*t*

Nickel, as auxiliary reducing reagent, 314

Nickel(II), as a reagent, complex formation titrations, 250*t*

Nitrate(s), extraction of, 502; organic, absorption of electromagnetic radiation by, 465*t*

Nitrate ion, analysis for, 236; as cathode depolarizer, 396; elimination by evaporation, 529

Nitrite(s), analysis for, 236; iodometric determination of, 340*t*

Nitrobenzene, use in Volhard method, 179, **581**

Nitrogen, analysis for, 233, 237*t* (*see also* Kjeldahl method); as mobile phase, gas-liquid chromatography, 519

Nitrogen dioxide, gas-sensing electrode for, 370

Nitrogen oxides, elimination by evaporation, 529; formation of, during Kjeldahl digestion, 235

Nitro group, as chromophore, 465*t*; polarographic behavior of, 420

5-Nitro-1, 10-phenanthroline iron(II) complex, transition range for, 307*t*

Nitroso group, as chromophore, 465*t*; polarographic behavior of, 420

Noise, in measurement of absorbance, 451, 472*t*

Nonflame atomizer(s), atomic absorption spectroscopy, 489

Nonisomorphous inclusion(s), 130

Normal calomel electrode, potential of, 355*t*

Normal error curve, 60

Normal hydrogen electrode, 275 (*see also* Standard hydrogen electrode)

Normality, *N*, 17*t*, 148; calculation of, 155

Notebook, laboratory, **568**

Nucleation, 124

Number of theoretical plates, *N*, 507

Occlusion, coprecipitation by, 130

Oesper's salt, 336 [*see also* Iron(II)]

Ohmic potential, 389

Olefin(s), bromination of, 329 (*see also* Alkenes)

Optical density, *D*, 440*t*

Optical lever(s), for analytical balances, 533

Organic compound(s), absorption in ultraviolet and visible regions by, 464; coulometric analysis of, 406

Organic functional group analysis, 239; gravimetric, 138; polarographic, 420; volumetric, 239

Organic precipitating reagent(s), for gravimetric analysis, 136

Orthophenanthroline(s), as oxidation-reduction indicators, 307*t* [*see also* 1, 10-Phenanthroline-iron(II) complex]

Orthophenathroline iron(II) complex, as indicator for permanganate titrations, 318

Osmium tetroxide, as catalyst for cerium(IV) titrations, 326, **601**

Ostwald-Folin pipet(s), 558*t*

Outlying result, rejection criteria for, 72–74

Oven drying, 133, 544

Overall equilibrium constant, β, 248

Overvoltage, 393 (*see also* Kinetic polarization)

Oxalate ion, homogeneous generation of, 132*t*; as titrant for lead(II), 421

Oxidant(s), 268 (*see also* Oxidizing agents)

Oxidation, definition, 268

Oxidation potential(s), 278

Oxidation-reduction indicator(s), 306, 307*t*

Oxidation-reduction reaction(s), equilibrium in, 41, 268 (*see also* Oxidation-reduction titrations)

Oxidation-reduction titration(s), 268–308; amperometric end point for, 424*t*, **617**; applications of, 313–343, **593–611** (*see also* under specific reagents); coulometric, 403, **615**; curves for, 296, 382; equivalence point potential for, 297–300; equivalent weight for, 150; experimental variables affecting, 305; indicator electrodes for, 359

Oxide(s), precipitation of, 498

Oxidizing agent(s), 268; auxiliary, 316; definition of, 268; titration of, 327, 336

Oxine, 137 (*see also* 8-Hydroxyquinoline)

Oxygen, coulometric analysis for, 407; effect of, on stability of thiosulfate solutions, 338; iodometric determination of, 340*t*, 342 (*see also* Winkler method); overvoltage effects with, 394; polarographic behavior of, 418

Ozone, iodometric determination of, 340*t*

Packed column(s), for gas-liquid chromatography, 520

Pan(s), for analytical balance, 533

Pan arrest(s), 534

Paper, as filtering medium, 548, **551**

Paper chromatography, 503*t*, 516

Parallax, error due to, 559

Particle growth, 124

Particle size, of precipitates, 123

Particulate properties, of electromagnetic radiation, 435

Partition chromatography, 503*t*, 510

Partition coefficient, 41, 499

Partition ratio(s), *K*, 503

Parts per billion, concentration in, 13

Parts per million, concentration in, 13

Peak area(s), evaluation of, gas-liquid chromatography, 524

Peptization, of colloidal precipitates, 128

Percentage, concentration in, 14

Perchloric acid, as volumetric reagent, 228

Periodate ion, iodometric determination of, 340*t*

Permanganate ion, 317 (*see also* Potassium permanganate)

Peroxide(s), as auxiliary oxidizing reagents, 317; iodometric determination of, 340*t*; polarographic behavior of, 420

Personal error(s), 53

pH, direct potentiometric measurement of, 371, 375; errors in measurement of, with glass electrode, 376; measurement with glass electrode, 360; separations based on control of, 100 (*see also* Hydrous oxides); spectrophotometric determination of, **620**

pH calculation(s), for solutions containing; conjugate acid-base pairs, 194 (*see also* Buffer solutions), mixtures of acids or bases, 206, polyfunctional acids or bases, 208–218, salts of the type NaHA, 209, strong acids and strong bases, 190, weak acids of the type HA, 201, H_2A, 208, weak bases, 206

pH effect(s), on composition of EDTA solutions, 251 (*see also* EDTA), on particle size of precipitates, 124, on solubility of precipitates, 91–95, on stability of thiosulfate solutions, 337, upon titrations with iodine, 332

pH meter(s), 354

1,10-Phenanthroline, photometric determination of iron with, **618**

1,10-Phenanthroline-iron(II) complex(es), charge-transfer absorption by, 466; as oxidation-reduction indicators, 307*t*, 325; preparation of, **600**

Phenol, bromination of, 329, **609**

Phenolphthalein, preparation of solution, **583**; transition range for, 189*t*

Phenol red, preparation of solution, **583**; transition range for, 189*t*

Phenosafranine, transition range for, 307*t*

Phosphate(s), potentiometric titration of, **613**

Phosphate ion, homogeneous generation of, 132*t*

Phosphorescence, relaxation by, 438

Phosphorus, elemental analysis for, 237*t*

Phosphorous pentoxide, as desiccant, 545

Photocell(s), 451

Photochemical reaction(s), 438

Photodecomposition, of cerium(IV) solutions, 325, of permanganate solutions, 319; of silver halides, 180, 573; of thiosulfate solutions, 338

Photoelectric colorimeter(s), 454 (*see also* Photometers)

Photometer(s), 444, 454, 455–458; for flame emission spectroscopy, 494

Photometric analysis, applications, **618–621**; calibration curves for, 470; filter selection for, 457

Photometric error, 471–475

Photomultiplier tube(s), 453, 491

Photon(s), 435

Phototube(s), 452

Photovoltaic cell(s), 451

Picric acid, coulometric titration of, 406

Pipet(s), 556 (*see also* Volumetric pipets)

Plasma, atomic, at high temperatures, 482

Planck constant, *h*, 436

Plate height, chromatographic analysis, 506

Plate theory, elution chromatography, 506

Platinum electrode(s), for electrogravimetric analysis, 394; for polarographic analysis, 422; for potentiometric titrations, 359

Pointer, for equal-arm balance, 533

Polarization, of electrochemical cells, 390 (*see also* Concentration polarization, Kinetic polarization); of microelectrodes, in amperometric titrations, 425, in polarographic analysis, 414

Polarogram(s), 408 (*see also* Polarographic waves)

Polarographic analysis, 408–421; applications, 418, 420, **616**; electrodes for, 408–410; of mixtures, 416, **616**; with rotating platinum electrode, 422; treatment of data, 420

Polarographic wave(s), 408, 411; effect, of complex-forming species upon, 415; of temperature upon, 418; interpretation of, 412; for irreversible reactions, 416; maxima in, 419, for mixtures, 416, **616**

Polarography, 408 (*see also* Polarographic analysis)

Polyethylene glycol, as stationary phase, gas-liquid chromatography, 521

Polyfunctional acids and bases, α-values for, 218–220; pH calculations for, 208; titration curves for, 208–218

Pooled data, standard deviation for, 66

Porcelain filtering crucible(s), 548

Position of equilibrium, 25

Potassium antimony(III) tartrate, as primary standard for iodine solutions, 334

Potassium bromate, as oxidizing reagent, 318*t*, 327, **608**; as source of bromine, 328, **609**

Potassium bromate solution(s), amperometric titration of arsenic(III) with, **617**; indicators for, 328; preparation, **608**

Potassium dichromate, applications, 327, **602**; as oxidizing reagent, 318*t*, 326; preparation of standard solutions, 327, **602**; as primary standard for thiosulfate solutions, 339, **605**

Potassium ferricyanide, as primary standard for thiosulfate solutions, 339

Potassium hydrogen iodate, as primary standard, for bases, 232, for thiosulfate solutions, 339

Potassium hydrogen phthalate, as primary standard for bases, 232, **585**

Potassium iodate, as oxidizing reagent, 330, **609**; as primary standard for thiosulfate solutions, 339, **605**

Potassium iodate solution(s), indicators for, 331; preparation of, **609**; titration of iodine-iodide mixtures with, **610**

Potassium iodide, as primary standard for permanganate solutions, 321

Potassium ion, liquid-membrane electrode for, 367

Potassium permanganate, as oxidizing reagent, 317–324; 318*t*, 322*t*

Potassium permanganate solution(s), preparation and standardization of, 319–321, **594**; stability of, 319; titration of calcium in limestone with, 323, **598–600**; titration of iron(II) with, 321, **595**

Potassium sulfate, use in Kjeldahl analysis, 234

Potassium thiocyanate, as indicator for iron(III), 306 (*see also* Thiocyanate ion)

Potassium thiocyanate solution(s), preparation and standardization of, 579 (*see also* Volhard method)

Potential, asymmetry, 363; calculation of, 287 (*see also* Nernst equation); changes in, during cell discharge, 289; decomposition, 412; electrode, 273 (*see also* Electrode potential); equivalence point, 297 (*see also* Equivalence point potential); formal, 294; half-wave, 412 (*see also* Half-wave potential, Polarographic waves); of hydrogen electrodes, 274; liquid junction, 287 (*see also* Liquid junction potential); measurement of, 350–354; ohmic, 389; standard electrode, 279 (*see also* Standard electrode potential)

Potentiometer(s), 351–354

Potentiometric method(s), 350–384 (*see also* Direct potentiometric measurements, Potentiometric titrations)

Potentiometric titration(s), 376–384; automatic, 384; complex formation, 381; differential, 383; end point detection for, 378; general instructions for, **611**; neutralization, 382, **612**; oxidation-reduction, 382; precipitation reactions, 379, **611**

Power, *P*, of electromagnetic radiation, 435

Power supply, for coulometric titrations, 400

Precipitate(s), colloidal, 125 (*see also* Colloidal precipitates); crystalline, 129–131; drying and ignition of, 133, **547**, **561**; factors affecting solubility of, 87–113, common ion, 30, complexing reagents, 95–99, electrolyte concentration, 103–111, pH, 91–95, solvent composition, 112; filtration and washing of, **550**; mechanism of formation, 124; physical properties of, 122–131; rate of formation, 112

Precipitate electrode(s), 359, 368

Precipitating reagent(s), 134, 135*t*, 138*t*; effect on electrode potentials, 285

Precipitation, analytical separations based upon, 99–103, 497

Precipitation from homogeneous solution, 131; determination of aluminum by, **576**

Precipitation titration(s), 167–182; amperometric, 424*t*, **617**; applications of, 180, 181*t*, 182*t*, 403*t*, **578–583**; coulometric, 402; curves for, 167–174, 380; end points for, 174–180, factors affecting, 170–174; equivalent weight for, 151; Fajans method, 179 (*see also* Fajans method); of halide mixtures, 172, 380, **611**; Mohr method, 176 (*see also* Mohr method); potentiometric, 379, **611**; Volhard method, 177 (*see also* Volhard method)

Precision estimate(s), 48, 61

Premix burner(s), for atomic absorption spectroscopy, 489

Pressure, effect on chemical equilibrium, 25

Pressure broadening, 486

Primary adsorption layer, 127

Primary standard(s), 144, 527 (*see also* under specific reaction types, reagents)

Prism monochromator(s), 447

Product(s), propagation of errors in, 75

Proportional error(s), 54

Purity, of precipitates, 122

p-Value(s), 12, 146 (*see also* pH); significant figures in, 79

Pyrophosphate, as masking agent for iron(III), 341

Q test, 73

Quadratic equation, 651

Quadrivalent cerium, as volumetric oxidizing reagent, 324 [*see also* Cerium(IV)]

Qualitative analysis, based, on absorption spectroscopy, 466, on chromatographic data, 518, 523

Quanta, of electromagnetic radiation, 435

Quantity of electricity, q, 397; measurement of, 399, 405

Quartz, as dispersing agent for electromagnetic radiation, 448

Quartz, filtering crucibles, 548

Quotient(s), propagation of errors in, 75

Radial development, paper chromatography, 516

Radiant intensity, I, 435, 440t

Radiant power, P, 435, 440t

Radiation, electromagnetic, 433 (*see also* Electromagnetic radiation); monochromatic, 438

Radiation detector(s), for absorption spectroscopy, 451–454

Radiation source(s), for absorption spectroscopy, 444, 487

Range, as precision estimate, 48

Rate of reaction, and electrode potentials, 295; of precipitate formation, 112

Rate theory, elution chromatography, 508

Reactant concentration(s), effect, on concentration polarization, 393, on end points, 170, on particle size of precipitates, 124, 130

Reagent chemical(s), 527; rules for use of, **528**

Redox reaction(s), 268 (*see also* Oxidation-reduction titrations)

Reducing agent(s), 268; auxiliary, 313; for gravimetric analysis, 136; volumetric, 336t

Reduction, definition, 268

Reductor(s), 314 (*see also* Jones reductor, Walden reductor)

Reference electrode(s), 354; calomel, 355; for polarographic analysis, 410

Reference standard(s), 527

Reflection grating(s), 449

Reflection losses, at air-glass interfaces, 439

Refractive index, effect on Beer's law, 441

Rejection criteria, for outlying result, 72–74

Rejection quotient, Q, 72

Relative electrode potential(s), 277 (*see also* Electrode potential)

Relative error, 49; in absorption analysis, 472; propagation of, in computations, 74–77

Relative half-cell potential(s), 273 (*see also* Electrode potential)

Relative precision, 49

Relative supersaturation, 123

Relaxation, of excited atoms and molecules, 436, 437

Replica grating(s), 449

Reprecipitation, of colloidal precipitates, 129; of hydrous ferric oxide, **574**

Residual current, correction for, 419; polarographic wave, 412

Resistance, of glass electrodes, 363

Resonance line(s), 486

Rest point, of equal-arm balance, 535

Retardation factor, R_F, 506; paper chromatography, 516

Retention time(s), t, 506; gas-liquid chromatography, 520

Reversibility, electrochemical, 271; polarographic, 412

Rider, for equal-arm balance, 537

Rotating platinum electrode(s), 422; amperometric titrations with, **617**

Rotational absorption spectra, 462

Rubber policeman, 550

Salt bridge(s), 269, 287; effect on direct potentiometric measurements, 371

Sample container(s), for absorption analysis, 450

Sample size, effect on constant error, 33, 56

Saturated calomel electrode, SCE, 355; for polarographic analysis, 410

Secondary standard(s), for volumetric analysis, 145

Second-order electrode(s), 358

Selenium, as catalyst for Kjeldahl oxidation, 235

Self-dissociation, of solvents, 6, 27

Semimicro balance(s), capacity of, 531

Sensitivity, of an analytical balance, 533, 534; evaluation of, **539**

Separation(s), analytical, 497 (*see also* Precipitation, Chromatographic separations, Extraction)

Serological pipet(s), 558t

Shot noise, 474

Short swings, method of weighing, **538**

Sign convention, for electrode potentials, 276; for electrodes, 272

Significant figure(s), 77–79; in logarithms, 79; in titration-curve calculations, 169

Silica gel, as solid chromatographic support, 511, 518

Silver, as auxiliary reducing reagent, 314 (*see also* Walden reductor); as indicator electrode, 357

Silver chloride, photodecomposition of, 573

Silver nitrate solution(s), gravimetric

determination of chloride with, **572**; precipitation titrations with, 167, 181*t*, 425, **578–583**, **611**; preparation of, 578; as reagent, for complex formation titrations, 250*t*

Silver-silver chloride electrode(s), 356

Silver(II), coulometric generation of, 403, 404*t*

Single-beam instrument(s), for absorption analysis, 454, 456, 458

Single-electrode potential(s), 273 (*see also* Electrode potential)

Single-pan balance(s), 531, 535; weighing with, **536**

Sintered glass crucible(s), 547

Slit width, effect on absorption spectra, 466; for spectrophotometers, 448, 459

Sodium bismuthate, as auxiliary oxidizing reagent, 316

Sodium carbonate, as primary standard for acids, 228, **585**; titration of, **587**, in mixtures, 237, **587**, **612**; titration curves for, 217

Sodium hydrogen carbonate, titration of, 237, **587**, **612**

Sodium hydroxide, titration of, 238*t*, **584**, in the presence of sodium carbonate, 237, **588**

Sodium hydroxide solution(s), preparation and standardization, 229–232, **584**, **585**

Sodium oxalate, as primary standard, for cerium(IV) solutions, 326, **601**, for permanganate solutions, 320, **594**

Sodium peroxide, as auxiliary oxidizing reagent, 317

Sodium tetraborate, as primary standard for acids, 229

Sodium tetraphenylboron, as precipitating reagent, 136

Sodium thiosulfate, as reducing reagent, 336*t*

Sodium thiosulfate solution(s), preparation of, **605**; reaction with iodine, 337 (*see also* Iodometric methods of analysis); stability of, 337; standardization of, against potassium bromate, **608**, against potassium dichromate, **605**, against potassium iodate, 339, **605**, **610**

Solid(s), role, in chemical equilibrium, 29

Solid-state electrode(s), 359, 368

Solid support(s), for partition chromatography, 511

Soluble starch, 333 (*see also* Starch)

Solubility, of metal hydroxides, 94

Solubility of precipitates, factors affecting, common ion, 30, 98, competing equilibria, 88–99, complexing reagents, 95–99, electrolyte concentration, 103–111, pH, 91–95, solvent composition, 112, temperature, 112; separations based upon differences in, 99–103, 497

Solubility-product constant(s), K_{sp}, 29, 87–98,

660; evaluation of, from electrochemical measurements, 292

Solute(s), migration of, through a chromatographic column, 504 (*see also* Chromatographic columns)

Solution(s), chemical composition of, 4; methods for expressing concentration of, 10–15, 17*t*, 154

Solution-diluent ratio(s), expression of concentration in, 15

Solvent(s), amphiprotic, 5; autoprotolysis of, 6; effect, on acid-base strength, 7, on solubility, 112

Source(s), for atomic absorption spectroscopy, 487; of electromagnetic radiation, 444, indeterminate error due to, 472–475

Spark spectroscopy, 483*t*

Specific gravity, of solutions, 13

Specific indicator(s), oxidation-reduction, 306

Specific surface, of precipitates, 130

Spectra, absorption, 442 (*see also* Absorption spectra); electromagnetic, 436; excitation of, in flames, 483, 484

Spectral band width, effect on absorption spectrum, 466

Spectrophotometer(s), 444, 454, 458–460, atomic absorption, 489, flame emission, 494

Spectrophotometric analysis, of mixtures, 470, **620**; scope of, 468; standard addition method, 470, **620**; wavelength selection for, 469

Spread, *w*, as precision estimate, 48

Sputtering, in hollow cathode lamp, 487

Squalene, as stationary phase, gas-liquid chromatography, 521

Stability, of an analytical balance, 534

Standard acid solution(s), preparation and standardization of, 228, 516, **585**, **585**

Standard addition method, for absorption analysis, 470, **620**; for atomic absorption spectroscopy, 493; for direct potentiometric measurement, 374; for flame emission spectroscopy, 495; for polarographic analysis, 420

Standard base solution(s), preparation and standardization of, 229, 516, **584**, **585**

Standard cell(s), 352

Standard deviation, σ, 61–65; estimate of, *s*, 64; of pooled data, 66

Standard electrode potential, E^0, 279, 281, 656; calculation of equilibrium constants from, 292; limitations to use of, 294

Standard hydrogen electrode, SHE, 273

Standard samples, for detection of determinate error, 55

Standard solution(s), for volumetric analysis, 143, 145 (*see also* under specific reaction types, reagents); calculation of concentration for, 157

Starch, as indicator for iodine, 306, 333, 337, **603**

Static charge(s), effect on weighing operation, 542

Stationary phase(s), 502; paper chromatography, 516; gas-liquid chromatography, 521; types of, 503*t*

Statistics, treatment of data with, 60–74

Steel, determination of manganese in, **619**

Stepwise equilibria, 41 (*see also* Acid-base equilibria, Complex formation)

Stibnite, determination of antimony in, 334, **604**

Stirring, effect, on coagulation of colloids, 127, on electrolytic deposits, 395, on particle size of precipitates, 130

Stirring rod(s), 550

Stirrup(s), for analytical balance, 533

Stopcock(s), 559, **562**

Stray radiation, in monochromators, 450

Strong acid(s), 7; coulometric titration of, 401; pH calculations for solutions of, 191; titration curves for, 190 (*see also* Titration curves)

Strong base(s), 7; coulometric titration of, 402; pH calculations for solutions of, 191; titration curves for, 193 (*see also* Titration curves)

Strong electrolyte(s), 4, 5*t*

Substitution reaction(s), amperometric end point for, 424*t*; analytical, 329, **609**

Substitution weighing, 535

Subtraction(s), propagation of errors in, 74

Successive approximations, method of, 652

Sulfate ion, amperometric titration of, **617**; homogeneous generation of, 132*t*; as titrant for lead(II), 421

Sulfide ion, homogeneous generation of, 132*t*

Sulfide separations, 101, 498

Sulfonic acid group(s), titration of, 239

Sulfur, elemental analysis for, 236, 237*t*

Sulfur dioxide, as auxiliary reducing reagent, 315; gas-sensing electrode for, 368

Sulfuric acid, as volumetric reagent, 228

Sum(s), propagation or errors in, 74

Supersaturation, relative, 123

Support(s), for gas-liquid chromatography, 521

Supporting electrolyte(s), for polarographic analysis, 412, 413

Surface adsorption, 125

Syringe pipet(s), 558*t*

Technical grade, designation for chemicals, 526

Temperature, effect, on atomic absorption and emission, 484, 485, on chemical equilibrium, 25, on coagulation of colloids, 127, on

electrolytic deposits, 396, on kinetic polarization, 393, on K_w, 28*t*, on particle size of precipitates, 124, 130, on performance of chromatographic column, 521, on polarographic wave, 418, on solubility of precipitates, 112, on volume measurements, 556, 566*t*, on weighing data, 542; estimation from color, 550*t*

THAM *tris* (hydroxymethyl) aminomethane, as primary standard for acids, 229

Theoretical plate(s), number of, *N*, 507

Theoretical potential(s), of electrochemical cells, 287

Thermal conductivity detector(s), gas-liquid chromatography, 522

Thermobalance, 133

Thermodynamic equilibrium constant expression(s), 26, 110

Thermodynamic potential(s), of electrochemical cells, 287

Thin-layer chromatography, 503*t*, 518

Thiocyanate ion, charge-transfer complexes with iron(III), 466; as indicator for iron(III), 306; role, in iodometric determination of copper, 431; separation of iron(III) from aluminum(III) by, 515, **576**; titrations with, 177 (*see also* Volhard method)

Thiosulfate, reaction of, with iodine, 337 (*see also* Sodium thiosulfate solutions)

Thymol blue, preparation of solution, **583**; transition range for, 189*t*

Thymolphthalein, preparation of solution, **583**; transition range for, 189*t*

Time, measurement of, for coulometric titrations, 398, 405

Tin(II) chloride, reduction of iron(III) with, 323, **597**

Tirrill burner(s), 549

Titanium(III), coulometric generation of, 404*t*; as reducing reagent, 336*t*

Titer, of solutions, 12, 154

Titration(s), 143 (*see also* under specific reaction types, reagents); amperometric, 421 (*see also* Amperometric titrations); calculation of results from, 158; concentration changes during, 146 (*see also* Titration curves); coulometric, 399 (*see also* Coulometric titrations); differential, 383; directions for performing, **563**; potentiometric, 376 (*see also* Potentiometric titrations)

Titration curve(s), 146; amperometric, 421; for complex formation reactions, 254 (*see also* EDTA titrations); oxidation-reduction, 296 (*see also* Oxidation-reduction titrations); potentiometric, 378; precipitation, 167 (*see also* Precipitation titrations); significant figures in calculation of, 169; for solutions containing, mixtures of acids or bases, 206, polyfunctional acids or bases, 212–218, strong acids and strong bases, 190–193, weak acids and weak bases, 193–218

Titration error, 144; with acid-base indicators, 189

Titrator(s), automatic, 384

Titrimetric methods, 143 (*see also* Volumetric analysis)

Top-loading balance(s), 543

Total consumption burner(s), for atomic absorption spectroscopy, 489

Total salt concentration, determination by ion-exchange chromatography, 515

Trace(s), concentration by ion exchange, 515

Transducer, 444

Transfer pipet(s), 557 (*see also* Volumetric pipets)

Transfer of precipitates, **550**

Transition element(s), absorption of electromagnetic radiation by, 465

Transition range(s), for acid-base indicators, 187, 189*t* (*see also* Acid-base indicators); for oxidation-reduction indicators, 306, 307*t*

Transmission grating(s), 449

Transmittance, T, 440; errors in measurement of, 472–475

Trichloroacetic acid, coulometric titration of, 406

Triiodide ion, 332 (*see also* Iodine)

Triple-beam balance(s), 543

Triple bond(s), absorption of electromagnetic radiation by, 464, 465*t*

Triton-X, as maximum suppressor, 419

True oxidation-reduction indicator(s), 306 (*see also* Oxidation-reduction indicators)

Tungsten filament lamp(s), as source for visible radiation, 445

t, values for, 70

Two-dimensional paper chromatography, 517

Ultraviolet radiation, absorption of, 401, 463; sources for, 445, species absorbing in, 464

Uranium(IV), coulometric generation of, 404*t*

Urea, as source of hydroxide ion, 131, 132*t*, **576**

USP grade, designation for chemicals, 527

Vacuum train, for filtering crucibles, 548, 554

Vacuum ultraviolet region, 464

Variable current coulometry, 405 (*see also* Coulometric titrations)

Variance, 63

Velocity of electromagnetic radiation, c, 434

Vibrational absorption spectra, 462

Vinegar, determination of acid content in, **586**

Visible radiation, absorption of, 461, 463; sources for, 445; species absorbing in, 464

Volatilization, gravimetric methods based on, 117

Volhard method, 177, 181*t*; determination of chloride, **581**

Voltage divider(s), 351

Voltaic cell(s), 271

Voltammetry, 408 (*see also* Polarographic analysis, Amperometric titrations)

Volume, measurement of, 555–570

Volume correction(s), for amperometric titration data, 422

Volume percent (v/v), expression of concentration in, 14

Volumetric analysis, 143–160, **578–611** (*see also* under specific reaction types, reagents); amperometric, 420 (*see also* Amperometric titrations); calculations for, 148–160; coulometric, 398 (*see also* Coulometric titrations)

Volumetric flask(s), 559; calibration of, **567**; use of, **564**

Volumetric pipet(s), 557; calibration of, **567**; use of, **560**

Von Weimarn, P. P., 123

Walden reductor, 315, 316*t*

Washing, of precipitates, 129, **550**

Water, analysis for, in a crystalline hydrate, **571**; autoprotolysis of, 6; determination of hardness in, 263, **593**; determination of iron in, **618**; ion product constant, K_w, for, 27; role of, in chemical equilibrium, 27, 279, in functioning of glass electrode, 362; volume of, at various temperatures, 566*t*

Wave(s), polarographic, 408 (*see also* Polarographic waves)

Wavelength, λ, of electromagnetic radiation, 434; selection of, for absorption analysis, 454, 469

Wave number, 435

Wave properties, of electromagnetic radiation, 434

Weak acid(s), 7; analysis of, 239, 382, **586**, **612**; coulometric titration of, 401; dissociation constants for, 32, 663; pH calculations for solutions of, 33 (*see also* pH); titration curves for, 193, 201 (*see also* Titration curves)

Weak base(s), 7; analysis of, 233 (*see also* Kjeldahl method, Winkler method); coulometric titration of, 402; dissociation constants for, 33, 666; pH calculations for solutions of, 37 (*see also* pH); titration curves for, 193, 206 (*see also* Titration curves)

Weak electrolyte(s), 5

Weighing, by difference, 546; with equal-arm balance, **538–540**; of hygroscopic substances, 546; of liquids, 546; with a single-pan balance, **536**; by substitution, 535

Weighing bottle(s), 543, **545**

Weight, distinction from mass, 530; units of, 8, 16*t*

Weight percent (w/w), expression of concentration in, 14

Weights, for an analytical balance, 533, 537

Weight/volume percent (w/v), expression of concentration in, 14

Weston cell(s), 352

Winkler method, for carbonate-hydroxide mixtures, 239, **558**; for determination of dissolved oxygen, 342, **607**

Witt plate(s), 555

Working electrode(s), coulometric analysis, 399 (*see also* Generator electrode)

X-ray(s), absorption of, 461

X-ray absorption spectroscopy, 483*t*

X-ray fluorescence spectrometry, 483*t*

Zero point, for analytical balances, 535; determination of, **538**

Zimmermann-Reinhardt reagent, 323; preparation of, **597**

Zinc, as auxiliary reducing reagent, 314 (*see also* Jones reductor)

Zinc(II), coulometric titration of, 403*t*

Zone shape, elution chromatography, 508; variables affecting, 509

1-2
1-2
1-3
1-3
3-1
3-1
12-2
4-2
4-1
7-2
7-1